AF868636

Fatigue and Fracture of Nanostructured Materials

Pasquale Cavaliere

Fatigue and Fracture of Nanostructured Materials

Springer

Pasquale Cavaliere
Department of Innovation Engineering
University of Salento
Lecce, Italy

ISBN 978-3-030-58087-2 ISBN 978-3-030-58088-9 (eBook)
https://doi.org/10.1007/978-3-030-58088-9

© Springer Nature Switzerland AG 2021
This work is subject to copyright. All rights are reserved by the Publisher, whether the whole or part of the material is concerned, specifically the rights of translation, reprinting, reuse of illustrations, recitation, broadcasting, reproduction on microfilms or in any other physical way, and transmission or information storage and retrieval, electronic adaptation, computer software, or by similar or dissimilar methodology now known or hereafter developed.
The use of general descriptive names, registered names, trademarks, service marks, etc. in this publication does not imply, even in the absence of a specific statement, that such names are exempt from the relevant protective laws and regulations and therefore free for general use.
The publisher, the authors, and the editors are safe to assume that the advice and information in this book are believed to be true and accurate at the date of publication. Neither the publisher nor the authors or the editors give a warranty, expressed or implied, with respect to the material contained herein or for any errors or omissions that may have been made. The publisher remains neutral with regard to jurisdictional claims in published maps and institutional affiliations.

This Springer imprint is published by the registered company Springer Nature Switzerland AG
The registered company address is: Gewerbestrasse 11, 6330 Cham, Switzerland

Preface

Nanostructured materials represent a possible alternative for a broader range of applications, outperforming many of today's engineering materials. As new nanomaterials are rapidly developing and many applications exist, mainly within fields such as medicine, communication, consumer goods, and engineering, it is necessary to identify what special properties this fairly new material group can offer. All engineering materials based on nanotechnology, involving the understanding of physical properties and how they change with material dimensions, are to be considered alternatives in existing products.

Metals with a grain size in the range of 100–1000 nm are classified as ultrafine grain; grain sizes less than 100 nm are considered to be in the nanocrystalline domain. The altered response of such properties is a direct consequence of the nanoscale microstructural arrangements of the atoms themselves. Regarding engineering design, these metals pose significant promise as next-generation structural materials due to reported increases in ultimate strength, resistance to fatigue, and wear resistance.

The abnormally high volume fractions of noncrystalline material exaggerate the importance of the grain boundaries, ultimately leading to a shift in the physical plasticity mechanisms which take place during deformation. At the smallest grain sizes traditional intragranular dislocation-based mechanisms begin to shut off which leads to the dominance of grain boundary-mediated mechanisms.

The main routes employed to produce UFG and NC materials are presented by underlying the pros and cons of the application of each production from mechanical alloying and severe plastic deformation to electrodeposition, sputtering, and surface modification. All those mechanisms, active at the nanoscale, governing the crack initiation and growth behavior as well as the crack-grain-microstructural features, are theoretically described. The strength-ductility effect on the fatigue life of nanocrystalline materials with particular attention to cyclic plastic strain is discussed. The effect of grain boundary strengthening through alloying with the related consequences on the fatigue life and fatigue mechanisms acting during cyclic deformation is underlined. The different behaviors due to the various mechanisms acting during creep in nanocrystalline metals and alloys have been described. The effect of nanostructuring on the creep and superplastic behavior of metals and alloys is also shown. The different mechanical behavior of thin films due to the reduced thickness and the confined deformation mechanisms is described as well. How mechanical properties of thin films differ from those of bulk materials is underlined. The wear behavior related to the effect of grain refinement and the consequent grain-mediated deformation mechanisms is shown. The contact fatigue and fretting mechanisms acting in nanostructured metals and alloys are also described. The cycle sliding-activating fatigue mechanisms in nanostructured metals and alloys are shown too.

My special thanks to the professionalism of the editorial office manager and assistants. I would like to dedicate this book to my son Paolo.

Lecce, Italy

Pasquale Cavaliere

Contents

Abbreviations

AA	Aluminum alloy
AD	Axial direction
AGG	Abnormal grain growth
AIF	Amorphous intergranular film
ALD	Atomic layer deposition
ALG	Abnormally large grain
AM	AMorphous
ARB	Accumulative roll-bonding
C/T	Compression/tension
CCC	Cylinder-covered compression
CF	Corrosion fatigue
CGB	Clean grain boundary
CGBS	Cooperative grain boundary sliding
CGP	Constrained groove pressing
CIP	Cold isostatic pressure
CO	Coble
CR	Cold rolling
CRSS	Critical resolved shear stress
CSL	Coincidence site lattice
CSS	Cyclic stress strain
CTB	Coherent twin boundaries
CVD	Chemical vapor deposition
DLC	Diamond-like carbon
DSC	Displacement shift complete
DSI	Depth-sensing indentation
DUFG	Deformed ultrafine grain
EAC	Environmentally assisted cracking
ECAP	Equal-channel angular pressing
EIS	Electrochemical impedance spectroscopy
F	Ferrite

FCB	Fatigue crack behavior
FCG	Fatigue crack growth
FEM	Finite element modeling
FGM	Functionally graded materials
FSP	Friction stir processing
GA	Grain agglomerates
GB	Grain boundary
GBAZ	Grain boundary-affected zone
GBCD	Grain boundary character distribution
GBD	Grain boundary dislocation
GBP	Grain boundary process
GBS	Grain boundary sliding
GNB	Geometrically necessary boundary
GNC	Gradient nanocrystalline
GND	Geometrically necessary dislocations
H	Hardness
HAGB	High-angle grain boundary
HCF	High-cycle fatigue
HE	Hydrogen embrittlement
HEA	High-entropy alloy
HESP	High-energy shot peening
HIP	Hot isostatic pressure
HNM	Heterogeneous nanostructured materials
H-P	Hall-Petch
HPT	High-pressure torsion
HSR	High strain rate
IDB	Incidental dislocation boundary
IGSCC	Intergranular stress corrosion cracking
IHPE	Inverse Hall-Petch effect
II	Ion implantation
ISE	Indentation size effect
ITB	Incoherent twin boundary
LAGB	Low-angle grain boundary
LBQ	Laser beam quenching
LCF	Low-cycle fatigue
LEDS	Low-energy dislocation structures
LP	Leading partial
LSNT	Large spacing nanotwinned
LSP	Laser shock peening
M	Martensite
MC	Microcrystalline
MD	Molecular dynamic
MDS	Molecular dynamic simulation
MEMS	Microelectromechanical systems

nc	Nanocrystalline
NEMS	Nanoelectromechanical systems
NH	Nabarro–Herring
NRD	Nanoscale rotational deformation
NS	Nanostructured
NT	Nanotwinned
OGBC	Ordered grain boundary complexion
PD	Partial dislocations
PGM	Plastically graded material
PIII	Plasma immersion ion implantation
P-N	Peierls–Nabarro
PN	Plasma nitriding
PSB	Persistent slip band
PZ	Plastic zone
RCS	Repetitive corrugation and straightening
RD	Radial direction
RHAGB	Random high-angle grain boundary
ROM	Rule of mixture
RPZ	Reversible plastic zone
S^2PD	Surface severe plastic deformation
SB	Slip bands
SCC	Stress corrosion cracking
SF	Stacking fault
SFE	Stacking fault energy
SFSP	Submerged friction stir processing
SIF	Stress intensity factor
SMAT	Surface mechanical attrition treatment
SMC	Submicrocrystalline
SMGT	Surface mechanical grinding treatment
SNH	Surface nano-crystallization and hardening
SP	Shot peening
SPD	Severe plastic deformation
SPS	Spark plasma sintering
SRS	Strain rate sensitivity
SS	Stainless steel
SSNT	Small spacing nanotwinned
SSP	Severe shot peening
STZ	Shear transformation zone
TB	Twin boundary
TD	Tangential directions
TJ	Triple junction
TL	Triple line
TP	Twin plane
TTS	Tribologically transformed structure

UDUFG	Undeformed ultrafine grain
UHP	Ultrahigh pressures
UNSM	Ultrasonic nanocrystal surface modification
UP	Ultrasonic peening
USP	Ultrasonic shot peening
USRP	Ultrasonic surface rolling processing
WIF	Wear induced by fretting

Chapter 1
Nanostructuring of Metals, Alloys, and Composites

1.1 Introduction

Nanostructured materials are the most potential and exciting candidates in many fields for revolutionizing traditional material designs. Nanostructured metals and alloys are a class of materials exhibiting novel characteristics across a wide range of properties including increased hardness, superplasticity, and electrical conductivity (Schaefer 2010). Here, the difference between ultrafine and nanocrystalline metals needs a special mention. Conventionally, metals with a grain size in the range of 100–1000 nm are classified as ultrafine grain; grain sizes less than 100 nm are considered to be in the nanocrystalline domain (Gleiter 1989, 1993).

The altered response of such properties is a direct consequence of the nanoscale microstructural arrangements of the atoms themselves (Murr 2015). Regarding engineering design, these metals pose significant promise as next-generation structural materials due to reported increases in ultimate strength, resistance to fatigue, and wear resistance (Mittemeijer 2010).

Nanostructured materials represent a possible alternative for a broader range of applications, outperforming many of today's engineering materials. As new nanomaterials are rapidly developing and many applications exist, mainly within fields such as medicine, communication, consumer goods, and engineering, it is necessary to identify what special properties this fairly new material group can offer. All engineering materials based on nanotechnology, involving the understanding of physical properties and how they change with material dimensions, are to be considered alternatives in existing products.

Through following the Hall-Petch equation

$$\sigma_y = \sigma_0 + \frac{k_y}{\sqrt{D}}$$

© Springer Nature Switzerland AG 2021

P. Cavaliere, *Fatigue and Fracture of Nanostructured Materials*,
https://doi.org/10.1007/978-3-030-58088-9_1

where σ_y is the yield strength, σ_0 is a material constant (i.e., 25 for Cu, 70 for Fe, and 80 for Ti), k_y represents the resistance of grain boundaries to dislocation mobility (transmission through grain boundaries) giving the quantification of the grain boundary strengthening (i.e., 0.11 for Cu, 0.74 for Fe, and 0.4 for Ti), and D is the mean grain size. A high density of grain boundaries limits the length of dislocation pileups and, due to restricted dislocation motion, extraordinarily high yield stress values can be observed (Gutkin and Ovid'ko 2004). However, the extraordinary mechanical properties of NC metals are limited by mechanisms related to the competing length scales (Anderson et al. 2014; Mohamed 2016).

The Hall-Petch relationship essentially describes grain boundary strengthening, a process by which grain boundaries, or regions between crystallites of different lattice orientation, act as physical barriers for continued dislocation movement within the material (Armstrong 2014). For more traditional, course-grained materials consisting of average grain sizes ranging from 100 nm upwards of several microns, gliding dislocations act as the main carriers of plasticity. For this reason, as grain size is decreased, significant strengthening can be expected (Brandl 2019). The core of this idea is that grain boundaries hinder dislocations, which accumulate and cause stress concentrations (Suryanarayana and Koch 2000). This resistance is thought to be proportional to the misalignment between the meeting slip systems and the magnitude of the burger's vector of the residual grain boundary dislocation that is created by transmission. The reliance on a planar dislocation pileup arrangement has led to criticism of this theory. In fact, it was revealed that the length of these pileups has not been correlated with grain size, nor are they likely to form in materials where cross-slip to other planes is relatively easy. Instead, a different explanation was proposed in which grain boundaries serve as nucleation sites for dislocations. The idea is that a greater grain boundary area will provide more dislocation sources and lead to a higher dislocation content at a given strain. Strength is known to depend on the square root of dislocation density. Another alternate explanation of the Hall-Petch effect has been offered by proposing that strain gradients imposed by compatibility requirements between grains increase the dislocation density. These strain gradients become larger as grain size decreases, and so does the number of geometrically necessary dislocation needed to support them.

A strong effort has been devoted to metal and alloy nanostructuring in order to reach the material's theoretical maximum strength (G/10). The pioneering view of Gleiter (1989) pushed the research on nanostructured metals in the recent decades. In the paper, nanocrystalline materials are defined as single- or multiphase polycrystals with nanoscale grain size (1–250 nm). As upper limit, "ultrafine grain" is used to indicate grain size in the range of 250–1000 nm. First of all, the properties of these materials are related to the increase of the grain boundary volume fraction as the mean grain size decreases. The abnormally high volume fractions of noncrystalline material exaggerate the importance of the grain boundaries, ultimately leading to a shift in the physical plasticity mechanisms which take place during deformation (Meyers et al. 2006a). At the smallest grain sizes traditional intragranular dislocation-based mechanisms begin to shut off which leads to the dominance of grain boundary-mediated mechanisms.

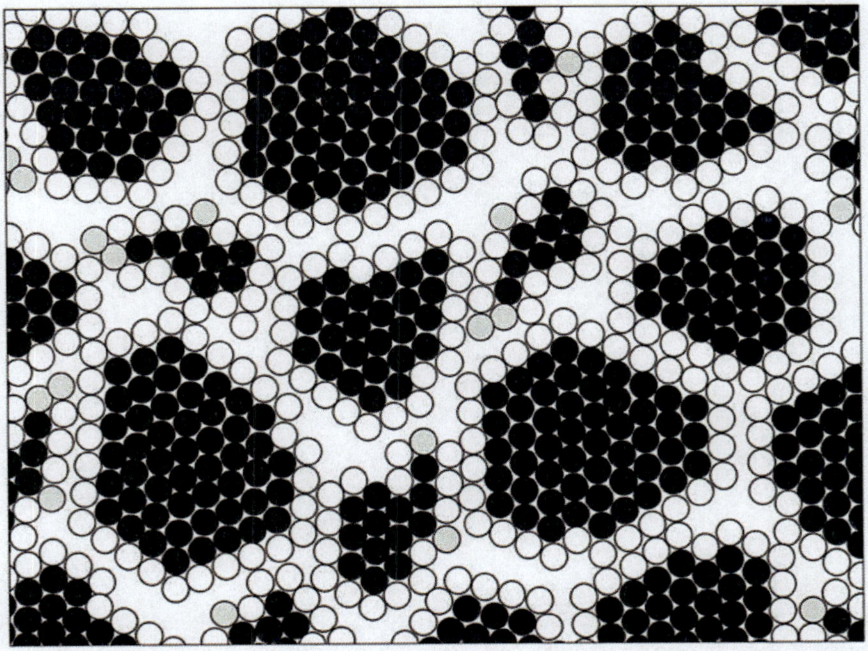

Fig. 1.1 2-D model of nanostructured materials

The altered response of NC material properties is a direct consequence of the nanoscale microstructural arrangements of the atoms themselves. The atoms in the crystal interior are closely packed in an ordered configuration; those in the grain boundaries are in a more chaotic high-energy configuration with a different spacing (Fig. 1.1).

As the grain size decreases, the nanocrystalline material interfaces contain higher atom fractions. For grain size in the order of magnitude of 5 nm the fraction is in the order of 50%; for 10 nm, the grain boundary volume fraction is around 30%; for ultrafine-grained materials it is around 2% while for microcrystalline metals and alloys it is orders of magnitude lower (Fig. 1.2).

Unfortunately, due to the high volume fractions of grain boundaries, the driving force for nanocrystalline grain growth is amplified (Andrievski and Khatchoyan 2016). Therefore, minimizing grain growth and maintaining their beneficial properties prove extremely challenging.

In nanocrystalline materials, the intercrystalline volume fraction is found to comprise as much as 50% of the total crystal volume. These solids are assumed to have a different kind of atomic structure: a crystalline structure with long-range order for all the atoms far from the grain boundaries and a disordered structure with some short-range order at the interfacial region. Hence, the mechanical properties of these nanocrystalline materials are expected to be different as compared to their equal polycrystalline material. There are two main strategies that have garnered the most

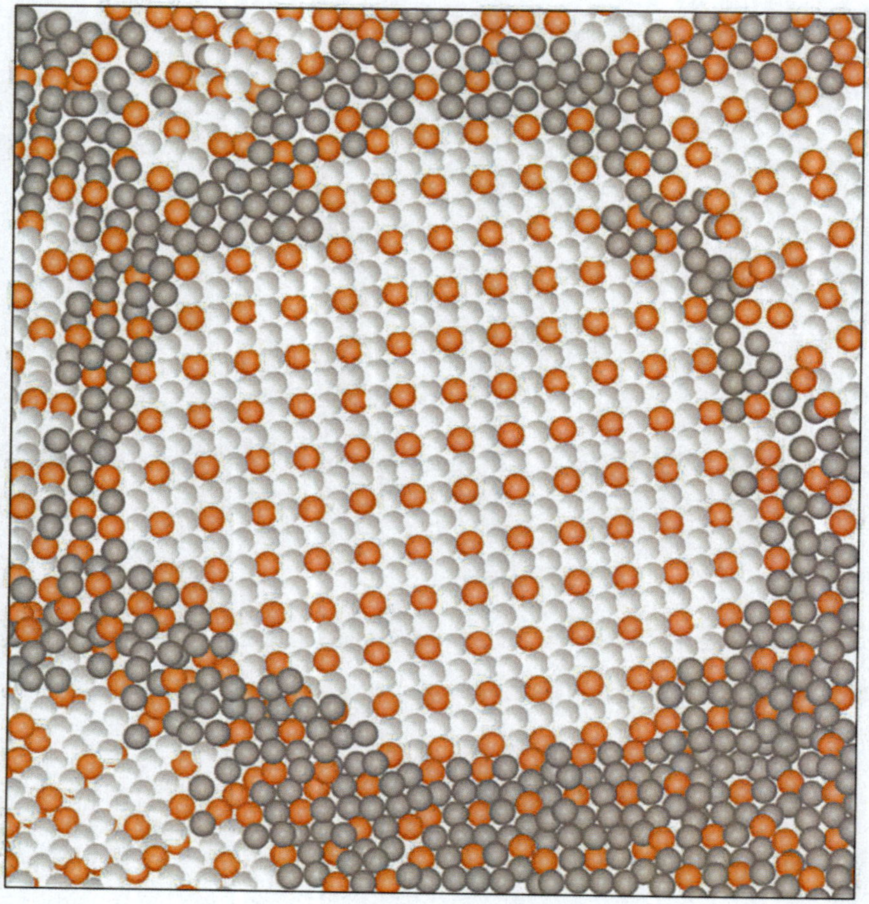

Fig. 1.2 MD model of nanocrystalline Ni-Fe with 5 nm grain size

attention. The first strategy utilizes kinetic mechanisms that decrease grain boundary mobility, such as solute drag, reduced diffusivity, particle drag, and chemical ordering. The second strategy utilizes solute segregation to reduce the grain boundary energy and consequently the driving force (Razumov 2014).

The grain boundary energy is given by the following equation:

$$\gamma = \gamma_0 - \Gamma(\Delta H^{seg} + RT \ln X)$$

where γ is the grain boundary energy, γ_0 is the original grain boundary energy of the pure metal, Γ is the grain boundary solute excess, ΔH_{seg} is the enthalpy of segregation, R is the Boltzmann constant, T is the temperature, and X is the global dopant composition. The equation ultimately means that segregation of dopant to grain

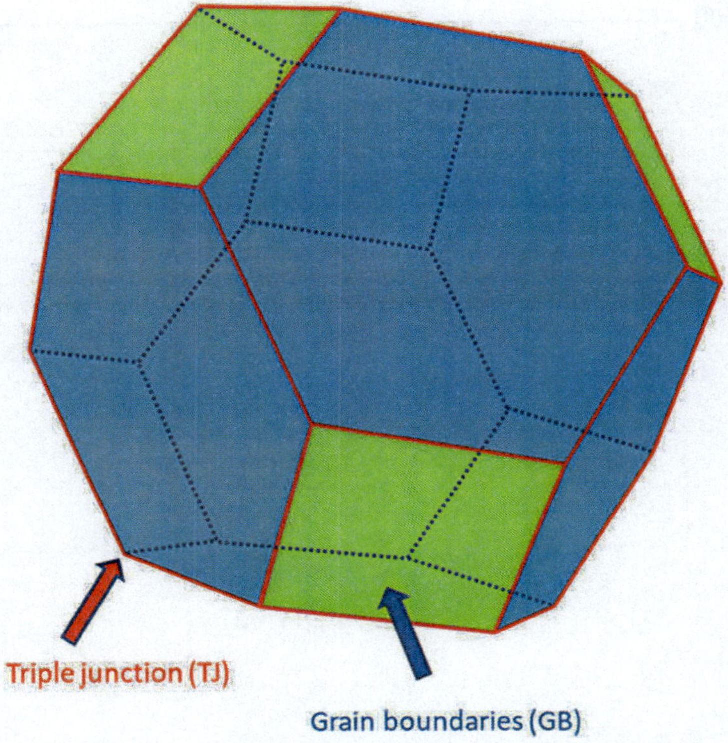

Fig. 1.3 Ideal tetrakaidecahedron grains with a grain boundary thickness of 1 nm

boundaries can lower the grain boundary energy and consequently change its behavior.

By analyzing this aspect through the employment of an ideal tetrakaidecahedron grain structure with a grain boundary thickness of 1 nm (Fig. 1.3), it is possible to plot the relationship between the grain size and the volume fraction.

So, both the volume fraction and the triple junctions increase as the mean grain size decreases (Fig. 1.4).

The dotted lines show the evolution of grain boundary thickness and its effect as a function of grain size (range 1–0.1 nm).

Experimental evidences show that GB size is around 0.5 nm for NC materials with FCC crystal structure and 1 nm for NC BCC crystal structure. From Fig. 1.5 it can be observed that for a grain size of 5 nm approximately 40% of the atoms lie in the grain boundary (Siegel and Thomas 1992; Spearot and McDowell 2009).

The volume fraction as a function of grain size distribution was measured for many Ni and Ni-Fe electrodeposited nanostructured materials (Fig. 1.6).

So, the overall behavior of nanocrystalline materials is dependent on not only the grain size but also the nature of grain boundary structures. Generally, these interfaces are responsible for the strength behavior of metals and alloys. The atom orientation

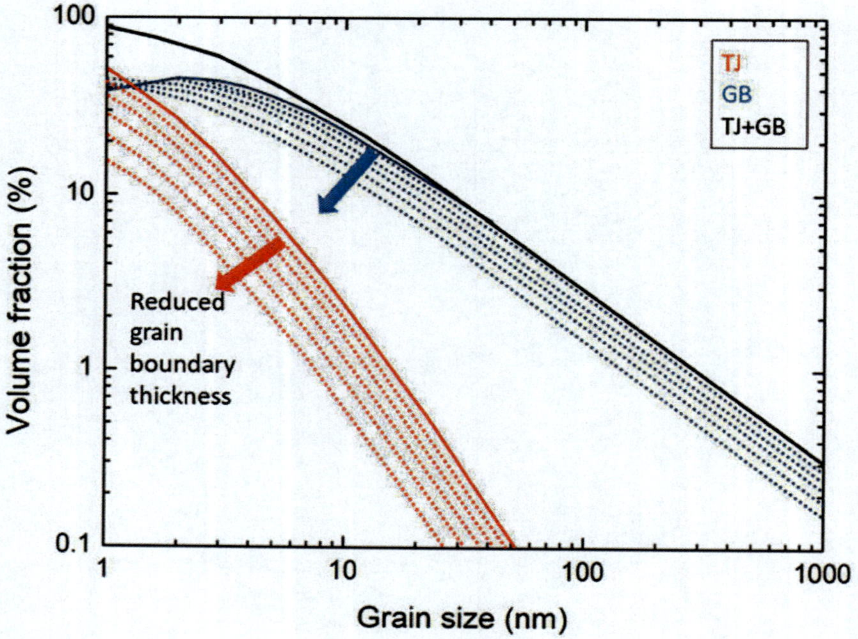

Fig. 1.4 Volume fraction and triple junctions as a function of the grain size

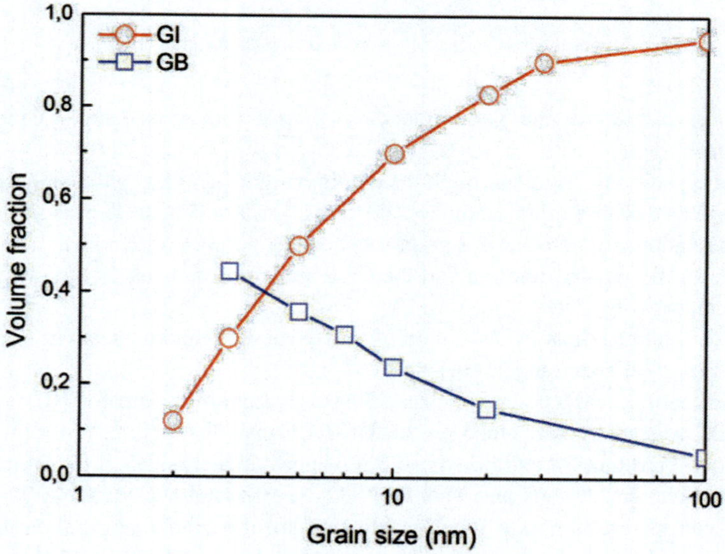

Fig. 1.5 Brain boundary vs. grain interior volume fraction

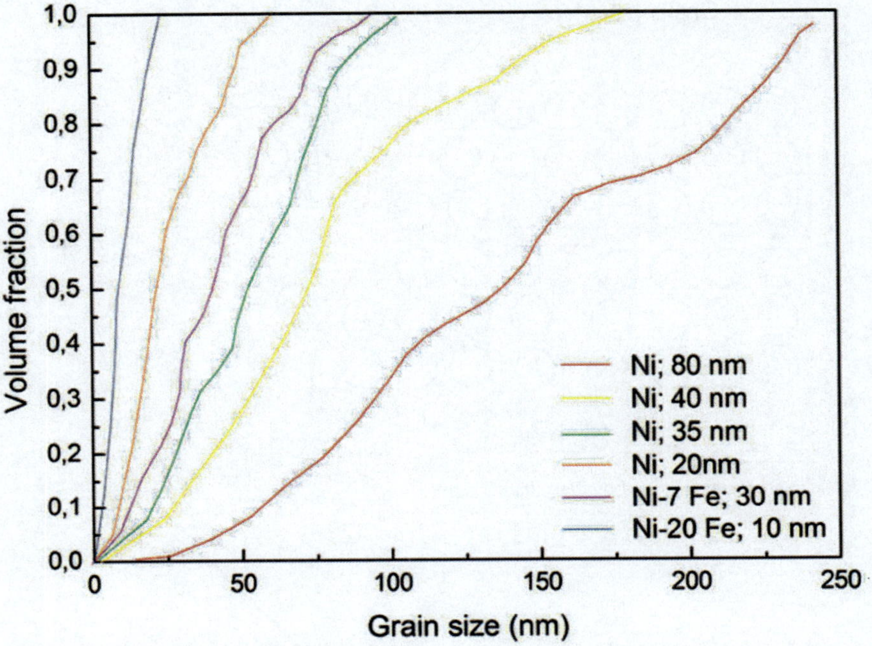

Fig. 1.6 Volume fraction vs. grain size distribution for different Ni and Ni-Fe

in adjacent grain is different; in addition, the atoms in the grain boundaries are in a disordered state. These factors lead to a difficulty for a dislocation to continuously slip in all the material volume because it must overcome the grain boundaries with a change in the slip direction (Fig. 1.7).

This hinders the material plasticity with a direct effect on the yield strength increase of the material (Sun 2014). The stress applied to the crystal (σ) generates a shear stress (τ_a); the crystal opposes a resistance (τ_0) to the generated shear stress (Fig. 1.8).

So, the effective stress (τ_{eff}) for the dislocation sliding is

$$\tau_{\text{eff}} = \tau_a - \tau_0$$

This is the model describing the situation in the grain interior. By approaching the grain boundary, the dislocation must have sufficient energy to overcome the grain boundary; otherwise it is pinned by the grain boundary. In this model the shear stress at grain boundary (τ_{gb}) is given by

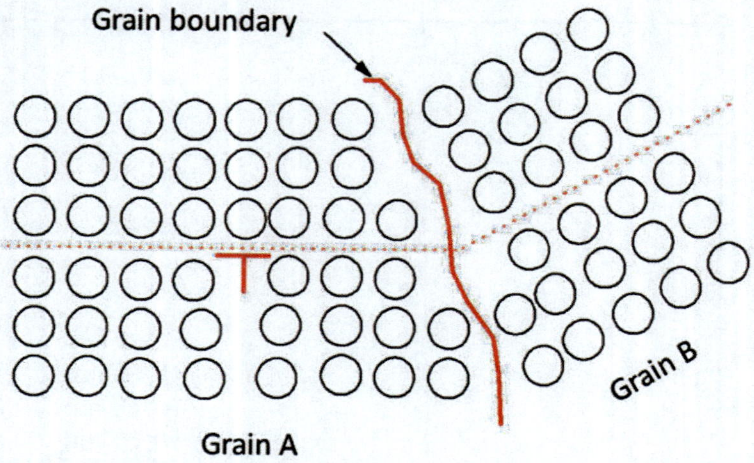

Fig. 1.7 Dislocation-grain boundary interaction

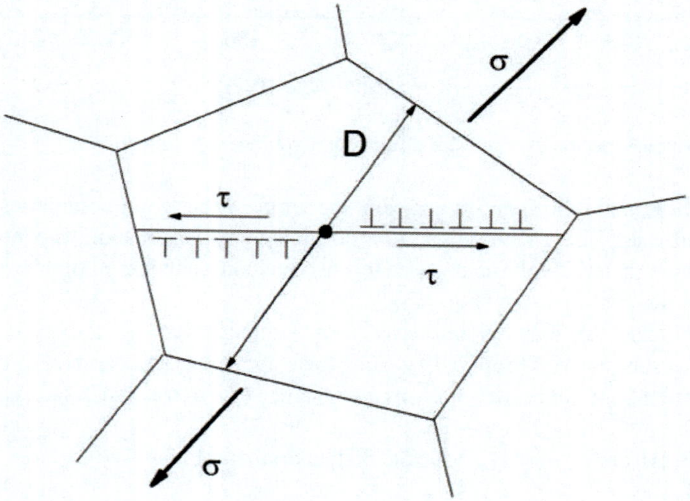

Fig. 1.8 Stress leading to dislocation sliding inside the grain

$$\tau_{bg} = \tau_{eff}\sqrt{\frac{D}{4r}} = \tau_a - \tau_0\sqrt{\frac{D}{4r}}$$

where r is the distance from the point of dislocation generation and the grain boundary. Finally, the shear stress necessary for the dislocation movement taking into account the crystal interior and the grain boundary resistances is given by

$$\tau_a = \tau_0 + \tau_{bg}\sqrt{\frac{4r}{D}}$$

that is the Hall-Petch relationship in terms of shear.

Some researchers observed that the capacity of generating dislocations inside the grain is related to the parameter

$$\mu = \frac{\text{emitted dislocation length}}{\text{grain area}}$$

that is related to the dislocation density (ρ) by

$$\rho = \frac{3\mu}{D}$$

Finally, the Hall-Petch relationship can be expressed by

$$\tau = \tau_0 + \alpha G b\sqrt{\frac{3\mu}{D}} = \tau_0 + k_y' D^{-1/2}$$

where α is a parameter depending on the crystal structure, G is the shear modulus, and b is the Burgers vector.

Now, the most common dislocation generation mechanism in the grain interior is given by the Frank-Read source. By reducing the grain size the dislocation loop dimension is reduced up to disappearance for nanosized grains (Fig. 1.9).

So, by decreasing the grain size, the material strength increases because of the limited dislocation generation and the limited space for sliding before the interaction with the grain boundary (Fig. 1.10).

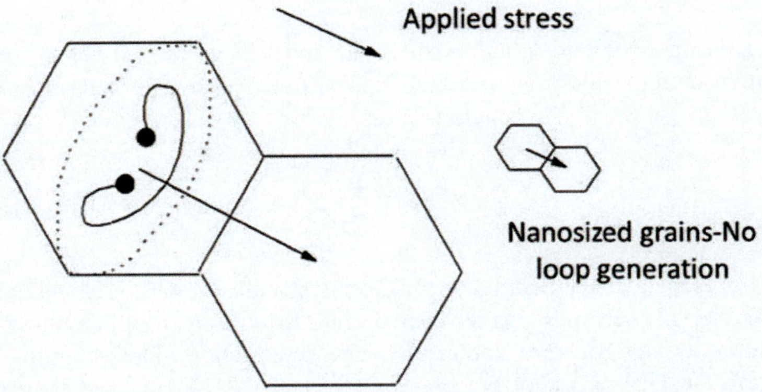

Fig. 1.9 Dislocation generation inside the grains

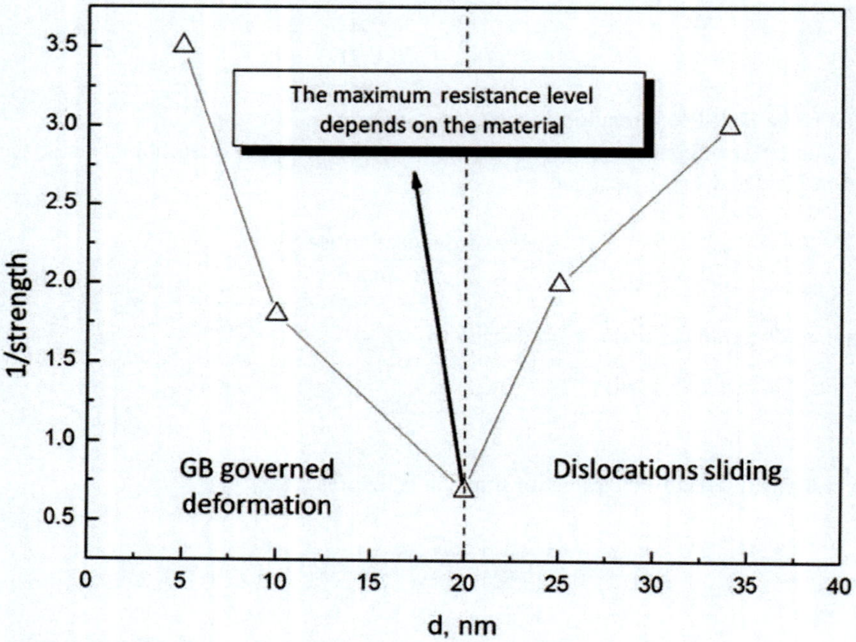

Fig. 1.10 Strength variation vs. grain size in pure Cu

Dislocation activity is grain size dependent. Full dislocation activity may be inhibited at very small grain sizes as partial dislocation activity becomes more favorable. In terms of the Orowan relation, the resolved shear stress required for expansion of a dislocation loop with a diameter of D is

$$\tau_{\mathrm{RSS}} = \frac{\mu b}{D}$$

where μ is the shear modulus and b is the Burgers vector of dislocation. Full dislocation multiplication requires that Frank-Read-type sources have a minimum grain size (D^*) at the yield strength (σ_y) of

$$D^* = \frac{\mu b}{\frac{\sigma D^*}{m}}$$

where $1/m$ is the average Schmid factor for polycrystals ($m = 3$). This would mean that as grain size becomes smaller than D^* full dislocation is inhibited and partial dislocation activity becomes dominant in the deformation of most grains, hence enhancing the GB relaxation process (Zhang et al. 2018). As the GB relaxation process is triggered by plastic deformation, it may not happen in the nanograined metals which possess high GB energy and hence poor thermal stability (Zhou et al. 2018; Wrobel 2012).

According to the results of kinetic modeling, carried out within the developed dislocation model, considering the wholeness of all the possible deformation mechanisms and presenting the development of the well-known composite models, the temperature alteration of the plastic deformation brings to the activation or suppression of some deformation processes in GBs of Cu with different microstructures. On the one hand, the temperature increases as well as the hydrostatic pressure growth activates the work of the Frank-Read sources. The efficiency of GBs as the sinks for dislocations increases. Both the annihilation of the screw dislocations during their double cross-slip and their annihilation during the nonconservative motion increase (Alexandrov and Chembarisova 2012).

In the intermediate nanocrystalline grain size range above approximately 10 nm and below 100 nm, it is likely that there exists competition between conventional lattice dislocation slip and diffusional deformation, with the relative contributions of these deformation modes being dependent upon the distribution of grain sizes (Gapontsev and Kondrat'ev 2002).

If the deformation is governed by the dislocation sliding, the strength increases as the grain size decreases. This continues up to a grain size limit, then the deformation is governed by the grain boundary deformation, and the strength starts to decrease as the grain size decreases; this is commonly known as the Hall-Petch inversion (Fig. 1.11).

The Hall-Petch inversion has been observed for many nanocrystalline metals produced via different routes (Bober et al. 2016). The shear deformation process of

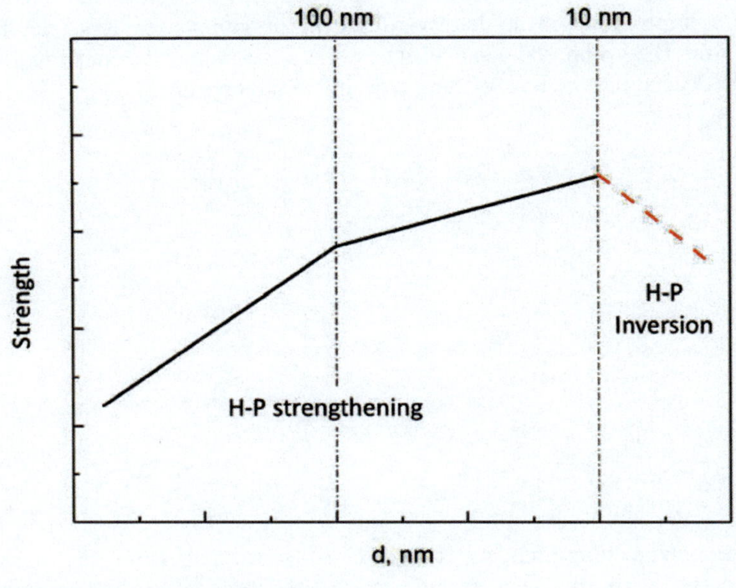

Fig. 1.11 Qualitative crystal strength vs. grain size

nanocrystals based on assuming an athermal deformation has been modeled. In accordance with them the shear propagation is composed of the grain boundary sliding and the shear deformation of crystalline matrix occurring mostly near the grain boundary plane. The latter deformation may be a twinning deformation, partial dislocation glide, or a stress-induced amorphization leading to a grain boundary thickening (Sitek and Degmová 2006). Because of the distribution of the angles between the boundary planes and the shear plane its width varies from a boundary to another. If the thickness of the boundary is t, the average width of the boundary region in the shear plane is αt, where α is about 2. If the stress necessary to propagate the shear front in the boundary region is σ_b and that in the crystalline matrix is σ_c, then $\sigma_c > \sigma_b$, and hence as the grain boundary region increases the deformation stress decreases, leading to an inverse H-P relation. Along the shear front, the fraction of the boundary region is $\alpha t/d$ and that of the crystalline region is $1-\alpha t/d$. Thus, the stress necessary to propagate the shear front is written as

$$\sigma = \left(1 - \frac{\alpha t}{d}\right)\sigma_c + \left(\frac{\alpha t}{d}\right)\sigma_b$$

The value of σ_c is material dependent while the value σ_b depends on the material, its purity, the processing to produce the nanocrystal, and the history of the treatment, because the structure of the boundary which affects σ_b should be sensitive to the purity and processing.

The quantitative variation of yield strength as a function of grain size for pure Ni, Cu, Fe, and Ti is shown in Fig. 1.12.

In Zhao et al. (2003) it is shown how the melting temperature of the nanostructured crystals decreases with decreasing particle size; the Hall-Petch relationship becomes limited and is no longer sufficient for grain sizes less than around 15–30 nm. They proposed a model where there is a numerical maximum whose location depends on the bulk melting enthalpy of the crystals:

$$\sigma(d) = \sigma_0 + \left[k_t + k_d D^{-\frac{1}{2}}\right] \exp\left[-\frac{1}{T_d}\frac{H_m/3R}{\frac{D}{6h} - 1}\right]$$

where

$$k_t = k_t' \exp\left[\frac{T_{m0}}{2T_d}\right]$$

$$k_d = k_d'' \exp\left[\frac{T_{m0}}{2T_d}\right]$$

$$H_m = T_{m0} S_m$$

If the bulk melting enthalpy is applied to nanostructured Ni (Fig. 1.13), the transition grain size can be calculated.

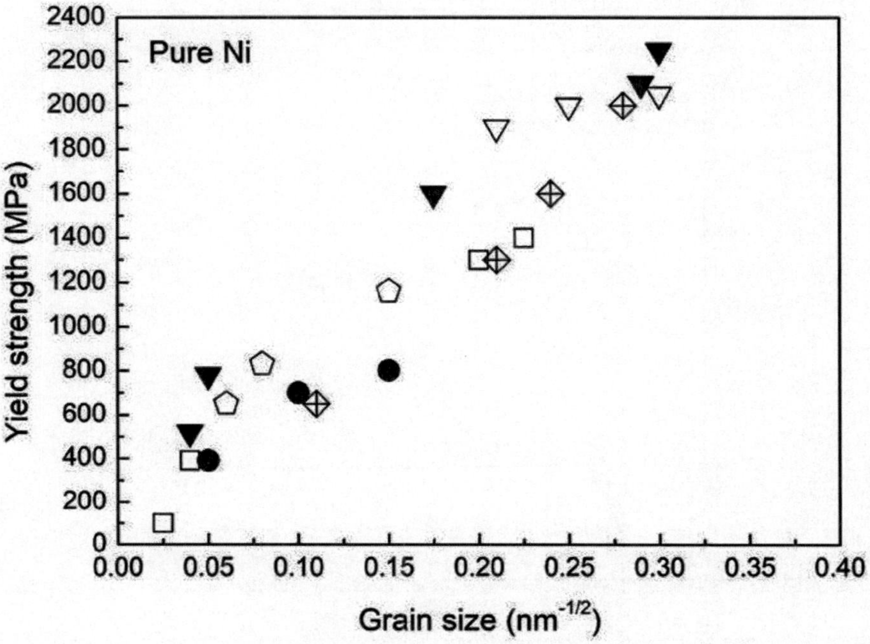

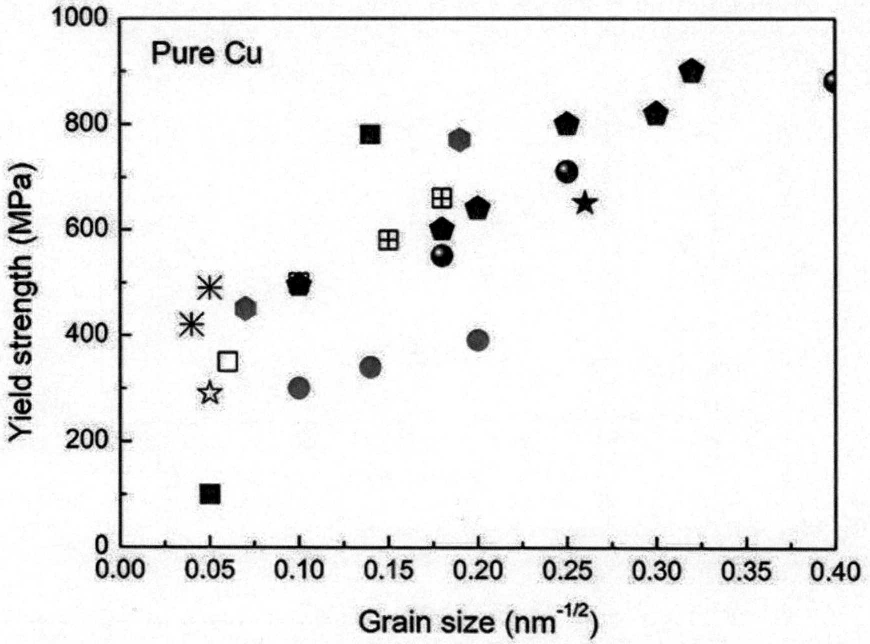

Fig. 1.12 Yield strength vs. grain size for pure Ni, Cu, Fe, and Ti

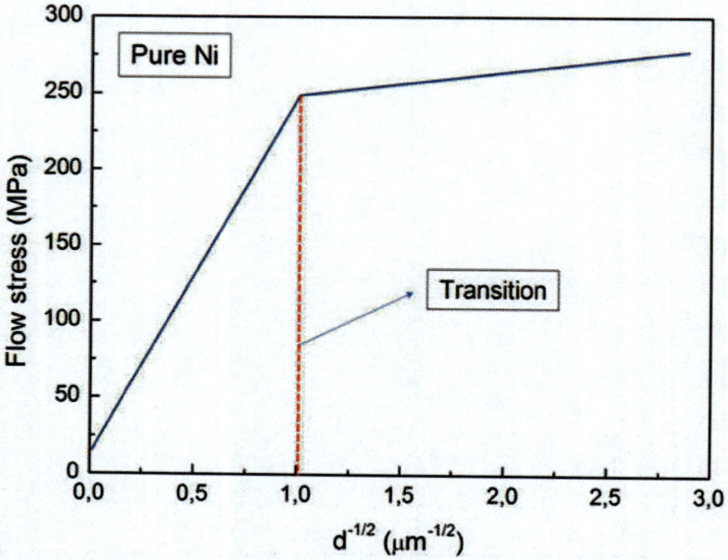

Fig. 1.13 Flow stress vs. grain size based on the bulk melting enthalpy

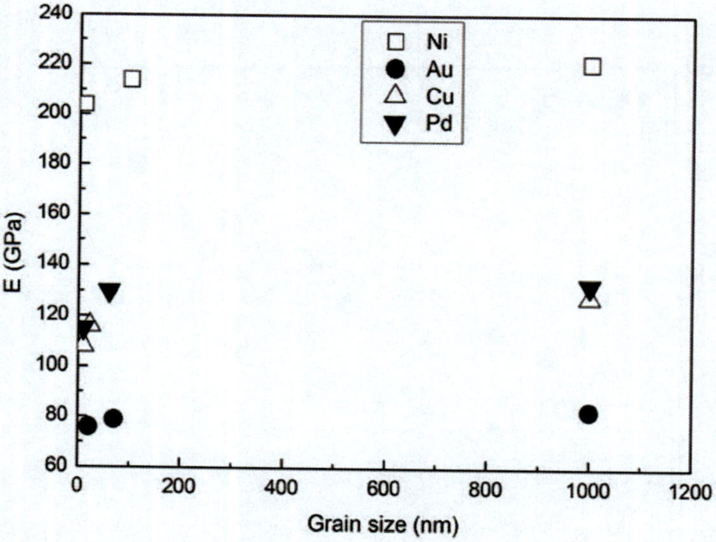

Fig. 1.14 Young's modulus variation with grain size

Another physical property to vary as the grain size is reduced is the Young's modulus. In Fig. 1.14 the E variation with grain size for different electrodeposited pure metals is shown.

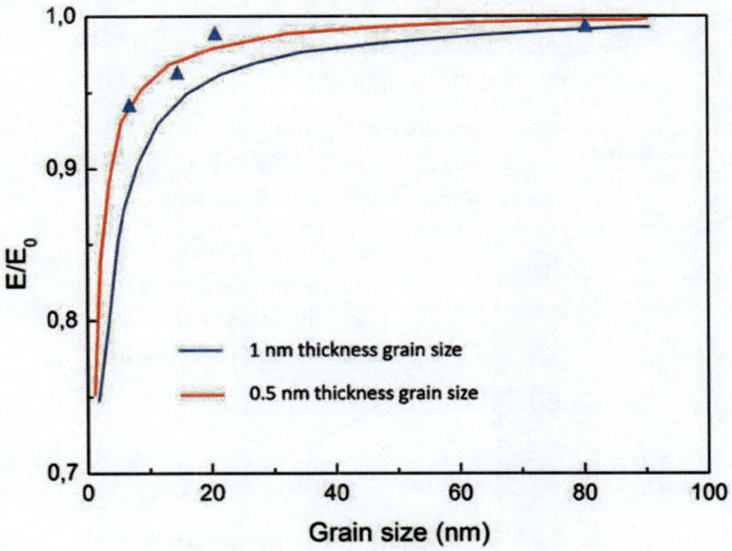

Fig. 1.15 Young's modulus variation with grain size for nanocrystalline Fe

Given that the values of a material's elastic constants reflect the bonding nature of its constituent atoms, it seems logical to expect that nanocrystalline materials would exhibit different moduli of elasticity compared to coarse-grained polycrystalline solids because of the high volume fraction of atoms located at or near the grain boundaries, triple junctions, and quadruple nodes (Ramesh 2009). In particular, since the degree of atomic structural disorder is greater within a grain boundary as compared to the crystal lattice, the average atomic distance within it is generally known to be larger. It could then be concluded that the grain boundary as a whole exhibits a lower bond strength and, therefore, has local elastic moduli values lower than those of the lattice (Zhu and Zheng 2010).

Different calculations and experimental analyses of pure Fe revealed a remarkable difference in Young's modulus with grain size (Fig. 1.15).

It is known that ultrafine-grained materials follow the Hall-Petch grain size strengthening behavior into the nanocrystalline regime. On the other hand, at some grain size transition from the regular to inverse Hall-Petch has also been reported in literature. The dependency of yield strength or hardness on grain size may become weaker, and even reverse at extremely fine-grain sizes. This phenomenon is known as the "inverse Hall-Petch effect" or "softening effect" (Carlton and Ferreira 2007). The reasons for the transition are a changing balance between competing deformation mechanisms (Zhou and Qu 2019). When conventional grain size metals are deformed at room temperature strain is carried exclusively by dislocations. Under these circumstances, grain boundaries influence deformation but they are not carriers of it. At the nanoscale, grain boundaries can mediate deformation more directly by sliding and shear-coupled migration or serving as dislocation sources and sinks.

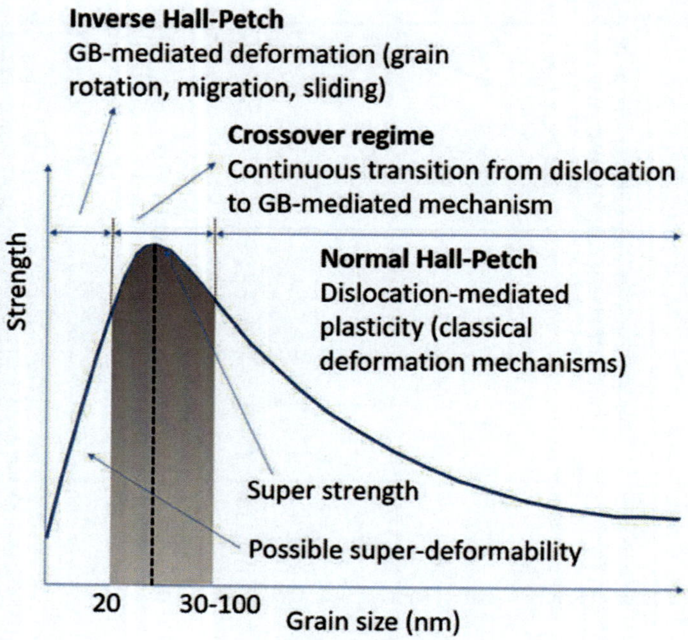

Fig. 1.16 Hall-Petch behavior at different length scales

There is strong evidence that applied shear stresses can drive rapid, diffusionless motion of nanocrystalline grain boundaries via shear coupling. Grain boundary sliding can also permit grains to slide past their neighbors, with accommodation provided by atomic shuffling. It has even been proposed that this leads to large grain rotations. At the same time, typical dislocation-based mechanisms become difficult to operate. From an atomistic point of view, there is a probability for atoms on a dislocation core to be absorbed by the grain boundary. The larger the grain size, the larger the number of atoms absorbed by the grain boundary, if the dislocation is also to be absorbed. Thus, for larger grain sizes, the probability for dislocation absorption by the grain boundary is lowered.

The most accepted general behavior is schematically shown in Fig. 1.16.

Below about 100 nm, there simply is not the space inside grains to activate conventional sources, or form pileups or dislocation tangles. This has the remarkable consequence that dislocation storage does not occur. Instead, dislocations are likely to be emitted from the boundaries, cross the grain, and be absorbed at the opposite side, without ever encountering another dislocation (Zhang et al. 2020).

The dislocation-based model (Yamakov et al. 2002) is based on the principle that dislocations are responsible for the flow in nanocrystalline materials and the dislocation energy is reduced as the grain size decreases. Anyway, by reaching a threshold grain size value twinning and partial dislocation generation lead to an increase in flow stress reducing the material strength (Zhu et al. 2008). Another diffusion-based model explains IHPE as the effect of unexpected Coble creep, combination of dislocation sliding and creep, or thermally activated grain boundary shearing (Conrad and Narayan 2000). Another model considers that the grain boundary shearing is the dominant mechanism in IHPE because of the domination on the dislocation motion by approaching very low grain sizes (Takeuchi 2001). An alternative model considers the nanocrystalline materials as composites of two phases with different yield strengths (the grain interior and the grain boundary). The grain interior deforms elastically under external stresses, while the plastic deformation of the grain boundary layer is governed by a Maxwell's equation. Based on this model, it is shown that the strength of a NC material decreases linearly with decreasing grain size, when the grain size is below a certain threshold (Fan et al. 2005). Carlton and Ferreira (2007) proposed a very precise analytical model for the IHPE. By considering dislocations with a certain length (l), it becomes increasingly important as decreasing the grain size. The structure and nature of grain boundaries do not change in the nanocrystalline regime (Van Swygenhoven et al. 2000). The modified Hall-Petch equation results to be

$$\sigma_y = \sigma_0 + k\left[\frac{1 - P_{\text{dis}}}{d}\right]^{\frac{1}{2}}$$

where P_{dis} is the probability of a dislocation being absorbed by the grain boundary and k is given by

$$k = \left[\frac{\sigma_1 \mu b}{\cos\theta \cos\varphi}\right]^{\frac{1}{2}}$$

σ_1 is the critical stress at the leading dislocation to activate a dislocation source on an adjacent grain, θ is the angle between the tensile axis and the normal to the slip plane, and ϕ is the angle between the tensile axis and the slip direction (cosqcosf Schmid factor). The equation describing the yield stress of a material as a function of grain size, taking into account the probability of grain boundary absorption, reverts to the classical Hall-Petch equation when the boundary is rigid and $P_{\text{dis}} = 0$. The essential effect of grain boundary dislocation absorption is to reduce the number of dislocations, n, in the pileup, which decreases the stress at the leading dislocation. If all moving dislocations are absorbed by the existing grain boundaries ($P_{\text{dis}} = 1$), then the equation is reduced to $\sigma_y = \sigma_0$; that is, grain boundaries do not play any role in strengthening the material. In addition, as the term P_{dis} is negative, reducing the Hall-Petch coefficient, k, is simply an inversion of the Hall-Petch relation.

1.2 Nanostructured Material Synthesis

It is well known that grain size significantly affects the properties of materials. Hence, it has led to many efforts and extensive research in investigating the best technique possible in producing the nanostructured material. A relatively large number of synthesis routes have been used to produce nanocrystalline materials resulting in a huge diversity of structures from these processing methods. Two main technological approaches are used to synthesize nanostructured materials: bottom-up (obtaining the structure by arranging the nanostructure atom by atom and layer by layer) and top-down (nanostructure obtained by breaking down the original microstructure). The various processing routes employable to obtain nanostructured materials (Meyers et al. 2006b) will be described in the following: mechanical alloying, severe plastic deformation, and electrodeposition as the main routes to produce bulk nanostructural metals and alloys (Koch et al. 2007, 2010). The current understanding of nanocrystalline metals has been primarily built around average grain size, *d*, driven by the past success of the Hall-Petch relation. At fine grain sizes where the Hall-Petch relationship breaks down, it has been replaced by new scaling rules that again relate strength to grain size. The transition from one scaling rule to another occurs at critical grain sizes where the dominant deformation mechanisms change. The first grain size threshold is 100 nm, below which dislocations nucleate at grain boundaries, sweeping through entire grains without interacting with each other and forming tangles. At even smaller grain sizes, around 10 nm, grain boundary sliding and rotation supplant dislocations as carriers of plasticity, eventually leading to an inverse Hall-Petch slope. Similar grain size-based relationships have been applied to other mechanical properties, like wear and fatigue resistance, and functional properties, like magnetic coercivity and permeability (Wolf et al. 2001; Harin et al. 2018). The common theme to the deformation mechanisms above is that grain boundary sites become increasingly important, yet characterization of nanostructured materials rarely focuses on the boundary itself. Expanding the characterization of nanocrystalline microstructures to include more grain boundary information may help address unanswered questions about structure-property relations and also open the door for control of such features in the future. Work on conventional, coarse-grained metals has demonstrated that grain boundary networks can control a wide range of properties, from fracture to corrosion. In ultrafine-grained (UFG) metals, or those metals with grain size larger than 100 nm but smaller than 1000 nm, the fraction of high angle boundaries has been implicated as a possible key to enhancing ductility (Ma 2004). Such effects are expected to be exaggerated at nanoscale grain sizes, where a large fraction of atoms resides in the grain boundary region.

Primary among various parameters that characterize the mechanical behavior of NC materials are strength and ductility. In general, strength and ductility are often inversely related: the stronger a material, the less ductile it is and vice versa (Ma 2005). This tendency is no different in case of NC materials in that although NC materials always possess higher strength than their coarse-grained counterparts, they usually exhibit relatively low ductility (Mohamed 2020). For engineering applications, low ductility is undesirable because this characteristic not only hinders the process of forming the material into useful products but also impairs the material's ability to absorb overload, a condition that reduces the margin of safety in design. A primary reason for the poor ductility in NC metals is the lack of work hardening. In general, work hardening in large-grained materials arises from the multiplication and interaction of dislocations in the interiors of the grains. However, such a process does not occur in NC materials due to the very small grain sizes (Firstov et al. 2018). Dislocations are generated from a grain boundary, move in the interior of the grain, and then are annihilated at another. This scheme results in the absence of dislocation accumulation and interactions; that is, nanograins are not able to sustain arrays of dislocations. The lack of dislocation accumulation leads to a loss of work hardening, which in turn results in low ductility (Liu et al. 2014). It has been suggested that ductility can be improved by introducing substructural features that not only impede the motion of the generated dislocations to boundaries but also enhance dislocation interactions. These features include the introduction of nanoparticles, a bimodal grain size distribution, deformation twins, annealing twins, or a combination of bimodal grain size distribution and twins. The role of bimodal grain size distributions and twins in achieving high ductility while maintaining or even increasing strength can be attributed to two factors. First, the presence of large grains enhances strain hardening, which prevents excessively large local strains and suppresses the propagation of cracks nucleated in the surrounding regions of the very small grains. Second, twins in the interiors of the large grains in the bimodal grain size distributions can block moving dislocations, a process that results in dislocation accumulation (Qian et al. 2014; Ratman et al. 2017). Dislocation accumulation in turn promotes strain hardening.

Some of the only nanocrystalline work to explicitly consider grain boundary type has been investigations into the unexpected ductility of nanotwinned materials (this will be specified in the following chapters). Since different nanocrystalline processing methods are controlled by a variety of physical mechanisms, there is reason to expect that these techniques will produce materials with different grain boundary networks. Large deformations subdivide grains through the accumulation of dislocations that form into low-energy dislocation structures (LEDS). Continued deformation causes the misorientation across LEDS to increase, forming low-angle boundaries, and eventually high-angle ones. Twin fragmentation is a complementary mechanism which has been proposed for nanoscale refinement, where deformation twins form within existing grains and narrow twins are then subdivided by LEDS. Continued deformation drives nanocrystalline grain rotation, transforming low-angle boundaries into high-angle ones. The extent to which these competing mechanisms may operate is sure to affect the grain boundary network. On the other hand, materials produced by deposition methods must be understood within a different framework. For both physical vapor deposition and electrodeposition, films form as atoms bond to the growth surface. This commonality causes similar structural development, even though one process is purely physical and the other is electrochemical (Jung et al. 2015). As new atoms deposit, they briefly undergo surface diffusion before being confined within the bulk. In nanocrystalline growth, adatoms are restricted to small rearrangements and cluster with only their immediate neighbors. Clusters grow outward until contacting adjacent grains, with grain size determined by the relative rates of nucleation and growth. As a result, the microstructure is determined by the kinetics of growth surface phenomena. In the absence of deformation or recrystallization, grain orientation is fixed at nucleation, i.e., before a grain has formed boundaries with most of its neighbors. As a general conclusion, different processes can produce very different grain boundary character distributions (GBCDs).

UFG material synthesis can be broadly divided into four different categories. They are (a) inert gas condensation, (b) mechanical alloying, (c) crystallization from amorphous solids, and (d) severe plastic deformation (SPD) (Lowe et al. 2000). The SPD route again comprises different processes which include (a) equal channel angular pressing (ECAP), (b) high-pressure torsion (HPT), (c) accumulative roll bonding (ARB), and (d) friction stir processing (FSP). Some of them will be largely described in the following.

1.2.1 Mechanical Alloying

Mechanical alloying is the repeated deformation of coarse-grained powders into a ball mill leading to the disintegration of the microcrystalline microstructure by obtaining nanostructured materials (Gupta et al. 2017). The process is normally acted on pure elemental powders at cryogenic temperatures (cryomilling) in order to induce a more severe refining and the desired composition for further processing (Witkin and Lavernia 2006). The consolidation of cryomilled powder provides a potential pathway towards large-scale manufacturing of nanostructured metallic materials (Han et al. 2007). The nanostructuring process is schematically described in Fig. 1.17.

The sequence of nanostructuring follows the increase of dislocation density with the deformation enhancement; these dislocations locate into shear bands that evolve

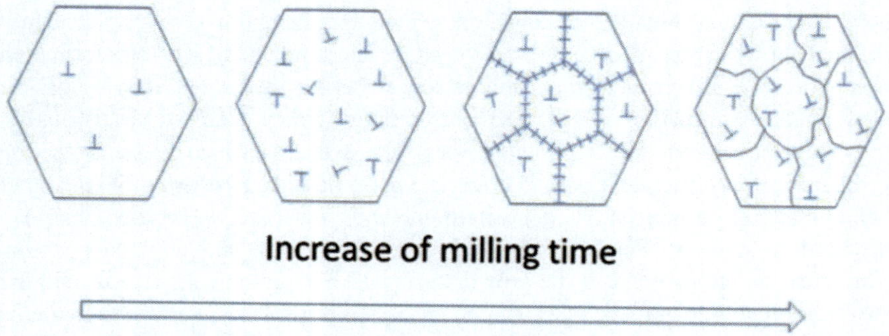

Fig. 1.17 Nanostructured grain formation during ball milling

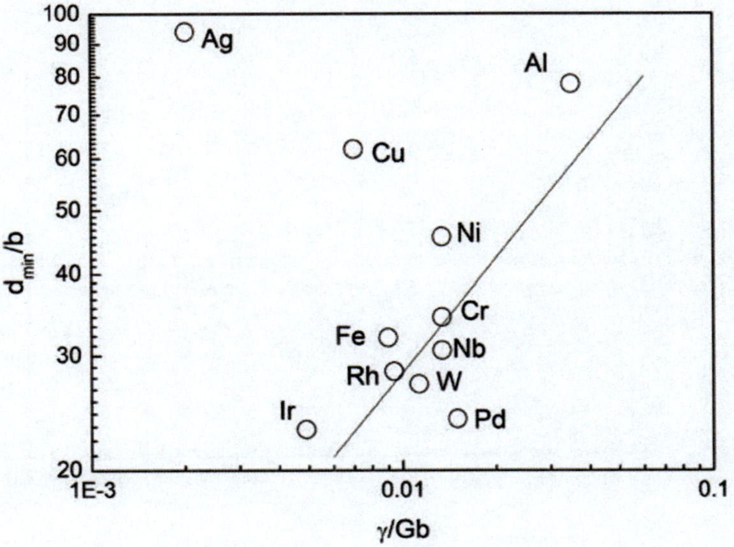

Fig. 1.18 Minimum grain size vs. stacking fault energy for milled materials

into a nanoscale subgrain through the annihilation and recombination mechanisms. The minimum grain size achievable through cryomilling is shown to be dependent on the melting temperature and on the crystal structure (Mohamed 2003). In the proposed model, the minimum grain size (d_{min}) is governed by a balance between the hardening rate introduced by dislocation generation and the recovery rate arising from dislocation annihilation and recombination. It depends on hardness, stacking fault energy, and an exponential function of the activation energy for recovery. By taking into account the parameter d_{min}/b, the dependence on the stacking fault energy (γ/Gb) is shown in Fig. 1.18.

The dependence on the melting temperature (Tm) and on the bulk modulus (B) is shown in Fig. 1.19.

The grain refinement efficiency strongly depends on the milling time (Fig. 1.20).

Lavernia et al. (2008) calculated the minimum achievable grain size with a complex equation taking into account the involved physical and mechanical parameters:

$$\frac{d_{min}}{b} = A_3 \exp\left(\frac{-\beta Q}{4RT}\right)\left(\frac{D_{PO}Gb^2}{\nu_0 k_B T}\right)^{0.25}\left(\frac{\gamma}{Gb}\right)^{0.5}\left(\frac{G}{\sigma}\right)^{1.25}$$

where b is the magnitude of the Burgers vector, A_3 a dimensionless constant, β constant, Q the self-diffusion activation energy, R the gas constant, T the absolute temperature, D_{PO} the diffusion coefficient, G the shear modulus, ν_0 the initial dislocation velocity, k_B Boltzmann's constant, γ the stacking fault energy, and σ the applied stress. The model predicts that the minimum grain size scales inversely

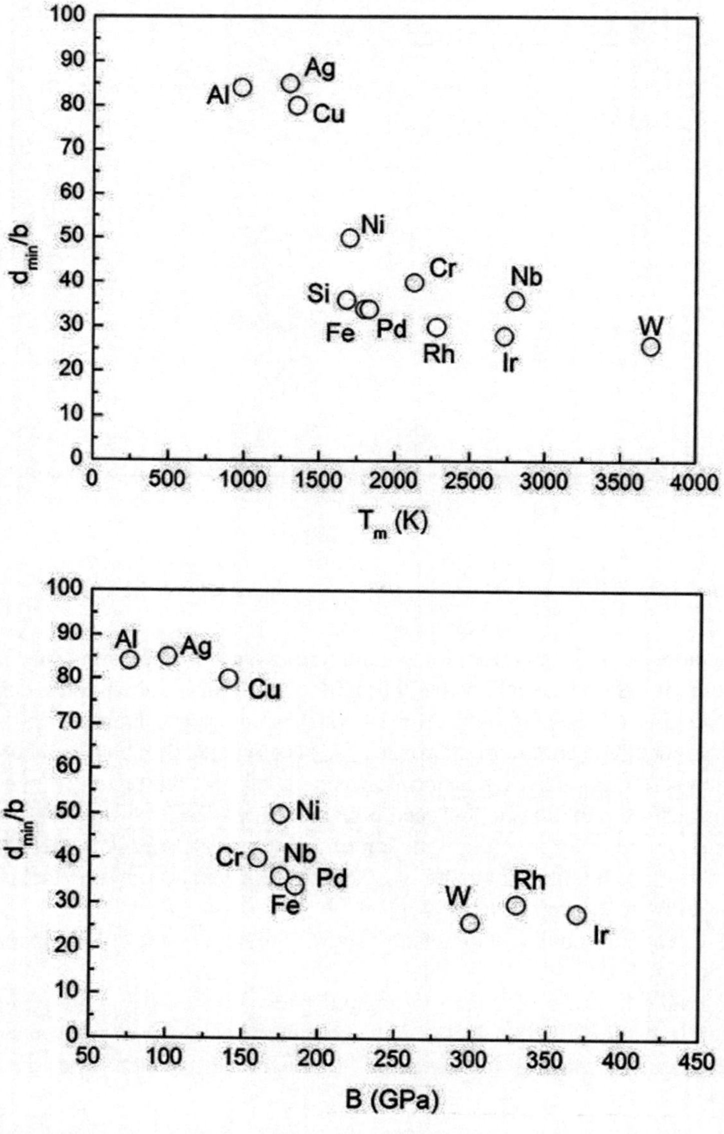

Fig. 1.19 Minimum grain size vs. melting temperature (**a**) and bulk modulus (**b**) for milled materials

with hardness, proportionally with the stacking fault energy and exponentially with the activation energy for recovery.

After milling, efficient consolidation technologies are required to compact the nanograined particles. The most employed approaches are hot isostatic pressing (HIP), Ceracon™ forging, cold isostatic pressing (CIP), spark plasma sintering

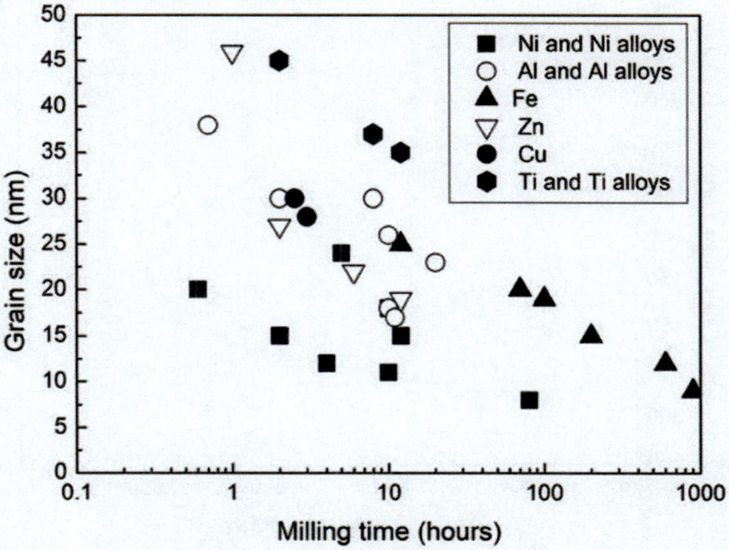

Fig. 1.20 Grain size as a function of cryomilling time

(SPS), ultrahigh pressures (UHP) during pressing, and shock consolidation, as well as others.

Obviously, the mechanical properties of the consolidated components are first of all related to the retention of the nanostructure of the milled powders (Enayati et al. 2007). As a matter of fact, Fig. 1.21 shows the grain growth of 5083 cryomilled powders with initial grain size of 305 nm after annealing for 50 h at different temperatures.

The grain growth is described by the Beck's equation:

$$D^n - D_0{}^n = kt$$

where D is the average instantaneous grain size, D_0 the initial grain size, n the grain growth exponent, t the annealing time, and k a rate constant that depends on the temperature but is insensitive to the grain size. k is expressed by an Arrhenius-type equation:

$$k = k_0 \exp\left(\frac{-Q}{RT}\right)$$

where k_0 is a constant and Q is the activation energy for grain growth. By differentiating the Beck's equation

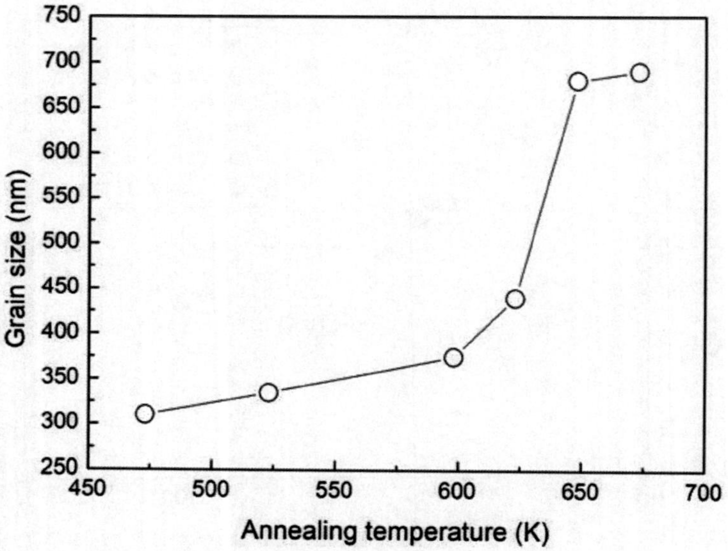

Fig. 1.21 Grain growth after annealing AA5083 for 50 h

$$\frac{dD}{dt} = \frac{k}{n}\left(\frac{1}{D}\right)^{n-1}$$

The grain growth exponent n can be calculated by the slope of log(dD/dt)−log ($1/D$).

Burke described a model of grain growth for materials containing dispersion particles. He considered that the grain growth rate is controlled by the decreasing difference between the ultimate limiting grain size and the changing value of the instantaneous grain size, rather than by the instantaneous grain size. The model is expressed by the following equation:

$$\frac{D_0 - D}{D_m} + \ln\left(\frac{D_m - D_0}{D_m - D}\right) = \frac{k_0 t}{{D_m}^2}\exp\left(\frac{-Q}{RT}\right)$$

where D_m is the limiting ultimate grain size for the particular annealing temperature. By differentiating the previous equation, the following growth rate equation is obtained:

$$\frac{dD}{dt} = k\left(\frac{1}{D} - \frac{1}{D_m}\right)$$

From the linear plot of $dD/dt - 1/D$, the value of slope (k) at the different annealing temperature can be determined.

Another approach derived the classic law of grain growth assuming that grain boundaries migrate because of curved interfaces:

$$v = MxF = M\gamma_b k$$

where the grain boundary velocity (ν) depends on the mobility (M) and the driving force (F). The driving force can be further described as the product of the grain boundary energy (γb) and curvature of the grains (κ). The grain boundary mobility obeys Arrhenius behavior:

$$M = M_0 \exp\left(-\frac{Q}{k_B T}\right)$$

where the mobility depends on the pre-exponential mobility ($MMoo$) multiplied by the exponential of the activation energy (Q) divided by Boltzmann's constant (k_B) and temperature (T). Therefore, temperature has a strong influence on grain boundary velocity. The global driving force is to reduce the total grain boundary volume. Grain boundaries are usually the most energetic components of a microstructure and it is energetically favorable to remove them. In addition, curved grain boundaries induce local tensile and compressive stresses on the crystal lattice, so the local driving force for grain growth is to decrease grain curvature. The proposed mechanism for grain growth is that atoms hop across grain boundaries to relieve the local pressure differences. Macroscopically, curvature-driven grain growth dictates that the smallest grains shrink and the largest grains grow.

For a single-phase microstructure that has a given average grain size, the grain growth rate is proportional to the average grain boundary velocity:

$$\frac{d\overline{G}}{dt} = \alpha\overline{v}_b = \frac{D_b^{\perp} 2\gamma_b V_m}{\beta RT\overline{G}\omega}$$

where the grain growth rate ($d\overline{G}/dt$) depends on the diffusivity across the grain boundary ($D_b^{\perp}$), grain boundary energy (γ_b), molar volume (V_m), proportionality constant (β), universal gas constant (R), temperature (T), average grain radius ($\overline{G}$), and grain boundary thickness (ω). If the equation is integrated from t to t_0, the difference between the final grain size and the original grain size is

$$\overline{G}^2 - \overline{G}_0^2 = \left[\frac{4D_b^{\perp}\gamma_b V_m}{\beta RT\omega}\right] t = kt$$

where the difference of the final grain size ($\overline{G}$) minus the original grain size ($\overline{G}_2$) depends on the variables described in the equation multiplied by time (t). The variables in the square brackets are assumed to be constant, and are combined into a single rate constant (k). Now, it is showed that grain boundary velocity depends on grain boundary mobility and driving force. Thus, there are two viable routes to reduce grain boundary velocity: reduce grain boundary mobility with kinetic barriers or lessen the thermodynamic driving force with solute segregation (Panin and Egorushkin 2010).

1.2.2 Severe Plastic Deformation

The top-down approach is finalized to induce UFG microstructures through heavy straining or shock loading (Valiev and Langdon 2006). The imposed strain is necessary for the conversion of the coarse-grained material into an ultrafined one through the introduction of a high density of dislocations and their rearrangement into an array of grain boundaries. Now, severe plastic deformation (SPD) may be defined as those metal-forming procedures in which a very high strain is imposed on a bulk solid without the introduction of any significant change in the overall dimensions of the solid and leading to the production of exceptional grain refinement so that, typically, the processed bulk solids have 1000 or more grains in any section (Tschopp et al. 2014). The most common severe plastic deformation techniques are equal-channel angular pressing (ECAP), high-pressure torsion (HPT), multidirectional forging, twist extrusion, cyclic extrusion–compression, reciprocating extrusion, repetitive corrugation and straightening (RCS), constrained groove pressing (CGP), cylinder-covered compression (CCC), accumulative roll bonding (ARB), friction stir processing (FSP), and submerged friction stir processing (SFSP). Among all these techniques, ECAP, HPT, ARB, and FSP are well established for the production of UFG metal alloys and composites (Viswanathan et al. 2006). Although the grain sizes of metallic materials can be substantially reduced from microscale to nanoscale through the application of various SPD techniques, an unlimited refinement process cannot occur (Sob et al. 2017). Saturated or limiting grain sizes can be achieved due to the balance between the grain refinement governed by the appropriate plastic deformation mechanisms and the grain growth induced by GB migration and dynamic recovery. It has been proposed that many factors, mainly including intrinsic parameters, e.g., SFE, melting point, activation energy, bulk modulus, and external factors, i.e., loading modes, temperature, and strain rate, crucially control the saturated grain size of single-phase materials (Borodin and Mayer 2012). A model to predict the saturated grain size ($d$min) of NS metallic materials is given by

$$\frac{d_{\min}}{b} = A\left(\frac{\mathrm{SFE}}{Gb}\right)^{q}$$

where b is the Burgers vector, G is the shear modulus, and A is a dimensionless constant. The change of q values originates from the transformation of the refinement process from a dislocation slip-dominated mechanism to one of deformation twinning. It is well established that the limiting grain size is closely related to the dislocation density and the grain refinement mechanism. Decreasing the SFE crucially restricts the recovery process, leading to an accumulation of more dislocations during SPD, while the introduction of solute atoms distributed randomly in the host lattice facilitates the generation and accumulation of dislocations induced by local free energy effects. The saturated grain size decreases with a reduction in SFE regardless of the SPD technique applied, while the precise dependence of limiting grain size on the SFE is governed at least in part by the external loading conditions. For pure metal, the theoretical and experimental investigations revealed that the steady-state grain size decreased with a lowering of the SFE. However, for alloys whose SFE is tuned by introducing solid-solution atoms into the matrix, the solute-dislocation interactions alone can actually reduce the limiting grain size, which makes the relative significance of SFE and alloying on the grain size debatable. Both factors are remarkably influential for grain refinement, and it is very challenging to separate their individual effects in the engineering alloys. In alloy systems where the introduction of alloying elements cannot effectively decrease the SFE, the effects of solute-matrix atomic-size mismatch and modulus interaction on the dislocation activities in terms of their multiplication, mobility, accumulation, and annihilation provide extra grain refinement to reduce the saturation grain size. In alloy systems where the SFE is prominently lowered by introducing solid atoms into the matrix, grain refinement is efficiently promoted when the deformation mechanism is transformed from dislocation slip to deformation twinning with a decrease in the SFE (Jang 2016). Therefore, depending on the solute dislocation interactions and effectiveness in lowering the SFE when introducing alloying elements into a matrix, both alloying and decreasing the SFE are essential methods to crucially reduce the limiting grain size obtained by SPD processing.

In ECAP a road or a bar is pushed into a die (Fig. 1.22).

As a general behavior, the equivalent imposed strain is defined by the following equation:

$$\varepsilon_N = \frac{N}{\sqrt{3}}\left[2\cot\left(\frac{\Phi}{2}+\frac{\Psi}{2}\right) + \Psi\cos ec\left(\frac{\Phi}{2}+\frac{\Psi}{2}\right)\right]$$

where N is the number of passes. The authors gave a graphical representation of this general equation allowing to immediately calculate the imposed strain as a function of the experimental layout per each single pass. This is shown in Fig. 1.23.

The main conclusion is that the effect of the arc of curvature on the imposed strain is low; in addition, for the most conventional configuration with a channel angle of 90°, the imposed strain per pass is ~1.

Now, depending on the chosen processing route different slip systems are activated during the deformation. In this way, a direct influence on the obtained microstructure will be obtained. Firstly, the processing route A is known as a route where the sample is pressed into the die without rotation (Fig. 1.24a). The corresponding slip systems associated with the route are shown in Fig. 1.24b. The sample is not rotated at each pass with two shearing planes intersecting at 90°.

The processing route B_A and the associated slip systems are shown in Fig. 1.25.

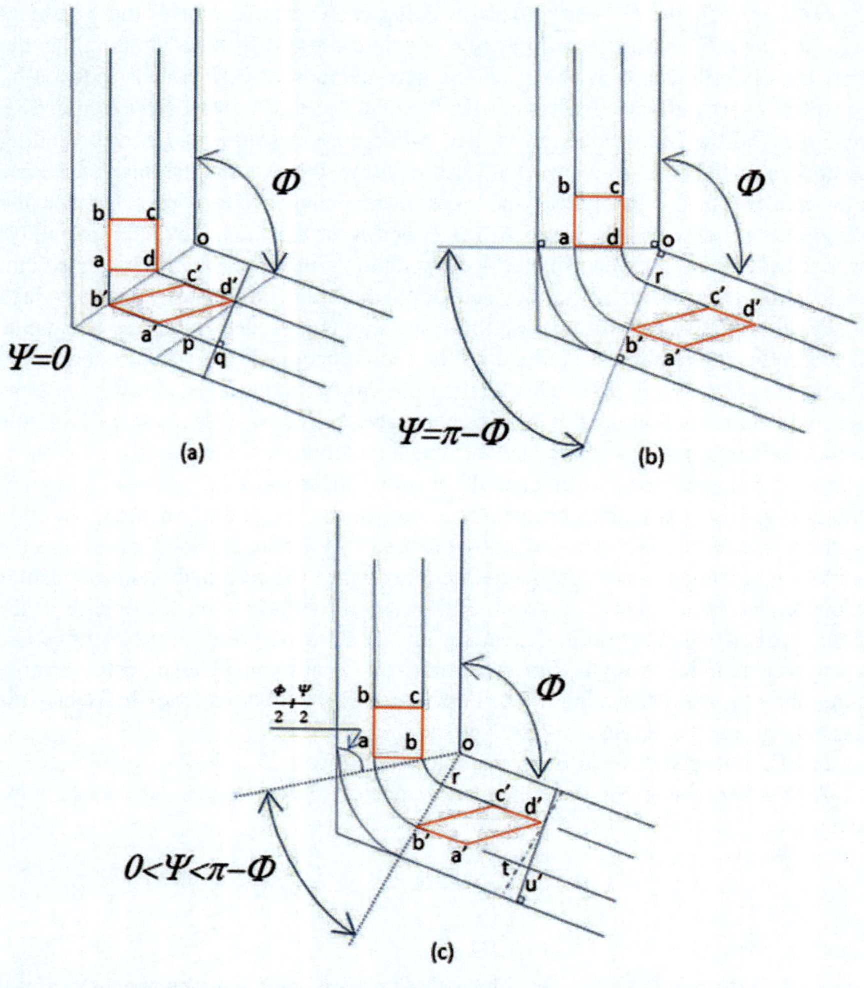

Fig. 1.22 Principle of ECAP, Φ, is the angle of intersection of the two channels and ψ is the angle subtended by the arc of curvature at the point of intersection

Here, the sample is rotated 90° in alternate mode at each pass. Here, four sharing different planes intersect at 120°.

The so-called route Bc and the corresponding slip systems are shown in Fig. 1.26.

It is apparent that route B_C is a redundant strain process because slip in the first pass is cancelled by slip in the third pass and slip in the second pass is cancelled by slip in the fourth pass.

Finally, the so-called route C and the corresponding slip planes are shown in Fig. 1.27.

In route C, the shearings continue on the same plane in each consecutive passage through the die but the direction of shear is reversed on each pass: thus, route C is

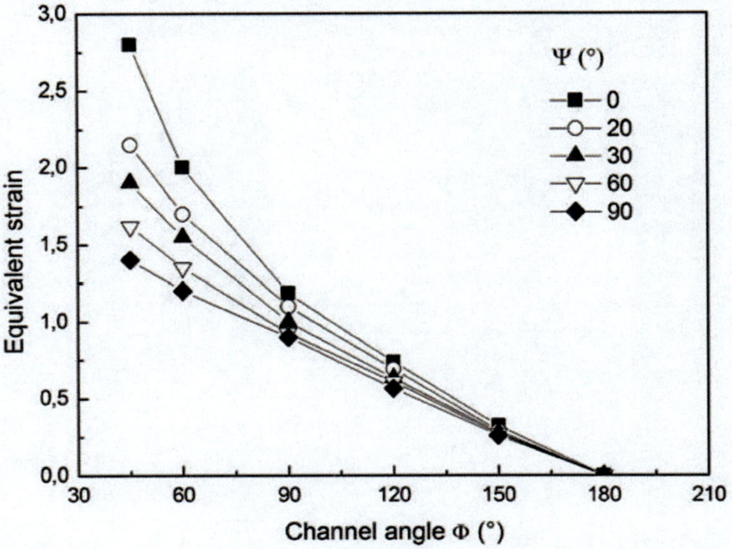

Fig. 1.23 Variation of the equivalent strain per each single pass as a function of the channel angle for different values of the arc of curvature

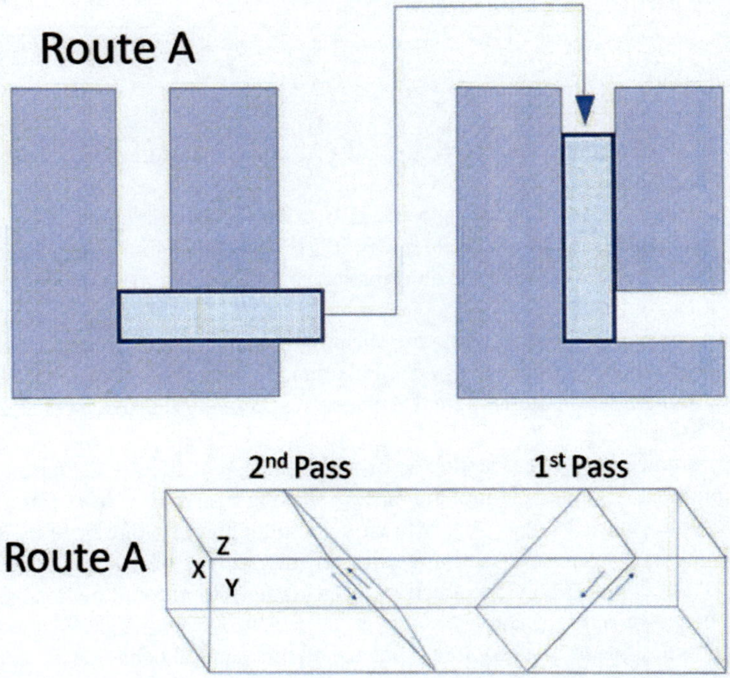

Fig. 1.24 Route A without rotation of the sample and corresponding slip systems

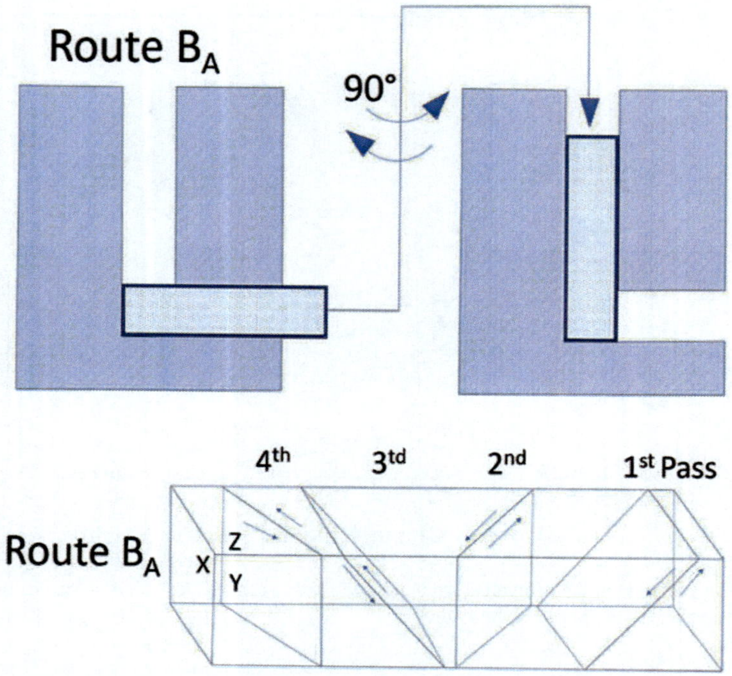

Fig. 1.25 Route B_A with 90° of alternating rotation at each pass of the sample and corresponding slip systems

termed as a redundant strain process and the strain is restored after every even number of passes.

The most employed processing route is B_C. This is because taking into consideration the deformation of a cubic element, it is restored after four passes (Furukawa et al. 1998). The schematic of the deformation of the cubic element during route B_C is shown in Fig. 1.28.

This s confirmed by considering the shearing patterns of each route; it is demonstrated that route B_C yields the largest angular range with values of η of 90°, 63°, and 63° after four passes on the X, Y, and Z planes, respectively (Furukawa et al. 2002)). This is shown in Fig. 1.29.

Many studies have been conducted on the influence of the die geometry on the ECAP process. All conclude that the most promising approach is to construct a die with a channel angle of $\Phi = 90°$, with an outer angle of curvature of $\Psi = 20°$ and with no arc of curvature at the inner point of intersection of the two parts of the channel (Han et al. 2009). The effective strain taking into account the tool material friction is shown in Fig. 1.30.

The pressing speed has very low influence on the material behavior, especially in route B_C that saturates after four passes. The pressing temperature has strong influence on ECAP because of the involvement of many microstructural features.

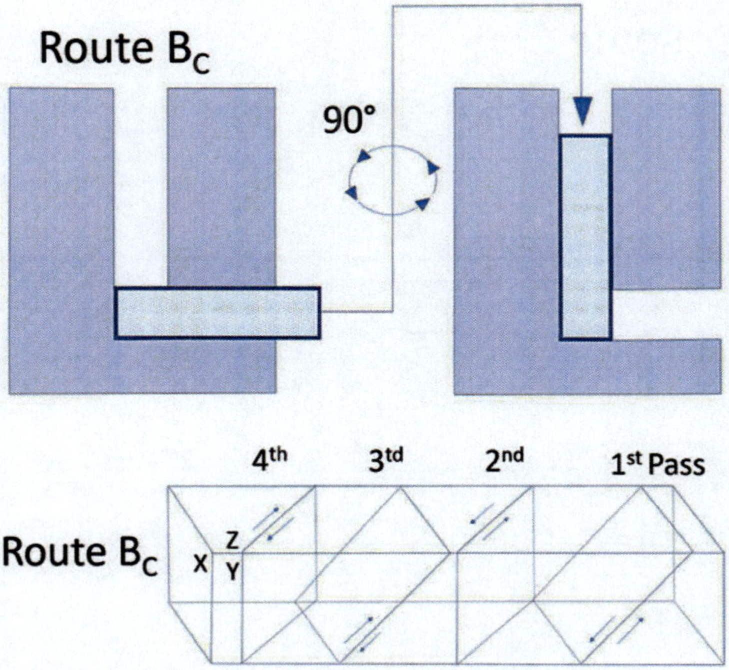

Fig. 1.26 Route B_C with 90° of rotation at each pass of the sample and corresponding slip systems

First of all, temperature increase leads to grain growth. In addition, recovery processes are accelerated with an increase of the dislocation annihilation leading to an increase of the fraction of low-angle grain boundaries. A summary belonging to different experimental evidences is shown in Fig. 1.31. All the results belong to ECAP performed with route BC and the grain size is measured after four passes.

From an industrial point of view, the employment of back pressure during pressing attracted large attention (Stolyarov et al. 2003). The employment of back pressure at the exit channel is demonstrated to be beneficial to retard cracking of the billets after many ECAP passes. In this way it is possible to achieve larger strains and deep grain refinement in the processed materials. The effect of back pressure is to increase the resultant dislocation density in both the cell walls and the cell interiors. A concomitant increase of the cell wall thickness and cell wall volume fraction and, notably, a decrease in the resultant average cell size were also revealed (Mckenzie et al. 2007).

High-pressure torsion (HPT) is a severe plastic deformation technique where samples are subjected to a compressive force and concurrent torsional straining (Zhilyaev and Langdon 2008). The schematic of the process is shown in Fig. 1.32.

The sample straining has been evaluated in Kuznetsov et al. (2015) as described in the following equation:

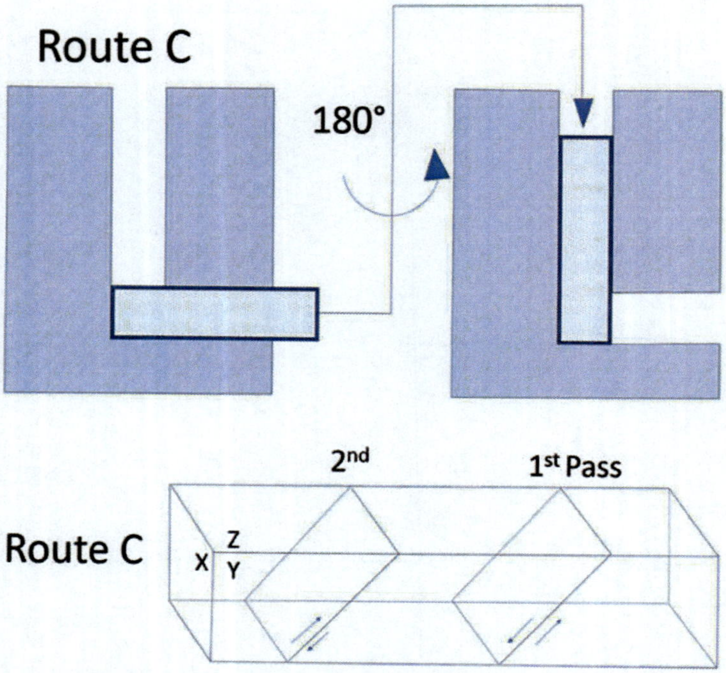

Fig. 1.27 Route C with 180° of rotation at each pass of the sample and corresponding slip systems

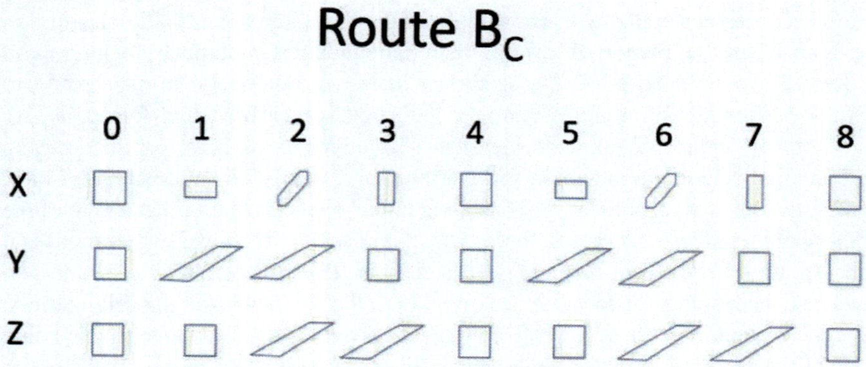

Fig. 1.28 Cubic element distortion at each pass on every plane during route B_C

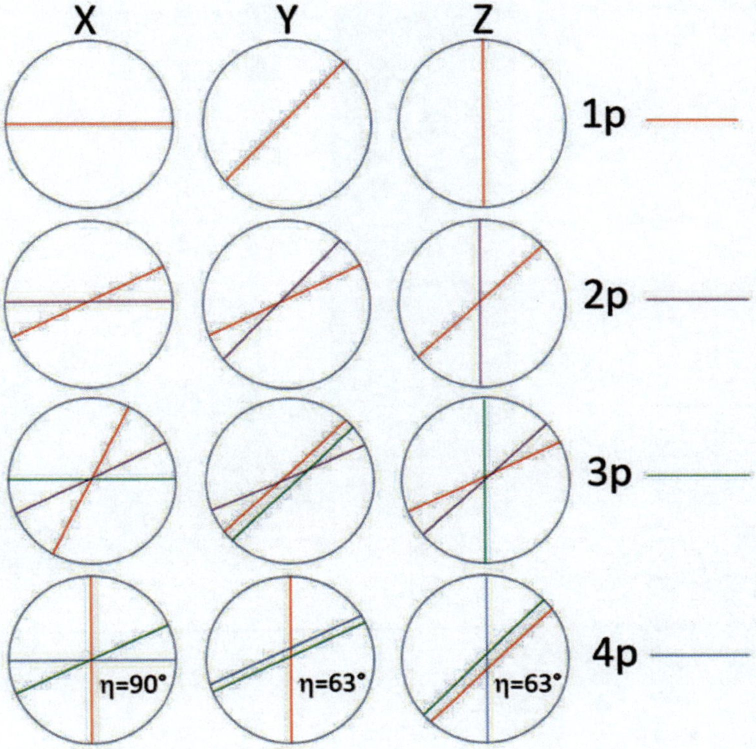

Fig. 1.29 Shearing patterns for route B_C

$$\varepsilon = \ln\left(1 + \frac{\varphi^2 r^2}{h^2}\right)^{\frac{1}{2}}$$

where ϕ is the rotation angle during torsion, and r and h are the disk radius thickness, respectively. Taking into account that $\phi r/h >> 1$ and that $\phi = 2\pi N$ with N number of rotations

$$\varepsilon = \ln\left(\frac{\varphi \cdot r}{h}\right) = \ln\left(\frac{2\pi N \cdot r}{h}\right)$$

As a matter of fact, the shear strain via HPT results in elongation along the shear direction and reduction of the smallest grain dimension (Fig. 1.33).

The basic reason for the use of this technology is that a brittle material as cast iron is capable of exhibiting very large torsional strains once compression is applied during torsional straining (Fig. 1.34).

The first observation to be underlined is that the deformation increases from the disk center to the sample edge. For this reason, the material microstructure is

Fig. 1.30 Effective strain during ECAP

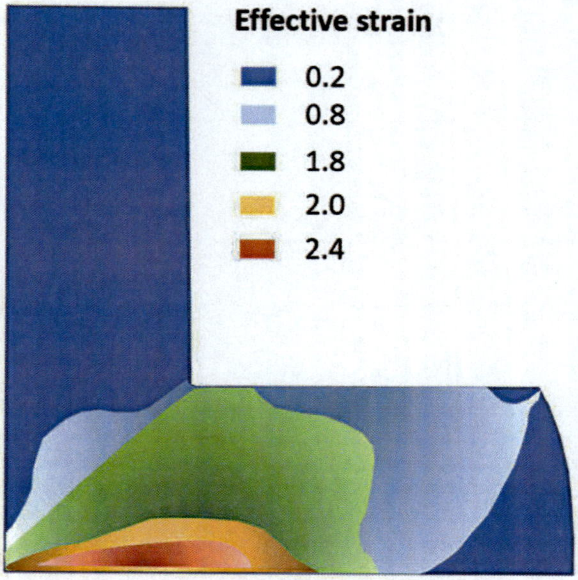

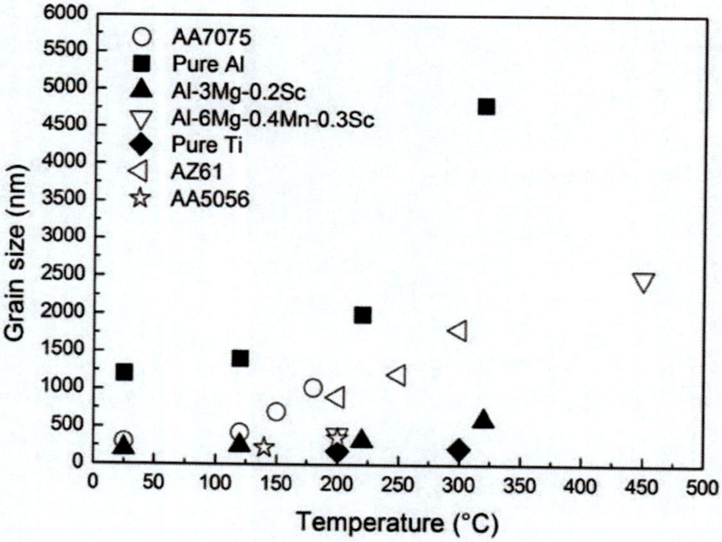

Fig. 1.31 Effect of pressing temperature on the grain size of ECAPed materials

normally nonhomogeneous. It was demonstrated that more heterogeneous plastic deformation characteristics were exhibited following the friction coefficient increase (Song et al. 2018). Anyway, microstructure and mechanical properties homogenize as the straining and compression load increase (Zhilyaev et al. 2001). Also, the

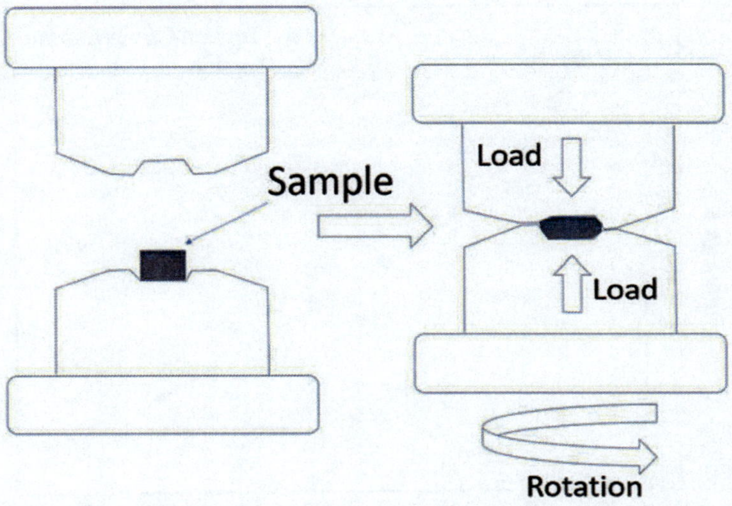

Fig. 1.32 Schematic of high-pressure torsion

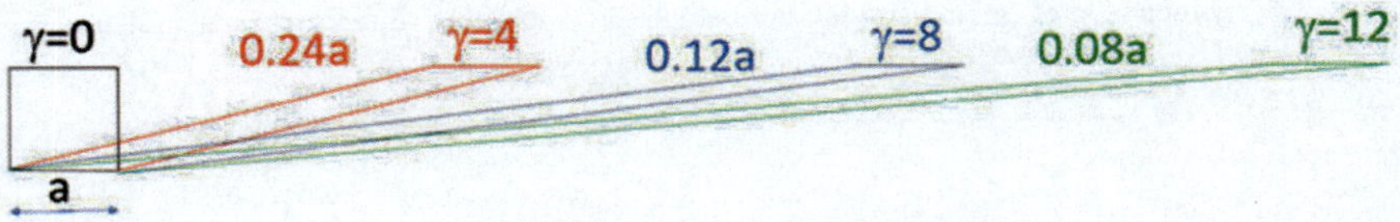

Fig. 1.33 Deformation of a cubic element as a function of shear strain

dislocation evolution was evaluated for HPT of pure metals; the results describe the dislocation density with straining:

$$\rho = \frac{2\sqrt{3}\varepsilon}{d_c \boldsymbol{b}}$$

where ρ is the dislocation density, ε is the imposed strain, d_c is the cell size, and $\boldsymbol{b}$ is the Burgers vector. Calculations performed on pure Ni allowed to measure the dislocation density at different distances from the center of HPTed material (Yang and Welzel 2005). The results are shown in Fig. 1.35.

As expected, strain decreases from the disk center to the edge, and the dislocation density follows the same trend; as a consequence, grain refinement is more pronounced by moving towards the peripheral region.

Simulations performed on HPTed materials (Fig. 1.36) show the stress distribution during torsion.

By increasing the torsional strain at constant pressure the accumulated plastic strain increases (Fig. 1.37).

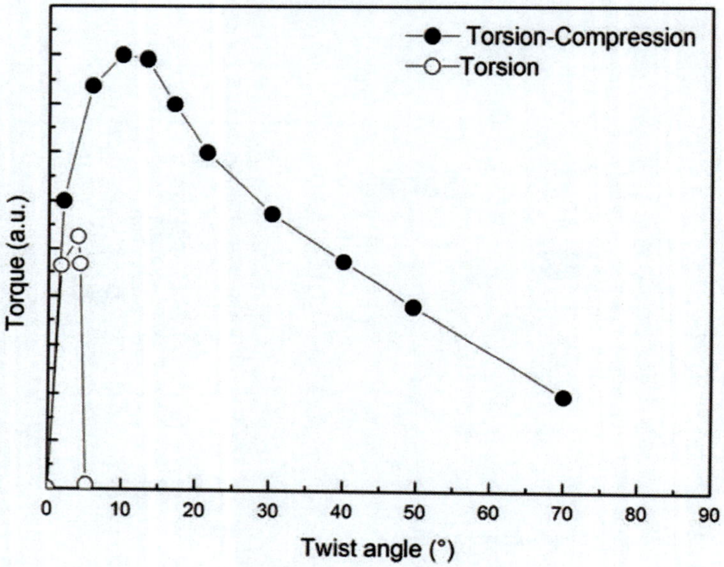

Fig. 1.34 Comparison of torsion and torsion-compression straining of the same material

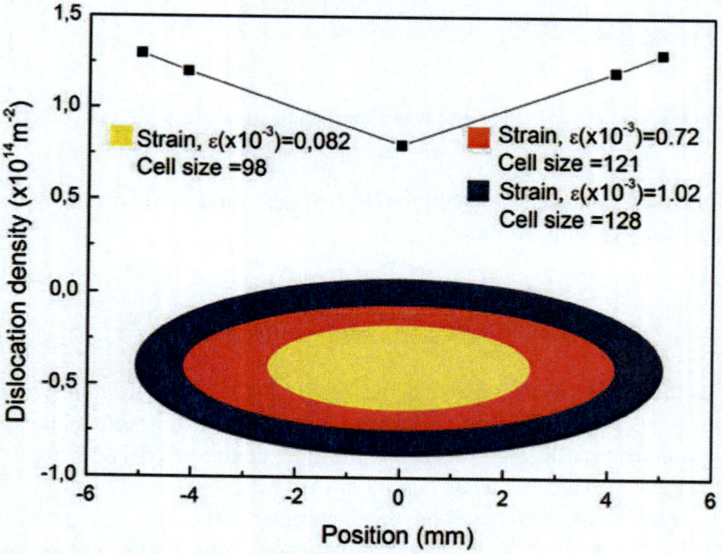

Fig. 1.35 Dislocation density as a function of the distance from the disk center for HPTed pure Ni

Estrin et al. (2008) calculated the accumulated plastic strain during HPT demonstrating that the straining becomes more uniform as the torsional turns increase (Fig. 1.38).

Fig. 1.36 FEM simulations of HPT

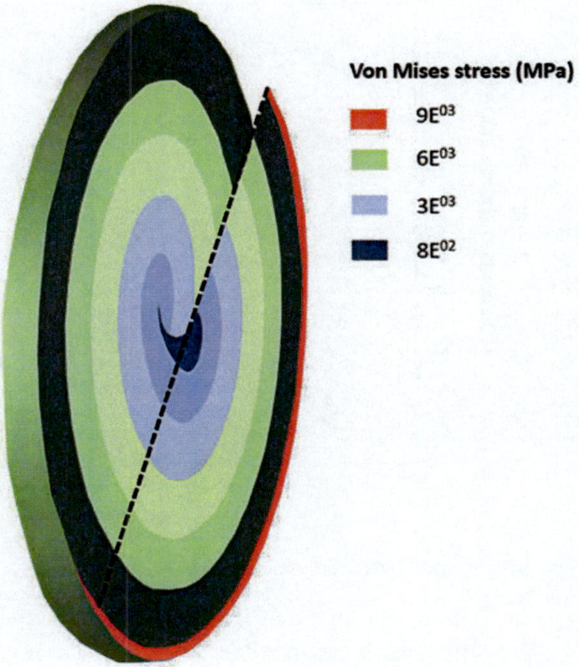

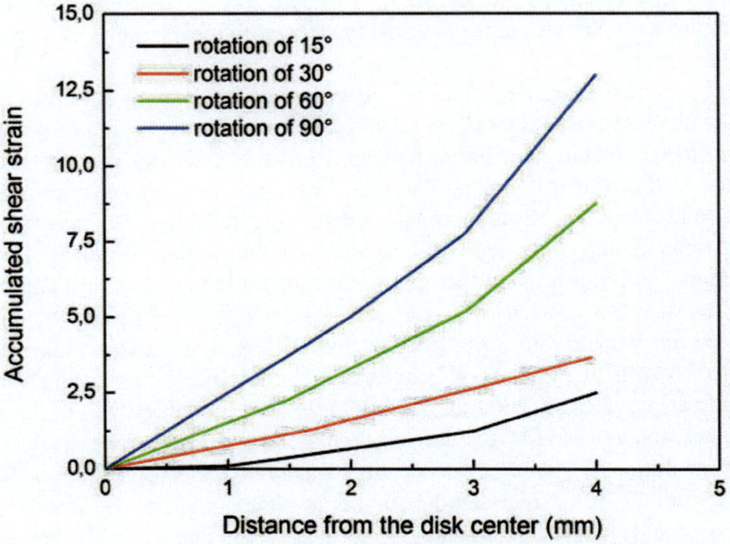

Fig. 1.37 Accumulated shear strain as a function of the torsion degrees during HPT

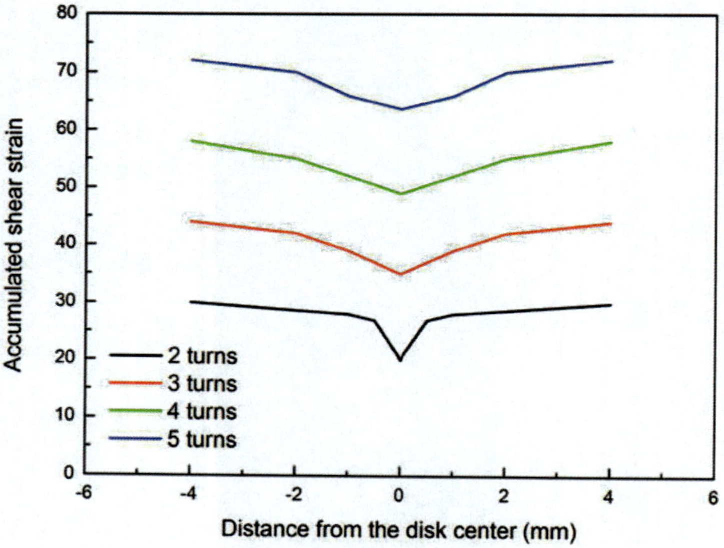

Fig. 1.38 Accumulated shear strain as a function of the torsion turns during HPT

By considering the grain refinement effect, it was demonstrated that pure materials reach a steady-state grain size depending on their crystal structure and physical parameters (Hedalati and Horita 2012). The pure metals were processed at room temperature with different processing parameters up to the obtaining of lowest grain size (Fig. 1.39).

Being a severe shear deformation, the grain size tends to be finer as the shear modulus increases. The authors revealed that the steady-state grain sizes are at the submicrometer level in elements with metallic bonding and at the nanometer level in elements with covalent bonding. The grain size decreases with an increase in the atomic bond energy and the homologous temperature. Khajouei-Nezhad et al. (2017) showed that in pure aluminum disk center and the number of turns. However, between two and four turns further grain refinement was not observed. The smallest grain size with the value of 0.41 μm was achieved at the periphery of the disks processed for two and four turns. The fraction of HAGBs increases with increasing the distance from the disk center; however considerable difference between the HAGB fractions determined at the half radius and the periphery was not observed. The highest fraction of HAGBs was achieved at the periphery of the disk processed by four revolutions (~73%). The dislocation density increased by increasing both the distance from the disk center and the number of HPT turns. The dislocation density is a monotonously increasing function of the shear strain applied in HPT processing. The maximum dislocation density was 6.8×10^{14} m^{-2} which was achieved at the periphery of the sample processed by four turns.

High-pressure torsion (HPT) processing was applied to cast pure magnesium, and the effects of the deformation on the microstructure were examined by electron

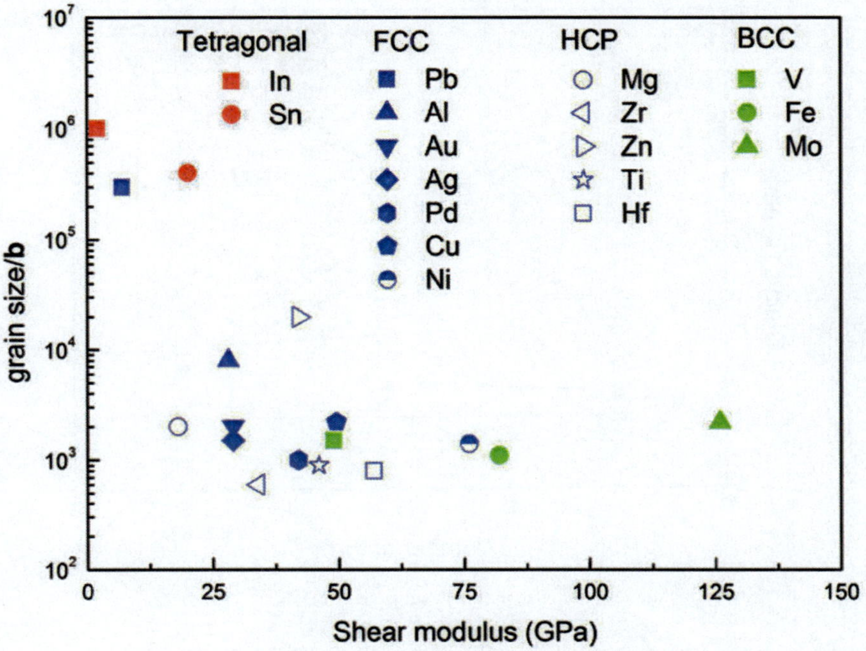

Fig. 1.39 Grain refinement for different pure metals subjected to HTP at room temperature

backscatter diffraction. Measurements showed that the grain size was effectively refined and the basal texture was intense even after processing through only one turn. Nevertheless, the microstructure became more homogeneous throughout the disks by increasing the HPT processing to five turns (Ahmadkhaniha et al. 2018). The authors calculated the evolution of HAGB and LAGBs as a function of the torsion straining (Fig. 1.40).

The refined grain structure also exhibited recovered grains with low-angle grain boundaries at the surface layer and a few large recovered grains at the bottom layer of the disk in the center region after one turn.

These series of steps that describe the microstructures resulting from SPD processes are commonly referred to as the formation of geometrically necessary boundaries and also as continuous recrystallization (Fig. 1.41).

SPD induces a uniform dislocation distribution in the deformed material (a); dislocations distribute onto a cell structure as the deformation proceeds (b); dislocations are locked by subgrain boundaries (c); elongated subgrains break up (d); subgrains reorient forming equiaxed recrystallized grains (e) (Bagherpour et al. 2019; Podrezov 2006).

A group of surface treatment techniques based on surface severe plastic deformation (S^2PD) have been developed to further improve the fatigue resistance of metallic components (Dai and Shaw 2008). There are several variants of S^2PD; however, S^2PD with high-energy ball impacts has received most of the attention

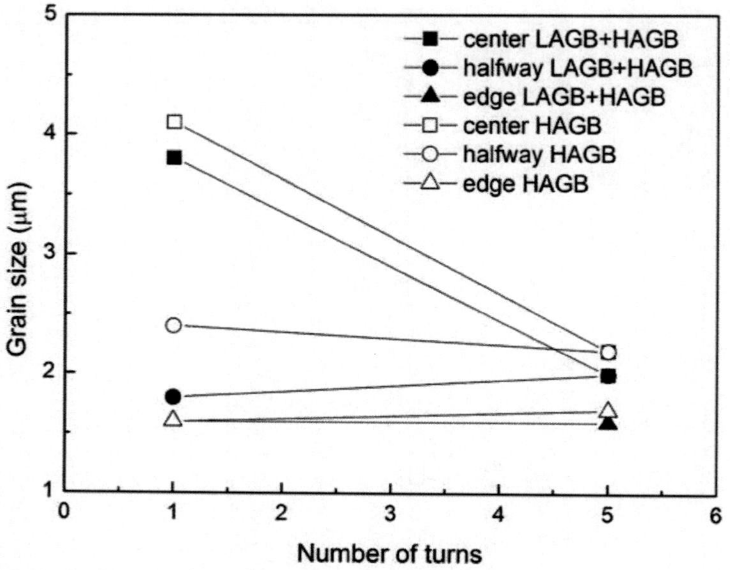

Fig. 1.40 Microstructural evolution in pure Mg subjected to HPT

because of their flexibility in processing complex-shaped objects. An example of this group of S^2PD processes is the surface nanocrystallization and hardening (SNH) process, which has been shown to be capable of enhancing the fatigue limit of a single-phase materials as well as of more complex alloys (Shaw 2013). By analyzing the strain condition at the tensile side of the sample under bending, it is concluded that the improved fatigue limit of SNH-processed samples is due to the combined effects of the NC surface layer, residual compressive stresses, and work-hardened region.

Many alternative S^2PD methods have been previously explored to improve the surface properties of bulk materials. Ultrasonic peening (UP), laser shock peening (LSP), surface mechanical attrition treatment (SMAT), ultrasonic shot peening (USP), ultrasonic surface rolling processing (USRP), surface mechanical grinding treatment (SMGT), severe shot peening (SSP), high-energy shot peening (HESP), surface nanocrystallization and hardening (SNH), and ultrasonic nanocrystal surface modification (UNSM) are some of these. Among these techniques, the principles at work in their process vary, and different grain refined layers with varying properties, such as surface topography, depth of processing layer, and size of refined grains, are produced. All S^2PD-based processes produce different deformation rates to repeat the impacts of the workpiece surface (Wu et al. 2014). Surface nanocrystallization mechanisms depend strongly on the crystal structure and stacking fault energy (SFE) of the material. For materials with high stacking fault energies, grain refinement is dominated by dislocation activities, entailing generation of high dislocation densities, formation of subgrains, and evolution of subgrain boundaries to highly

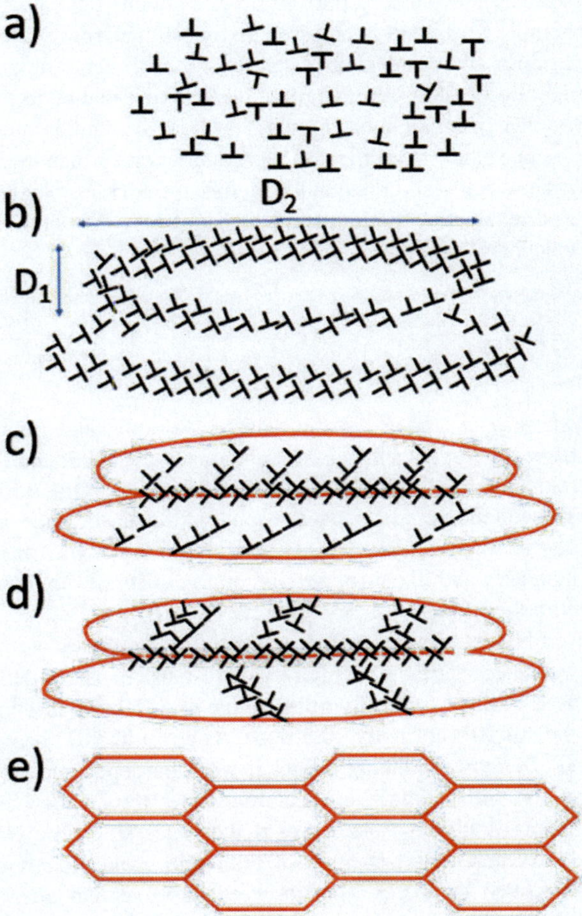

Fig. 1.41 Microstructural evolution during SPD

misoriented grain boundaries (Beyerlein and Knezevic 2020). Also the microstructure refinement during large deformation is associated with the creation of deformation-induced boundaries, including (1) geometrically necessary boundaries (GNBs), which separate regions that deform by different combinations of slip systems, and (2) incidental dislocation boundaries, consisting of ordinary cell boundaries that form by the trapping of glide dislocations. The formation of geometrically necessary boundaries with moderate-to-high misorientation requires a relatively high level of strain (Dutta et al. 2015).

Surface mechanical attrition treatment (SMAT) is a process which can transform the coarse-grained surface layer of a material into nanosized grains by severe plastic deformation (SPD). SMAT is based on mechanical impacts of material by metallic balls with high strain rate. It allows introducing a large number of dislocations and/or deformation twins in order to obtain refined grains down to nanometer scale at the surface. The nanocrystallized surface obtained by SMAT is coupled with significant compressive residual stresses and work hardening. The grain refinement and high compressive residual stress are expected to delay fatigue crack initiation and propagation of SMATed material and thus to enhance its fatigue strength. SMAT thus involves higher kinetics energies than SP, and is therefore expected to result in thicker nanocrystalline and work-hardened surface layers as well as deeper surface regions with larger residual compressive stresses. Neglecting work hardening, strain rate sensitivity, friction, and thermal effects, the depth of the plastic zone (h) can be calculated as follows:

$$h = \frac{3\pi^{\frac{1}{3}}}{(1.145)^{\frac{1}{4}}\left(\frac{2}{3}\right)^{\frac{1}{12}}(2.5)^{\frac{1}{20}}} \frac{\rho^{\frac{1}{5}}v^{\frac{3}{20}}R^{\frac{2}{3}}}{E_H^{\frac{1}{6}}E_S^{\frac{1}{30}}}$$

where ρ, v, and R are the density, velocity, and radius of the impacting sphere, respectively; E_S is the elastic modulus of the solid; and E_H is the equivalent modulus related to the Young's moduli and Poisson's ratios of the ball and solid materials. The equation predicts that the depth of the plastic zone, and therefore the thicknesses of the work-hardened layer and of the layer with residual compressive stresses, increases with the size, density, and velocity of the impacting bodies, and therefore with their kinetic energy (Ortiz et al. 2010).

SNH is similar to SP in that both processes entail repeated impacts of the workpiece surface by high-velocity balls or shots. However, the impacting balls used in SNH (typically in the range of 5–10 mm in diameter) are much larger than shots in SP (normally ~0.2 mm). As a result, SNH has a higher kinetic energy than SP, thereby producing a thicker work-hardened surface layer and a deeper surface region with larger residual compressive stresses than SP. SNH can also generate a nanocrystalline surface layer if the processing time is long enough. It is demonstrated that fatigue strengths in both high-cycle and low-cycle fatigue regimes can be improved by S^2PD. The observed improvement after S^2PD processing has been attributed to the presence of a nanocrystalline surface layer and high dislocation densities in the work-hardened region, both of which increase the resistance to fatigue crack initiation. It is well known that SP generates compressive residual stresses on the surface region of the workpiece. This is also true for S^2PD which also produces a thin NC surface layer. It was demonstrated that in S^2PD the NC surface layer and work-hardened surface region are more effective than residual compressive stresses in improving the fatigue limit.

As one of the S^2PD methods that improve material surface properties, UNSM has already been used as a treatment for many materials. The properties of wear and fatigue of materials can be improved through UNSM, as it produces surface hardness and compressive residual stress. With an increase in strike number, higher fatigue strength is obtained and small fatigue cracks are restrained under the compressive residual stress produced by UNSM treatment. Force analysis between strike pin and specimen surface using a strike tip is shown in Fig. 1.42.

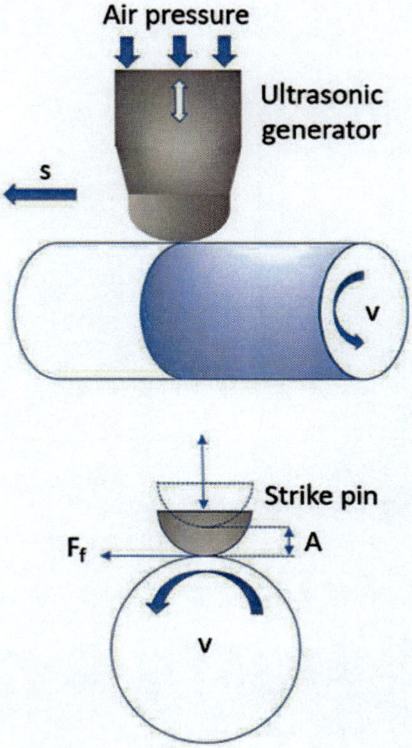

Fig. 1.42 UNSM process schematic

The frictional force between the strike pin and specimen surface increases as the static load grows. At the same time, the amplitude of ultrasonic process decreases under high static load. The total force between the strike pin and specimen surface is expressed by

$$\boldsymbol{F_m} = \boldsymbol{F_s} + \boldsymbol{F_a} \sin \omega t + \boldsymbol{F_f}$$

where F_s is the static loading produced by air compressor devices, F_a is the loading produced by ultrasonic generator which changes with processing time (ω is $2\pi f$, f is processing frequency), and F_f is the friction force between strike tip and workpiece surface. The friction force is smaller than F_m and F_m plays an important role in producing plastic deformation. As the static loading increases, F_m increases exponentially. Deeper and more refined grains in the surface layer are produced. However, as the static loading increases, the friction force between the strike pin and treated surface is enhanced along with a decrease in amplitude and increase in contact time. The increase of friction force and contact time, which is not in a vibrational direction, produces instability in the ultrasonic system working in a resonance state. This force perpendicular to the striking direction destroys the resonance produced by the ultrasonic system, and the UNSM system cannot work

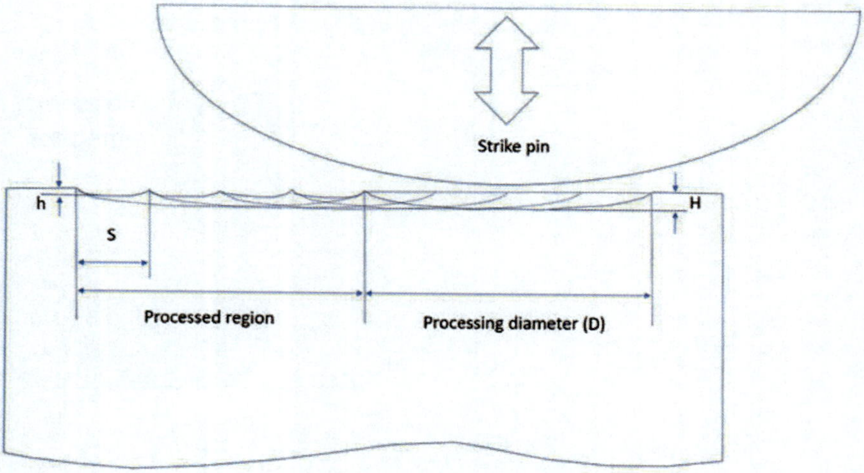

Fig. 1.43 UNSM process schematic

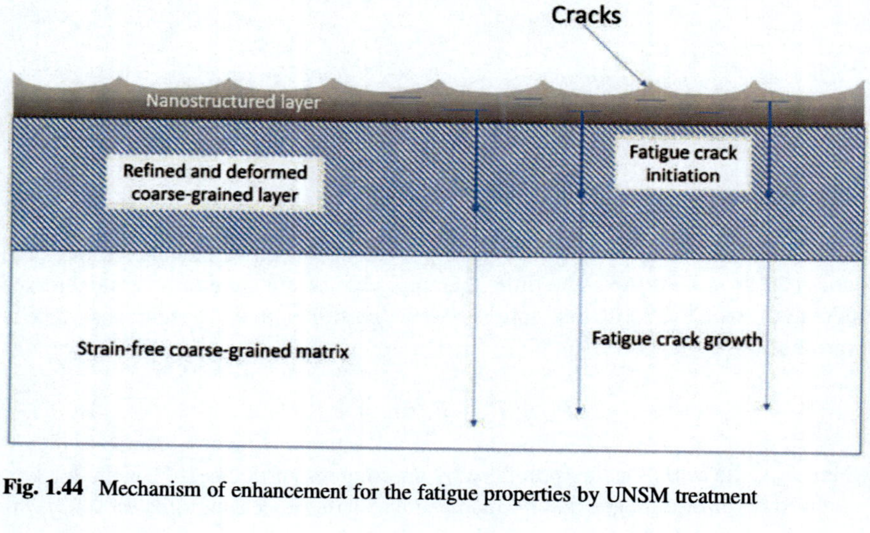

Fig. 1.44 Mechanism of enhancement for the fatigue properties by UNSM treatment

in a stable state. During the process of UNSM, the strike pin moved along an axial direction by guide unit. Uniform processing is obtained as the specimen rotates on the metal surface. The UNSM process along an axial direction is shown in Fig. 1.43.

The refined and deformed layers play a primary role in the enhancement of fatigue properties (Fig. 1.44).

Another bulk severe plastic deformation technology is the accumulative roll bonding. After heating at a suitable temperature, two or more metal sheets are rolled to a given thickness. So, a bonding interface is created between the sheet surfaces.

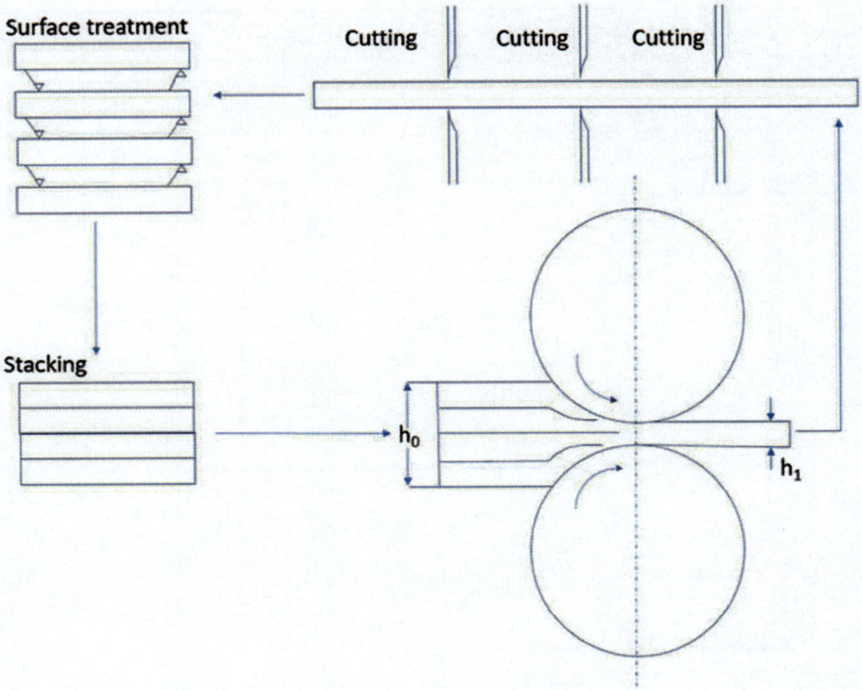

Fig. 1.45 Schematic of ARB

The product of the first rolling is cut into two or more sheets that are overlapped and rolled again (Fig. 1.45).

As the number of rolling passes increases, the grain size of the samples gradually decreases. The imposed strain is given by the following equation:

$$\varepsilon = \frac{2N}{\sqrt{3}} \ln \frac{h_0}{h_1}$$

It is calculated that the ARB can induce the largest strain in the samples (~1.6). This implies that the process requires a smaller number of passes to obtain the same total equivalent strain in the workpiece (Yu et al. 2014). First of all, the interfacial bond strength is fundamental for the quality of the final sheet. In Fig. 1.46 the bond strength between two similar or dissimilar sheets is shown as a function of the deformation.

In order to achieve optimal bonding, a reduction of 70% must be produced.

Brailovski et al. (2011) showed that in the strain intensity-deformation temperature-grain size space (Fig. 1.47), there is the grain refinement potential of HPT and ECAP techniques, when they are applied to Ti–Ni shape memory alloys.

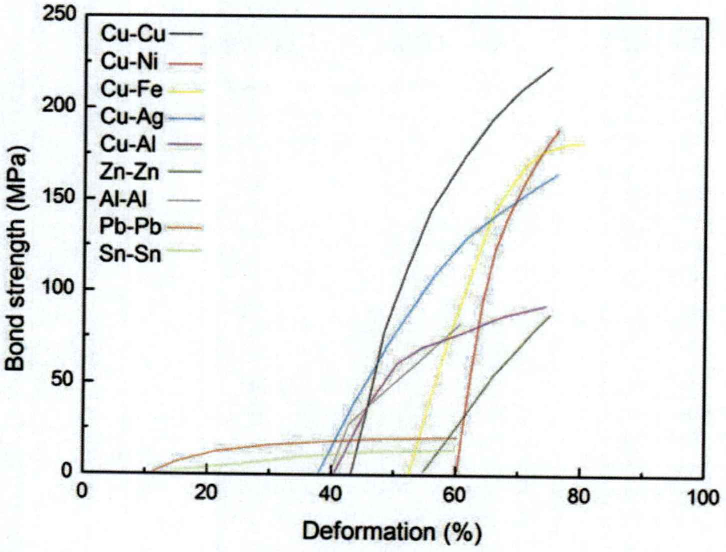

Fig. 1.46 Bond strength as a function of deformation during ARB

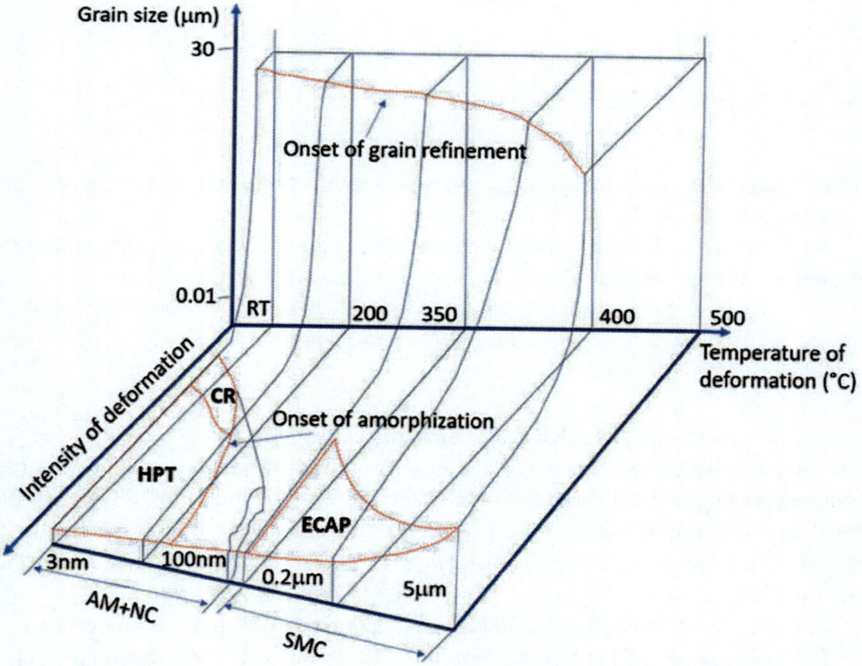

Fig. 1.47 SPD techniques' grain refinement potential in the strain intensity-deformation temperature-grain size space

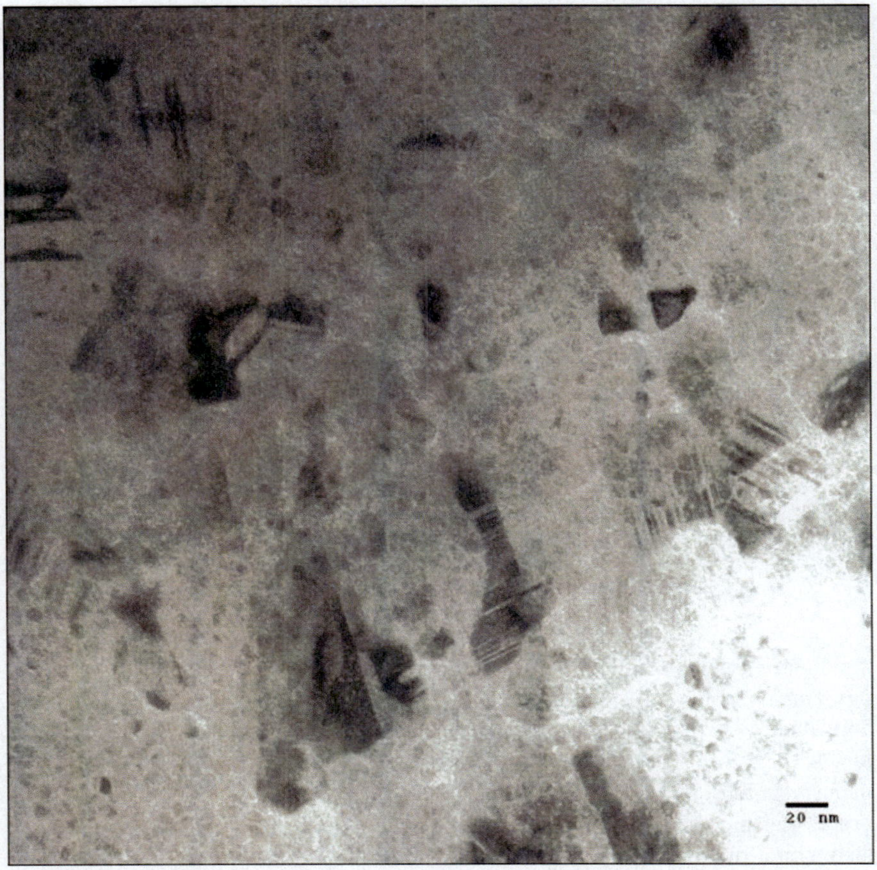

Fig. 1.48 Mixed amorphous–nanocrystalline microstructure

HPT processing below 300 °C (true deformation strain e from 1.5 to 6. 0.8) of Ti–Ni alloys results in a mixed amorphous–nanocrystalline structure (AM + NC). Homogeneous nanostructure can then be created using post-deformation annealing in the 350–450 °C temperature range (Fig. 1.48).

However, HPT is appropriate for exclusively fundamental investigations, being limited to an individual processing of thin disk shape specimens with 3–10 mm (maximum 20 mm) of diameter. ECAP is much more efficient than HPT from both the productivity and product size points of view. However, ECAP technology in its actual state does not allow creating true nanostructured Ti–Ni alloys. Given low deformability of these alloys, the lowest possible ECAP deformation temperature corresponds to 350 °C, which results in ECAP formation of submicrocrystalline structure (SMC) (grain size >100–200 nm)—instead of true nanostructure.

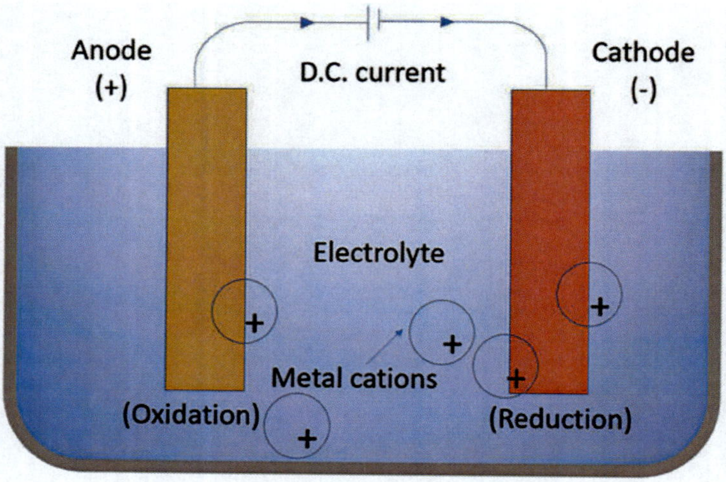

Fig. 1.49 Electrodeposition cell

1.2.3 Electrodeposition

Among the available bottom-up technologies, electrodeposition is the main technology employed for the production of nanocrystalline metals and alloys (Protsenko and Danilov 2020). Metals from the iron group (Ni, Fe, Co); some of the precious metals, e.g., Au, Pt, and Rh; and a number of others can be electrodeposited in nanocrystalline form through proper control of the chemical and electrical processing parameters. During electrodeposition, the metal or the alloy structure grows on a substrate thanks to the electrochemical reaction of ions belonging to an electrolyte (Gamburg and Zangari 2011). Metal films are deposited via the electrochemical reduction of the corresponding metal cations from the electrolyte. Depending on the electrochemical conditions (mainly temperature and current) the grain size of the deposit can be precisely tuned (Agarwal et al. 2010). In order for an electrodeposition to take place, the major components of the electroplating cell consist of the electrolytic plating solution, two electrodes where one acts as the anode and the other cathode, and a power source (Fig. 1.49).

The cathode acts as the site for metal deposition. Metallic films are formed at the cathode when electrons are consumed which reduces the aqueous metal ions.

The equilibrium potential (E^{eq}) on the ion activity (a_{Me+}) is described by the well-known Nernst equation:

$$E^{eq} = E_0^{eq} + \frac{RT}{zF} \ln \left(a_{Me+}\right)$$

where T is the absolute temperature, R is the gas constant, and F is the Faraday constant.

In the case of pure Ni electrodeposition, the most common cathodic reactions are

$$Ni^{2+} + 2e^- \rightarrow Ni$$

or

$$Ni^{2+} + H_2O \rightarrow NiOH^+ + H^+$$

$$NiOH^+ + e^- \rightarrow (NiOH)_{ads}$$

$$(NiOH)_{ads} + H^+ + e^- \rightarrow Ni + H_2O$$

Current is supplied by an external power source and the circuit is completed by the anode-cathode electrode couple immersed in the electrolyte. In principle, the negatively polarized cathode may be any electrically conductive material that is chemically resistant to the plating solution, while the positively charged anode may be either a nonconsumable dimensionally stable anode such as Pt-clad Nb or a consumable anode that oxidizes as electrodeposition proceeds. The applied current density has two important effects on electrocrystallization, namely it increases the rate of metal ion reduction on the surface and reduces the critical crystal nucleation size by increasing the applied overpotential. The critical crystal nucleation size, r_c, is inversely proportional to overpotential, η/current density, I, as shown by the Gibbs-Kelvin equation:

$$r_c = \frac{2\Phi M}{\rho z F \eta}$$

where Φ is the interfacial tension of the metal/solution interface, M the molecular weight, ρ the density, and zF the molar charge. Higher current density results in smaller crystal sizes. In addition to current density effects, electrolyte additives can have a drastic effect on the surface diffusion behavior of the adatom once it has deposited onto the cathode surface. In other words, bath additives designed to discourage adatoms from diffusing to "join" a preexisting crystal on the deposit surface in lieu of nucleating a new crystal are powerful tools with which to influence the overall extent of microstructural refinement. Electrical parameters that maximize the driving force for grain nucleation throughout electrocrystallization, combined with inhibition of adatom surface diffusion by chemical means, represent the methodology for the production of nanocrystalline materials by electrodeposition (Fig. 1.50).

Electrodeposition is considered to be the most attractive method in synthesizing thin, protective, nanostructured coatings as well as magnetic materials (Erb et al. 2000). Electrodeposition provides many advantages as previously discussed

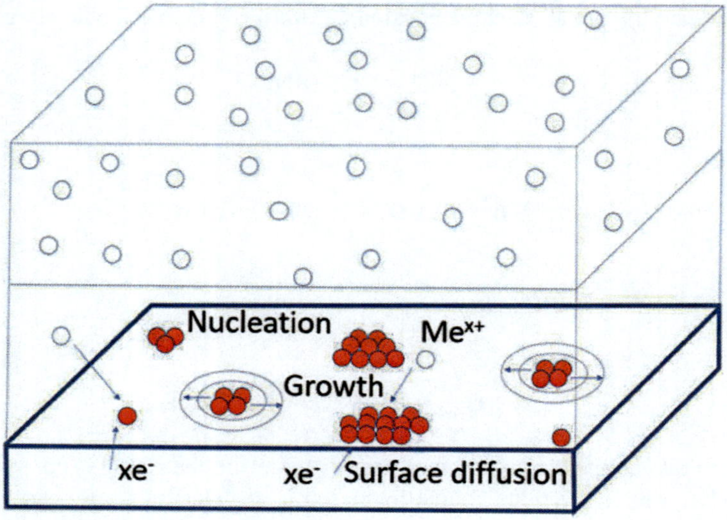

Fig. 1.50 Nucleation, growth, and surface diffusion during electrocrystallization

compared to other processing techniques. In addition, electrodeposition allows the thickness and microstructure of the films to be controlled by adjusting the deposition parameters such as the electrolyte composition, pH, temperature, current density, and agitation (Xu et al. 2009). Now, in pure metal systems a physical limit fixes the minimum obtainable grain size at around 10 nm. In order to produce nanocrystalline materials with finer grain sizes, two or more elements can be deposited by varying the bath temperature and the pulsed current (Liu and Kirchheim 2004). In this way, the solid solution elements segregating at grain boundaries allow for the production of materials with extremely fine grains (Fig. 1.51).

In addition, by varying the bath temperature and the current density during deposition, it is possible to produce a material with a graded microstructure along the growth direction. In engineering applications, property gradation offers possibilities to control and optimize material response through redistribution of stresses, either mechanical or thermal, as well as relaxation of stress concentration zones, and control of local crack driving force. Such materials can find applications in diverse fields such as optimization of thermomechanical stresses in aircraft parts and space vehicles, damage-resistant surfaces in armored plates and bulletproof vests, and barrier coatings for structural components and industrial tools (Clement et al. 2012). The described gradient can be achieved through the variation in the elastic properties (Young's modulus) or in the plastic ones (yield strength) as shown in Fig. 1.52 (Cavaliere 2009).

As previously described, the addition of solid solution elements allows for the obtaining of a broad range of grain sizes and in addition it is possible to tune the electrodeposition processing parameters in order to obtain samples in which the grain size varies linearly along the thickness. Such possibility allows that the grade

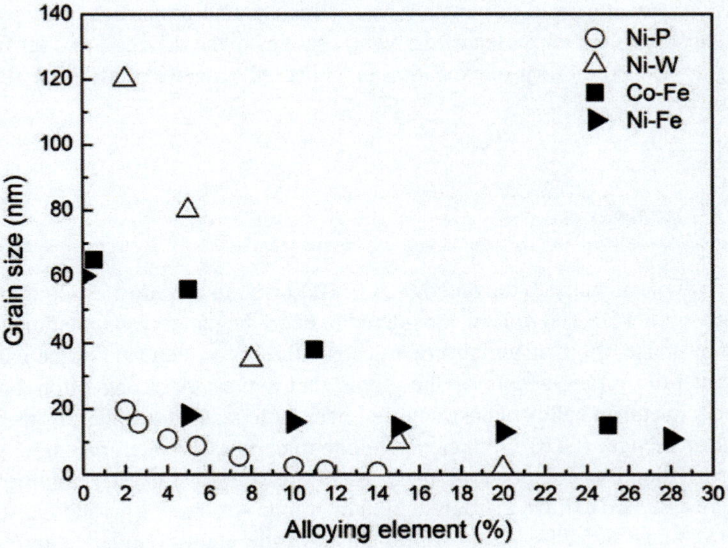

Fig. 1.51 Grain size variation in Ni-based electrodeposited alloys

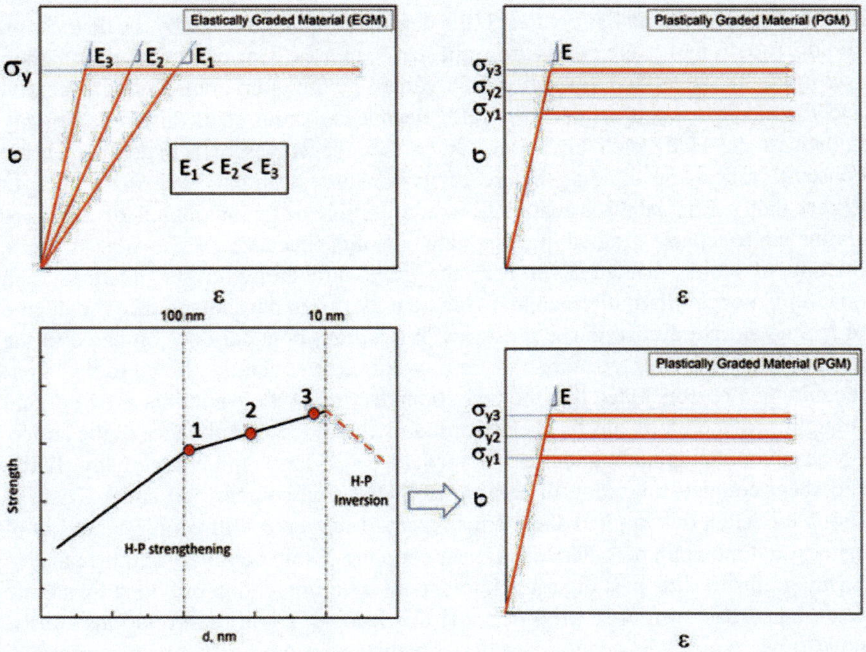

Fig. 1.52 Gradient in elastic and plastic properties of materials as a function of different location and yield strength variation with grain size as described by the Hall-Petch relationship

in the microstructure and consequent plastic mechanical properties of nanocrystalline electrodeposited alloys leads to several changes in the driving force for fracture, leading to the possibility of new design philosophy against failure of structural components.

1.3 Conclusions

Metals with a grain size in the range of 100–1000 nm are classified as ultrafine grain; grain sizes less than 100 nm are considered to be in the nanocrystalline domain. The altered response of NC material properties is a direct consequence of the nanoscale microstructural arrangements of the atoms themselves. A strong effort has been devoted to metal and alloy nanostructuring in order to reach the material's theoretical maximum strength (G/10). In nanocrystalline materials, the intercrystalline volume fraction is found to comprise as much as 50% of the total crystal volume. These solids are assumed to have a different kind of atomic structure: a crystalline structure with long-range order for all the atoms far from the grain boundaries and a disordered structure with some short-range order at the interfacial region. Hence, the mechanical properties of these nanocrystalline materials are expected to be different as compared to their equal polycrystalline material. So, the overall behavior of nanocrystalline materials is dependent not only on the grain size but also on the nature of grain boundary structures. If the deformation is governed by the dislocation sliding, the strength increases as the grain size decreases. This continues up to a grain size limit, and then the deformation is governed by the grain boundary deformation and the strength starts to decrease with decreasing grain size; this is commonly known as the Hall-Petch inversion. Given that the values of a material's elastic constants reflect the bonding nature of its constituent atoms, it seems logical to expect that nanocrystalline materials would exhibit different moduli of elasticity compared to coarse-grained polycrystalline solids because of the high volume fraction of atoms located at or near the grain boundaries, triple junctions, and quadruple nodes (Shvindlerman and Gottstein 2005). In particular, since the degree of atomic structural disorder is greater within a grain boundary as compared to the crystal lattice, the average atomic distance within it is generally known to be larger. It could then be concluded that the grain boundary as a whole exhibits a lower bond strength and, therefore, has local elastic moduli values lower than those of the lattice. At the nanoscale, grain boundaries can mediate deformation more directly by sliding and shear-coupled migration or serving as dislocation sources and sinks. There is strong evidence that applied shear stresses can drive rapid, diffusionless motion of nanocrystalline grain boundaries via shear coupling. Grain boundary sliding can also permit grains to slide past their neighbors, with accommodation provided by atomic shuffling. It has even been proposed that this leads to large grain rotations. At the same time, typical dislocation-based mechanisms become difficult to operate. A relatively large number of synthesis routes have been used to produce nanocrystalline materials resulting in a huge diversity of structures from these processing methods. Two main technological approaches are used to synthesize nanostructured materials: bottom-up (obtaining the structure by arranging the nanostructure atom by atom and layer by layer) and top-down (nanostructure obtained by breaking down the original microstructure).

UFG material synthesis can be broadly divided into four different categories. They are (a) inert gas condensation, (b) mechanical alloying, (c) crystallization from amorphous solids, and (d) severe plastic deformation (SPD). The SPD route again comprises different processes which include (a) equal channel angular pressing (ECAP), (b) high-pressure torsion (HPT), (c) accumulative roll bonding (ARB), and (d) friction stir processing (FSP). The top-down approach is finalized to induce UFG microstructures through heavy straining or shock loading (Kawasaki 2011). The imposed strain is necessary for the conversion of the coarse-grained material into an ultrafine one through the introduction of a high density of dislocations and their rearrangement into an array of grain boundaries. Now, severe plastic deformation (SPD) may be defined as those metal-forming procedures in which a very high strain is imposed on a bulk solid without the introduction of any significant change in the overall dimensions of the solid and leading to the production of exceptional grain refinement so that, typically, the processed bulk solids have 1000 or more grains in any section. Among the available bottom-up technologies, electrodeposition is the main technology employed for the production of nanocrystalline metals and alloys. Metals from the iron group (Ni, Fe, Co); some of the precious metals, e.g., Au, Pt, and Rh; and a number of others can be electrodeposited in nanocrystalline form through proper control of the chemical and electrical processing parameters. During electrodeposition, the metal or the alloy structure grows on a substrate thanks to the electrochemical reaction of ions belonging to an electrolyte. Metal films are deposited via the electrochemical reduction of the corresponding metal cations from the electrolyte. Depending on the electrochemical conditions (mainly temperature and current) the grain size of the deposit can be precisely tuned. Electrodeposition is considered to be the most attractive method in synthesizing thin, protective, nanostructured coatings as well as magnetic materials. Electrodeposition provides many advantages as previously discussed compared to other processing techniques. In addition, electrodeposition allows the thickness and microstructure of the films to be controlled by adjusting the deposition parameters such as the electrolyte composition, pH, temperature, current density, and agitation. Now, in pure metal systems a physical limit fixes the minimum obtainable grain size at around 10 nm. In order to produce nanocrystalline materials with finer grain sizes, two or more elements can be deposited by varying the bath temperature and the pulsed current.

References

Agarwal M, Kumar V, Malladi SRK et al (2010) Effect of current density on the pulsed co-electrodeposition of nanocrystalline nickel-copper alloys. JOM 62:88–92. https://doi.org/10.1007/s11837-010-0095-6

Ahmadkhaniha D, Huang Y, Jaskari M, Järvenpää A, Heydarzadeh Sohi M, Zanella C, Pentti Karjalainen L, Langdon TG (2018) Effect of high-pressure torsion on microstructure, mechanical properties and corrosion resistance of cast pure Mg. J Mater Sci 53:16585–16597. https://doi.org/10.1007/s10853-018-2779-1

Alexandrov IV, Chembarisova RG (2012) State and processes in grain boundaries of Cu subjected to severe plastic deformation. Rev Adv Mater Sci 31:91–99

Anderson PM, Carpenter JS, Gram MD, Li L (2014) Mechanical properties of nanostructured metals. In: Bhushan B, Luo D, Schricker S, Sigmund W, Zauscher S (eds) Handbook of nanomaterials properties. Springer, Berlin. https://doi.org/10.1007/978-3-642-31107-9_20

Andrievski RA, Khatchoyan AV (2016) Grain growth and nanomaterials behavior at high temperatures. In: Nanomaterials in extreme environments, Springer series in materials science, vol 230. Springer, Cham. https://doi.org/10.1007/978-3-319-25331-2_2

Armstrong RW (2014) Engineering science aspects of the Hall–Petch relation. Acta Mech 225:1013–1028. https://doi.org/10.1007/s00707-013-1048-2

Bagherpour E, Pardis N, Reihanian M et al (2019) An overview on severe plastic deformation: research status, techniques classification, microstructure evolution, and applications. Int J Adv Manuf Technol 100:1647–1694. https://doi.org/10.1007/s00170-018-2652-z

Beyerlein IJ, Knezevic M (2020) Mesoscale, microstructure-sensitive modeling for Interface-dominated, nanostructured materials. In: Andreoni W, Yip S (eds) Handbook of materials modeling. Springer, Cham. https://doi.org/10.1007/978-3-319-44677-6_82

Bober DB, Khalajhedayati A, Kumar M et al (2016) Grain boundary character distributions in nanocrystalline metals produced by different processing routes. Metal Mater Trans A 47:1389–1403. https://doi.org/10.1007/s11661-015-3274-9

Borodin EN, Mayer AE (2012) Yield strength of nanocrystalline materials under high-rate plastic deformation. Phys Solid State 54:808–815. https://doi.org/10.1134/S1063783412040038

Brailovski V, Prokoshkin S, Inaekyan K, Demers V (2011) Functional properties of nanocrystalline, submicrocrystalline and polygonized Ti-Ni alloys processed by cold rolling and post-deformation annealing. J Alloys Compd 509(5):2066–2075. https://doi.org/10.1016/j.jallcom.2010.10.142

Brandl C (2019) Molecular dynamics simulations of nanopolycrystals. In: Hsueh CH et al (eds) Handbook of mechanics of materials. Springer, Singapore. https://doi.org/10.1007/978-981-10-6884-3_12

Carlton CE, Ferreira PJ (2007) What is behind the inverse Hall–Petch effect in nanocrystalline materials? Acta Mater 55:3749–3756. https://doi.org/10.1016/j.actamat.2007.02.021

Cavaliere P (2009) Mechanical properties of nanocrystalline metals and alloys studied via multi-step nanoindentation and finite element calculations. Mater Sci Eng A512:1–9. https://doi.org/10.1016/j.msea.2009.03.008

Clement K, Iseli A, Karote D, Cremer J, Rajagopalan S (2012) Nanostructured materials: industrial applications. In: Kent J (ed) Handbook of industrial chemistry and biotechnology. Springer, Boston. https://doi.org/10.1007/978-1-4614-4259-2_9

Conrad H, Narayan J (2000) On the grain size softening in nanocrystalline materials. Scr Mater 42:1025–1030. https://doi.org/10.1016/S1359-6462(00)00320-1

Dai K, Shaw L (2008) Analysis of fatigue resistance improvements via surface severe plastic deformation. Int J Fatigue 30:1398–1408. https://doi.org/10.1016/j.ijfatigue.2007.10.010

Dutta RK, Malet L, Hermans MJM, Godet S, Richardson IM (2015) Formation of nanostructures in severely deformed high-strength steel induced by high-frequency ultrasonic impact treatment. Met Trans 46A:813–830. https://doi.org/10.1007/s11661-014-2658-6

Edalati K, Horita Z (2012) Parametrs influencing steady-state grain size of pure metals processed by high-pressure torsion. Mater Sci Forum 706:3034–3039. https://doi.org/10.4028/www.scientific.net/MSF.706-709.3034

Enayati MH, Bafandeh MR, Nosohian S (2007) Ball milling of stainless steel scrap chips to produce nanocrystalline powder. J Mater Sci 42:2844–2848. https://doi.org/10.1007/s10853-006-1371-2

Erb U, Palumbo G, Aust KT (2000) Electrodeposited nanostructured films and coatings: synthesis, structure, properties and applications. In: Chow GM, Ovid'ko IA, Tsakalakos T (eds) Nanostructured films and coatings, NATO science series (Series 3: high technology), vol 78. Springer, Dordrecht. https://doi.org/10.1007/978-94-011-4052-2_2

Estrin Y, Molotnikov A, Huw C et al (2008) Strain gradient plasticity modelling of high-pressure torsion. J Mech Phys Solids 56(4):1186–1202. https://doi.org/10.1016/j.jmps.2007.10.004

Fan GJ, Choo H, Liaw PK, Lavernia EJ (2005) A model for the inverse Hall–Petch relation of nanocrystalline materials. Mater Sci Eng A 409:243–248. https://doi.org/10.1016/j.msea.2005.06.073

Firstov SA, Rogul TG, Shut OA (2018) Hardening in the transition to nanocrystalline state in pure metals and solid solutions (ultimate hardening). Powder Metal Met Ceram 57:161–174. https://doi.org/10.1007/s11106-018-9964-2

Furukawa M, Iwahashi Y, Horita Z, Nemoto M, Langdon TG (1998) The shearing characteristics associated with equal-channel angular pressing. Mater Sci Eng A257:328–332. https://doi.org/10.1016/S0921-5093(98)00750-3

Furukawa M, Horita Z, Langdon TG (2002) Factors influencing the shearing patterns in equal-channel angular pressing. Mater Sci Eng A332:97–109. https://doi.org/10.1016/S0921-5093(01)01716-6

Gamburg YD, Zangari G (2011) Technologies for the electrodeposition of metals and alloys: electrolytes and processes. In: Theory and practice of metal electrodeposition. Springer, New York. https://doi.org/10.1007/978-1-4419-9669-5_13

Gapontsev VL, Kondrat'ev VV (2002) Diffusion phase transformations in nanocrystalline alloys under severe plastic deformation. Dokl Phys 47:576–579. https://doi.org/10.1134/1.1505513

Gleiter H (1989) Nanocrystalline materials. Prog Mater Sci 33:223–315. https://doi.org/10.1016/0079-6425(89)90001-7

Gleiter H (1993) Nanostructured materials. In: Nastasi M, Parkin DM, Gleiter H (eds) Mechanical properties and deformation behavior of materials having ultra-fine microstructures, NATO ASI series (Series E: applied sciences), vol 233. Springer, Dordrecht. https://doi.org/10.1007/978-94-011-1765-4_1

Gupta RK, Murty B, Birbilis N (2017) Mechanical properties of high-energy ball milled nanocrystalline Al alloys. In: An overview of high-energy ball milled nanocrystalline aluminum alloys, Springerbriefs in materials. Springer, Cham. https://doi.org/10.1007/978-3-319-57031-0_4

Gutkin MY, Ovid'ko IA (2004) Yield stress of nanocrystalline materials. In: Plastic deformation in nanocrystalline materials, Springer series in materials science, vol 74. Springer, Berlin. https://doi.org/10.1007/978-3-662-09374-0_2

Han BQ, Ye J, Tang F et al (2007) Processing and behavior of nanostructured metallic alloys and composites by cryomilling. J Mater Sci 42:1660–1672. https://doi.org/10.1007/s10853-006-0907-9

Han J-H, Chang H-J, Lee K-K, Oh K-H (2009) Effects of die geometry on variation of the deformation rate in equal channel angular pressing. Met Mater Int 15:439–445. https://doi.org/10.1007/s12540-009-0439-3

Harin EV, Sheftel EN, Tedzhetov VA (2018) Approximation of the influence of effective magnetic parameters on the coercivity of Iron-based nanocrystalline films. Tech Phys Lett 44:420–423. https://doi.org/10.1134/S106378501805019X

Jang D (2016) Nanosized nanocrystalline and nanotwinned metals. In: Bhushan B (ed) Encyclopedia of nanotechnology. Springer, Dordrecht. https://doi.org/10.1007/978-94-017-9780-1

Jung A, Weinmann M, Natter H (2015) Electroforming and electrodeposition on complex 3D geometries: special requirements and new methods. In: Aliofkhazraei M, Makhlouf A (eds) Handbook of Nanoelectrochemistry. Springer, Cham. https://doi.org/10.1007/978-3-319-15207-3_46-1

Kawasaki M (2011) Processing of ultrafine-grained materials through the application of severe plastic deformation. Metal Mater Trans A 42:3035–3045. https://doi.org/10.1007/s11661-010-0501-2

Khajouei-Nezhad M, Paydar MH, Ebrahimi R, Jenei P, Nagy P, Gubicza J (2017) Microstructure and mechanical properties of ultrafine-grained aluminum consolidated by high-pressure torsion. Mater Sci Eng A682:501–508. https://doi.org/10.1016/j.msea.2016.11.076

Koch CC, Scattergood RO, Murty KL (2007) The mechanical behavior of multiphase nanocrystalline materials. JOM 59:66–70. https://doi.org/10.1007/s11837-007-0042-3

Koch CC, Scattergood RO, Youssef KM et al (2010) Nanostructured materials by mechanical alloying: new results on property enhancement. J Mater Sci 45:4725–4732. https://doi.org/10.1007/s10853-010-4252-7

Kuznetsov PV, Mironov YP, Tolmachev AI et al (2015) Positron spettroscopy of defects in submicrocrystalline nichel after low-temperature annealing. Phys Solid State 57:219–228. https://doi.org/10.1134/S1063783415020225

Lavernia EJ, Han BQ, Shoenung LM (2008) Cryomilled nanostructured materials: processing and properties. Mater Sci Eng A493(1–2):207–214. https://doi.org/10.1016/j.msea.2007.06.099

Liu F, Kirchheim R (2004) Nano-scale grain growth inhibited by reducing grain boundary energy through solute segregation. J Cryst Growth 264:385–391. https://doi.org/10.1016/j.jcrysgro.2003.12.021

Liu H, Zhou J, Zhang S et al (2014) Modeling the strain-hardening effect and plastic deformation of nanocrystalline FCC metals dispersed with the finest grains. J Nanopart Res 16:2449. https://doi.org/10.1007/s11051-014-2449-4

Lowe TC, Zhu YT, Semiatin SL, Berg DR (2000) Overview and outlook for materials processed by severe plastic deformation. In: Lowe TC, Valiev RZ (eds) Investigations and applications of severe plastic deformation, NATO science series (Series 3. High technology), vol 80. Springer, Dordrecht. https://doi.org/10.1007/978-94-011-4062-1_45

Ma E (2004) Recent progress in improving ductility of ultra-high strength nanostructured metals. Met Mater Int 10:527–531. https://doi.org/10.1007/BF03027414

Ma E (2005) Four approaches to improve the tensile ductility of high-strength nanocrystalline metals. J Mater Eng Perform 14:430–434. https://doi.org/10.1361/105994905X56179

Mckenzie PWJ, LApokov R, Estrin Y (2007) The influence of back pressure on ECAP processed AA 6016: modeling and experiment. Acta Mater 55(9):2985–2993. https://doi.org/10.1016/j.actamat.2006.12.038

Meyers MA, Mishra A, Benson DJ (2006a) The deformation physics of nanocrystalline metals: experiments, analysis, and computations. JOM 58:41–48. https://doi.org/10.1007/s11837-006-0214-6

Meyers MA, Mishra A, Benson DJ (2006b) Mechanical properties of nanocrystalline materials. Prog Mater Sci 51:427–556. https://doi.org/10.1016/j.pmatsci.2005.08.003

Mittemeijer EJ (2010) Mechanical strength of materials. In: Fundamentals of materials science. Springer, Berlin. https://doi.org/10.1007/978-3-642-10500-5_11

Mohamed FA (2003) A dislocation model for the minimum grain size obtainable by milling. Acta Mater 51:4107–4119. https://doi.org/10.1016/S1359-6454(03)00230-1

Mohamed FA (2016) Mechanical properties of nanocrystalline metals. In: Bhushan B (ed) Encyclopedia of nanotechnology. Springer, Dordrecht. https://doi.org/10.1007/978-94-017-9780-1

Mohamed FA (2020) Creep and superplasticity: evolution and rationalization. Adv Eng Mater 22:1900532. https://doi.org/10.1002/adem.201900532

Murr LE (2015) Mechanical and other properties of nanocrystalline materials. In: Handbook of materials structures, properties, processing and performance. Springer, Cham. https://doi.org/10.1007/978-3-319-01815-7_47

Ortiz AL, Tian J-W, Shaw LL, Liaw PK (2010) Experimental study of the microstructure and stress state of shot peened and surface mechanical attrition treated nickel alloys. Scripta Mater 62:129–132. https://doi.org/10.1016/j.scriptamat.2009.10.015

Panin VE, Egorushkin VE (2010) Nanostructural states in solids. Phys Metals Metallogr 110:464–473. https://doi.org/10.1134/S0031918X10110062

Podrezov YM (2006) Nanocrystalline structure formation under severe plastic deformation and its influence on mechanical properties. In: Zhu YT, Varyukhin V (eds) Nanostructured materials by high-pressure severe plastic deformation, NATO science series (II: Mathematics, physics and chemistry), vol 212. Springer, Dordrecht. https://doi.org/10.1007/1-4020-3923-9_22

Protsenko VS, Danilov FI (2020) Current trends in electrodeposition of electrocatalytic coatings. In: Inamuddin, Boddula R, Asiri A (eds) Methods for electrocatalysis. Springer, Cham. https://doi.org/10.1007/978-3-030-27161-9_11

Qian T, Karaman I, Marx M (2014) Mechanical properties of nanocrystalline and ultrafine- grained nickel with bimodal microstructure. Adv Eng Mater 16(11):1323–1339. https://doi.org/10.1002/adem.201300570

Ramesh K (2009) Mechanical properties: density and elasticity. In: Nanomaterials. Springer, Boston. https://doi.org/10.1007/978-0-387-09783-1_4

Ratman D, Marx M, Motz C (2017) Crack propagation and mechanical properties of electrodeposited nickel with bimodal microstructures in the nanocrystalline and ultrafine grained regime. J Mater Res 32(24):4573–4582. https://doi.org/10.1557/jmr.2017.353

Razumov IK (2014) Grain boundary segregations in nanocrystalline alloys. Russ J Phys Chem 88:494–502. https://doi.org/10.1134/S0036024414030212

Schaefer HE (2010) Nanocrystalline materials. In: Nanoscience. Springer, Berlin. https://doi.org/10.1007/978-3-642-10559-3_6

Shaw LL (2013) Surface nanocrystallization and hardening (SNH). In: Wang QJ, Chung YW (eds) Encyclopedia of tribology. Springer, Boston. https://doi.org/10.1007/978-0-387-92897-5

Shvindlerman LS, Gottstein G (2005) Cornerstones of grain structure evolution and stability: vacancies, boundaries, triple junctions. J Mater Sci 40:819–839. https://doi.org/10.1007/s10853-005-6498-z

Siegel RW, Thomas GJ (1992) Grain boundaries in nanophase materials. Ultramicroscopy 40 (3):376–384. https://doi.org/10.1016/0304-3991(92)90135-7

Sitek J, Degmová J (2006) Aluminium based amorphous and nanocrystalline structure. In: Lippens PE, Jumas JC, Génin JMR (eds) ICAME 2005. Springer, Berlin. https://doi.org/10.1007/978-3-540-49850-6_17

Sob PB, Alugongo AA, Tengen TB (2017) Adv Mater Phys Chem 7:58–69. https://doi.org/10.4236/ampc.2017.72006

Song Y, Zhang Z, Gao D, Zhang H, Kim H-S (2018) Effects of friction on heterogeneous deformation characteristics of interstitial-free steels during the torsion stage of high pressure torsion. In 3rd international conference on control, automation and artificial intelligence (CAAI 2018), pp 186–189

Spearot DE, McDowell DL (2009) Atomistic modeling of grain boundaries and dislocation processes in metallic polycrystalline materials. J Eng Mater Technol 131(4):041204. https://doi.org/10.1115/1.3183776

Stolyarov VV, Lapovok R, Brodova IG, Thomson PF (2003) Ultrafine-grained Al–5 wt.% Fe alloy processed by ECAP with backpressure. Mater Sci Eng A357(1–2):159–167. https://doi.org/10.1016/S0921-5093(03)00215-6

Sun CQ (2014) Nanograins: II. Plasticity and yield stress. In: Relaxation of the chemical bond, Springer series in chemical physics, vol 108. Springer, Singapore. https://doi.org/10.1007/978-981-4585-21-7_28

Suryanarayana C, Koch C (2000) Nanocrystalline materials—current research and future directions. Hyperfine Interact 130:5. https://doi.org/10.1023/A:1011026900989

Takeuchi S (2001) The mechanism of the inverse Hall–Petch relation in nanocrystals. Scr Mater 44:1483–1487. https://doi.org/10.1016/S1359-6462(01)00713-8

Tschopp MA, Murdoch HA, Kecskes LJ et al (2014) “Bulk” nanocrystalline metals: review of the current state of the art and future opportunities for copper and copper alloys. JOM 66:1000–1019. https://doi.org/10.1007/s11837-014-0978-z

Valiev RZ, Langdon TG (2006) Principles of equal-channel angular pressing as a processing tool for grain refinement. Prog Mater Sci 51(7):881–981. https://doi.org/10.1016/j.pmatsci.2006.02.003

Van Swygenhoven H, Farkas D, Caro A (2000) Grain boundary structures in polycrystalline metals at the nanoscale. Phys Rev B 62:831–838. https://doi.org/10.1103/PhysRevB.62.831

Viswanathan V, Laha T, Balani K, Agarwal A, Seal S (2006) Challenges and advances in nanocomposite processing techniques. Mater Sci Eng R54:121–285. https://doi.org/10.1016/j.mser.2006.11.002

Witkin DB, Lavernia EJ (2006) Synthesis and mechanical behavior of nanostructured materials via cryomilling. Prog Mater Sci 51:1–60. https://doi.org/10.1016/j.pmatsci.2005.04.004

Wolf H, Guan Z, Li X et al (2001) Magnetic properties of grain boundaries of nanocrystalline Ni and of Ni precipitates in Nanocrystalline NiCu alloys. Hyperfine Interact 136:281–287. https://doi.org/10.1023/A:1020583105443

Wrobel RJ (2012) Thermal stability of nanocrystalline iron. Mater Sci-Pol 30:63–69. https://doi.org/10.2478/s13536-012-0009-0

Wu B, Zhang J, Zhang L, Pyoung Y-S, Murakami R-I (2014) Effect of ultrasonic nanocrystal surface modification on surface and fatigue properties of quenching and tempering S45C steel. Appl Surf Sci 321:318–330. https://doi.org/10.1016/j.apsusc.2014.09.068

Xu L, Du J, Ge S et al (2009) Preparation and characterization of nanocrystalline Fe–Ni–Cr alloy electrodeposits on Fe substrate. J Appl Electrochem 39:713–717. https://doi.org/10.1007/s10800-008-9714-0

Yamakov V, Wolf D, Phillpot SR, Muckherjee AK (2002) Dislocation processes in the deformation of nanocrystalline aluminium by molecular-dynamics simulation. Nat Mater 1:45–49. https://doi.org/10.1038/nmat700

Yang Z, Welzel U (2005) Microstructure-microhardness relation of nanostructured Ni produced by high-pressure torsion. Mater Lett 59:3406–3409. https://doi.org/10.1016/j.matlet.2005.05.077

Yu HL, Lu C, Tieu AK, Kong C (2014) Fabrication of nanostructured aluminum sheets using four-layer accumulative roll bonding. Mater Manuf Proc 29:448–453. https://doi.org/10.1080/10426914.2013.872259

Zhang W, Gao Y, Nieh TG (2018) Competing grain boundary and interior deformation mechanisms with varying sizes. In: Hsueh CH et al (eds) Handbook of mechanics of materials. Springer, Singapore. https://doi.org/10.1007/978-981-10-6855-3_75-1

Zhang Z, Ódor É, Farkas D et al (2020) Dislocations in grain boundary regions: the origin of heterogeneous microstrains in nanocrystalline materials. Metal Mater Trans A 51:513–530. https://doi.org/10.1007/s11661-019-05492-7

Zhao M, Li JC, Jiang Q (2003) Hall–Petch relationship in nanometer size range. J Alloys Comp 361:160–164. https://doi.org/10.1016/S0925-8388(03)00415-8

Zhilyaev AP, Langdon TG (2008) Using high-pressure torsion for metal processing: fundamentals and applications. Prog Mater Sci 53:893–979. https://doi.org/10.1016/j.pmatsci.2008.03.002

Zhilyaev AP, Lee S, Nurislamova GV, Valiev RZ, Langdon TG (2001) Microhardness and microstructural evolution in pure nickel during high-pressure torsion. Scripta Mater 44:2753. https://doi.org/10.1016/S1359-6462(01)00955-1

Zhou H, Qu S (2019) Mechanical properties of nanostructured metals: molecular dynamics studies. In: Hsueh CH et al (eds) Handbook of mechanics of materials. Springer, Singapore. https://doi.org/10.1007/978-981-10-6884-3_19

Zhou X, Li XY, Lu K (2018) Enhanced thermal stability of nanograined metals below a critical grain size. Science 360:526–530. https://doi.org/10.1126/science.aar6941

Zhu L, Zheng X (2010) Influence of interface energy and grain boundary on the elastic modulus of nanocrystalline materials. Acta Mech 213:223–234. https://doi.org/10.1007/s00707-009-0263-3

Zhu YT, Liao XZ, Wu XL (2008) Deformation twinning in bulk nanocrystalline metals: experimental observations. JOM 60:60. https://doi.org/10.1007/s11837-008-0120-1

Chapter 2
Cyclic Deformation of Metal Alloys and Composites

2.1 Introduction

In polycrystalline metals and alloys, grain boundaries are first of all obstacles to dislocations because of different slip systems in adjacent grains. Deformation mechanisms that typically operate in CG metals and their alloys, e.g., the nucleation, migration, interaction, and annihilation of lattice dislocations in crystalline grains, are severely limited in NC metals; instead, interface-mediated mechanisms such as GB sliding/migration, dislocation nucleation from GB sources, and twinning tend to dominate (Konakov et al. 2017). Since grain boundaries are more abundant and more important in nanocrystalline systems, increased attention has been focused on studying the atomic structure of these interfaces (Beyerlein and Knezevic 2018). Figure 2.1 schematically shows various transitions in deformation mechanism encountered as the grain size is reduced from microns (traditional) down to extremely NC grained (Nano-1). In the CG grain size regime, a decrease in grain size is associated with an increase in strength (the Hall-Petch relation). In the NC regime, the onset of GB sliding processes at grain sizes below 20 nm leads to weakening, also known as the inverse Hall-Petch effect (Naik and Walley 2020).

The transition to nonclassical, interface-mediated deformation mechanisms at NC grain sizes has profound implications regarding the predictive modeling of plastic deformation in these materials (Wolf 2005). First, traditional theories of crystal plasticity do not apply at nanoscale grain sizes, since these do not account for GB-mediated processes (McDowell 2019). Second, in the regime of submicron grain sizes, experimental and computational studies have reported strength asymmetry in tension vs. compression. The plastic deformation is mainly carried by dislocations—line defects of the regular crystal lattice—within the individual grains (Panin et al. 2018). Dislocations can move through the crystal grains and can interact with each other. When they meet grain boundaries (GBs), GBs often hinder the transmission of dislocations, creating dislocation pileups at the boundaries and thereby making the material harder to deform. Here, strengthening is due to the

© Springer Nature Switzerland AG 2021

P. Cavaliere, *Fatigue and Fracture of Nanostructured Materials*,
https://doi.org/10.1007/978-3-030-58088-9_2

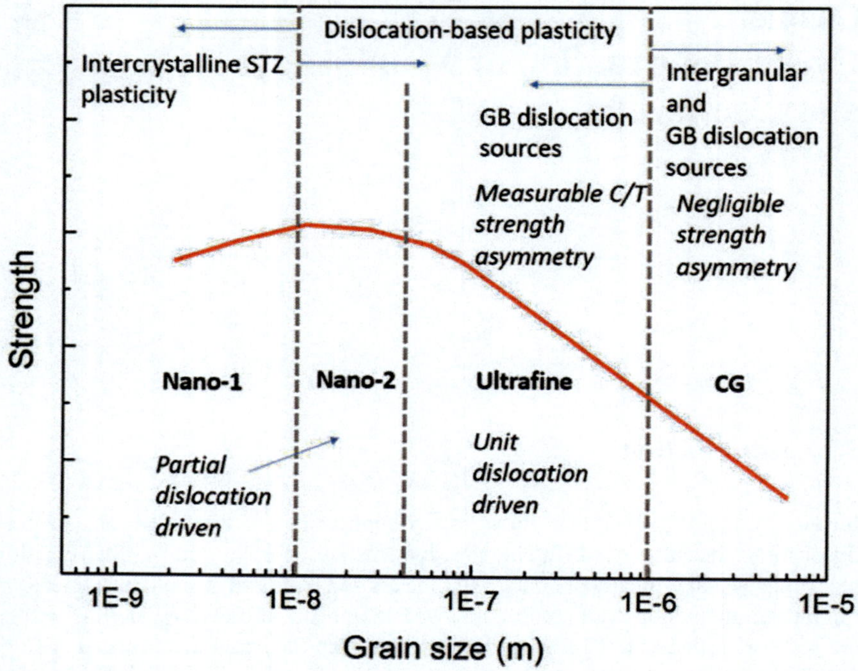

Fig. 2.1 Scaling of nominal strength with grain size *d* in metallic materials

dislocation pileups at the grain boundaries. With further grain refinement, when the grain sizes are reduced to nanoscale, there is a large fraction of atoms sitting at the grain boundaries and these atoms are in a nonequilibrium state. The grain boundary (GB) activities become active (Wang et al. 2017a), and the dislocation activities within grains may become difficult and even cease, i.e., the dislocation-based theory established in the coarse-grained and ultrafine-grained materials may not persist in nanocrystalline materials any more (Scattergood et al. 2008). Strengthening mechanisms that are well documented for coarse-grained metals and alloys can also operate in nanocrystalline metals and alloys, although modified by the nanoscale grain size and nonequilibrium microstructures (Cherkaoui and Capolungo 2009). Grain boundaries, and especially the presence of grain boundary segregates, appear to play a unique role in nanocrystalline microstructures, in terms of both potential changes in properties such as grain boundary energy and cohesive strength and effects on mechanical properties (Murr 2016).

Plastic deformation of NC grains is limited by the interface-mediated nucleation and pinning of dislocations; therefore, classical strengthening mechanisms that are based on dislocation multiplication + storage and dislocation-obstacle interactions in the lattice may not apply to NC metals.

A number of studies have reported that nanocrystalline metals often contain nonequilibrium grain boundaries, characterized by excess free volume or grain boundary dislocations, in their as-prepared state (Fultz and Frase 2000). Perhaps not surprisingly since boundaries are so important for nanocrystalline plasticity, reports have shown that mechanical strength is highly dependent on this grain boundary structural state (McDowell 2018). As an example, grain boundary relaxation results in a significant increase in hardness, even though grain size was unchanged. It is well known that the material resistance depends on grain size with increased plastic resistance as the grain size decreases. The basic principles of this phenomenon were first described in the middle of the twentieth century and summarized in the well-known Hall-Petch equation (Hall 1951; Petch 1953). The phenomenon was demonstrated for low-carbon steel where the stress concentration due to dislocation pileup in the soft grain interior stopped at the grain boundary.

2.2 Elastoplastic Behavior in Nanostructured Metals and Alloys

2.2.1 Dislocations and Plasticity

The best available model describing the effect of grain size dimension of the metal strength considers a double-ended dislocation pileup moving on a slip system in the grain interior of a single grain (with a given diameter d) of a polycrystalline material during tensile stress σ. This is shown in Fig. 2.2.

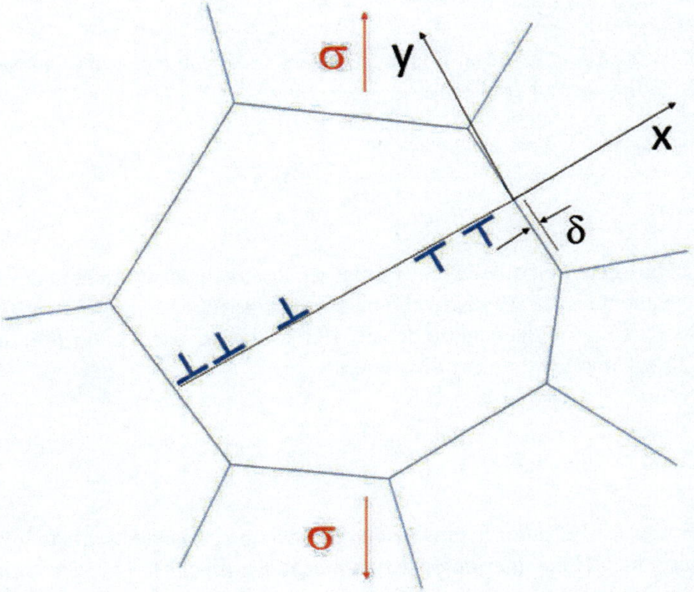

Fig. 2.2 Model of the deformation of a single grain subjected to tensile stress

The stress concentrating at a distance d from the grain boundary σ_{xy} $(x = \delta)$ felt by the adjacent grain is given by

$$\sigma_{xy}(x = \delta) = (\sigma - \sigma_0)m\sqrt{\frac{d}{4\delta}}$$

Here σ_0 is the critical tensile stress to be applied for initiating the dislocation motion across the slip plane in the grain interior and m is the Schmid factor of the slip plane with respect to the tensile axis. The grain yields $(\sigma = \sigma_y)$ when σ_{xy} $(x = \delta)$ reaches a critical shear stress (σ_{sc}) leading to the percolation of plastic behavior; this gives the well-known Hall-Petch equation:

$$\sigma_y = \sigma_0 + \frac{k_y}{\sqrt{d}}$$

k_y is known as friction stress and is given by

$$k_y = \frac{\sigma_{sc}}{m}\sqrt{4\delta}$$

Naturani and Takamura (1991) defined the equation describing the flow stress dependence on the dislocation density:

$$\left(\frac{\sigma - \sigma_0}{\alpha\mu b}\right)^2 = \frac{\left(\xi_d(\varepsilon) + \frac{A\varepsilon}{d}\right)}{\eta}$$

where σ_0 is the friction stress, b is the Burgers vector, and μ is the isotropic shear modulus. In the case of small strains

$$\sigma \cong \sigma_0 + \frac{\alpha\mu b\sqrt{\frac{A\varepsilon}{\eta}}}{\sqrt{d}}$$

with a Hall-Petch-type behavior. For larger strains the flow stress is described by a term dependent on the grain size and on a grain size-independent factor. The model takes into account the dislocation density due to the imposed strain $\rho(\varepsilon)$ leading to the final following equation for flow stress:

$$\left(\frac{\sigma - \sigma_0}{\alpha\mu b}\right)^2 = \rho(\varepsilon) + \frac{\beta\varepsilon}{d}$$

The flow stress behavior begins to change once the grain size starts to fall into the nanometer range. Here other deformation mechanisms start to play prevailing roles. Many experimental and numerical evidences show that in the nanometer range the

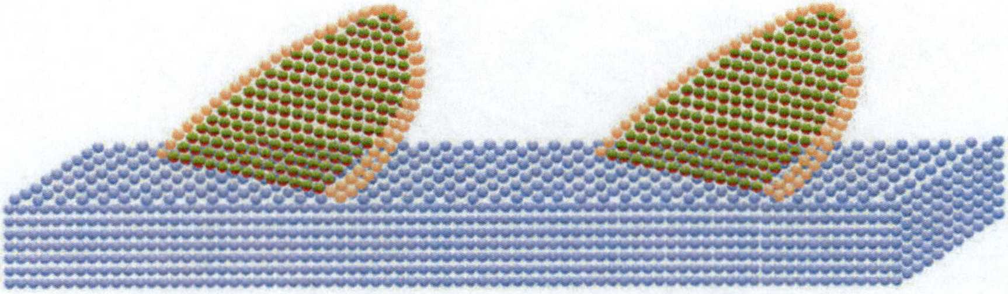

Fig. 2.3 MDS of partial dislocation emissions at grain boundary of NC Ni

development of dislocations during plastic flow becomes shorter (Yamakov et al. 2003; Schiots and Jacobsen 2003). They all showed that plastic flow depends on grain boundary shear and emissions of full or partial dislocations from grain boundaries (Fig. 2.3).

As the leading partial dislocation enters the grain it produces partial segments along the side boundaries (Dao et al. 2007). Dislocation nucleation and propagation are limited by activity at grain boundary sites. These dislocations often become pinned at grain boundary ledges as they move through the grain, giving the spacing between boundary pinning points as the characteristic length scale of the mechanism. Experimental data could be well described by a model that invokes Orowan-type pinning of dislocations with the grain size taken as the distance between obstacles (Koch et al. 2008). Atomistic simulations have suggested that dislocations in NC metals nucleate from GB sources (Farkas and Selinger 2005). A snapshot from such a simulation is shown in Fig. 2.4.

The snapshot was captured following the successive nucleation of two partial dislocations—a leading partial (LP) and a trailing partial (TP) separated by a stacking fault (red), from the GB network (blue atoms show GB atoms that lie in the dislocation slip plane). Due to the small size of grains in nanocrystalline materials, these GB sources tend to be small compared to Frank-Read sources in CG metals. Further, the figure shows that the dislocations LP and TP nucleate from different sites on the GB network. This is the result of small-scale stress variations in the GB atoms, which not only influence the nucleation of dislocations, but also serve to pin dislocation loops as they thread through the grain (Gutkin and Ovid'ko 2007). At grain sizes on the lower end of the NC regime (below 20 nm), GBs can have a non-negligible contribution to strain in the form of several GB processes (GBPs), e.g., GB sliding, GB migration, and grain rotation (Ovid'ko and Sheinerman 2016). Extensive studies have been made to explore the microstructures and microscopic deformation mechanisms in NC metals and alloys (Rekhi 2002). Two main deformation mechanisms have been suggested. One is GB-mediated processes (e.g., GB sliding/GB migration/GB diffusion, grain rotation, and grain growth) (Herzig and Mishin 2005). For example, where most of the plastic deformation is due to a large number of small "sliding" events of atomic plane at the grain boundaries, with only a minor part being caused by dislocation activity in the grains, the softening that we

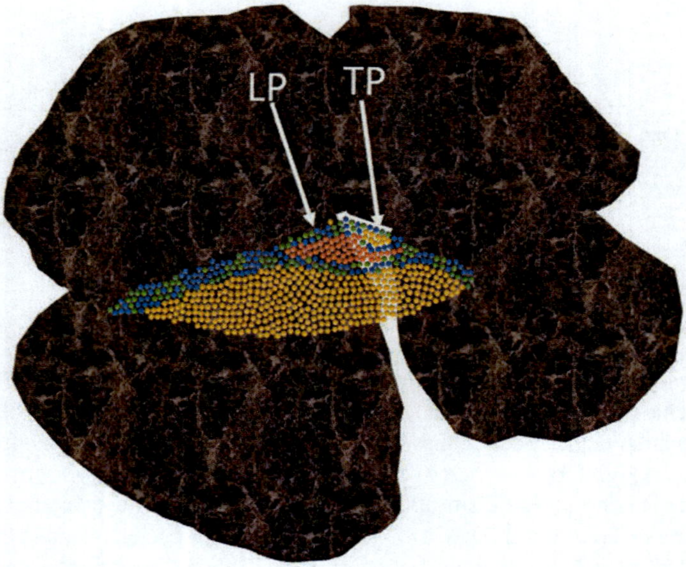

Fig. 2.4 MD simulation of NC Al, showing a pair of leading partial (LP) and trailing partial (TP) dislocations separated by a stacking fault

see at small grain sizes is therefore due to the large fraction of atoms at grain boundaries. The NC materials possessing high-angle GBs change to low-angle GBs after the deformations. With plastic deformation, a relative shear between grains along their boundaries will be conducted by gliding grain boundary dislocations (GBDs), whereas the crystal lattice or plane orientation in the neighboring grains will subsequently occur by climbing GBDs, which results from the splitting of gliding GBD at triple junction into two climbing GBDs. GB sliding transforming into crystal lattice rotation in neighboring grain is suggested as the deformation mechanism in the NC materials. GB and triple-junction (TJ) nonequilibrium state plays a crucial role in the extent of plastic deformation. The other deformation mechanism is dislocation/twinning-based plasticity (Zhu 2005). The emission of lattice dislocations from GBs or partials from GBDs has been suggested to be nucleated and absorbed in the GB with the very small temporal and spatial dimensions in the nanocrystalline regime (Gutkin et al. 2005). With conventional dislocation theory in mind, three possible inter-grain deformation mechanisms are assumed: the first by means of partial dislocations that extend across the entire grain creating stacking fault defects; the second by means of full dislocations: a leading and a trailing partial on the same slip plane leaving no dislocation debris; and the third by means of mechanical twinning: a leading and twinning partial on two adjacent planes nucleating a two-layer micro-twin which is taken as the precursor for deformation twining (Misra 2008). Large-scale MD simulations have provided evidence for a transition with decreasing grain size from intergranular plastic deformation based on grain boundary accommodation to a mixture of intergranular and intragranular

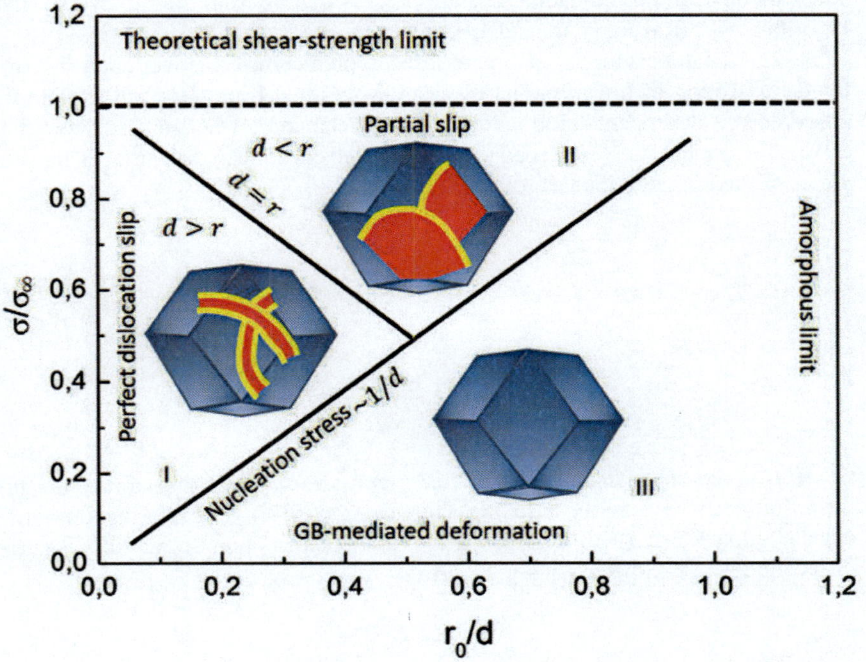

Fig. 2.5 Deformation mechanism map incorporating the role of the stacking fault energy in the deformation behavior

processes where the grain boundary acts as a source of imperfect or partial dislocations (Zhou and Qu 2018).

Figure 2.5 presents a typical work predicting the deformation map to elucidate the transition from a dislocation-based to a GB-based deformation mechanism in nanocrystalline metals.

The map is expressed in reduced units of the stress ($\sigma/\sigma\infty$) and inverse grain size (r_0/d). The parameters $\sigma\infty$ and r_0 are functions of the stacking fault energy (SFE) and the elastic properties of the material. r and d are the partial dislocation separation and the average grain size, respectively. In region I, characteristic of a large grain size (d) and/or a high-SFE (small r) metal, slip deformation prevails as the grains are larger than partial dislocation separation, r, and complete, extended dislocations nucleated from the GBs can propagate across the grains. In region II, characteristic of a small grain size and/or a low SFE (large r) metal, only incomplete dislocations can be nucleated; the grains are therefore transected by stacking faults that inhibit dislocation propagation, thus giving rise to strain hardening. Region III characterizes the very small grain size or low-stress regime in which no dislocations are present at any stress, and the deformation is therefore controlled by GB-mediated processes. These GB-mediated processes, i.e., GB sliding/migration/diffusion, grain rotation, and sometimes the combination of these processes, may result in the inverse Hall-Petch behavior (Razumov et al. 2019).

Several theoretical models were proposed to predict the crossover grain size (d_C) for the transition of full to partial dislocations emitted from GBs and explain the observed size-dependent deformation twinning behavior in FCC metals (Cizek et al. 2016). The critical stresses needed to move a full dislocation and a partial dislocation are, respectively, described as follows:

$$\sigma_{\mathrm{Full}} = \frac{\mu b}{d}$$

and

$$\sigma_{\mathrm{Partial}} = \frac{\alpha - 1}{\alpha} \frac{\mu b}{d} + \frac{1}{3} \frac{\gamma_{\mathrm{SF}}}{b}$$

where μ is the shear modulus, b is the magnitude of Burgers vector of the full dislocation, and γ_{SF} is the stacking fault energy (SFE). Considering the stress concentration effect (with a factor n of ~2–4), a transition from full to partial dislocation is thus expected at a grain size d_C:

$$d_C \approx \frac{3m - 1}{m} \frac{\mu b^2}{\gamma_{\mathrm{SF}}}$$

At small grain sizes $d < d_C$, emission of Shockley partials in lieu of full dislocations (when $d > d_C$) from GBs becomes favorable, which in turn produces deformation twins (and SFs) to accommodate plastic deformation of NC metals (Zhang et al. 2020).

As the grain size decreases below 5 nm, the material becomes nearly amorphous, and a transition to glasslike deformation mediated by shear transformation zones (STZs) is expected. Now, the change in mechanical response due to the formation of amorphous intergranular film formed during annealing can be reconciled by the elastic contrast between the grain interior and the grain boundary (Miglierini and Grenèche 1999). From studies on NC Al-Ni-Ce amorphous alloys it is clear that annealing drives strong chemical segregation of Ni and Ce to the grain boundaries. This is expected to energetically relax the grain boundaries as well as mechanically stiffen them, making them less prone to sliding or localized atomic shuffling (Balbus et al. 2020). The model described in Fig. 2.6 was proposed to describe the formation of amorphous grain boundary during annealing.

The effects of the amorphous intergranular film are shown in Fig. 2.7.

A dislocation traversing the grain must overcome the barrier for nucleation (τ_n), the barrier due to pinning forces at the grain boundary (τ_{pin}), as well as the image forces from the stiff grain boundary region (τ_{gb}) in order to move under the applied stress (τ). The formation of a stiff, amorphous intergranular film, in contrast with a compliant grain interior, rationalizes the enhanced hardness of the annealed material.

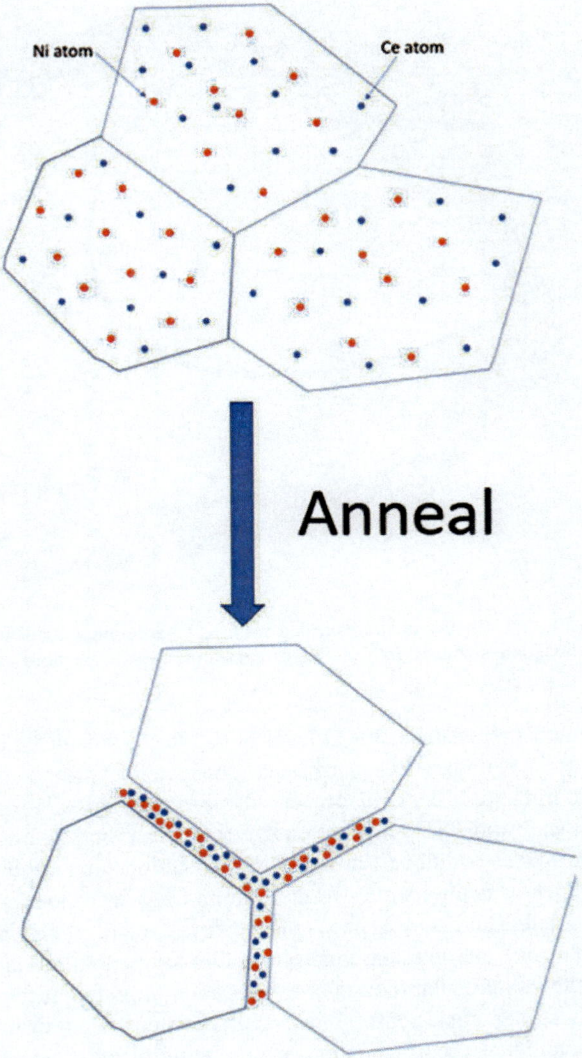

Fig. 2.6 Amorphous GB formation during annealing

Additional implications on the mechanical response due to the presence of an amorphous intergranular film include the suppression of shear localization (Xu et al. 2008). The presence of an amorphous grain boundary enables the absorption of several dislocations prior to failure (Schuh et al. 2007). Von Mises stress in neighboring grains is nearly unaffected by the absorption of dislocations at the amorphous grain boundary, whereas the absorption of a dislocation at an atomically sharp grain boundary leads to high stresses and crack formation. The formation of amorphous intergranular film may be responsible for the suppression of long-range localization. In conjunction with the suppression of grain boundary-

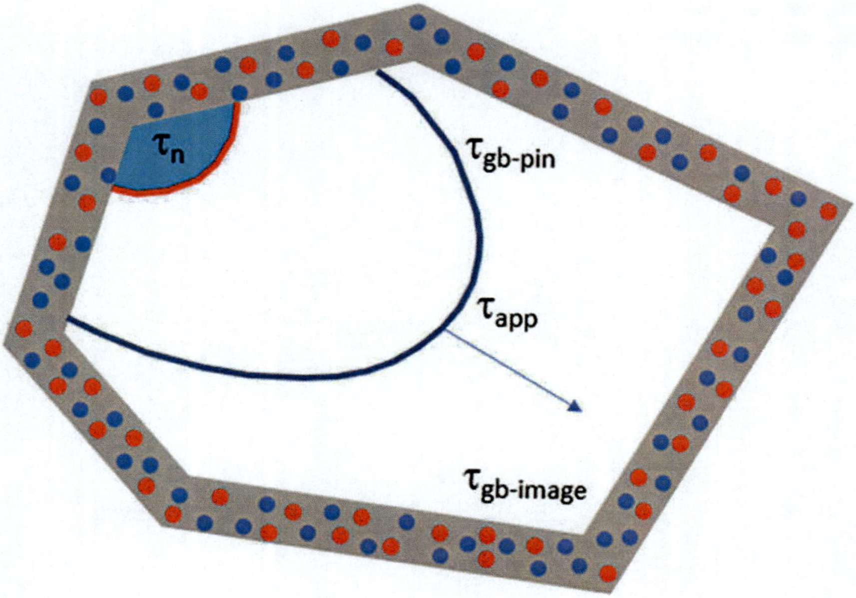

Fig. 2.7 Ability of the annealed material to accommodate dislocations at the amorphous grain boundaries without transmitting or nucleating new dislocations in neighboring grains

mediated mechanisms in favor of dislocation-mediated plasticity, the tendency for a grain boundary to absorb several dislocations prior to initiating plasticity in neighboring grains would prevent the formation of a percolating path necessary for localization. The segregation-driven increase in stiffness of the grain boundary likely provides additional screening of dislocations that impinge on the grain boundaries. Such screening would further inhibit long-range localization, promoting both high strength and homogeneous plastic flow (Gutkin and Ovid'ko 2004). Thus, the role of the stiff intergranular amorphous film serves three primary purposes: (1) to impede the intragranular motion of a dislocation providing increased strength, (2) to serve as a strong obstacle for intergranular dislocation transmission, and (3) to provide additional accommodation of impinging dislocations at the grain boundary, all of which assist in mitigating the propensity for localization.

Besides the aforementioned mechanisms, the abundance of interfaces in NC metals increases the propensity of diffusive rearrangement in GB regions, and Coble creep could be a significant contributor to plasticity. Recent experimental observations based on in situ microscopy have confirmed several mechanisms suggested by prior atomistic simulation studies, namely the activation of GB dislocation sources and significant GB migration.

Dislocation nucleation and emission from GBs are important strain accommodation mechanisms in NC metals. Instead of traditional Frank-Read sources, GBs become the source for both partial and full dislocations in NC materials (Fig. 2.8).

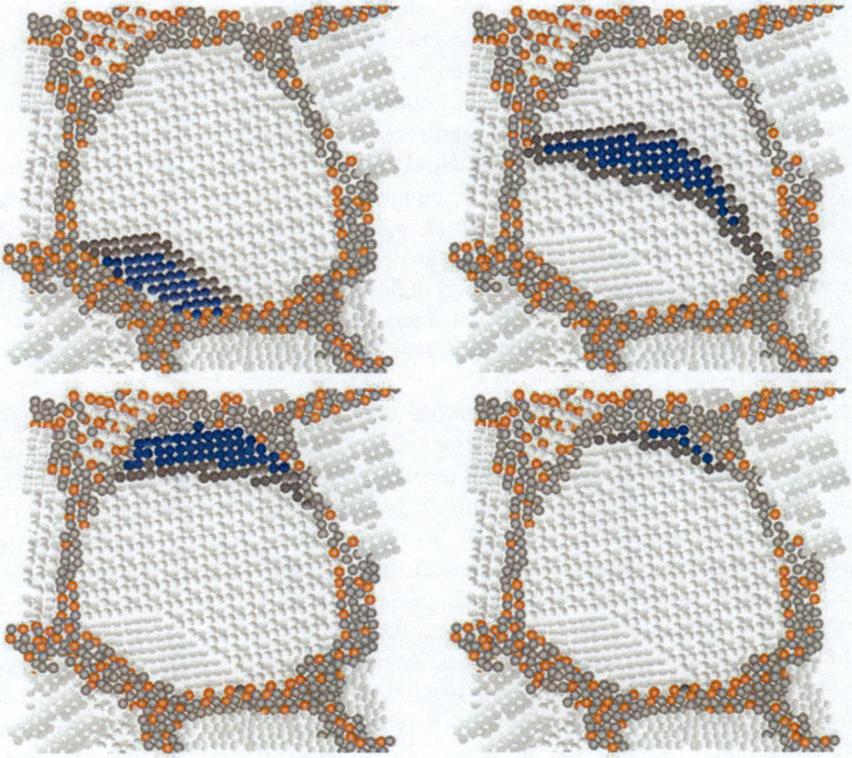

Fig. 2.8 Nucleation/emission of a full dislocation from a GB/TJ region and traversing the entire grain being absorbed into another GB/TJ region

Due to the small grain size, the emitted dislocations tend to glide across the grain and become absorbed in other GBs, sometimes leaving behind a stacking fault for the case of partial dislocation emission. Prior to dislocation nucleation in NC materials, atomistic simulations have shown that free volume migration occurs within the boundary and nearby TJs. Excess GB free volume has been noted to be a good measure of the degree of "nonequilibrium" state, and is defined as the additional amount of free volume as compared to that present in the equilibrium GB structure. Excess free volume correlates with higher interfacial energy and atomic misfit, and is therefore a key physical attribute directly affecting many important GB properties, such as sliding, migration, and dislocation mediation processes. Excess free volume within certain GB regions promotes the formation of the Burgers vector required for dislocation nucleation under applied load. The atomic structure, energy, and free volume have all been found to be influential in determining GB behavior and dislocation nucleation. Free volume migration has also been noted in NC metals as an accompanying process to dislocation nucleation and interfacial reordering. Atomic shuffling and GB sliding are also important mechanisms in NC metals, and free volume has been noted to influence these mechanisms as well. It is clear that free volume plays a crucial role in GB properties, and that the mechanical deformation in nanostructured materials is therefore influenced (Thangadurai et al. 2020). Following dislocation nucleation, interfacial atoms in specific regions undergo coordinated migration in the GB period direction towards areas in the GB where dislocations nucleate (nucleation regions), while free volume migrates in the opposite direction towards regions where dislocation nucleation has not occurred (non-nucleation regions). This process is the interfacial activation event for partial dislocation nucleation, and produces a change in both structure and free volume. Immediately following partial dislocation nucleation, a significant drop in the free volume concentration is observed in nucleation regions. This is attributed to atomic rearrangements (or structural changes) occurring within the GB.

All these mechanisms of plastic deformation not only cause relaxation of high-concentrated stresses, thereby hampering the nucleation of nanoscale cracks, but can also slow down or arrest the growth of formed cracks. It has been proposed that these deformation modes can play an important role in the toughening of nanocrystalline and ultrafine-grained materials. Of particular interest is the experimentally documented phenomenon of the nanoscale amorphization (the formation of nanoscale amorphous regions) in deformed nanocrystalline and ultrafine-grained materials. Furthermore, computer simulations and theoretical models have provided convincing evidence for the fact that nanoscale amorphization can effectively occur in nanocrystalline and polycrystalline materials. Local amorphization can occur in the vicinity of crack tips and thereby influence crack growth in conventional coarse-grained materials. Nanoscale amorphization can effectively occur at GBs and their triple junctions as a process driven by relaxation of the elastic energy of GB disclinations (line defects associated with abrupt changes/gaps in GB misorientation in the absence of any external mechanical load) (Feng et al. 2019). Nanoscale amorphization can occur at GB disclination dipole and serve as a special mode of local plastic deformation near crack tips. It is well known that dislocation emission from crack tips is one of the most fundamental processes for understanding crack blunting in nanocrystalline and ultrafine-grained materials. Once dislocations are emitted, they move out of the crack tip area leaving behind a dislocation-free zone. An internal back stress due to the dislocations emitted from crack tip accommodates the stress intensity factor to the applied load, causing an increase fracture toughness of materials. Edge dislocations emitted from cracks are stopped at GBs, resulting in blunting of cracks. Both crack blunting and stress field of the arrested dislocations hamper further dislocation emission from cracks. As a result, grain size reduction causes nanocrystalline and ultrafine-grained materials to show a brittle behavior. Nevertheless, the effect of the specific plastic deformation, which plays an important role in the toughening of nanocrystalline and ultrafine-grained materials, on the emission criteria of dislocations is not considered. In fact, the dislocation emission criterion deeply affects the number of the dislocations emitted from the crack tip as well as fracture toughening of nanocrystalline and ultrafine-grained materials. High local stresses operating near the crack tip can initiate GB sliding. Then, the nanoscale

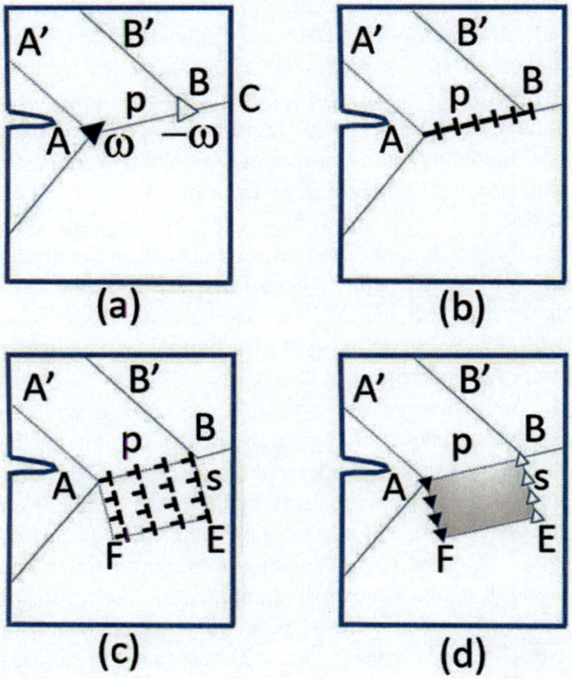

Fig. 2.9 Formation of an amorphous region around a GB disclination dipole near a crack tip in a nanocrystalline solid deformed by GB sliding

amorphization can occur through the splitting transformation of GB disclinations which are produced by GB sliding (Ovid'ko and Skiba 2014a). The influence of the nanoscale amorphization, which is an important deformation mode in nanocrystalline and ultrafine-grained materials, on the emission criterion of the dislocation from the elliptical blunt crack tip was theoretically described by Feng et al. (2019). The schematic model is shown in Fig. 2.9.

Consider a deformed nanocrystalline or ultrafine-grained solid consisting of nanoscale grains divided by GBs and containing an elliptical blunt crack. The external load concentrated at the crack tip can initiate GB sliding along one GB adjacent to the crack tip, which can induce the nanoscale amorphization. Within the model, the amorphization with a rectangular region occurs through the splitting transformation of GB disclinations which are produced by GB sliding. More specifically, GB sliding along the high-angle boundary AC makes a high-angle tilt boundary transferred from position AA′ to BB′ and results in the formation of a GB disclination dipole AB characterized by the strengths $\pm\omega$ and the arm (the distance between disclinations) p (Ovid'ko and Skiba 2014b). The magnitude ω represents the tilt misorientation of the high-angle boundary. Then, according to the theory of defects in solids, such a disclination dipole is equivalent to a finite wall of GB dislocations distributed between the junctions A and B (Fig. 2.9b). The relaxation of mechanical stresses in loaded solids occurs by means of plastic deformation and/or failure processes. Therefore, it is thought that the GB dislocations move during their splitting transformations and thereby carry local plastic deformation to relax the stresses created by the disclination dipole. Since the GB dislocations are not lattice ones, their movement is accompanied by formation of a disordered region ABEF in the wake of the GB dislocations and this disordered region is logically treated as the amorphous region. It is assumed that dislocations are continuously distributed over the amorphous region with a constant density. Then, the continuous uniform distribution of edge dislocations (Fig. 2.9c) can be considered as an array of dislocation walls located between the opposite boundaries, AF and BE, of the amorphous region. The uniform dislocation distribution over the amorphous region is equivalent to the two uniform distributions of wedge disclinations over the boundaries AF and BE. That is, the amorphization within a rectangular region can be modeled as two uniform distributions of disclinations over two boundaries of the amorphous region (Fig. 2.9d).

The nanoscale amorphization can either enhance or weaken the critical applied SIFs for dislocation emission, depending on the emission angle, the radius of curvature of the elliptical blunt crack, and the distance between the nanoscale amorphization and the crack tip. There is a critical crack-junction distance making the dislocation emission most difficult and at this critical crack-junction distance, the stress release effect of the nanoscale amorphization is strongest. The location and the strength of the nanoscale amorphization, as well as the radius of curvature of the elliptical blunt crack, have great influence on the most probable angle for dislocation emission. Meanwhile, when the strength of the nanoscale amorphization is relatively strong, the most probable emission angle for the mode I loading is the same as that for the mode II under the same parameters. There is a critical crack length making the dislocation emission easiest. This critical crack length increases with the increasing radius of curvature, but decreases with the increment of the emission angle.

The behavior of nanocrystalline materials with grain sizes blow ~10–20 nm has been attributed to the emergence of grain boundary sliding and rotation as the dominant carriers of plastic deformation. A nanocrystalline aggregate model including dislocation and grain boundary sliding mechanisms of deformation is shown in Fig. 2.10.

This is due to the fact that grain boundaries' shear does not allow for the compatibility of each grain, so stress concentrates and dislocations are emitted from these regions of stress concentration in order to satisfy the compatibility. As a general behavior, as the grain size decreases the phenomenon is more pronounced.

A very realistic scenario of dislocation emissions from a grain boundary traveling towards the opposite grain boundary was presented by Meyers et al. (2006). The dislocation is emitted from the grain boundary AD (Fig. 2.11), leaving two screw dislocations AB and CD (a cubic grain of length d is considered).

The 3-D model of this behavior is shown in Fig. 2.12.

The left side of the grain is deformed by shear of γ. The increase of deformation work due to the formation of the screw dislocations AB and CD is

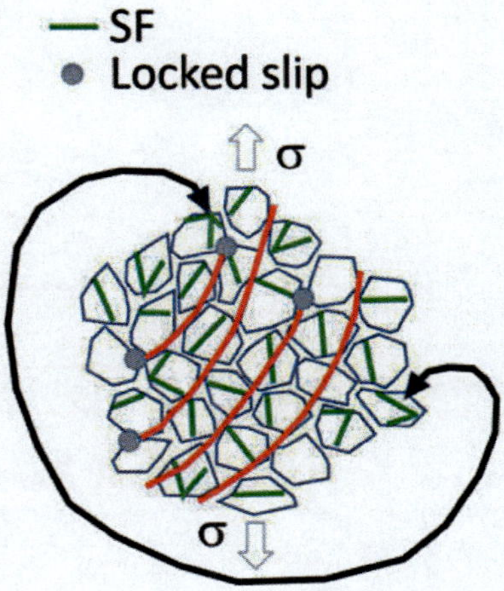

Fig. 2.10 Deformation by dislocation/SF emission

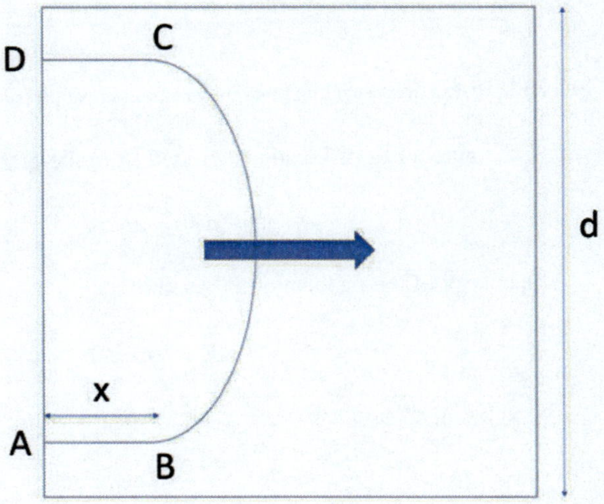

Fig. 2.11 Dislocation emission from the grain boundary

$$dW = \tau\gamma V$$

where V is the deformed volume:

$$V = xd^2$$

And the shear strain is

$$\gamma = \frac{nb}{d}$$

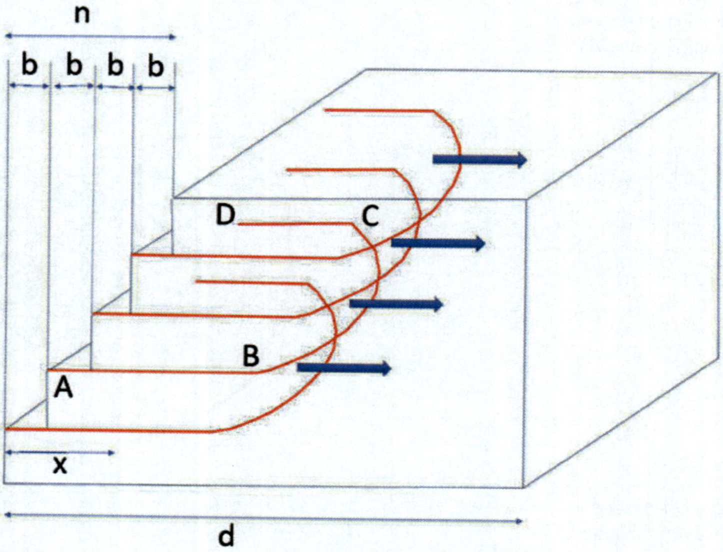

Fig. 2.12 Dislocations traveling through nanograin in parallel planes and creating a shear strain γ

with n the number of dislocations emitted from the grain boundary. So

$$dW = nbx\tau d$$

The energy of the dislocation is given by

$$dE = 2nx\alpha Gb^2$$

By equating dE and dW

$$\tau = \frac{2\alpha Gb}{d}$$

If partial dislocations are emitted instead of perfect dislocations

$$b_p = \frac{b}{\sqrt{3}}$$

Thus

$$\tau = \frac{2\alpha Gb}{\sqrt{3}d}$$

The SFE is the measure of the separation of partials:

$$\mathrm{SFE} = \frac{Gb_1b_2}{2\pi\gamma}\left(\cos\theta_1\cos\theta_2 + \frac{\sin\theta_1\sin\theta_2}{1-\nu}\right)$$

where θ_1 and θ_2 are the angles between each arm of the partial and the dislocation line. Assuming $\theta_1 = \theta_2 = 30°$ and $b_1 = b_2 = b_p$

$$\mathrm{SFE} = \frac{Gb_p^2}{2.2\pi\gamma}$$

The expression for the flow stress based on partial emission is

$$\tau_p = \frac{1}{\sin\alpha}\left(\frac{Gb_p}{2\pi d}\ln\frac{d}{b_p\sqrt{3}} + \frac{\gamma}{b_p}\right)$$

The natural consequence of the saturation of plastic deformation processes is fracture. Thus, the achievable ductility of a material is an important scientific topic because it demarcates the capacity of the active plastic deformation mechanisms to support plastic flow (Kumar et al. 2016). Plasticity is grain boundary dependent, but also failure is directly related to grain boundary structure in nanocrystalline materials (Malygin 2007). As a matter of fact, the formation of nano-voids in the grain boundary and triple junctions in front of a moving crack is fundamental. These interfacial voids can join together to make a larger micron size crack and cause intergranular fracture (Fig. 2.13).

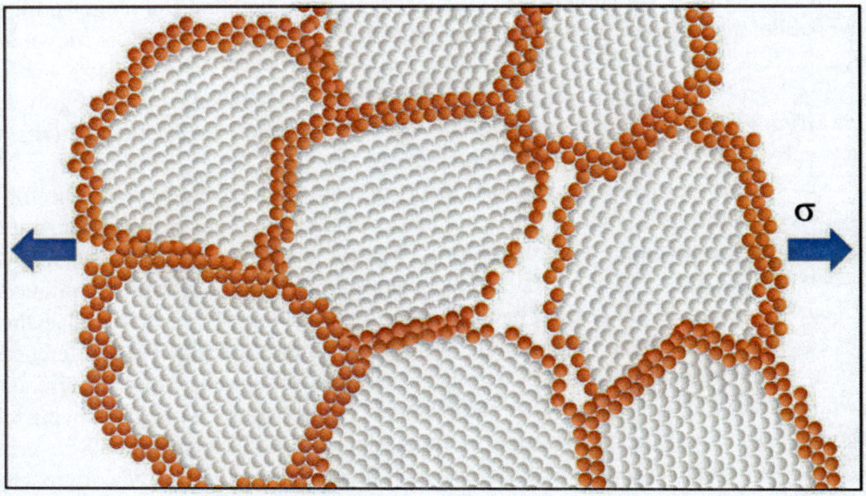

Fig. 2.13 Crack nucleation at triple junction in NC Ni

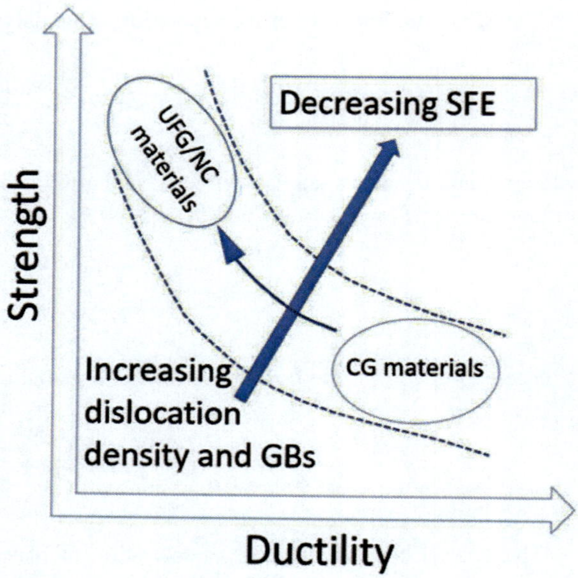

Fig. 2.14 Strength-ductility synergy vs. microstructure for FCC materials

In general, the increase in dislocation density and GBs enhances the material strength but goes to limit the ductility. By contrasting this behavior, a better strength-ductility synergy can be obtained by reducing the SFE (Fig. 2.14).

So, grain size and SFE can be tuned in order to obtain optimized synergy between strength and ductility. Analyses performed on Cu alloys allowed to model the strength variation as a function of the number of cycles during cyclic loading for a medium-SFE material (Fig. 2.15).

SFE can govern the choice of fatigue mechanisms in both the CG range and the nanoscale; the summary of the acting mechanisms as a function of SFE is shown in Fig. 2.16.

For high-SFE materials, the main observed micromechanisms are grain growth via GB migration and dislocation pattern. For low-SFE materials, local GB activities act such as atom shuffling and GB sliding/rotation.

The relaxation of internal stresses, recovery, and grain coarsening and the formation of sparse shear bands, which are narrower than the grain size and penetrate many grains, dominantly control the stage I cyclic softening process that mainly occurs in the matrix (An et al. 2019). The cyclic softening of the matrix is restrained gradually in stage II, while shear bands and grain growth mutually accommodate the cyclic plasticity and continue softening, resulting in an increasingly larger difference in the strength between the matrix region and shear banding areas. The reoccurrence of dramatic softening in stage III is ascribed to the loss of material integrity owing to the nucleation and the propagation of fatigue cracks. The reduction of SFE can significantly suppress the intrinsic softening feasibility of NS alloys which originates from changes of fundamental fatigue damage mechanisms (Pineau et al. 2016).

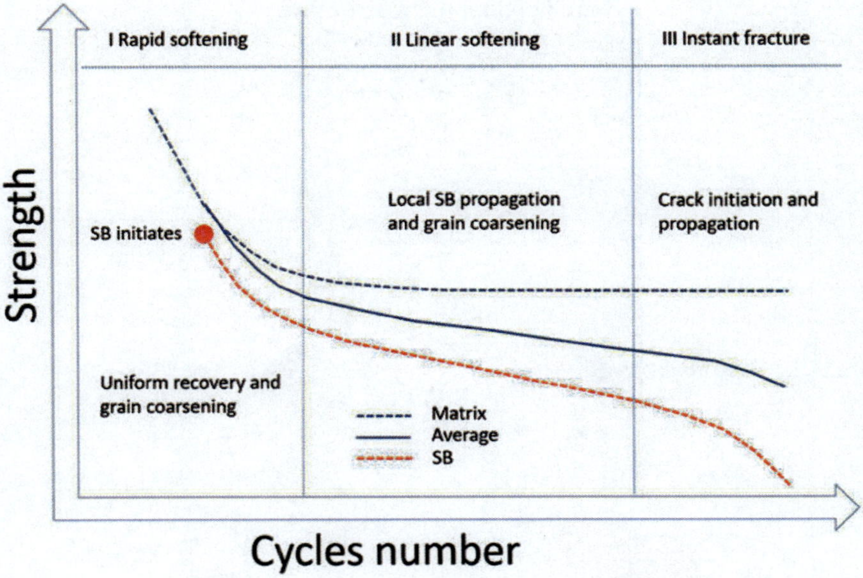

Fig. 2.15 Microscopic mechanisms during cyclic loading of NS materials

2.2.2 *Cyclic Behavior*

Plastic deformation plays a key role in the thermomechanics of engineering metals and alloys. Plasticity is of central concern in the ductile failure, surface wear, and deformation processing of metals. Additionally, mechanisms of fracture, fatigue, and damage typically involve plastic deformation (e.g., the blunting of cracks). At the atomic scale, plastic deformation in metals is the result of stress-driven irreversible rearrangement of atoms, i.e., bond switching (this should be contrasted with bond breaking, which leads to brittle fracture). Constitutive descriptions of plasticity attempt to establish relations between the kinetics (force-like quantities, e.g., stress) and kinematics (displacement-like quantities, e.g., strain) that emerge from such rearrangement events (Voyiadjis et al. 2003; Lu et al. 2014). Considerable research efforts were recently focused on the study of mechanical properties of submicron crystals. It was found that the deformation mechanisms, which we habitually associate with dislocation plasticity, change dramatically once the sample size is reduced below the micrometer range. Strength of such crystals was shown to be size dependent, with stress-strain response exhibiting pronounced intermittency and scale invariance over a wide range of scales, independently of crystal symmetry. Both, measured and computed, scaling exponents appear to be featuring continuous size dependence. Moreover, even though plasticity at macroscale is generally associated with ductility, crystal plasticity at submicron scales was shown to exhibit major stress drops or strain bursts reminiscent of brittle fracture (Ramesh 2009).

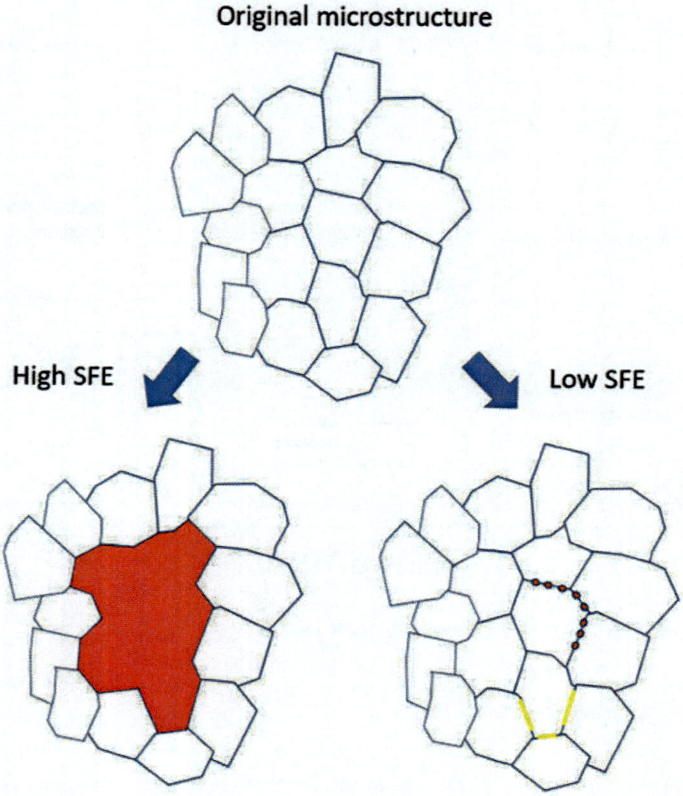

Fig. 2.16 Microscopic evolution after cyclic loading due to SFE

A defining characteristic of NS materials is the enhanced influence of interfaces such as grain boundaries (GBs) and/or free surfaces on the physical properties. In conventional engineering materials, internal length scales are typically on the order of microns or higher, and therefore, physical properties are controlled by the behavior of bulk phases(s) that constitute the material. However, when these length scales are reduced to nanometers, as is the case in NS solids, the population of atoms that belong to the interfaces and/or free surfaces becomes comparable to that of the bulk atoms. Consequently, material properties are strongly affected by interfacial structure, surface tension forces, impurity segregation at interfaces, and presence of disordered phases. From the standpoint of elastoplastic deformation, the significant differences between the CG microstructure and the compositionally similar NC microstructure arise from the suppression of lattice dislocation activity in the confined crystalline volumes inherent to NC materials. Unlike CG metals, NC metals do not have a significant concentration of either preexisting dislocations or dislocation sources. Additionally, extended dislocation pileups are unlikely to form owing to the small grain size.

The most suggested model to couple both grain boundary shear and dislocation motion takes into account both the contributions. First of all, the volume grain boundary shear strain rate is given by

$$\dot{\gamma}_{GB} = \gamma^{\mathrm{T}}\left(\frac{6\delta}{d}\right)\nu_D \exp\left[-\frac{Q_v}{kT}\left(1-\frac{\sigma}{\hat{\tau}_{\mathrm{is}}}\right)\right]$$

where δ is the grain boundary thickness where the flow is concentrated producing a transformation shear strain γ^{T}, ν_D is the atomic frequency, Q_v is the activation energy for viscous flow, and τ_{is} is the ideal shear strength of the disordered grain boundary. The contribution of dislocation motions is described through the shear rate due to dislocation plasticity:

$$\dot{\gamma}_D = \left(\frac{b}{d}\right)\nu_G \exp\left[-\frac{\Delta F_j}{kT}\left(1-\frac{\sigma}{\hat{\tau}}\right)\right]$$

Plasticity occurs by release of dislocations from pinning points associated with a forest of dislocations with a jog energy barrier ΔF_j. The exponential factor in the previous equations can be modified as

$$\exp\left[-\frac{Q_v}{kT}\left(1-\frac{\sigma}{\hat{\tau}_{is}}\right)\right] \rightarrow \left(\frac{\sigma}{\hat{\tau}_{\mathrm{is}}}\right)^{m_1}$$

$$\exp\left[-\frac{\Delta F_j}{kT}\left(1-\frac{\sigma}{\hat{\tau}}\right)\right] \rightarrow \left(\frac{\sigma}{\hat{\tau}}\right)^{m_2}$$

It can be demonstrated that $m_1 \approx 30$ and $m_2 \approx 60$; anyway they are normally considered as equal so that $m = m_1 = m_2 = 30$. In this way the overall strain rate is given by

$$\dot{\gamma} = f\dot{\gamma}_{GB} + (1-f)\dot{\gamma}_D$$

Taking into account that the volume fraction of grain boundary material $f = 6$ (b/d) and considering $b = d$

$$\dot{\gamma} = \gamma^{\mathrm{T}} f \nu_D \left(\frac{\sigma}{\hat{\tau}_{\mathrm{is}}}\right)^m + (1-f)\left(\frac{b}{d}\right)\upsilon_G\left(\frac{\sigma}{\hat{\tau}}\right)^m$$

By the calculations of Naturani and Takamura (1991) the threshold plastic resistance is related to grain size as

$$\widehat{\tau} = \alpha\mu\sqrt{\frac{b}{d}}$$

The shear strain rate becomes

$$\dot{\gamma} = \upsilon_D\left(\frac{\sigma}{\mu}\right)^m\left[\left(6\gamma_T\frac{b}{d}\right)\left(\frac{\mu}{\widehat{\tau}_{\mathrm{is}}}\right)^m + \left(1-6\left(\frac{b}{d}\right)\right)\left(\frac{\upsilon_G}{\upsilon_D}\right)\left(\frac{1}{\alpha}\right)^m\left(\frac{d}{b}\right)^{\frac{m}{2}}\right]$$

By taking a close look at the previous equation it can be observed that for a constant shear strain rate, the flow stress has a maximum for a certain grain size. Below this size, the first term of the previous equation dominates the shear deformation. Above this size, plasticity is dominant. The observation of new structure-property scaling laws in nanocrystalline materials is interesting scientifically because it signals the emergence of new physical mechanisms responsible for plasticity. As a matter of fact, systematic nanoindentation at different grain sizes shows that hardening occurs quickly and is grain size dependent. The change in strengthening slope observed below d ~ 100 nm is believed to be related to a shift from plastic deformation that is controlled by intragranular dislocation sources to grain boundary sites acting as sources and sinks for dislocation activity.

The deformation mechanism of the NC metals and alloys at nanoscale, especially at the length scale of less than 20 nm, is complex due to the competing dislocation-mediated and grain boundary-dominated activities. The strain rate sensitivity (SRS) can provide valuable insight into the deformation mechanism of the metallic alloys of ultrafine-grained (UFG) and/or NC-grained structure at multiple length scales (Mohebbi and Akbarzadeh 2017; Wei 2007) and further fingerprints the rate controlling mechanism during the thermally activated process of the UFG/NC materials.

Quasi-static and dynamics nanoindentation experiments have shown that nanocrystalline plasticity is likely due to the difference between the strain rates that are commonly applied in experimental and computational techniques. A maximum rate sensitivity of strength was found at a grain size of ~12 nm, meaning that high strain rates such as those used in molecular dynamics simulations would show a maximum strength at an intermediate grain size and inverse Hall-Petch scaling at smaller grain sizes.

Making some assumptions based on experimental and simulations results, for nanocrystalline copper the tensile peak flow stress is 2.3 GPa that gives rise to a shear peak stress of 1.33 GPa, with a normalized grain diameter d/b of 48. Considering $s = \sigma/\mu$ and $x = d/b$

$$\left(\frac{\partial s}{\partial x}\right)_{\dot{\gamma}} = 0$$

A general equation for the critical grain size x_p for the peak stress is

$$x^{\left(\frac{m}{2}\right)+1} - \left[\frac{6(m-2)}{m}\right]x^{\frac{m}{2}} - Q = 0$$

where

$$Q = 12\gamma^{\mathrm{T}} \frac{(\nu_G/\nu_D)(\alpha/(\widehat{\tau}_{\mathrm{is}}/\mu))^m}{m}$$

Now

$$s = \left(\frac{\dot{\gamma}}{\nu_D}\right)^{\frac{1}{m}} \left[\frac{A}{x} + Bx^{\frac{m}{2}} - 6Bx^{\left(\frac{m}{2}\right)-1}\right]^{-\frac{1}{m}}$$

where

$$A = \frac{6\gamma^{\mathrm{T}}}{\left(\frac{\widehat{\tau}_{\mathrm{is}}}{\mu}\right)^m}$$

and

$$B = \frac{\nu_G/\nu_D}{\alpha^m}$$

From the conclusions of Argon and Demkowicz (2006), $\gamma^{\mathrm{T}} = 0.015$, $\tau_{\mathrm{is}} = \mu/30$, $\alpha = 0.246$, $\nu_D = 10^{13}\ \mathrm{s}^{-1}$, and $\nu_G/\nu_D = 10^{-3}$; the global shear strain rate results to be $10^8\ \mathrm{s}^{-1}$. All this results in $Q = 6.6{*}10^{26}$ with $x_p = 47.8$ (for a threshold grain size of 12.2 nm for Cu). This gives a result for dimensionless peak stress $s_p = 0.0276$ (for a σ_p of 1.52 MPa). When $x >> x_p$ the dimensionless flow stress results as

$$s = \left(\frac{\dot{\gamma}/\nu_D}{B}\right)^{\frac{1}{m}} \sqrt{\frac{b}{d}} = 0.1\sqrt{\frac{b}{d}}$$

which is of the classical Hall-Petch character. For schematizing the analytically expressed behavior Fig. 2.17 has been plotted from Narutani and Takamura with the analytical graph.

Now, with decreasing grain size grain boundary activity is dominant on the dislocation activity. So, a variety of deformation mechanisms such as grain boundary sliding and grain rotation leading to the activation of local shear planes contribute to the material deformation. The multiple shear band formation leads to perfect plasticity deformation in contrast to the microcrystalline counterparts exhibiting hardening behavior due to the domination of the dislocation activity. In addition, the twin boundaries (Fig. 2.18) appearing and disappearing at almost the same location during the different stages of straining and after relaxation have been observed and

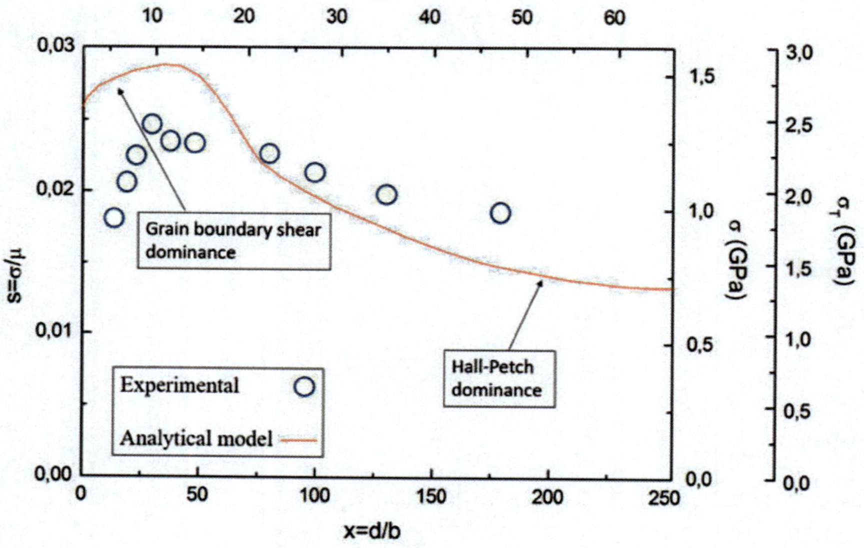

Fig. 2.17 Plastic behavior of polycrystalline Cu based on the presented model

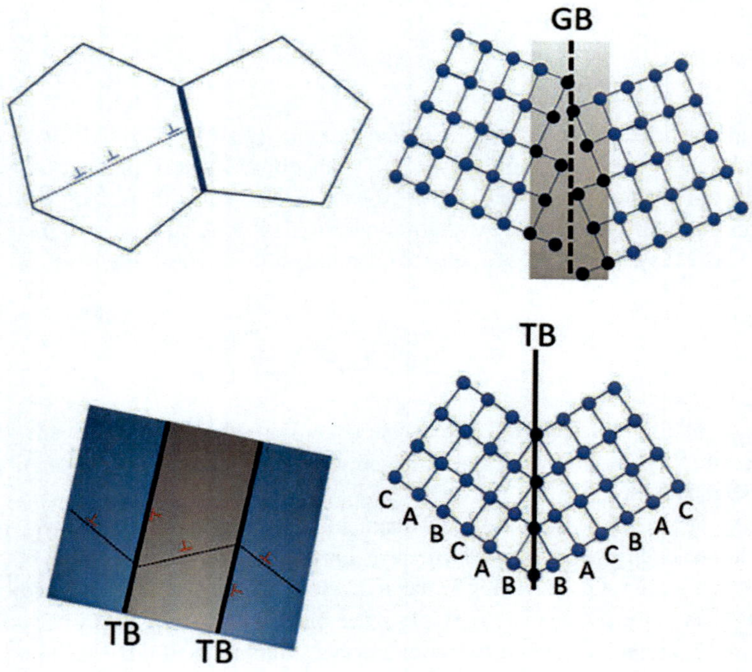

Fig. 2.18 Grain boundary and twin boundary

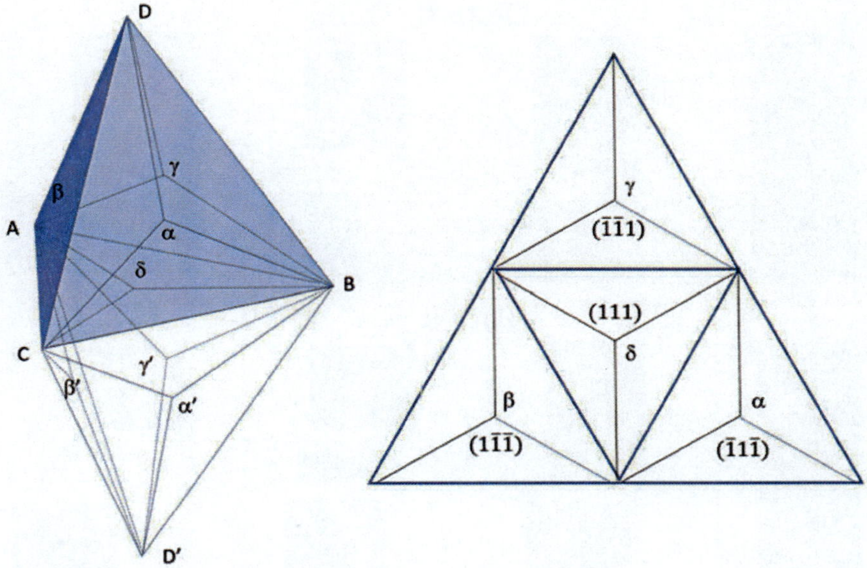

Fig. 2.19 Schematic representation of slip planes in FCC lattice

demonstrated to be due to a Bauschinger type effect as the different grains locally react very differently to the strain, leading to high stress accumulation at their interfaces, which apparently compensates for part of the plastic deformation during relaxation.

According to the different geometric and energetic considerations of twin boundaries compared to grain boundaries, one can expect to see different dislocation reactions (nucleation and propagation). However, both boundaries behave in a similar fashion to some degree. The development of nanostructured materials with high-density nanotwins is an optimal issue to improve the material ductility (Li et al. 2018, 2020). Although such materials are promising, they cannot be applied to a wide range of materials, since the formation of nanotwinned grains is only possible for materials with low stacking fault energy (You et al. 2020).

Simply, both boundaries act as an obstacle to dislocation motion/transmission at some degree. The easiest way to demonstrate the dislocation twin boundary interactions is using double Thompson tetrahedron (Fig. 2.19).

Now, twin nucleation in a face-centered cubic crystal is believed to be accomplished through the formation of twinning partial dislocations on consecutive atomic planes. Twinning should thus be highly unfavorable in face-centered cubic metals with high twin-fault energy barriers, such as Al, Ni, and Pt, but instead is often observed (Wang et al. 2017b). The proposed model for the twinning deformation is shown in Fig. 2.20.

In the traditional twinning route, twin nucleation involves the sequential nucleation of three partial dislocations on consecutive atomic layers. The highest energy

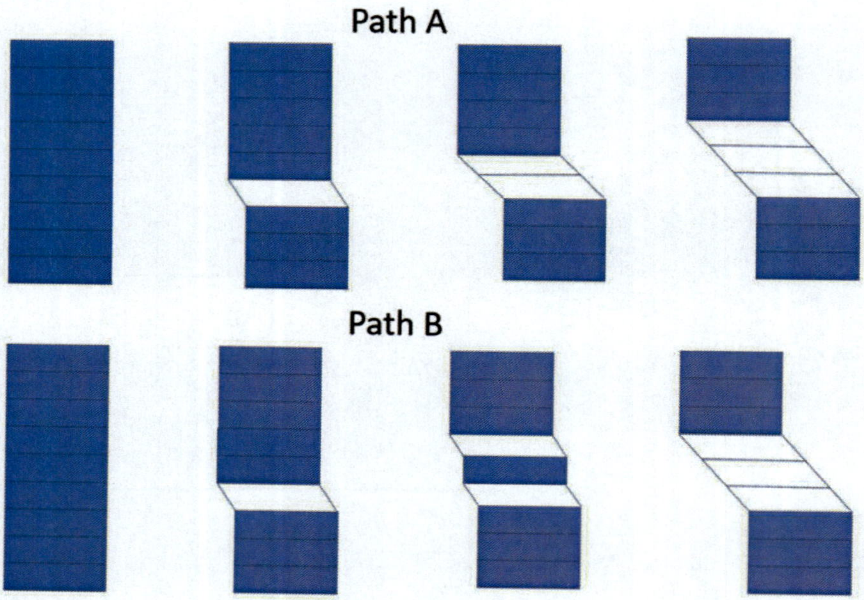

Fig. 2.20 Twin nucleation routes

barrier for path A is the unstable twin-fault energy. For nanocrystalline FCC metals, in the proposed model two partial dislocations are present, but they are separated with one atomic layer in between; then a third partial dislocation forms in between the two previously formed SFs, resulting in a three-layer twin. The highest energy barrier becomes the unstable middle twin-fault energy.

First of all, it should be considered that below the ultrafine regime (GS < 500 nm) the deformation behavior of materials must be considered as a superimposition of two contributes: the bulk with no dislocation slip and the grain boundary with dislocation-induced rotation.

In cyclic loading grains rotate towards the loading axis. The magnitude of lattice rotation can be calculated by knowing the amount of dislocation gliding through the slip planes (Fig. 2.21).

Assuming that the total number of dislocations produced during crack growth contributes to the total amount of displacement due to slip, an average rotation angle, θ, can be given as

$$\theta = \tan^{-1}\frac{nb}{l_w}$$

where n is the total number of dislocations, l_w is the average width of the grain, and b is the Burgers vector. Another factor that could contribute to grain rotation is higher grain boundary (GB) dislocation content (Goswami et al. 2017). The increase

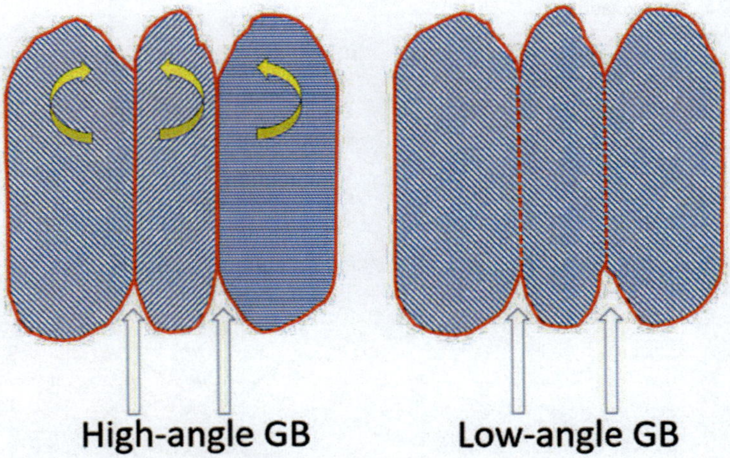

Fig. 2.21 Grain rotation-assisted grain coarsening

in grain boundary dislocation content alters the GB angle between neighboring grains due to the well-known Frank-Bilby equation. Fatigue crack growth in the material results in cycle-dependent plastic deformation due to dislocation motion, which gives higher magnitude of dislocation density and also indicates that dislocation motion is important. It is likely that the cyclic/reverse slip of dislocations near the crack tip can cause higher stresses and dislocation damage accumulation. This together with slip constraints leads to grain rotation and coalescence, thereby increasing the grain size (Fig. 2.22).

Randomly oriented grains rotate into each other's orientation until they are aligned. As they have no more different orientations, the grain coalescence and dislocation motion over several "former" grains are possible.

Obviously, once described the effect of both the contributions, the influence of defects such as nanocrack formation at the interfaces must be taken into account (Fig. 2.23).

The mechanism model of the crack formation at triple junction is shown in Fig. 2.24.

The crack has a length L and it is located in a direction perpendicular to the loading axis. The crack is at a distance d from the triple junction. The fatigue stresses generate a crack at the triple junction. The force F is defined as the elastic energy released when the crack moves over a unit distance:

$$F = \frac{\pi(1-v)l}{4G}\left(\overline{\sigma}_{yy}^2 + \overline{\sigma}_{xy}^2\right)$$

The mean weighted stress tensor components can be calculated as

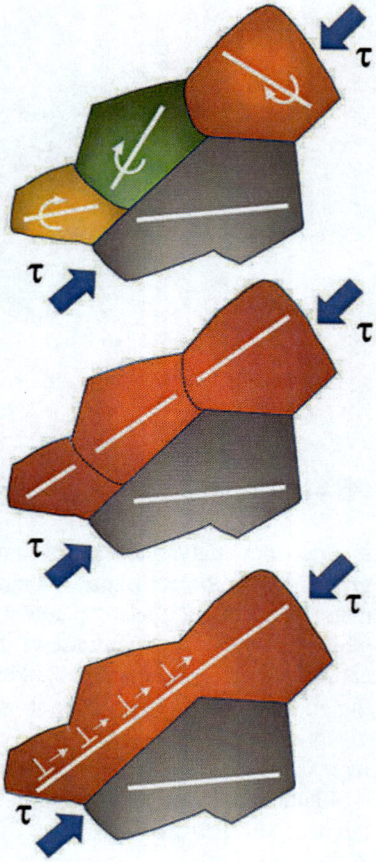

Fig. 2.22 Grain rotation and coarsening

$$\overline{\sigma}_{my} = \frac{2}{\pi L}\int_{0}^{l} \sigma_{my}(x, y = 0)\sqrt{\frac{x}{l - x}}dx$$

It is demonstrated that (Ovidko and Sheinerman 2004)

$$\overline{\sigma}_{yy} = \sigma_e + \frac{GB}{\pi(1 - v)l}$$

where

$$\sigma_e = \sigma_0\left(1 + \sqrt{L/4d}\right)$$

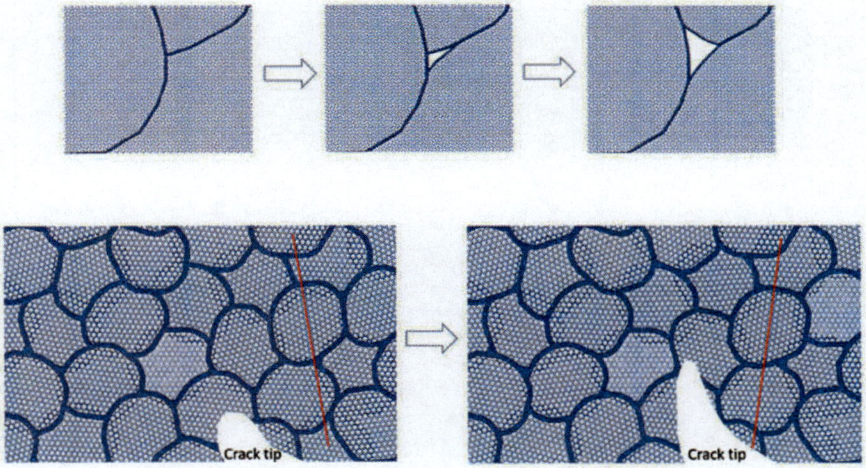

Fig. 2.23 Pore formation at triple junctions and grain rotation

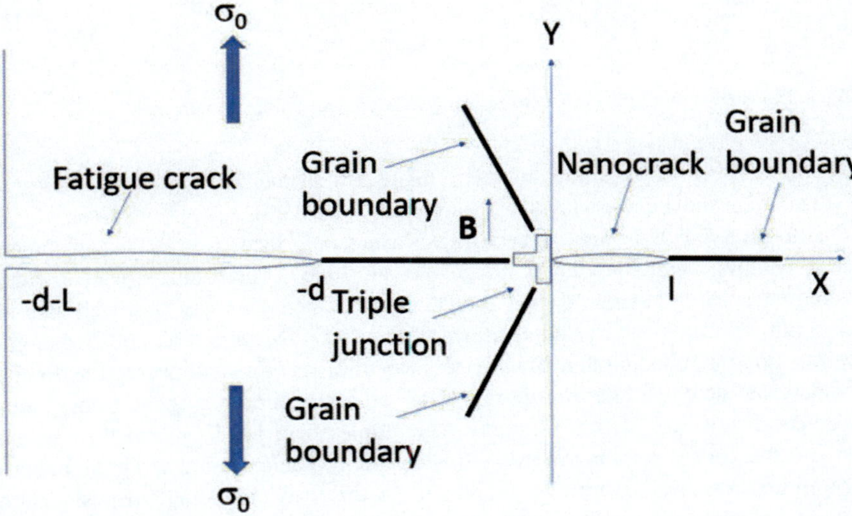

Fig. 2.24 Nucleation of nanocrack at triple junction

is the effective stress taking into account the stress concentration. When the fatigue crack is absent, $L = 0$ and $\sigma_e = \sigma_0$.

These mechanisms seem to be diffusion assisted without any dislocation activity at the nanoscale (He et al. 2017). Grain rotation is inhomogeneous with a relative rotation on the order of 5° for neighboring grains and in the order of 15° in proximity of the crack tip (Ovidko and Aifantis 2013). They described the tensile behavior of

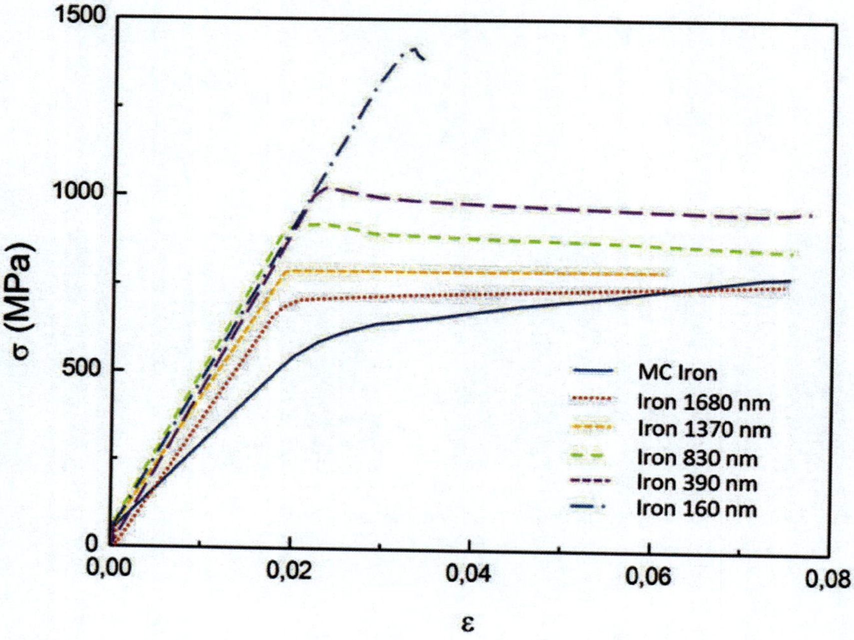

Fig. 2.25 Stress-strain behavior of electrodeposited iron with different grain sizes

ultrafine and nanocrystalline iron underlying that, with decreasing grain size in irons, the materials show decreasing strain hardening (Fig. 2.25).

The observed deformation mechanisms were described, first of all an intense shear banding at yield point (Gryaznov et al. 1993). These shear bands' width increases as the grain size decreases. This kind of behavior was observed also in other ultrafine and nanocrystalline alloys (Valiev 2004; Meyers et al. 2006; Malygin 2009). So, by reducing the grain size from the microcrystalline regime to the nanocrystalline one a first transition is observed at the ultrafine regime from homogeneous dislocation motion plasticity to multiple shear bending, and then below ~10 nm a second transition towards diffusion-induced deformation and grain boundary sliding/rotation starts to be dominant. In this way, reducing the dislocation motion as the grain size decreases leads to higher yield points in ultrafine and nanocrystalline materials. Ovidko and Aifantis (2013) calculated a critical strain rate at which dislocation-based plasticity due to grain boundary sources might occur:

$$\dot{\varepsilon} \approx \frac{10^4}{s}\left(\frac{b}{d}\right)^2 \frac{1}{4\pi^2}$$

In order to model the elastic and plastic behavior of NC and UFG materials, they are considered as a composite of bulk and grain boundary as two distinct phases. This will be very useful also in the modeling of crack nucleation and growth below the ultrafine regime. The two "phases" interact through an internal body force and so the equilibrium differential equations are

$$\mathrm{div}\boldsymbol{\sigma}_1 = \widehat{\boldsymbol{f}}$$

$$\mathrm{div}\boldsymbol{\sigma}_2 = -\widehat{\boldsymbol{f}}$$

And so

$$\mathrm{div}\boldsymbol{\sigma} = 0$$

with $\boldsymbol{\sigma}_1$ and $\boldsymbol{\sigma}_2$ the partial stress tensors for each phase and

$$\boldsymbol{\sigma} = \boldsymbol{\sigma}_1 + \boldsymbol{\sigma}_2$$

is the global stress tensor. It is proportional to the displacement:

$$\boldsymbol{\sigma}_k = \boldsymbol{L}_k \boldsymbol{u}_k$$

And

$$\widehat{\boldsymbol{f}} = \alpha(\boldsymbol{u}_1 - \boldsymbol{u}_2)$$

with

$$\boldsymbol{L}_k = \lambda_k \boldsymbol{G} + \mu_k \widehat{\nabla}$$

with

$$\boldsymbol{G} = \boldsymbol{I}\mathrm{div};\ \widehat{\nabla} = \nabla + \nabla^{\mathrm{T}}$$

where $\boldsymbol{I}$ is the identity tensor and div the divergence; the differential equation for the total displacement associated with the nanostructured material is

$$\mu\nabla^2\boldsymbol{u} + (\lambda + \mu)\mathrm{div}\boldsymbol{u} + c\nabla^2\left[c\nabla^2\boldsymbol{u} + (\lambda + \mu)\mathrm{div}\boldsymbol{u}\right] = 0$$

Finally, the following Hooke's law for nanoelasticity was defined:

$$\boldsymbol{\sigma} = \lambda(tr\boldsymbol{\varepsilon})\boldsymbol{I} + 2\mu\boldsymbol{\varepsilon} - c\nabla^2[\lambda(tr\boldsymbol{\varepsilon})\boldsymbol{I} + 2\mu\boldsymbol{\varepsilon}]$$

As for the elastic behavior, the flow stress τ is considered as the sum of two contributions, with τ_1 and τ_2 the bulk and grain boundary flow stress, respectively:

$$\tau = \tau_1 + \tau_2$$

In the case of simple shear, both the phases carry out the stress being in equilibrium:

$$\partial_x \tau_1 = \widehat{f};\, \partial_x \tau_2 = -\widehat{f} \rightarrow \partial_x \tau = 0$$

Considering τ_k and γ_k the flow stress and the shear strain per each phase

$$\tau_k = k_k \gamma_k; \widehat{f} = \alpha(\gamma_1 - \gamma_2) = \alpha\delta$$

Being $\gamma = \gamma_1 + \gamma_2$

$$\tau = k_1(\gamma + \delta) + k_2(\gamma - \delta) = k(\gamma) - \mu^*(\gamma)\delta$$

That in three dimensions becomes

$$\tau = k(\gamma) - c\nabla^2\gamma$$

Aifantis (2009) concluded that higher order strain gradients enter naturally into the standard equations of continuum mechanics as a consequence of the mechanical interaction between "bulk" and internal or external "surface" points. Since these interactions become more dominant, as the dimensions of the sample decrease, it may be argued that the resultant gradient-dependent equations of elasticity and plasticity may be used for describing the behavior at the micro- and nanoregime, in particular the mechanical response of ultrafine-grained materials and nanograined polycrystals. The effect of defects on deformation and crack will be described in the next chapter.

2.3 Nanoindentation

The probing of materials' mechanical properties through indentation is one of the most and fast methods employed in the research practice. The first property to be measured is the indentation hardness that is the resistance of the material to local plastic deformation:

$$H = \frac{F_{\max}}{A}$$

where $F_{\max}$ is the maximum applied load and A is the impression after unloading. The development of nanoindentation techniques allows for the measurement of mechanical properties of very small material volumes. The so-called depth sensing indentation (DSI) has been developed for the obtaining of mechanical properties of nanocrystalline materials. During instrumented nanoindentation a sharp indenter is pressed on the material surface and then removed. The displacement of the indenter (at a given load) is continuously recorded. Loads and depths are in the order of micronewtons and nanometers, respectively; the load-displacement (F-h) curve allows to calculate the material hardness and the Young's modulus of the material. Mechanical properties such as continuous stiffness, scratch resistance, film-substrate adhesion, residual stresses, time-dependent creep and relaxation properties, fracture toughness, and fatigue can be measured by nanoindentation technique, based on elastic contact theory and load-displacement data. The most employed procedure is known as Oliver-Pharr method and it is based on the assumption that unloading data arise from a purely elastic contact.

The main assumptions can be summarized as follows:

- After unloading the deformation is elastic.
- The compliance between the specimen and the indenter is related by

$$\frac{1}{E_r} = \frac{1-\nu_{\text{specimen}}^2}{E_{\text{specimen}}} + \frac{1-\nu_{\text{indenter}}^2}{E_{\text{indenter}}}$$

where E_r is the reduced modulus.

- The contact stiffness is given by

$$S = \frac{2}{\sqrt{\pi}} E_r \sqrt{A_c}$$

where A_c is the contact area.

The schematic of an indentation at maximum load and after unloading is shown in Fig. 2.26.

The indicated quantities of Fig. 2.26 refer to the F-h curve examples plotted in Fig. 2.27 (for comparison both the Oliver-Pharr and the Nix model are plotted).

The hardness is given by

$$H = \frac{F_{\max}}{A(h_c)}$$

where $A(h_c)$ is the projected contact area at the maximum load. The main difference with the classical microindentation measurements is that the area in the nanoindentation tests refers to the contact area at the maximum load while classical indentation considers the residual area after unloading.

A fundamental physical aspect of the nanoindentation tests is that the dislocation nucleation at the indenter tip is shown through the appearance of pop-ins in the load-displacement curve. The phenomenon is shown in Fig. 2.28 for a spherical indenter.

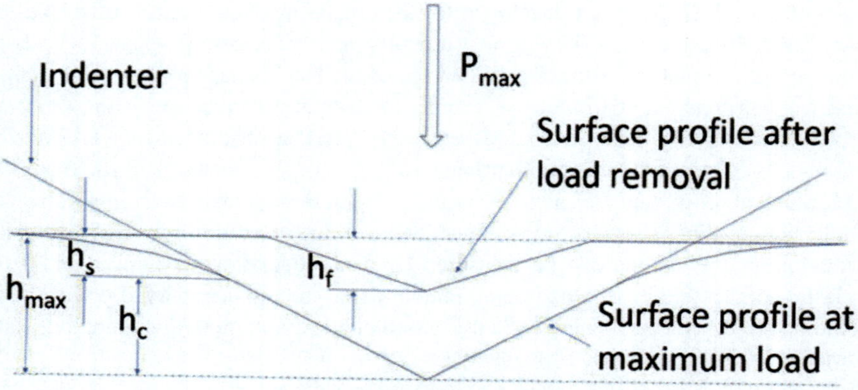

Fig. 2.26 Indentation profile at maximum load and after unloading

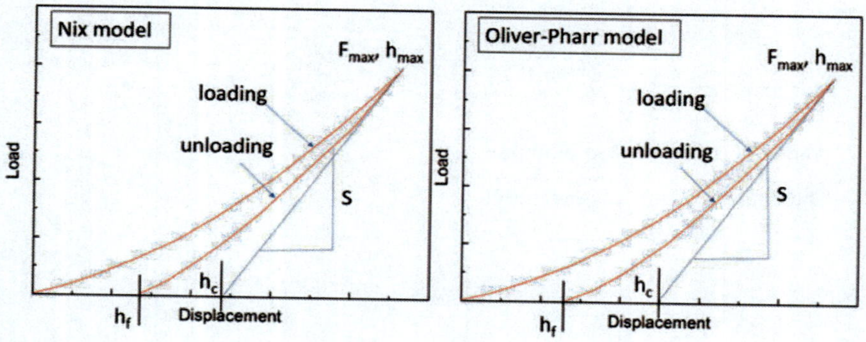

Fig. 2.27 Load-depth curves

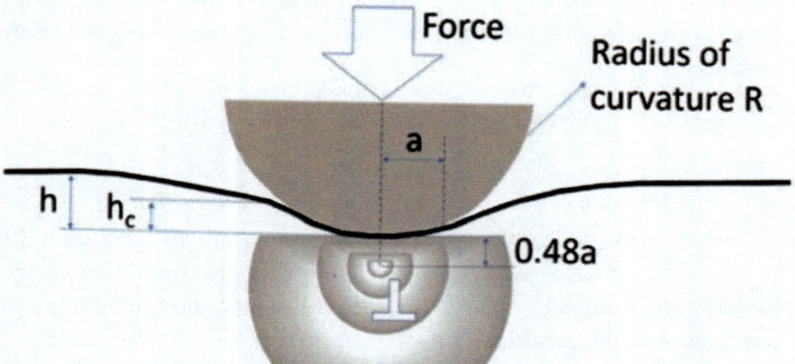

Fig. 2.28 Dislocation nucleation due to nanoindentation

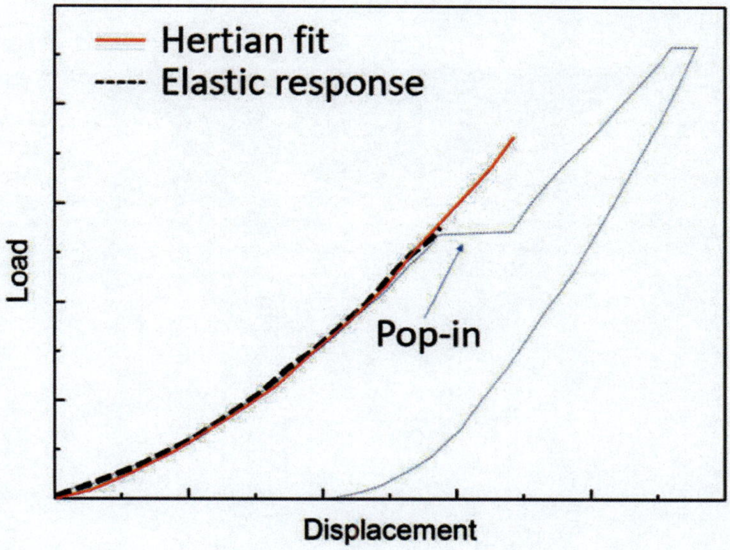

Fig. 2.29 Indentation curve with pop-in

The pop-in in the indentation curve is shown in Fig. 2.29.

Given that sometimes pop-ins can be due to the breaking of surface oxides or other types of defects, surface must be completely smooth and clean in order to observe pop-ins due to dislocation nucleation. As shown in the figure, the pop-in shifts the material behavior from pure elastic to plastic deformation. The relationship between the force and the indentation depth is given by

$$F = \frac{4}{3} E_r \sqrt{Rh^3}$$

The contact radius is expressed by

$$a^2 = Rh$$

It is demonstrated that, in this configuration, dislocation nucleation begins at a depth of 0.48a for the maximum shear stress:

$$\tau_{\max} = \frac{0.31}{\pi} \left(\frac{6FE_r^2}{R^2} \right)$$

Another important aspect to be considered is the so-called indentation size effect (ISE). The first effect of ISE is the decrease of hardness as the indentation size increases. This is physically due to the induced GNDs as the strain gradient increases. The most considered model assumes that indentation is accommodated

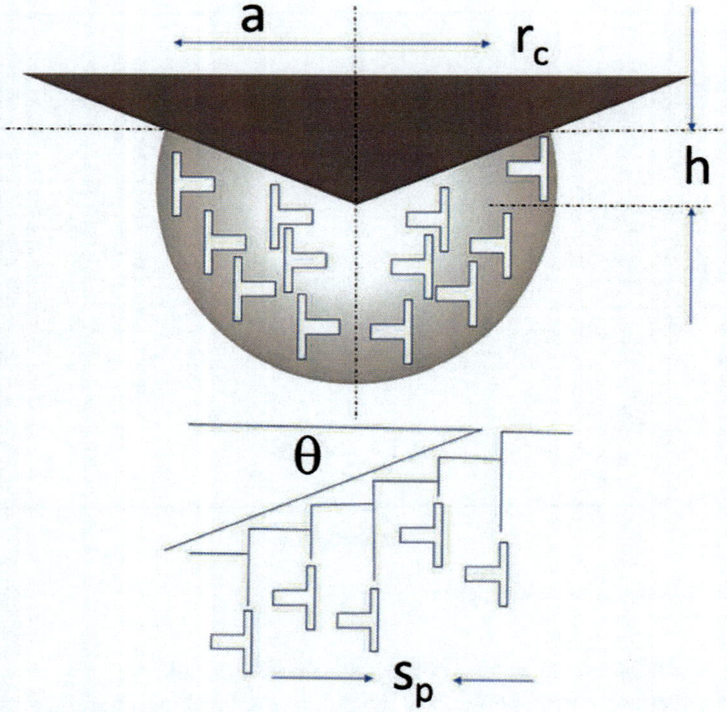

Fig. 2.30 GNDs formed by an indentation

by circular loops of GNDs with Burgers vector normal to the indentation surface (Fig. 2.30).

By considering that dislocation loops are equally spaced along the indentation surface (s_p)

$$\tan\theta = \frac{h}{a} = \frac{b}{s_p};\quad s_p = \frac{ba}{h}$$

The density of GNDs is given by

$$\rho_{\mathrm{GND}} = \frac{3h}{2ba^2} = \frac{3\tan^2\theta}{2bh}$$

Given that, the material hardness in the absence of GNDs is

$$H_0 = \alpha_1 \mu b \sqrt{\rho_s}$$

With α_1 being a constant, μ the shear modulus, and ρ_s the density of stored dislocations, the total hardness is

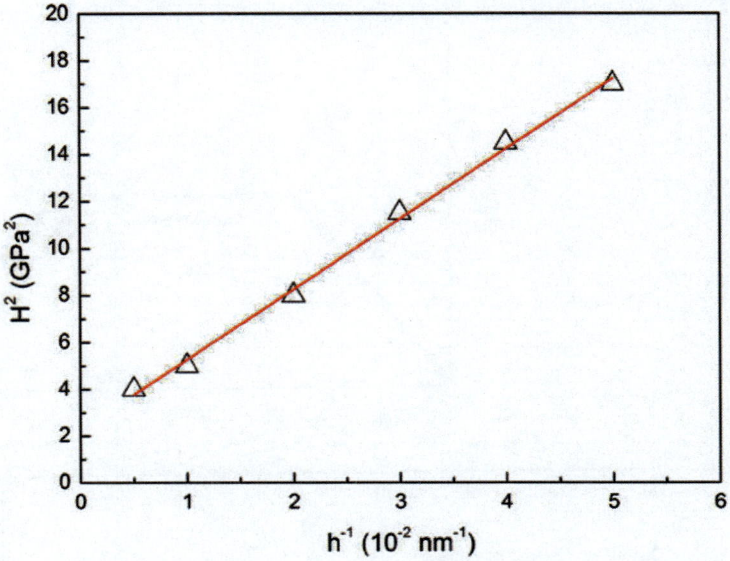

Fig. 2.31 Hardness dependence on the indentation depth for nanocrystalline Ni

$$H = 3\sqrt{3}\alpha_1 \mu b \sqrt{\rho_{\mathrm{GND}} + \rho_s}$$

In this context, a linear dependence of H on 1/h is demonstrated. An example for nanocrystalline Ni, with a mean grain size of 80 nm, is given in Fig. 2.31.

2.3.1 Cyclic Nanoindentation

The traditional load-time curve employed during nanoindentation is shown in Fig. 2.32.

After loading to the nominal maximum force, the load is maintained for a holding time in order to stabilize the elastoplastic behavior of the material, and then unloading phase is observed. Sometimes, continuous loading-unloading (without holding time) is used during the tests.

In order to analyze the dynamic behavior of nanostructured materials, cyclic loading or multistep nanoindentation experiments have been conducted in order to study the hardening–softening behavior of the materials as well as fatigue and crack behavior of NC metals and alloys (Fig. 2.33).

The cyclic loading functions contain a series of low-force loading and unloading cycles, the quantity and magnitude of which are varied based on fatigue and stress conditions, respectively.

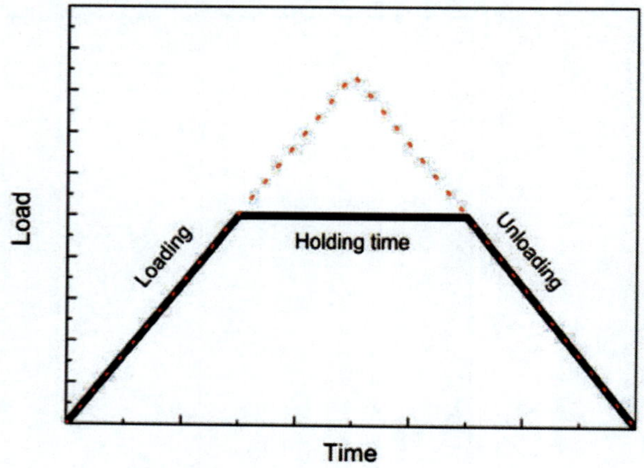

Fig. 2.32 Loading-time curve during nanoindentation

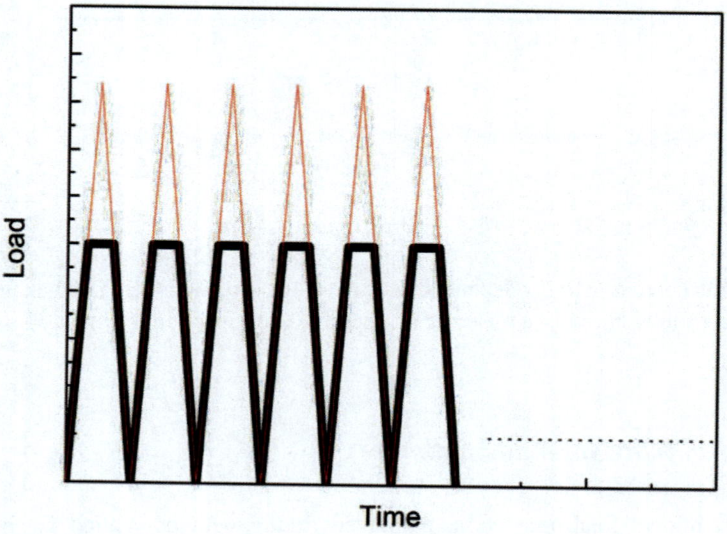

Fig. 2.33 Loading-time curve during cyclic nanoindentation

In other cases, cyclic loading can be performed starting from a certain minimum load in order to simulate fatigue behavior in different conditions (Fig. 2.34).

Many materials have a time-dependent behavior when placed under load and as a result conventional nanoindentation test methods may not provide an adequate estimation of material properties of interest (Charitidis et al. 2013). Two basic approaches appear in managing time-dependent behavior of nanoindentation testing, either application of an oscillatory displacement or load, in which the transfer function between the load and displacement provides a method of calculating the storage and loss modulus of the material, or application of a step load or displacement and subsequent measurement of displacement (creep) or load as a function of

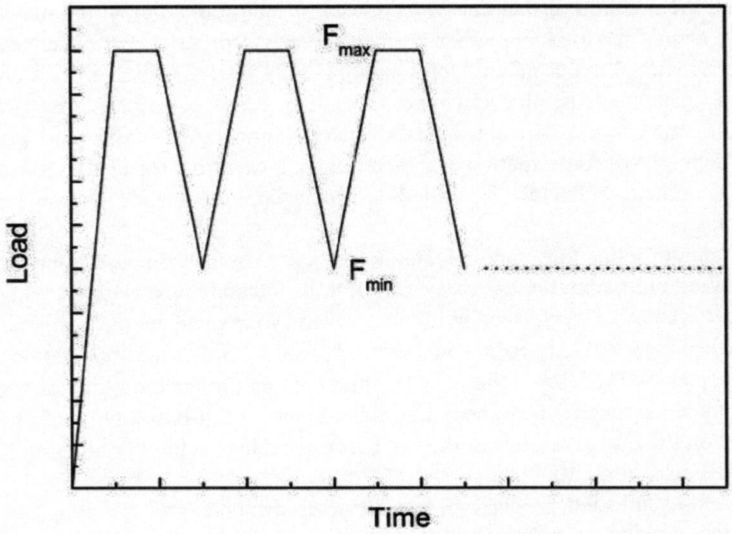

Fig. 2.34 Loading-time curve during cyclic nanoindentation around a fixed minimum load

time (relaxation). During nanoindentation, the applied load can be controlled at a constant value, whereas the penetration of the indenter tip into the sample surface is continuously recorded. This is often called constant-load indentation creep test, and it has been widely used to study the time-dependent properties of crystalline materials. The nanoindentation creep consists of two stages, transient (primary creep) and steady state (secondary creep). In a nanoindentation creep experiment, the tip is pushed into the surface at a constant rate of indentation until a prefixed load or penetration displacement is reached, and then the load is held constant while the indenter continues to creep into the material. With the indenter tip held fixed at that load or displacement, the material beneath the indenter tip continues to deform in time and finally the indenter tip is retracted from the material. Creep within a specimen occurs during the hold time of the loading phase of nanoindentation testing and manifests itself as a change of indentation displacement with the load kept constant. It is postulated that the stress fields in the material underneath the indenter develop a chemical potential gradient that leads to a thermally activated diffusional flux of atoms moving from below the indenter to the surface and along the interface between the indenter and the specimen, even under an elastic contact. In a nanoindentation test, creep and plastic deformation in the conventional sense, i.e., that occurs due to shear-driven slip, for example, should be regarded separately. Plasticity (yield or hardness) is commonly referred to as being an instantaneous event; however, creep may occur over time in an otherwise elastic deformation as a result of the diffusion and motion of atoms or movement of dislocations in the indentation stress field. Creep is often utilized to describe a delayed response to an applied stress or strain that may be a result of viscoelastic or viscoplastic deformation. In nanoindentation, the displacement recorded at each load increment will be, in general, the addition of that due to the elastic-plastic properties of the material and that occurring due to creep, either viscoelastic or viscoplastic. Permanent deformation in a material under indentation loading is thus seen as arising from a combination of instantaneous plasticity and creep (in addition, effects arise from the formation of cracks). An elastoplastic material undergoes elastic and non-time-dependent plastic deformation. A material that deforms elastically yet exhibits time-dependent behavior is called viscoelastic; otherwise, the material is viscoplastic.

The multiple-loading nanoindentation tests are useful to measure hardness and elastic modulus versus depth in the cases of thin-film and thick coatings, whereas the multicycle tests are performed at a constant load to investigate the dynamic indentation response to study localized fatigue behavior. Since the indentation fatigue damage starts with deformation, crack initiation, and propagation, the interpretation of the multiple-loading nanoindentation and nano-impact-based fatigue results will depend on the characteristics of the full loading cycle (loading–holding–unloading) history (Faisal et al. 2017).

A well-established property of nanostructured metals and alloys is the higher strain rate sensitivity with respect to their microcrystalline counterpart. Competing deformation mechanisms in nanocrystalline metals result in increased rate sensitivity which may not be uniform across all timescales: diffusion-controlled processes are important at slow strain rates and may dominate for large fractions of GBs, while intragranular mechanisms are dominant at high strain rates. The competition between the two processes is expected to control the strain rate sensitivity and activation volumes in nanocrystalline metals (Li and Weng 2013). Thus, probing the mechanical response of nanocrystalline metal films under different loading rates presents a means to capture the timescales at which different mechanisms play a rate-controlling role. The small grain size enhanced GB diffusion and concomitant inelastic deformation by increasing the net GB volume and contribution of GB-based deformation processes. The significantly increased role of the latter changed the relative importance of the inelastic deformation mechanisms, thus resulting in increased strain rate sensitivity and reduced apparent activation volumes. Obviously, this characteristic is first of all dependent on the crystal structure of the material (Fig. 2.35).

In FCC materials, strain rate sensitivity linearly increases with decreasing grain size. For BBC structures, m first decreases with decreasing grain size while it re-increases only at sizes around 30–50 nm.

2.4 Conclusions

Deformation mechanisms that typically operate in CG metals and their alloys, e.g., the nucleation, migration, interaction, and annihilation of lattice dislocations in crystalline grains, are severely limited in NC metals; instead, interface-mediated

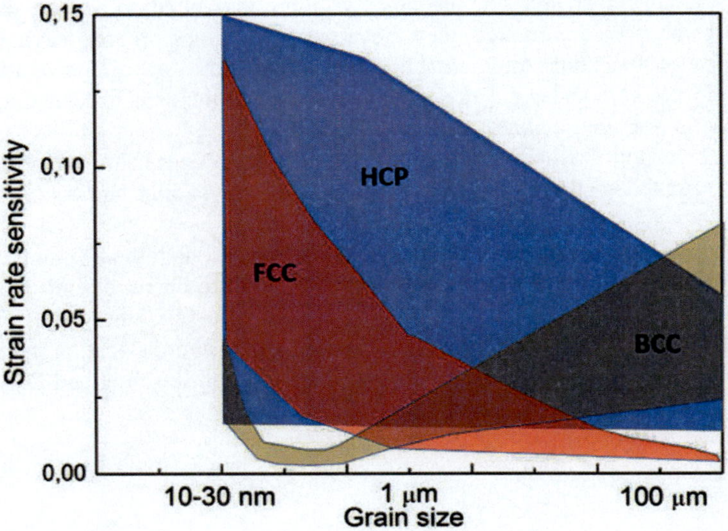

Fig. 2.35 Strain rate sensitivity vs. grain size for different crystal structures

mechanisms such as GB sliding/migration, dislocation nucleation from GB sources, and twinning tend to dominate.

Grain boundaries, and especially the presence of grain boundary segregates, appear to play a unique role in nanocrystalline microstructures, in terms of both potential changes in properties such as grain boundary energy and cohesive strength and effects on mechanical properties.

Plastic deformation of NC grains is limited by the interface-mediated nucleation and pinning of dislocations; therefore, classical strengthening mechanisms that are based on dislocation multiplication + storage and dislocation-obstacle interactions in the lattice may not apply to NC metals. The flow stress behavior begins to change once the grain size starts to fall into the nanometer range. Here other deformation mechanisms start to play prevailing roles. Many experimental and numerical evidences show that in the nanometer range the development of dislocations during plastic flow becomes shorter. Two main deformation mechanisms have been suggested. One is GB-mediated processes (e.g., GB sliding/GB migration/GB diffusion, grain rotation, and grain growth). The other deformation mechanism is dislocation/twinning-based plasticity.

Due to the small grain size, the emitted dislocations tend to glide across the grain and become absorbed in other GBs, sometimes leaving behind a stacking fault for the case of partial dislocation emission. Prior to dislocation nucleation in NC materials, atomistic simulations have shown that free volume migration occurs within the boundary and nearby TJs.

Nanoscale amorphization can occur at GB disclination dipole and serve as a special mode of local plastic deformation near crack tips. It is well known that dislocation emission from crack tips is one of the most fundamental processes for understanding crack blunting in nanocrystalline and ultrafine-grained materials. Once dislocations are emitted, they move out of the crack tip area leaving behind a dislocation-free zone. An internal back stress due to the dislocations emitted from crack tip accommodates the stress intensity factor to the applied load, causing an increase in fracture toughness of materials.

The behavior of nanocrystalline materials with grain sizes below ~10–20 nm has been attributed to the emergence of grain boundary sliding and rotation as the dominant carriers of plastic deformation.

SFE can govern the choice of fatigue mechanisms in both the CG range and the nanoscale. Grain size and SFE can be tuned in order to obtain optimized synergy between strength and ductility. For high-SFE materials, the main observed micromechanisms are grain growth via GB migration and dislocation pattern. For low-SFE materials, local GB activities act such as atom shuffling and GB sliding/rotation.

References

Aifantis EC (2009) Deformation and failure of bulk nanograined and ultrafine-grained materials. Mater Sci Eng A503(1–2):190–197. https://doi.org/10.1016/j.msea.2008.04.085

An XH, Wu SD, Wang ZG, Zhang ZF (2019) Significance of stacking fault energy in bulk nanostructured materials: insights from Cu and its binary alloys as model systems. Prog Mater Sci 101:1–45. https://doi.org/10.1016/j.pmatsci.2018.11.001

Argon AS, Demkowicz MJ (2006) Atomistic simulation and analysis of plasticity in amorphous silicon. Philos Mag 86(25–26):4153–4172. https://doi.org/10.1080/14786430600596852

Balbus GH, Wang F, Gianola DS (2020) Suppression of shear localization in nanocrystalline Al-Ni-Ce via segregation engineering. Acta Mater 188:63–78. https://doi.org/10.1016/j.actamat.2020.01.041

Beyerlein IJ, Knezevic M (2018) Mesoscale, microstructure-sensitive modeling for Interface-dominated, nanostructured materials. In: Andreoni W, Yip S (eds) Handbook of materials modeling. Springer, Cham. https://doi.org/10.1007/978-3-319-42913-7_82-1

Charitidis CA, Dragatogiannis DA, Koumoulos EP (2013) A study on time dependent properties of aluminum alloy by nanoindentation technique. Int J Struct Integr 4(1):33–54. https://doi.org/10.1108/17579861311303618

Cherkaoui M, Capolungo L (2009) Deformation mechanisms in nanocrystalline materials. In: Atomistic and continuum modeling of nanocrystalline materials, Springer series in materials science, vol 112. Springer, Boston. https://doi.org/10.1007/978-0-387-46771-9_6

Cizek P, Sankaran A, Rauch EF et al (2016) Microstructure and texture of electrodeposited nanocrystalline nickel in the As-deposited state and after in-situ and ex-situ annealing. Metal Mater Trans A 47:6655–6670. https://doi.org/10.1007/s11661-016-3810-2

Dao M, Lu L, Asaro RJ, De Hosson JTM, Ma E (2007) Toward a quantitative understanding of mechanical behaviour of nanocrystalline metals. Acta Mater 55:4041–4065. https://doi.org/10.1016/j.actamat.2007.01.038

Faisal NH, Prathuru AK, Goel S et al (2017) Cyclic Nanoindentation and Nano-impact fatigue mechanisms of functionally graded TiN/TiNi film. Shape Mem Superelasticity 3:149–167. https://doi.org/10.1007/s40830-017-0099-y

Farkas D, Selinger RLB (2005) Atomistics of fracture. In: Yip S (ed) Handbook of materials modeling. Springer, Dordrecht. https://doi.org/10.1007/978-1-4020-3286-8_44

Feng H, Tang J, Peng J, Wu H (2019) Nanoscale amorphization effect on dislocation emission from an elliptical blunt crack tip in deformed nanocrystalline and ultrafine-grained materials. Mech Mater 134:98–105. https://doi.org/10.1016/j.mechmat.2019.04.019

Fultz B, Frase H (2000) Grain boundaries of nanocrystalline materials—their widths, compositions, and internal structures. Hyperfine Interact 130:81–108. https://doi.org/10.1023/A:1011056927880

Goswami R, Feng CR, Qadri SB, Pande CS (2017) Fatigue-assisted grain growth in Al alloys. Sci Report 7:10179. https://doi.org/10.1038/s41598-017-10889-8

Gryaznov VG, Gutkin MY, Romanov AE et al (1993) On the yield stress of nanocrystals. J Mater Sci 28:4359–4365. https://doi.org/10.1007/BF01154943

Gutkin MY, Ovid'ko IA (2004) Localization of plastic flow in nanocrystalline materials. In: Plastic deformation in nanocrystalline materials, Springer series in materials science, vol 74. Springer, Berlin. https://doi.org/10.1007/978-3-662-09374-0_3

Gutkin MY, Ovid'ko IA (2007) Dislocation loop nucleation and plastic deformation of nanocrystalline materials. Mech Solids 42:271–282. https://doi.org/10.3103/S0025654407020124

Gutkin MY, Ovid'ko IA, Skiba NV (2005) Grain boundary sliding and lattice dislocation emission in nanocrystalline materials under plastic deformation. Phys Solid State 47:1662–1674. https://doi.org/10.1134/1.2045349

Hall EO (1951) The defoemation and ageing of mild steel: III discussion of results. Proc Phys Soc B64(9):747. https://doi.org/10.1088/0370-1301/64/9/303

He T, Xiao W, Zhang Y et al (2017) Effect of special rotational deformation on the dislocation emission from a branched crack tip in deformed nanocrystalline materials. Acta Mech 228:823–836. https://doi.org/10.1007/s00707-016-1742-y

Herzig C, Mishin Y (2005) Grain boundary diffusion in metals. In: Heitjans P, Kärger J (eds) Diffusion in condensed matter. Springer, Berlin. https://doi.org/10.1007/3-540-30970-5_8

Koch CC, Scattergood RO, Darling KA et al (2008) Stabilization of nanocrystalline grain sizes by solute additions. J Mater Sci 43:7264–7272. https://doi.org/10.1007/s10853-008-2870-0

Konakov YV, Ovidko IA, Sheinerman AG, Skiba NV (2017) Collective migration of low-angle tilt boundaries in nanocrystalline metals under fatigue loading. Rev Adv Mater Sci 52:113–120

Kumar P, Kawasaki M, Langdon TG (2016) Review: overcoming the paradox of strength and ductility in ultrafine-grained materials at low temperatures. J Mater Sci 51:7–18. https://doi.org/10.1007/s10853-015-9143-5

Li J, Weng GJ (2013) A micromechanical approach to the stress–strain relations, strain-rate sensitivity and activation volume of nanocrystalline materials. Int J Mech Mater Des 9:141–152. https://doi.org/10.1007/s10999-013-9214-1

Li X, Zhang X, Gao H (2018) Atomistic simulations of fracture and fatigue in nanotwinned and amorphous materials. In: Andreoni W, Yip S (eds) Handbook of materials modeling. Springer, Cham. https://doi.org/10.1007/978-3-319-50257-1_46-1

Li X, Zhang X, Gao H (2020) Atomistic simulations of fracture and fatigue in nanotwinned and amorphous materials. In: Andreoni W, Yip S (eds) Handbook of materials modeling. Springer, Cham. https://doi.org/10.1007/978-3-319-44680-6_46

Lu S, Li D, Brenner DW (2014) Molecular dynamics simulations of plastic damage in metals. In: Voyiadjis G (ed) Handbook of Damage Mechanics. Springer, New York. https://doi.org/10.1007/978-1-4614-8968-9_6-1

Malygin GA (2007) Plasticity and strength of micro- and nanocrystalline materials. Phys Solid State 49:1013–1033. https://doi.org/10.1134/S1063783407060017

Malygin GA (2009) Mechanism of the formation of shear microbands under plastic deformation of nanocrystalline materials. Phys Solid State 51:1814–1820. https://doi.org/10.1134/S1063783409090091

McDowell DL (2018) Multiscale crystalline plasticity for materials design. In: Shin D, Saal J (eds) Computational materials system design. Springer, Cham. https://doi.org/10.1007/978-3-319-68280-8_6

McDowell DL (2019) Multiscale modeling of interfaces, dislocations, and dislocation field plasticity. In: Mesarovic S, Forest S, Zbib H (eds) Mesoscale models, CISM International Centre for Mechanical Sciences (Courses and Lectures), vol 587. Springer, Cham. https://doi.org/10.1007/978-3-319-94186-8_5

Meyers MA, Mishra A, Benson DJ (2006) Mechanical properties of nanocrystalline materials. Prog Mater Sci 51(4):427–556. https://doi.org/10.1016/j.pmatsci.2005.08.003

Miglierini M, Grenèche J (1999) Temperature dependence of amorphous and interface phases in the Fe80Nb7Cu1B12 nanocrystalline alloy. Hyperfine Interact 122:121–128. https://doi.org/10.1023/A:1012645706470

Misra A (2008) Twinning in nanocrystalline metals. JOM 60:59. https://doi.org/10.1007/s11837-008-0119-7

Mohebbi MS, Akbarzadeh A (2017) Development of equations for strain rate sensitivity of UFG aluminum as a function of strain rate. Int J Plast 90:167–176. https://doi.org/10.1016/j.ijplas.2017.01.003

Murr LE (2016) Mechanical and other properties of nanocrystalline materials. In: Handbook of materials structures, properties, processing and performance. Springer, Cham. https://doi.org/10.1007/978-3-319-01905-5_47-2

Naik SN, Walley SM (2020) The Hall–Petch and inverse Hall–Petch relations and the hardness of nanocrystalline metals. J Mater Sci 55:2661–2681. https://doi.org/10.1007/s10853-019-04160-w

Naturani T, Takamura J (1991) Grain-size strengthening in terms of dislocation density measured by resistivity. Acta Metall Mater 39(8):2037–2049. https://doi.org/10.1016/0956-7151(91)90173-X

Ovid'ko IA, Sheinerman AG (2016) Free surface effects on rotational deformation in nanocrystalline materials. J Mater Sci 51:6444–6451. https://doi.org/10.1007/s10853-016-9942-3

Ovid'ko IA, Skiba NV (2014a) Nanotwins induced by grain boundary deformation processes in nanomaterials. Scr Mater 71:33–36. https://doi.org/10.1016/j.scriptamat.2013.09.028

Ovid'ko IA, Skiba NV (2014b) Generation of nanoscale deformation twins at locally distorted grain boundaries in nanomaterials. Int J Plast 62:50–71. https://doi.org/10.1016/j.ijplas.2014.06.005

Ovidko IA, Aifantis EC (2013) Nanocrystals and nanomechanics: mechanics and models. A selective review. Rev Adv Mater Sci 35:1–24

Ovidko IA, Sheinerman AG (2004) Triple junction nanocrack in fatigued nanocrystalline materials. Rev Adv Mater Sci 7:61–66

Panin VE, Surikova NS, Smirnova AS et al (2018) Mesoscopic structural states in plastically deformed nanostructured metal materials. Phys Mesomech 21:396–400. https://doi.org/10.1134/S102995991805003X

Petch NJ (1953) The orientation relationships between cementite and α-iron. Acta Cryst 6(1):96. https://doi.org/10.1107/S0365110X53000260

Pineau A, Benzerga AA, Pardoen T (2016) Failure of metals III: fracture and fatigue of nanostructured metallic materials. Acta Mater 107:508–544. https://doi.org/10.1016/j.actamat.2015.07.049

Ramesh K (2009) Fundamentals of mechanics of materials. In: Nanomaterials. Springer, Boston. https://doi.org/10.1007/978-0-387-09783-1_2

Razumov IK, Gornostyrev YN, Ermakov AE (2019) Nonequilibrium diffusional phase transformations in alloys induced by migration of grain boundaries and dislocations. Phys Solid State 61:214–224. https://doi.org/10.1134/S1063783419020215

Rekhi S (2002) Experimental and theoretical investigations on the compressibility of nanocrystalline nickel. J Mater Sci 37:2167. https://doi.org/10.1023/A:1017443007759

Scattergood RO, Koch CC, Murty KL, Brenner D (2008) Strengthening mechanisms in nanocrystalline alloys. Mater Sci Eng A493:3–11. https://doi.org/10.1016/j.msea.2007.04.132

Schiots J, Jacobsen KW (2003) A maximum in strength of nanocrystalline copper. Science 301(5638):1357–1359. https://doi.org/10.1126/science.1086636

Schuh CA, Hufnagel TC, Ramamurty U (2007) Mechanical behavior of amorphous alloys. Acta Mater 55:4067–4109. https://doi.org/10.1016/j.actamat.2007.01.052

Thangadurai TD, Manjubaashini N, Thomas S, Maria HJ (2020) Properties of nanostructured materials. In: Nanostructured materials, Engineering materials. Springer, Cham. https://doi.org/10.1007/978-3-030-26145-0_7

Valiev RZ (2004) Nanostructuring of metals by severe plastic deformation for advanced properties. Nat Mater 3:511–516. https://doi.org/10.1038/nmat1180

Voyiadjis GZ, Aifantis EC, Weber G (2003) Constitutive modeling of plasticity in nanostructured materials. In: Harik VM, Salas MD (eds) Trends in nanoscale mechanics, ICASE/LaRC interdisciplinary series in science and engineering, vol 9. Springer, Dordrecht. https://doi.org/10.1007/978-94-017-0385-7_5

Wang P, Yang X, Peng D (2017a) Effects of cyclic loading performance on grain boundary motion of nanocrystalline Ni. Metal Mater Trans A 48:4977–4989. https://doi.org/10.1007/s11661-017-4261-0

Wang L, Guan P, Teng J et al (2017b) New twinning route in face-centered cubic nanocrystalline metals. Nat Commun 8:2142. https://doi.org/10.1038/s41467-017-02393-4

Wei Q (2007) Strain rate effects in the ultrafine grain and nanocrystalline regimes—influence on some constitutive responses. J Mater Sci 42:1709–1727. https://doi.org/10.1007/s10853-006-0700-9

Wolf D (2005) Grain boundaries in nanocrystalline materials. In: Yip S (ed) Handbook of materials modeling. Springer, Dordrecht. https://doi.org/10.1007/978-1-4020-3286-8_106

Xu Y, Zhang J, Bai Y et al (2008) Shear localization in dynamic deformation: microstructural evolution. Metal Mater Trans A 39:811. https://doi.org/10.1007/s11661-007-9431-z

Yamakov V, Wolf D, Phillpot SR, Gleiter H (2003) Dislocation-dislocation and dislocation-twin reactions in nanocrystalline Al by molecular dynamics simulation. Acta Mater 51 (14):4135–4147. https://doi.org/10.1016/S1359-6454(03)00232-5

You Z, Luo S, Lu L (2020) Size effect of deformation nanotwin bundles on their strengthening and toughening in heterogeneous nanostructured Cu. Sci China Technol Sci. https://doi.org/10.1007/s11431-020-1584-6

Zhang J, Liu G, Sun J (2020) Deformation mechanisms, microstructural evolution and mechanical properties in small-scaled face-centered-cubic metallic thin films. Nano Mater Sci 2(1):58–65. https://doi.org/10.1016/j.nanoms.2019.11.002

Zhou H, Qu S (2018) Mechanical properties of nanostructured metals: molecular dynamics studies. In: Hsueh CH et al (eds) Handbook of Mechanics of Materials. Springer, Singapore. https://doi.org/10.1007/978-981-10-6855-3_19-1

Zhu YT (2005) Deformation twinning in nanocrystalline metals. J Mater Eng Perform 14:467–472. https://doi.org/10.1361/105994905X56269

Chapter 3
Crack Initiation and Growth in Metal Alloys and Composites

3.1 Introduction

The first phenomenological singularity of nanostructured metal and alloy deformation is an abnormal grain rotation if compared to their microcrystalline counterparts (Shan et al. 2008). In situ tensile straining transmission electron microscopy tests have been carried out on nanocrystalline Ni. Grain agglomerates (GAs) were found to form very frequently and rapidly ahead of an advancing crack with sizes much larger than the initial average grain size. High-resolution electron microscopy indicated that the GAs most probably consist of nanograins separated by low-angle grain boundaries. The formation of these agglomerates is thought to occur through transformations of high-angle grain boundaries into low-angle ones, due to crystal lattice rotations within nanograins under high local stresses near crack tips. Furthermore, both inter- and intra-GA fractures were observed. The observations suggest that these newly formed GAs may play an important role in the formation of the dimpled fracture surfaces of nanocrystalline materials. Cheng et al. (2010) concluded that grain coarsening occurs through crystal lattice reorientation under the high stress concentration near crack tips. Structural evolution inside of nanocrystalline shear bands shows that intense strain localization leads to grain growth and formation of a shear texture. The first type of evolution can be understood as the material being driven closer to an equilibrium state by high stresses during deformation (Greer et al. 2012). Atoms in the grain boundaries are defects and exist in a higher energy state than atoms found in the grain interiors (Fensin et al. 2019). Grain growth reduces the volume fraction of material located inside of grain boundaries, although the extent of grain growth will be a complex convolution of effects relating to the active deformation mechanisms and the driving force for growth. For starting grain size in the order of few nm, plasticity should be dominated by grain rotation and sliding, as well as grain boundary migration. These mechanisms are able to efficiently coarsen grain structures, with grain rotation capable of causing grain coalescence and boundary migration directly increasing

© Springer Nature Switzerland AG 2021

P. Cavaliere, *Fatigue and Fracture of Nanostructured Materials*,
https://doi.org/10.1007/978-3-030-58088-9_3

the size of one grain at the expense of another (Handwerker and Pollock 2014). As grains grow, there will be a gradual shift towards dislocation-based plasticity. Therefore, the grain growth will effectively shut itself off once grains are too large for collective grain boundary mechanisms to dominate plasticity. The observation of texture formation can be explained by a shift to dislocation-based mechanism through a combination of grain growth and high plastic strains. Plasticity due to grain boundary sliding and grain rotation randomizes texture, similar to observations of superplastic behavior in nanocrystalline metals. For larger grains, plasticity is dominated by dislocation glide, which has been shown to induce preferred orientation. The dominant deformation mechanism is also a function of plastic strain, with higher plastic strains increasing the importance of dislocation-based mechanisms.

A modulated structure is generated in a nanocrystalline Ni-Fe alloy under cyclic deformation. Substantial grain coarsening and loss of growth twins are observed in the path of fatigue cracks, while the grains away from the cracks remain largely unaffected. Statistical analyses suggest that the grain coarsening is realized through the grain lattice rotation and coalescence and the loss of growth twins may be related to the detwinning process near crack tip. Under the shear loading the detwinning deformation is related to the loading rate. The results show that there may be a critical shear rate. As the shear rate is sufficiently high the circle twin is found to be failed; as the shear rate is less than that rate, the size of circle twin becomes smaller and gradually approaches a constant value (Su and Tang 2013).

So, many experimental evidences show how nanoscale rotational mechanisms operate during the deformation of pre-cracked and crack-free NC materials (Panin et al. 2012). In Simokawa et al. (2005) the subject is analyzed for nanocrystalline Al with grain size from 5 to 80 nm deformed in tension. For grain size of 30 nm, a transition from grain size hardening to grain size softening is underlined. In the grain size hardening region, intergranular deformation is modeled with pileup of dislocations. For grain size lower than 30 nm the main deformation mechanism is intergranular deformation by grain sliding. Here, geometrical misfits by grain sliding are accommodated by grain rotation.

3.1.1 Dislocation Sources

A dislocation source showing pileup against grain size under an effective shear stress τ_e is described in Fig. 3.1:

$$\tau_e = \tau - \tau_0$$

where τ is the applied stress and τ_0 the contribution of internal friction and back stress. The stress at the head of pileup is calculated as

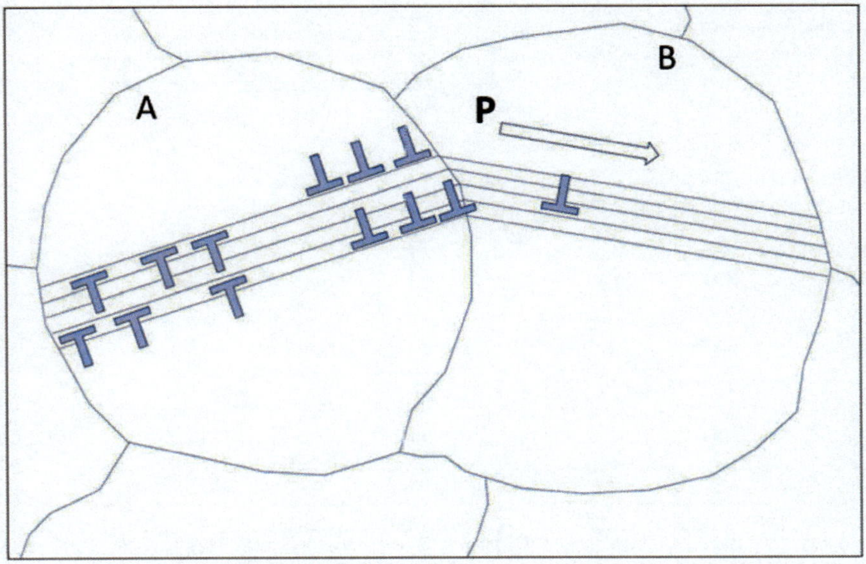

Fig. 3.1 Pileup of dislocations at grain boundary

$$\tau_p = n\tau_e$$

With n being the number of dislocations in the pileup depending on the pileup length L, the shear modulus μ is

$$n = \frac{q\pi L\tau_e}{\mu b}$$

The first model of pileup of dislocations at triple junctions was proposed by Fedorov et al. (2003). When a GBD pileup is stopped at a triple junction, different mechanisms can take place (Fig. 3.2).

The moving dislocation can split into two different dislocations moving along the adjacent grain boundaries b). This mechanism can be repeated for all the further dislocations moving along the same path c). Otherwise, an immobile dislocation with Burgers vector $\mathbf{b_1}$ stops at the triple junction and a mobile dislocation with Burgers vector $\mathbf{b_2}$ moves along the adjacent grain boundary.

Another mechanism can be the convergence of the dislocations into a dislocation with Burgers vector 2**b** (Fig. 3.3e) that then splits into a mobile dislocation moving in the adjacent grain interior with Burgers vector $\mathbf{b'_2}$ and a dislocation with Burgers vector $\mathbf{b'_1}$ stopping at the triple junction (Fig. 3.3f).

The case in which a stacking fault is formed is shown in Fig. 3.4. A dislocation splits into a partial dislocation stopping at the triple junction and a partial dislocation moving in the adjacent grain interior.

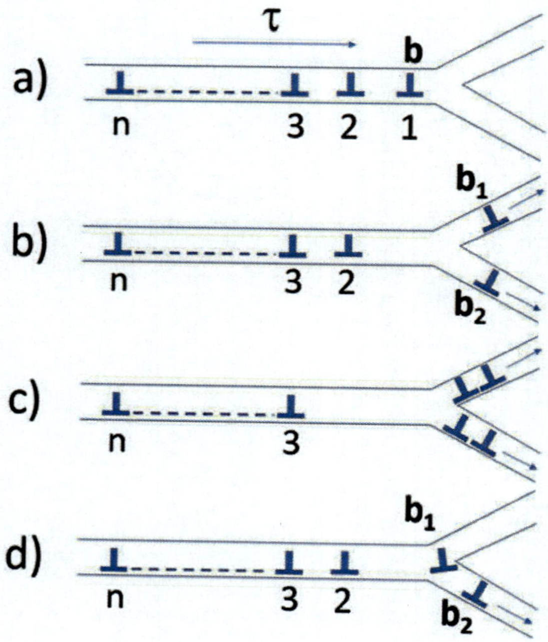

Fig. 3.2 Movement of dislocations at triple junctions

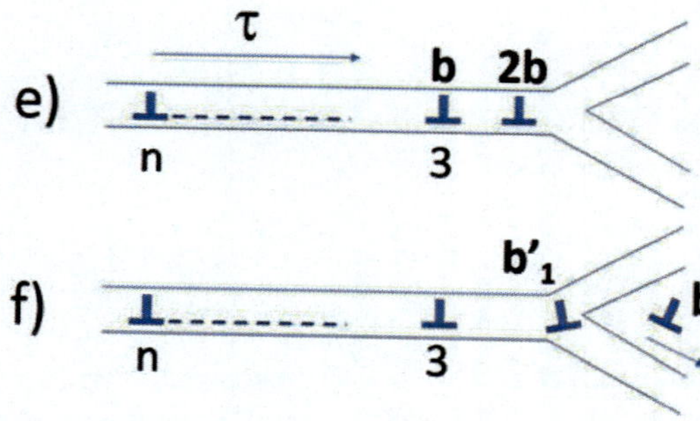

Fig. 3.3 Split of dislocations moving in the grain interior

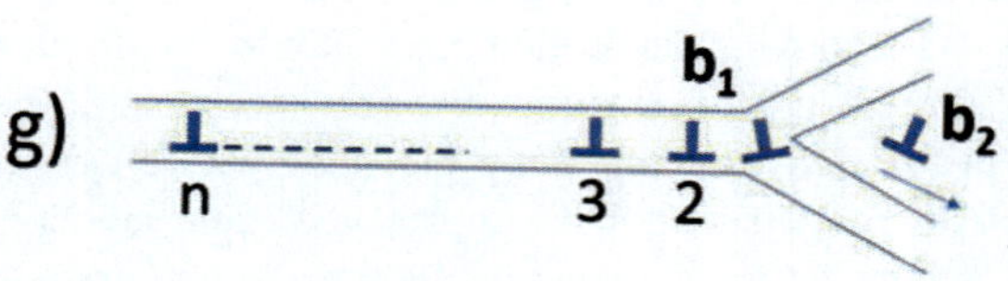

Fig. 3.4 Stacking fault formation

The mechanism of splitting of GBDs influences the grain boundary sliding during stress. Triple junctions are known as "soft" or "hard" depending on the occurring or not of the splitting. A high fraction of soft grain boundary favors grain boundary sliding as the main deformation mechanism.

3.1.2 GB Mechanisms

Grain boundary sliding has been widely observed in nanocrystalline materials (Shan et al. 2004). In addition, the model of the transition from grain boundary sliding to grain rotation was proposed by Gutkin et al. (2003). In the proposed model, moving dislocations at a triple junction splitting into climbing grain boundary dislocations occur repeatedly, resulting in the formation of two walls of dislocations and causing the crystal lattice rotation in the grain interior (Fig. 3.5).

By applying a shear stress to an undeformed crystal (a), grain boundary sliding acts as a consequence of dislocation motion along the grain boundary (b), and dislocations split, at the triple junction O, into climbing dislocations (c). The repeated gliding of dislocations leads to starting of the grain rotation (d). Dislocations reach the triple junction O′ starting to produce sliding in the adjacent grains (e).

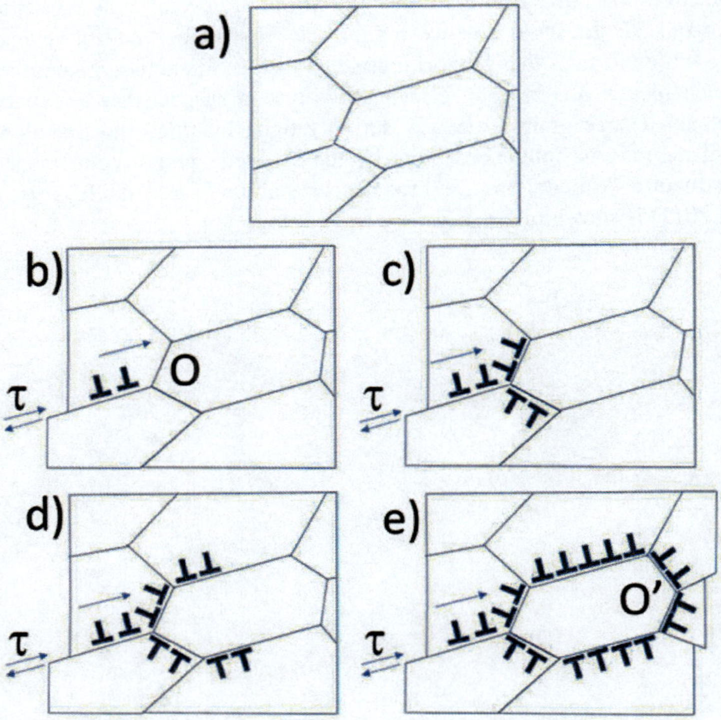

Fig. 3.5 Grain boundary sliding evolving into grain rotation

3.1.3 Grain Growth

Now, as described in Chap. 1, the volume fraction of atoms in the grain size increases with decreasing mean grain size. GBs in NC materials promote the total free energy of the system. The reduction of this excess free energy through the removal of grain boundary area represents a large driving force for the grain growth. Grain growth is due to the rotation and coalescence of adjacent grains (Gutkin et al. 2003). A MD simulation schematic of grain rotation and coarsening during tension loading of NC Al is shown in Fig. 3.6.

In the case of nanograined metals, analyses of subsurface crack initiation sites have indicated that the crack nucleation is associated with abnormally large grains (Furnish et al. 2017). The schematic of grain growth is shown in Fig. 3.7.

Aside from quantitative improvements in fatigue performance, NC metals also hold the possibility of new insight into the mechanisms responsible for traditional fatigue failure. As a matter of fact, the persistent slip mechanism responsible for conventional fatigue crack initiation may be suppressed when the grain size is below a certain threshold, likely on the order of 100 nm or several hundred nanometers. This length scale threshold is between grain sizes that support collective dislocation activity and grain sizes that support individual dislocation activity. Room-temperature mechanically driven grain growth leads one to suspect that NC metals may evolve such coarse grain structures during fatigue loading, and that the fatigue mechanisms may be influenced more by the evolved grain structure than by the initial structure. A model proposed for nanocrystalline Ni-based alloys (Boyce and Padilla 2011) is shown in Fig. 3.8.

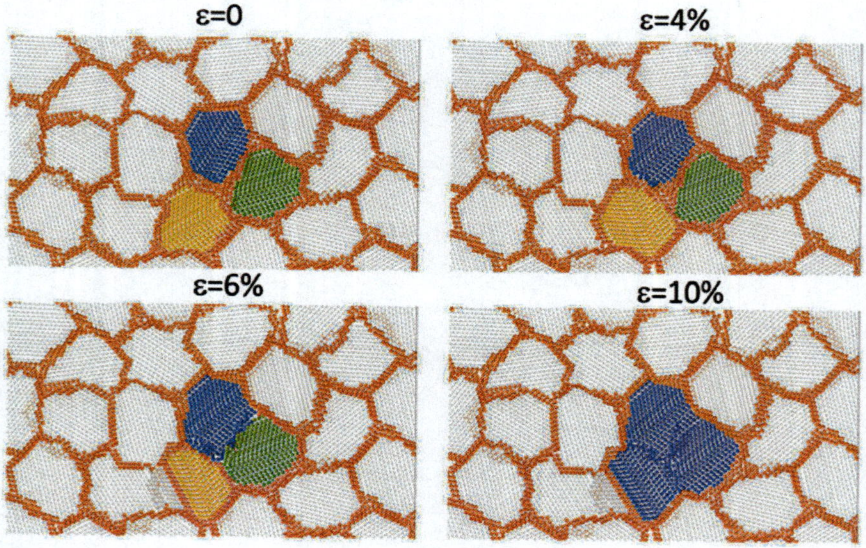

Fig. 3.6 Grain rotation and coarsening in NC Al during tension

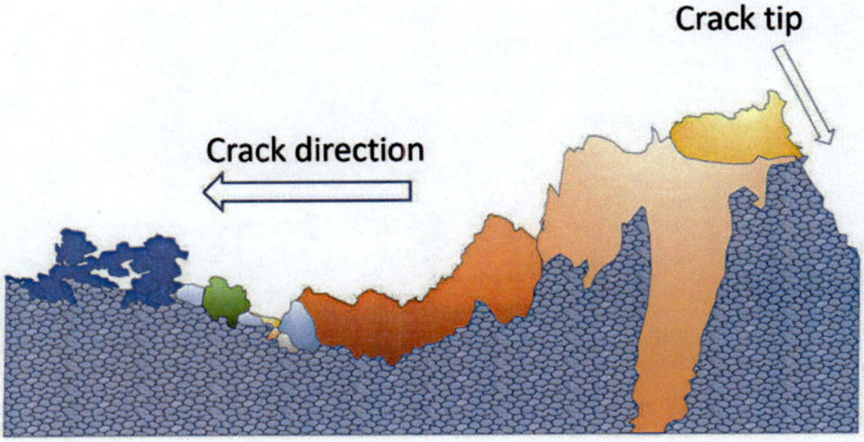

Fig. 3.7 Grain growth along the crack path

Cyclic loading is applied at a homogeneous NC Ni-based alloy microstructure (a); coarsened grains appear at the surface where the maximum stress is applied (b); persistent slip bands form inside the coarsened grains (c); crack starts in the coarsened grains and propagates into the NC material (d).

3.1.4 Dislocation Absorption

Another important mechanism in nanocrystalline materials is the absorption of dislocations at grain boundaries identified as one of the reasons of the strong increase of strain rate sensitivity as the grain size decreases (Fig. 3.9).

Gliding dislocations (green) and lattice dislocations (open signs) move under the effect of an applied shear stress. Climbing dislocations (small open signs) form due to the dipole configuration (Ovid'ko and Reizis 2001). Then climbing dislocations form at the triple junctions O and O′ interacting with the previously formed climbing dislocations leading to the annihilation of dislocations with opposite character and to the presence of only isolated climbing dislocations far from the triple junctions. This is the reason why plastic flow localization and fracture do not tend to act in nanocrystalline materials and also in high stresses (Ovid'ko 2007).

3.1.5 Strain Rate Sensitivity

Nanocrystalline metals also show high strain rate sensitivity if compared to microcrystalline materials (Hay et al. 2013). In addition, it has been widely recognized that NC metals have an increased sensitivity to loading changes with respect to their

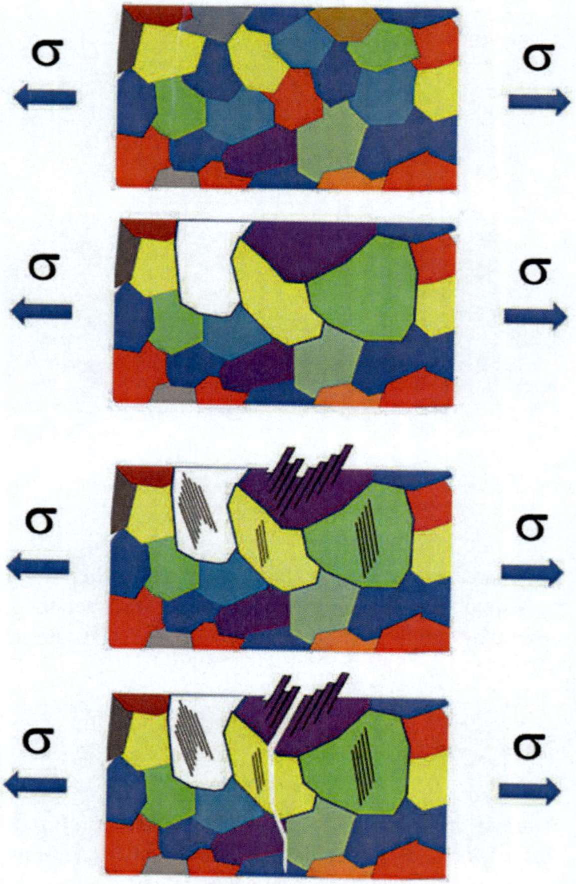

Fig. 3.8 Fatigue crack initiation mechanisms in NC Ni

microcrystalline counterparts (Peykov et al. 2012). The possibility to obtain the materials' response to strain rate changes is very useful in revealing many deformation mechanisms (Cavaliere 2008). The strain rate sensitivity of a material is defined as the variation of flow stress with strain rate at a given level of strain for a fixed temperature and it can be expressed as

$$m = \frac{\sqrt{3}kT}{\sigma v^*}$$

where k is Boltzmann constant, T the absolute temperature, σ the flow stress, and v^* the activation volume which can be considered as the derivative of the activation energy with respect to the effective shear stress (Maier-Kiener and Durst 2017). GB diffusion, GB sliding mediated by diffusion, and dislocation activity are the primary reasons for increased rate sensitivity, which is also corroborated by the small activation volumes. The larger activation volumes are a strong indication of

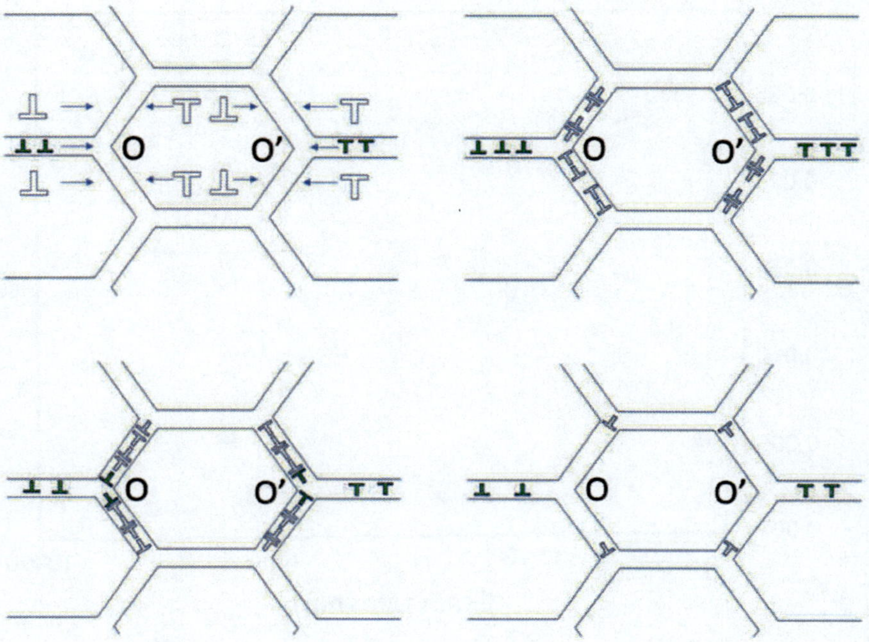

Fig. 3.9 Mechanism of dislocation absorption

intragranular dislocation-based plasticity, as opposed to the slower strain rates when GB processes dominate (Stegall and Elmustafa 2018).

By employing nanoindentation measurements, the flow stress can be related to the measured hardness ($H = 3\sigma$) and consequently the strain rate sensitivity is measured as

$$m = \frac{3\sqrt{3}kT}{Hv^*}$$

By measuring the nanoindentation hardness and calculating the activation volume, it becomes possible to obtain the strain rate sensitivity of the material. The grain refinement is largely known to lead to the exhibition of strain rate and loading rate sensitivity in metallic materials. It was explained via the existence of highly strained grain boundary-affected zone (GBAZ). Nanocrystalline alloys, on the other hand, are known to exhibit thermodynamically more stable states. Hence, it can be stated that grain refinement on nanoscale leads to high strengthening effects, thanks to the high grain boundary density creating obstacles to dislocations' motion. It is well established that in the first stage of deformation there are abundant dislocation sources quickly generating dislocations interacting with the grains and their boundaries. After a short while, the dislocation density continues to increase slowly and the phenomenon is accompanied with an increase in lattice rotation and the absorption

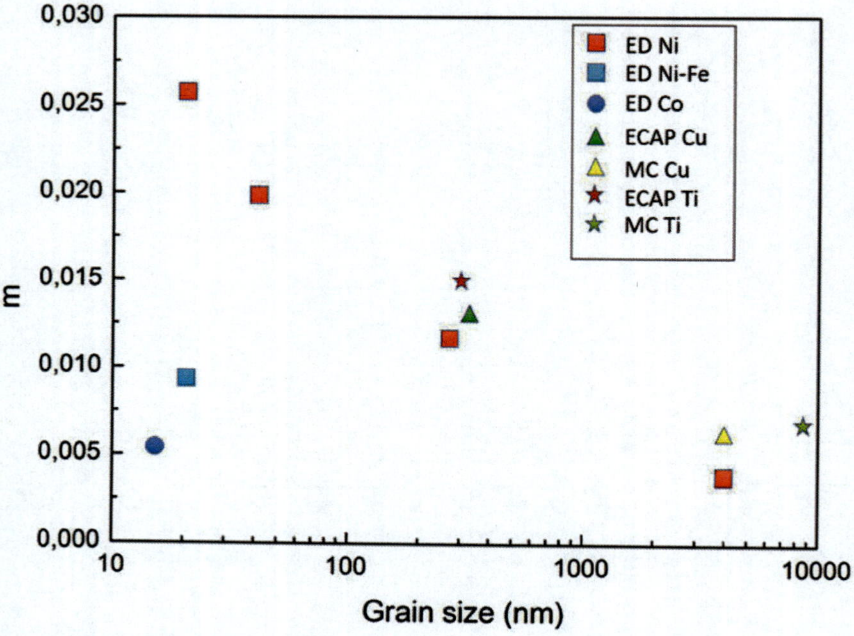

Fig. 3.10 Strain rate sensitivity as a function of grain size

of the dislocations as the strain increases. Such aspect is empathized as increasing the load rate leading also to a big strain hardening. A summary of the strain rate sensitivity measured for UFG and NC metals and alloys is shown in Fig. 3.10.

The same material is more sensitive to the strain rate variation when the grain size is decreased, demonstrating the strong effect of GBAZ on such mechanical property. Previous studies revealed, in fact, a sharpness of grain boundaries below 30 nm, showing also that grain boundary atoms and atoms up to 7–10 lattice parameters away from the grain boundary are heavily involved in plastic deformation. Deformation was mostly found to be taken up by atoms at and nearby grain boundaries. Therefore, it is well established that such a region close to grain boundaries is elastically strained despite the absence of defects and it is the region that largely contributes to deformation phenomena observed in NC metals. As a matter of fact, atoms within this GBAZ are more likely to be involved in the deformation process. This is very helpful indeed for the engineering application of these electrodeposited materials, in that the high rate sensitivity can have a strong effect in the delaying of necking during tensile deformation, leading to a significant increase in ductility (Ma 2006). In addition, the different behavior observed for UFG and NC metals can be explained in terms of activation volume. A small activation volume of dislocation mobility is responsible for the variation in strain rate sensitivity with decreasing mean grain size of metals. In addition, mechanisms of dislocation generation at grain boundaries coupled with grain rotation and migration are responsible for the whole plastic deformation in NC metals. Such mechanisms disappear by increasing the grain size from NC to UFG regime, thus decreasing the strain rate sensitivity of the materials at room temperature. Another explanation of the different behavior is related to the fact that many UFG metals are produced via severe plastic deformation, and in such cases the dislocation activity is completely different from the plastic deformation associated with the motion of dislocations generated during the processing route. In general, twin boundaries stop the dislocations' motion (Jang 2012), leading to increase in strength and to much higher structural and mechanical stability as shown by the reduced sensitivity to loading rates (Tian 2018). As a general behavior, the experimental results suggest that grain refinement to NC structures and ultrafine regime leads to a strong effect on the loading rate sensitivity. Very hard materials characterized by nanotwins appear more stable and less sensitive to changes in loading rates (Da et al. 2018). As a general trend, the decreasing of the grain size from microcrystalline to ultrafine to NC regime increases the number of grain boundaries acting as dislocation pinning. In this way, it decreases the rate-limiting process represented by the dislocation-grain boundary interaction. In NC metals and alloys, in fact, dislocations are generated at grain boundaries and this defect-assisted mechanism can be rate limiting in more stable microstructures. In addition, the steady-state density of dislocations is due to a dynamic balance between dislocation generation during plastic deformation and annihilation in the recovery process. Room-temperature recovery is an important process for NC/UFG metals, because the dislocations are generated from closely spaced sources (e.g., grain boundaries). They are packed in very close spaces and therefore the phenomenon of dislocation annihilation is prominent. Such a process balances the dislocation generation at grain boundary in a prominent way as the grain boundaries are stable, apparently leading to lack of strain rate sensitivity.

In order to analytically describe the grain boundary sliding let us consider two hexagonal grains as shown in Fig. 3.11.

Under the applied stress σ, the neighbor grains slide by a small quantity λ. This sliding is produced by the shear stress:

$$\tau = \sigma \sin \theta$$

with θ being the angle between the applied stress and the boundary. The work that is done by the shear stress on the sliding plane is

$$W_i = \tau l D \lambda = D \sigma l \lambda \sin \theta_i$$

where l is the length of the grain column. So, the total work for the overall sliding will be

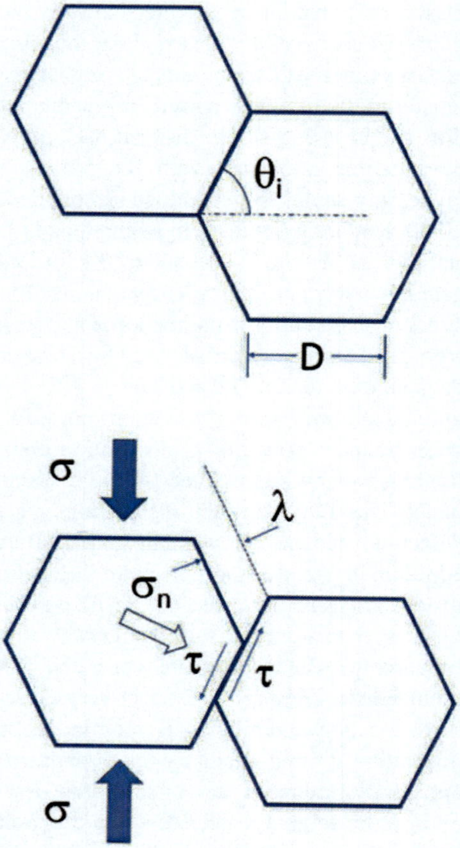

Fig. 3.11 Schematic of grain boundary sliding

$$W_{\text{total}} = \sum_i^n W_i = \sum_i^n D\sigma l\lambda \sin\theta_i = \frac{1}{\pi}\int_0^\pi [D\sigma l\lambda \sin\theta] d\theta = \frac{2D\sigma l\lambda}{\pi}$$

In the case of a perfect hexagon, $\theta = 60°$, the volume of the test sample is

$$V = \alpha l D^2$$

And

$$\Delta E = \frac{W_{\text{total}}}{V} = \frac{2\sigma\lambda}{\pi\alpha D} = \alpha'\sigma\left(\frac{\lambda}{D}\right) = \alpha'\sigma\varepsilon_{\text{gbs}}$$

where α' is a geometric constant and ε_{gbs} the sliding strain. The boundary sliding rate is expressed by

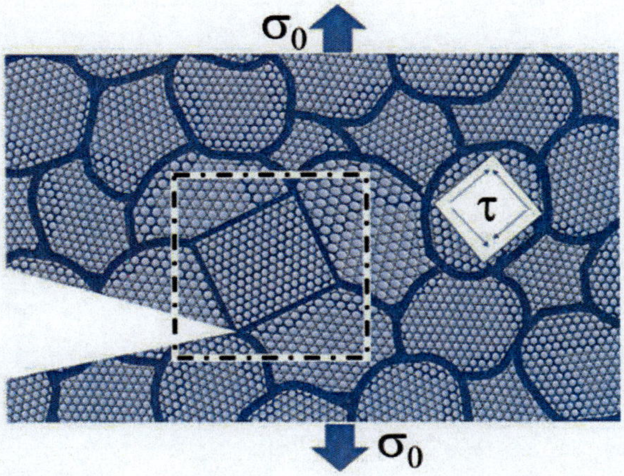

Fig. 3.12 Nanocrystalline material pre-cracked and tensioned

$$\dot{\varepsilon}_{\text{gbs}} = \frac{B\sigma^2}{d^p} \exp\left(-\frac{Q}{RT}\right)$$

B is a constant depending on the material, d is the grain size, and p is the grain size exponent; it is 2 when sliding is controlled by lattice diffusion and 3 when it is dependent on grain diffusion.

In the case of uniaxial tension or compression, the sliding strain will be

$$\varepsilon_{\text{gbs}} = \int_0^{\Delta t} \dot{\varepsilon}_{\text{gbs}} dt = \dot{\varepsilon}_{\text{gbs}} \Delta t = \frac{B\sigma^2 \Delta t}{d^p} \exp\left(-\frac{Q}{RT}\right) = B' \frac{\sigma^2 \Delta t}{d^p}$$

where B' is a material constant and Δt is the sliding time. So, the energy dissipation will be

$$\Delta E = \frac{W_{\text{total}}}{V} = \frac{2\sigma\lambda}{\pi\alpha D} = \alpha' \sigma \left(\frac{\lambda}{D}\right) = \alpha' \sigma \varepsilon_{\text{gbs}} = A \frac{\sigma^3 \Delta t}{d^p}$$

And the energy dissipation rate is

$$\frac{\Delta E}{\Delta t} = \dot{E} = A \frac{\sigma^3}{d^p}$$

The schematic describing the grain rotation in nanocrystalline metals associated with grain boundary sliding and grain boundary dislocation climbing is shown in Fig. 3.12.

The high stress concentration at the crack tip leads to the grain rotation (rectangular one). The rectangular grain can be considered as a quadrupole of immobile wedge disclinations whose strengths gradually increase during the formation process (Fig. 3.13) as described by Morozov et al. (2010).

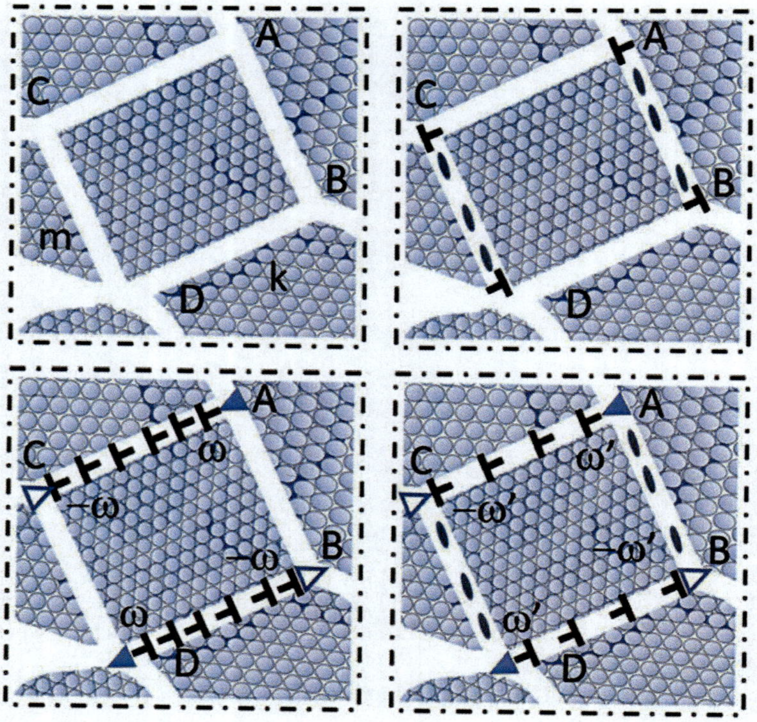

Fig. 3.13 Model of grain rotation

In this model the deformation acts as grain boundary sliding along grain boundaries AB and CD and diffusion-controlled climb of grain boundary dislocations along grain boundaries AC and BD. The sliding produces the formation of grain boundary dislocations at the junctions A, B, C, and D (Bobylev et al. 2009). The dislocation climb produces a special rotation leading to the formation of a quadrupole of wedge dislocations at the junctions. The formation of the quadrupole influences the crack growth through increased stress concentration.

The grain coalescence due to the grain rotation can be described as in Fig. 3.14.

The applied stress leads to grain translation due to the grain boundary dislocation glide. The triple junction impedes the motion of GB dislocations; with further stress the central grain rotates. The rotation leads to the reduction of misorientation angles with consequent grain coarsening. In addition, as the grain size is reduced the grain boundary migration appears as a competing mechanism with grain rotation and coarsening (Luo et al. 2016). The schematic of grain boundary migration is shown in Fig. 3.15.

The grain rotates as a consequence of grain boundary sliding, the misorientation with the adjacent grain is reduced during deformation, and the grain boundary migrates in a direction orthogonal to the grain rotation.

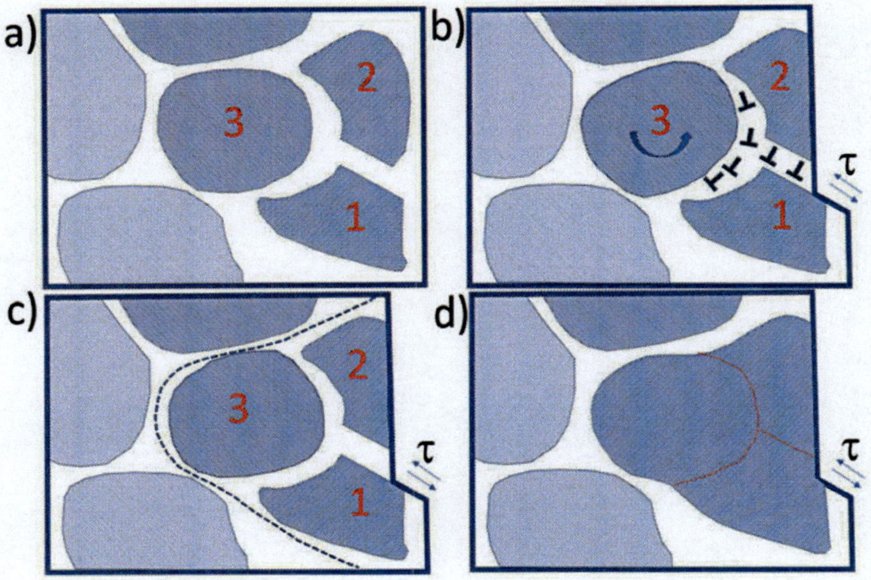

Fig. 3.14 Rotation-assisted grain coarsening

3.2 Grain Deformation at the Crack Tip

The grain rotation due to the formation of the quadrupole leading to the stress concentration influencing the crack growth is modeled through the introduction of a critical stress intensity factor K_{IC}. The crack propagates in a direction perpendicular to the tensile loading, and the formation of the quadrupole leads to a change in the value and behavior of the stress intensity factor. The disclination strength of the quadrupole is indicated as $\pm\omega$ (Fig. 3.13), with k and m being the quadrupole arms and α the angle between the crack plane and one of the quadrupole arms. So

$$K_{IC} = \frac{K^{\sigma}_{IC}}{\sqrt{(1 + A)^2 + B^2}}$$

with K^{σ}_{IC} being the stress intensity factor due to the tensile load and A and B geometry factor depending on the angle α and the sizes of the rotated grain facets $t = k/m$. A graphical description is given in Fig. 3.16.

The figure shows that the quadrupole formation leads to an average increase of the stress intensity factor.

If we consider that the damage propagates through the formation of multiple cracks with the formation of numerous quadrupoles, the critical stress intensity factor will be an average value depending on the inclination and characteristics of

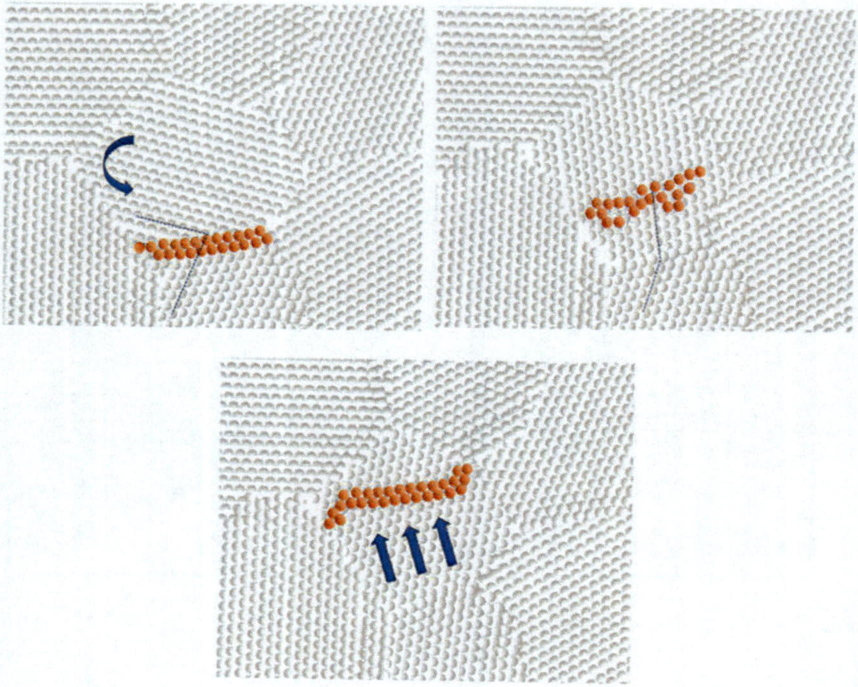

Fig. 3.15 Grain migration

all the quadrupoles. The angle a has random values described by a distribution function and t has a log-normal distribution:

$$\rho_\alpha(\alpha) = \frac{1}{2\pi}$$

$$\rho_t(t) = \frac{1}{t\sqrt{2\pi S_0^2}} \exp\left[-\frac{\ln^2 t}{2S_0^2}\right]$$

So, the average fracture toughness can be expressed as

$$\overline{K_{\mathrm{IC}}} = \int_{-\pi}^{\pi} \rho_\alpha(\alpha) d\alpha \int_0^\infty K_{\mathrm{IC}}(\alpha, t) \rho_t(t) dt$$

It is calculated that the grain rotation gives a contribution to the fracture toughness increases of ~10–15%.

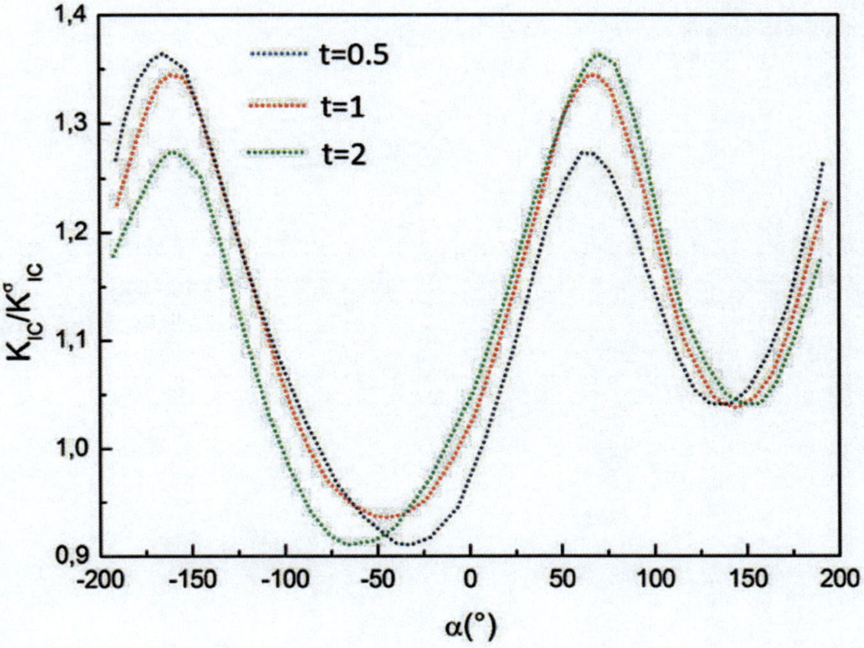

Fig. 3.16 Critical stress concentration behavior as a function of the quadrupole characteristics

3.3 Mechanisms of Cracking in NC Materials

During deformation in ultrafine and nanocrystalline grain regimes, specific mechanisms play different roles with respect to their microcrystalline counterparts because of the growing importance of grain boundary volumes. The plastic deformation mechanisms differ especially during high stress levels as in the case of cyclic and fatigue loading. So, it is fundamental to set the deformation mechanisms acting during high stress levels such as grain rotation during damage propagation.

3.3.1 Crack Behavior in NC Materials

Interest in nanocrystalline (NC) metals and alloys is motivated by the potential improvements these materials offer in terms of strength, hardness, and wear resistance over conventional coarse-grained (CG) materials. NC metals, by virtue of their small mean grain size (<100 nm), contain a significant volume fraction of interfaces in the form of grain boundaries (GBs), and, in some cases, twin boundaries. Such extreme refinement in grain size greatly reduces the mean free path of dislocation glide, suppressing both dislocation multiplication via Frank-Read sources and

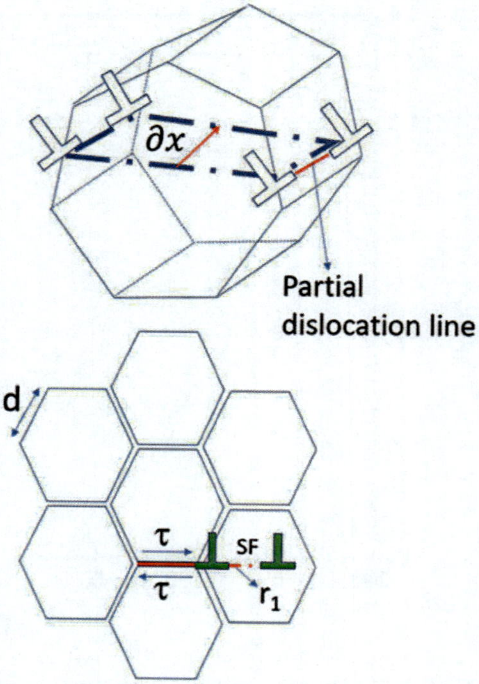

Fig. 3.17 Partial dislocation emission and stacking fault from grain boundary

strengthening due to dislocation pileups. Instead, inelastic deformation in NC metals is largely interface mediated, exhibiting mechanisms such as shear transformation zones (STZs) in the GBs, twinning, and dislocation nucleation from GB sources. Interface-mediated deformation mechanisms are believed to be the cause of anomalous behavior that occurs in an NC metal but not in the CG metal of similar composition, e.g., yield strength asymmetry under tension vs. compression (T/C asymmetry), non-Schmid effects in dislocation nucleation, and a nonlinear stress-strain response with recoverable "plasticity," otherwise known as an elasticity.

First of all, let us consider the nucleation of dislocations and stacking faults from grain boundaries. Basically, the conventional dislocation sources as Frank-Read loops disappear as the grain size decreases below the nanocrystalline regime because of the reduced volume at the grain interior as described in Chap. 1. Experimental evidences in FCC materials fix the critical size at 30 nm (Cheng et al. 2003). Below the critical grain size, grain boundary acts as source or sink for dislocations. Partial and perfect dislocations nucleate from grain boundary in the zones of stress concentration (Fig. 3.17).

Another deformation mechanism active as the grain size is below a critical value is the deformation twinning observed in nanocrystalline pure metals (Chen et al. 2003; Liao et al. 2004); an example of pure electrodeposited Ni is shown in Fig. 3.18.

The mechanism of twin formation is shown in Fig. 3.19.

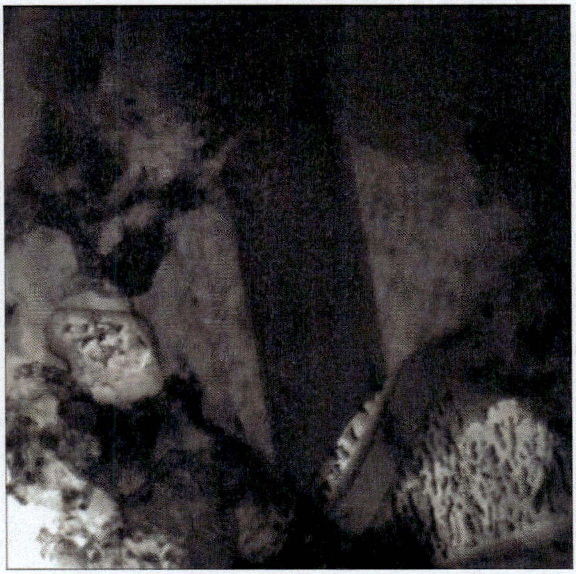

Fig. 3.18 Twins in NC pure electrodeposited Ni (20 nm grain size)

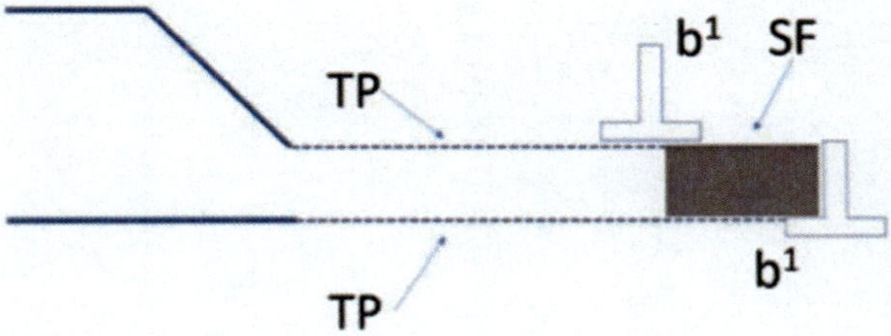

Fig. 3.19 Emission of deformation twin

A leading partial dislocation is emitted from a grain boundary; it is followed by a second leading partial dislocation with the same Burgers vector on an adjacent plane, forming a deformation twin. The gray area is the stacking fault. Deformation twinning acts when grain boundaries emit more partial dislocations than perfect ones.

Let us consider a polycrystalline material under loading in order to produce a shear stress in a region indicated by ABCD (Fig. 3.20).

One of the most suggested mechanisms of deformation is the formation of walls of dislocation dipoles due to the ideal shear stress; these dipoles are responsible for the grain rotation during the deformation. The shear stress acts on parallel slip planes generating walls of stacking faults (Fig. 3.21).

As the deformation proceeds, the magnitude of s increases reaching the value b (Fig. 3.22).

The plastic deformation produced by the walls of dislocations is accompanied with crystal rotation. Given d the nano-disturbance length, the energy change for its effect is

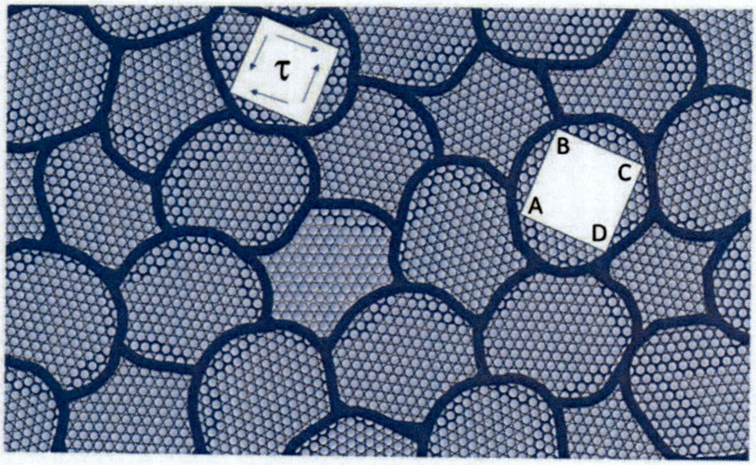

Fig. 3.20 Nanocrystalline material with shear stresses

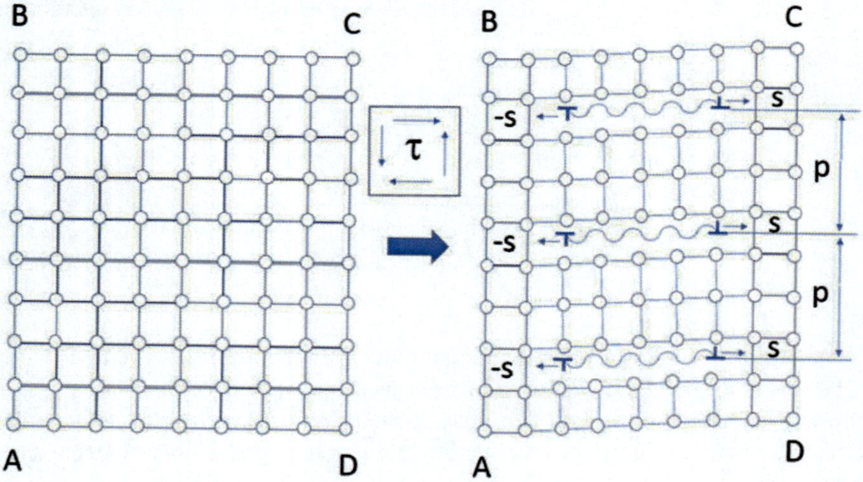

Fig. 3.21 Nanoscale rotational deformation on a single crystallographic plane

$$\Delta W(s) = \frac{Gs^2}{2\pi(1-\nu)}\left\{\ln\frac{d}{s} + 1 + \left(\frac{1}{n}\right)\sum_{k=1}^{n-1}(n-k)\left[\ln\frac{p^2k^2+d^2}{p^2k^2} - \frac{2d^2}{p^2k^2+d^2}\right]\right\} - \tau sd + \gamma(s)d$$

with $\gamma(s)$ the specific energy of a generalized stacking fault. $-\tau sd$ is the work of the external force to form the dislocation. $\gamma s(d)$ is the energy of a generalized stacking fault associated with the disturbance. The energy has a sinusoidal form as

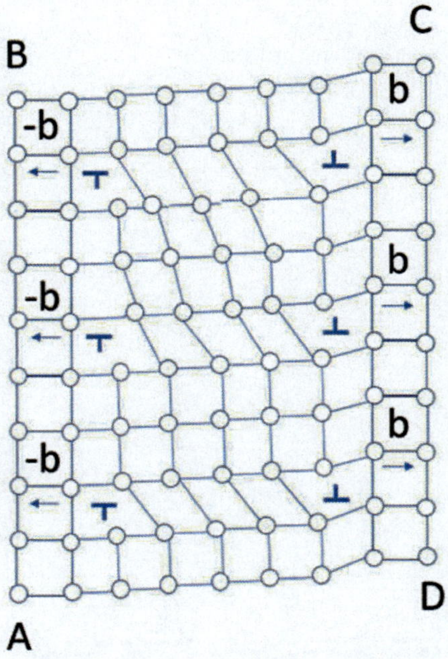

Fig. 3.22 The non-crystallographic dislocations transform into conventional dislocation with Burgers vector b

$$\gamma(s) = \begin{cases} \gamma_m \sin 2\pi u, & 0 < u < 1/4 \\ \frac{\gamma_m + \gamma_0}{2} - \frac{\gamma_m - \gamma_0}{2} \cos 4\pi u, & 1/4 < u < 3/4 \\ -\gamma_m \sin 2\pi u, & 3/4 < u < 1 \end{cases}$$

where $u = s/b$ and γ_m and γ_0 are the maximum and minimum values of the energy. Experimental evidences underline how for low shear stresses ΔW increases as s decreases. For very high stress levels, as in the case of nanocrystalline materials, the energy decreases as s increases leading to the so-called nanoscale rotational deformation (NRD). This critical stress level is quantified in

$$\tau_c = \frac{2\pi\gamma_m}{b}$$

For many nanocrystalline pure metals this is around G/20; as a matter of fact, the observation for pure Ni gives $\tau_c = 4.3$GPa (Ovidko and Sheinerman 2012a).

The mechanical behavior of materials is governed by crack nucleation and growth. This subject is very crucial for nanostructured materials. In this regard, cracks nucleate preferentially at grains' triple junctions. Kumar et al. (2003) showed how cracks nucleate at triple junctions of pure nanocrystalline nickel along the crack propagation path during tensile load. So, the mechanism is that voids nucleate at triple junctions and then they are absorbed by the main crack during propagation (Fig. 3.23).

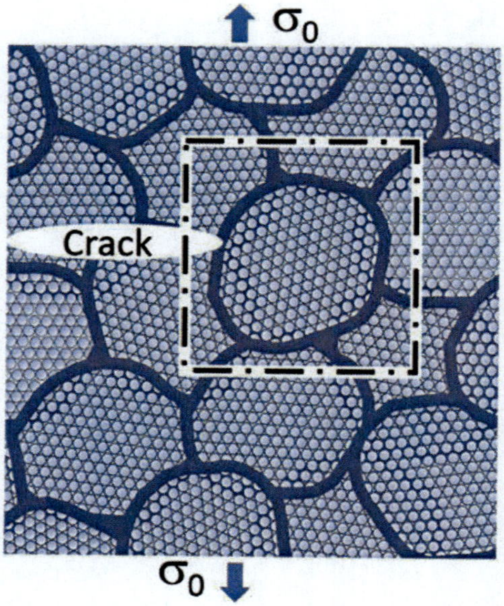

Fig. 3.23 Crack propagating in a nanocrystalline material under tensile loading

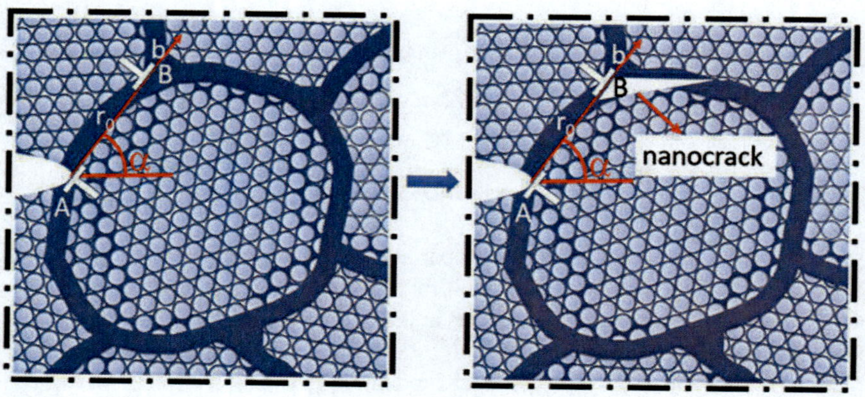

Fig. 3.24 Nanocrack formation at triple junction due to the sliding of an edge dislocation from the crack tip

Due to the sliding along the grain boundary AB, an edge dislocation forms a nanocrack at the triple junction B (Fig. 3.24) (Farkas et al. 2005).

The formation of nanocracks is described as to be dependent on external stresses and on internal defects such as dislocations, dislocation dipoles, and dislocation configuration (Ovidko and Sheinerman 2009). Additional studies from Ovidko and Sheinerman (2012b) have shown the dependence of the nanocrack formation on the grain size and on the crack geometry (modeled as an ellipse with a given radius of

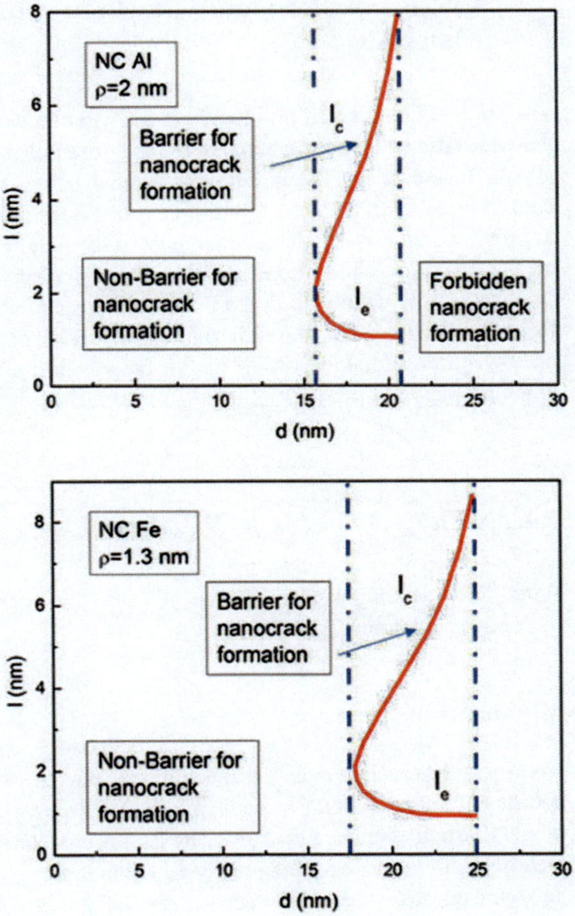

Fig. 3.25 State diagram for nanocrystalline Al and Fe with different crack geometries

curvature ρ, and a crack length l) for nanocrystalline Al and nanocrystalline Fe (Fig. 3.25).

Nanocrack generation and growth are enhanced in nanocrystalline solids by increasing the crack tip curvature radius ρ. By decreasing the grain size, the nanocrack formation is favored. For larger grain sizes, nanocracks form in the region $l < l_e$ and $l > l_c$. For intermediate grain sizes an energy barrier must be overcome. The analyses performed for both the materials demonstrate that the transition from a behavior to another happens in a very narrow grain size range.

3.4 Crack Initiation and Growth in Nanostructured Materials

The knowledge in crack initiation and growth evolved in the recent past because of the scientific evidences obtained by the research on nanostructured metals and alloys. In the past, fatigue life was treated as consisting of crack initiation and crack propagation periods, through an exact definition of transition from initiation to propagation. It was believed that most of the fatigue life of a component was spent on the crack initiation stage, and that the period of fatigue crack propagation was small. It is now widely accepted that fatigue crack initiation occurs early in life, and then the cracks grow through microstructural barriers. This growth period can occupy a considerable portion of the fatigue life, and it is essentially the fatigue crack propagation period.

3.4.1 Fatigue Cracks in NC Materials

From the pioneering work of Paris et al. (1961) the rate of crack growth is dependent on a power relationship:

$$\frac{da}{dN} = C(\Delta K)^m$$

where a is the crack length; N is the number of cycles; C and m are the so-called Paris coefficient and exponent, respectively; and $\Delta K = K_{max} - K_{min}$ is the stress intensity range. Plasticity-induced crack closure is the contact of plastically deformed residual material in the wake of a fatigue crack. Upon loading, crack surfaces separate when the opening stress level is reached, and only then can the crack advance. Thus, fatigue crack closure modifies the stress intensity factor range. The Paris relation is widely accepted for describing the fatigue crack growth behavior of materials, and the correction for crack closure provides more accurate description of the fatigue crack growth rate. By considering the effect of the crack closure during cyclic loading, the previous equation was modified:

$$\frac{da}{dN} = C(\Delta K_{eff})^m$$

where $\Delta K_{eff} = K_{max} - K_{clos}$.

In the field on nanostructured metals and alloys, the fundamental aspect is related to the relationships between the slipping driving the crack growth and microstructural barriers at the crack tip. In order to approach the problem, the following relationship was proposed to take into consideration the dimension of the plastic zone at the crack tip:

$$\frac{da}{dN} = 7.5\left(\frac{\varepsilon_y}{\varepsilon_f}\right)^2 \frac{(w^*)^2}{l}$$

where ε_f and ε_y are the fracture and yield strains and w^* is the extension of the plastic zone. It is clear how the crack growth rate is directly proportional to the extension of the plastic zone. From a microstructural point of view, it was demonstrated that the plastic zone increases as the SFE increases.

It was observed that, for stresses below the fatigue limit, cracks can be arrested by microstructural barriers. If the stress is increased at values higher than the fatigue limit they continue to grow. So, the crack growth can be divided into two contributions, the first one for short crack and the second one for long crack propagation:

$$\frac{da_i}{dN} = A\left(\Delta\gamma_p\right)^n (d_i - a)$$

$$\frac{da_i}{dN} = B\left(\Delta\gamma_p\right)^m (a - D)$$

d_i is the distance to the closest microstructural barrier, $\Delta\gamma_p$ is the plastic shear strain range, D is a threshold condition dependent on the level of applied stress-strain state, and A, B, n, and m are material constants. As the crack continues to grow, the radius of the plastic zone also grows, increasing the growth rate per cycle. The growth rate is inversely proportional to the yield strength; hence the rate is slower with higher yield strengths through various strengthening mechanisms. So, the generalized equation has been developed:

$$\frac{da}{dN} = \xi^{\frac{1}{b}}(2s)^{1-\frac{1}{b}}\left[\frac{\Delta K}{E}\right]^{\frac{2}{b}}$$

where

$$\xi = \frac{\mathrm{ES}}{4\sigma_y \varepsilon_f d}$$

S is the cell size and d the grain size. The smaller the cell size the higher the fatigue crack growth resistance; b varies between 0.5 and 1.

By considering the effect of crack closure

$$\Delta K_{\text{eff}} = 1,12(\sigma_{\max} - \sigma_{\text{cc}})\sqrt{2}a$$

where σ_{cc} is the plasticity-induced crack closure stress, with z being the distance between the crack tip and the next barrier:

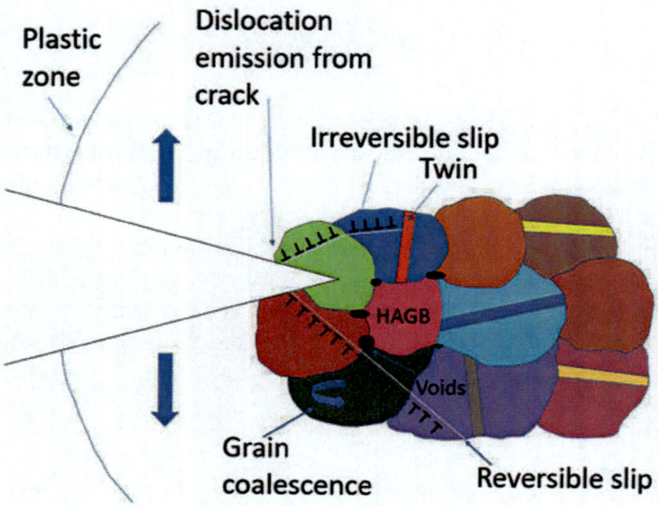

Fig. 3.26 Crack interaction with various metallurgical features (Sangid et al. 2011)

$$\sigma_{cc} = \frac{\alpha z}{2a}\sigma_{max}$$

Thus, the crack growth rate is

$$\frac{da}{dN} = C\Delta K^n\left[1 - \alpha z\left(\frac{1.12\sigma_{max}}{\Delta K}\right)^2\right]^n$$

This equation quantifies the effects of grain size, plastic stress amplitude, and microstructural barrier on the crack growth rate effectively (Fig. 3.26).

From the interaction between the dislocations emitted from the crack and grain boundaries or twins, residual dislocations within the twin or GB car remain. This leads to slip irreversibilities during fatigue. The mechanism is schematized in Fig. 3.27.

Upon loading, dislocations are emitted at the crack tip in the tensile semi-cycle (a); then the crack advances and the crack tip is blunted (b); during loading reversing, a different slip system is activated as dislocations are emitted (c); the emitted dislocations interact with microstructural barriers such as twins (d); the dislocation movements are impeded and the crack is retarded (e, f).

In this regard, it is interesting to underline the deformation behavior at the crack tip in the presence of bimodal NC/UFG microstructure (Fig. 3.28).

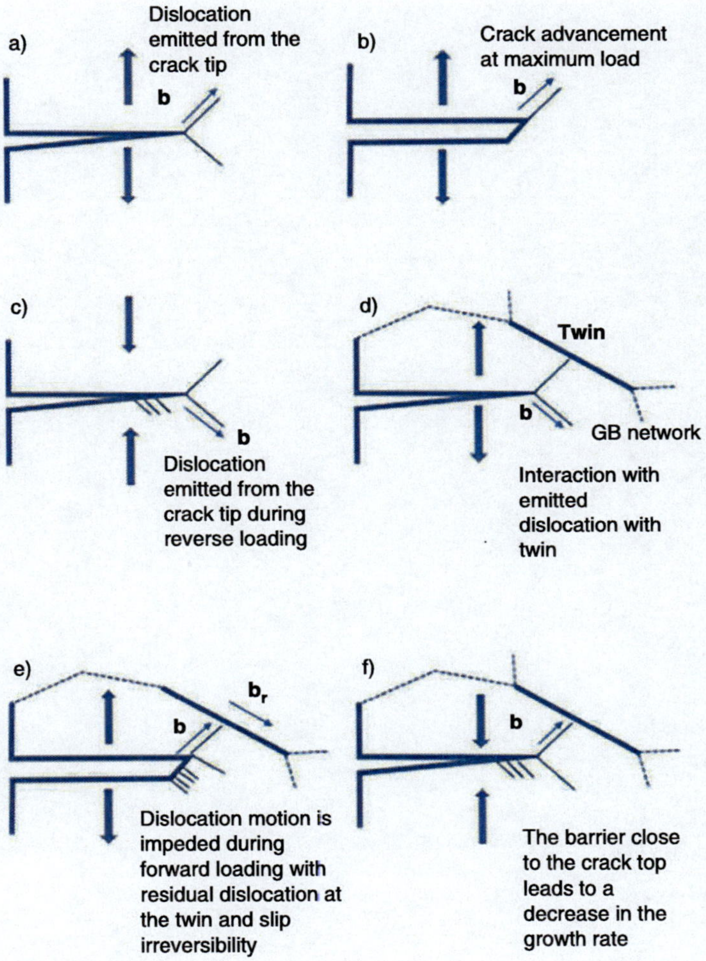

Fig. 3.27 FCG in the case of reversible slip

3.4.2 Crack Propagation

The problem of low ductility and low fracture toughness is directly coupled with many other issues in the design of nanocrystalline metals (Ramesh 2009). For example, nanocrystalline metals suffer from microstructure instability at low temperature and grain growth can hinder their use in real-life applications where they lose their original properties over time (Zheng and Peng 2018; Ramesh et al. 2007). It is clear that overcoming the main challenges in this class of materials such as low fracture toughness and low strain to failure becomes tightly bonded on the structure of the grain boundaries. The issue of limited ductility in nanocrystalline materials comes from the lack of inefficient accommodation of dislocation plasticity, leading

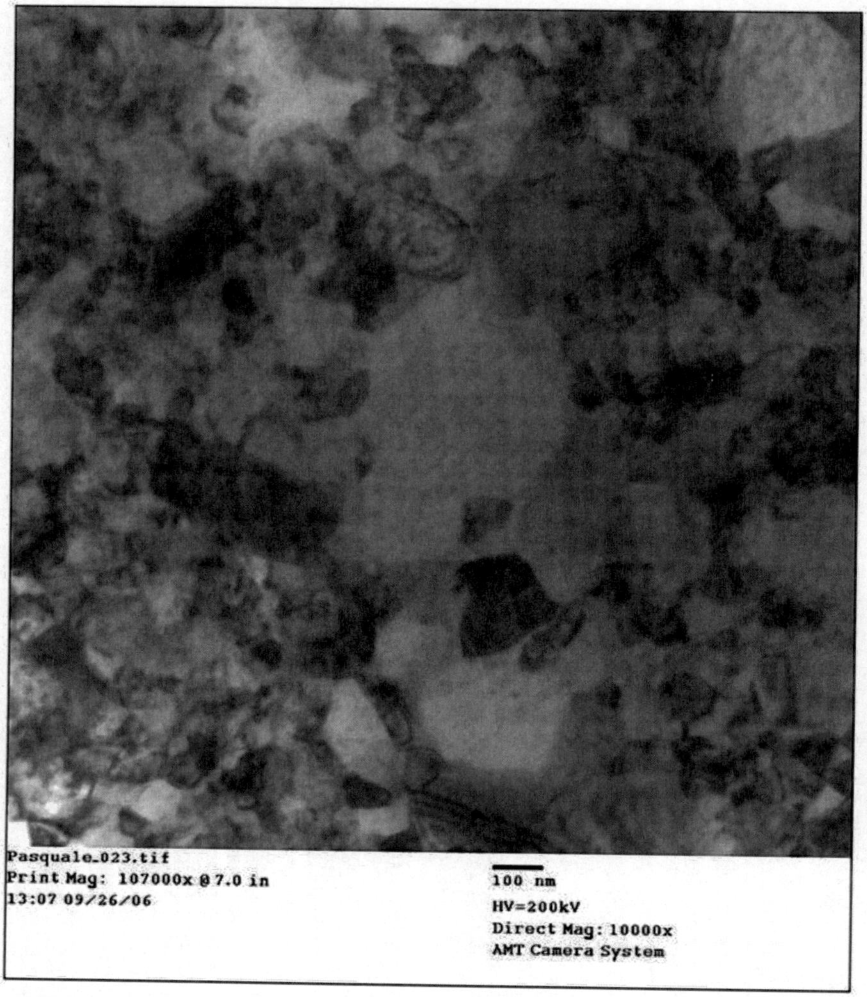

Fig. 3.28 Bimodal NC/UFG microstructure for Ni

to microcracks, or a lack of strain hardening. In recent years, several methods have been proposed to improve the toughness and ductility of nanocrystalline materials. One common strategy includes dispersing a larger grain size into the structure of a nanocrystalline material, which results in improved strain hardening ability. The lack of strain hardening and propensity for early fracture in nanocrystalline metals can also be related to the ability of the grain boundary in absorbing the mobile defects. To accommodate defects, interfaces are required to have excess free volume. In the absence of this free volume, dislocations cannot be absorbed and microcracks can form. By tuning the chemistry of the grain boundary and adding a stable disordered complexion, it may be possible to increase the ductility of nanocrystalline materials.

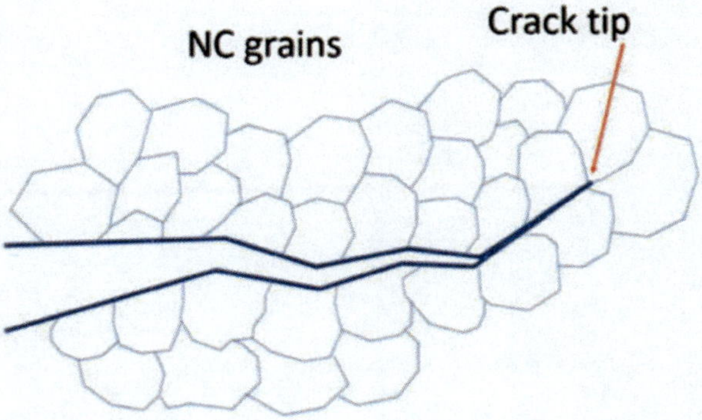

Fig. 3.29 Deformation mode in NC materials

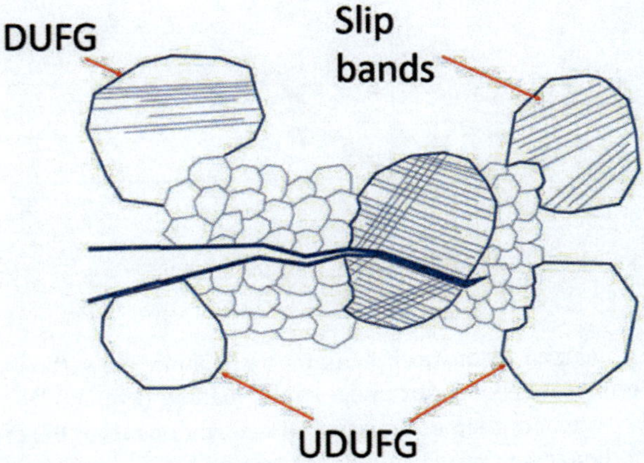

Fig. 3.30 Crack propagation in bimodal NC/UFG material

While this increases ductility, these materials with bimodal grain sizes do so at the expense of strength. Another method for improving ductility is the promotion of mechanically induced grain growth.

The schematic of the pure NC deformation mode is shown in Fig. 3.29.

The crack tip advances preferably in a transgranular way; in the case of presence of ultrafine grains on the crack path, these act as obstacles to the crack propagation (Fig. 3.30).

Fatigue crack growth is strongly influenced by both grain size and misorientation angles (Zhou et al. 2016). In NC materials the first emissive dislocation and grain boundary can hinder the next dislocation to emit. By considering the schematic in Fig. 3.31 and mode *I* crack opening, the resolved shear stress from the external forces on the crack tip acting on the *k*th dislocation is

$$\tau_a^k = \frac{K_I \sin\varphi \cos\left(\frac{\varphi}{2}\right)}{2\sqrt{2\pi r_k}}$$

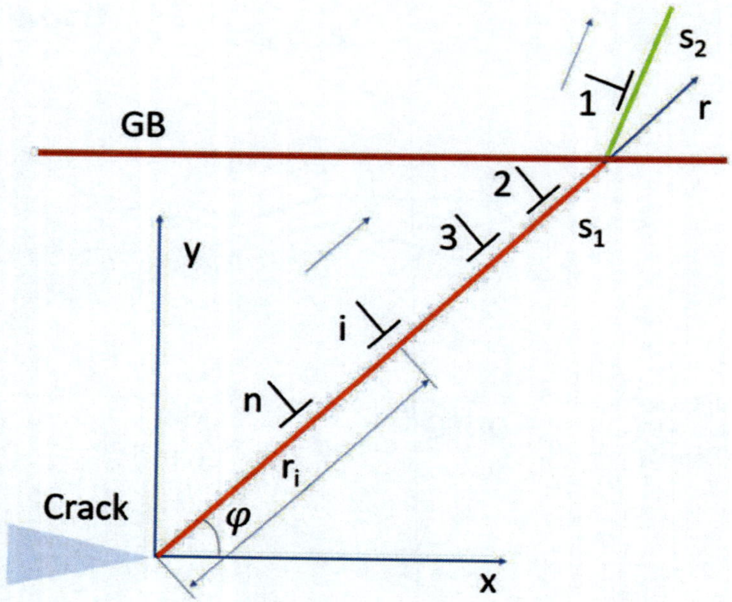

Fig. 3.31 Dislocation emission from the crack tip

where r_k is the position of the dislocation. Now, the crack tip emits a large number of screw dislocations under cyclic loading (Fig. 3.32).

With continuous emission of dislocations from the crack tip (Yu et al. 2018), the fatigue crack propagation is expressed by

$$\frac{da}{dN} = \frac{r_{\max}}{2G}\sum_{i=1}^{k}\tau_k$$

That combined with the previous equation will be

$$\frac{da}{dN} = \frac{r_{\max}}{2G}\frac{\Delta K_I \sin\varphi\cos\left(\frac{\varphi}{2}\right)}{2\sqrt{2\pi}}\sum_{i=1}^{k}\frac{1}{\sqrt{r_i}} - \frac{br_{\max}}{8\pi(1-\nu)}\sum_{i=1}^{k}\frac{1}{r_i} + \frac{br_{\max}}{4\pi}\sum_{i=1}^{k}\sum_{j\neq i}\sqrt{\frac{r_j}{r_i}} \times \frac{1}{r_j - r_i}$$

So, for a given grain size and misorientation angle, the corresponding crack growth at a given ΔK can be predicted. It is demonstrated that the crack growth rate increases with the decrease of the grain size and the misorientation angle increases with the crack growth rate decreasing. Because nanocrystalline metals with grain size above 10 nm plastically deform through grain boundary dislocation

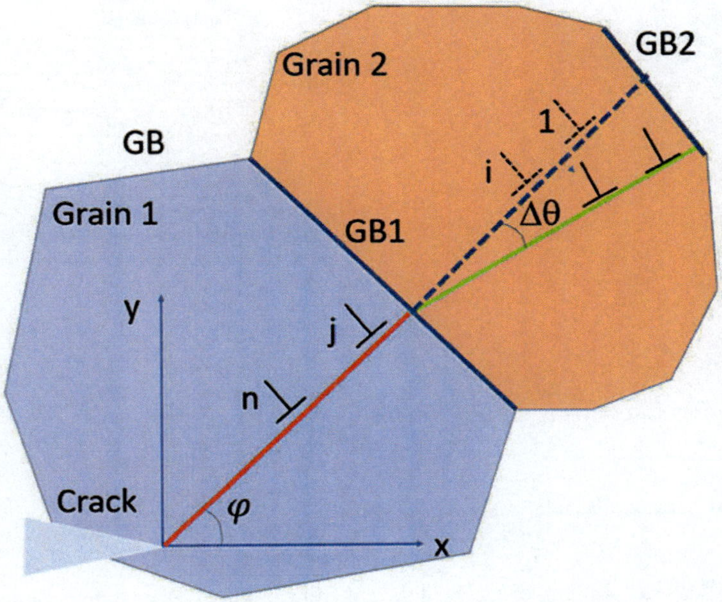

Fig. 3.32 Dislocations emitted from the crack tip and penetrating the GB1 are stopped at GB2

emission and absorption, grain boundary complexions should have a major impact on the ductility and toughness of these materials.

One very important factor influencing the growth of the fatigue crack is the interaction between the slip mechanism driving the fatigue crack and potential microstructural barriers that are ahead of the crack tip. The barriers mostly discussed in literature are grain boundaries and twin boundaries. The recent scientific literature demonstrates how nanotwinned NS materials exhibit high damage tolerance (Chowdhury et al. 2016). The concept relies on the fact that twin boundaries impart material strengthening in a fashion similar to conventional high-angle grain boundaries, namely by blocking dislocation motion. However, unlike conventional nanocrystalline materials, decreasing the twin lamellae thickness also results in an increase in work hardening capacity. Apparently, the mechanism for this increased work hardening capacity with increased twin density is that the twin boundaries act as locations for dislocation accumulation. Nanostructuring through severe plastic deformation or other techniques allows for the obtaining of high-strength materials. Anyway, their technological application is strongly related to the optimal synergy among strength-ductility-crack resistance (Fig. 3.33).

In FCC metals, there are two important twin boundary types: (1) Σ3 {111} coherent twin boundary (described above) and Σ3 {112} incoherent twin boundary. It is worth to mention that, depending on the material, other types of twin boundaries may also become important (Li et al. 2015). Both coherent and incoherent twin boundaries are energetically much different than grain boundaries. For example, in

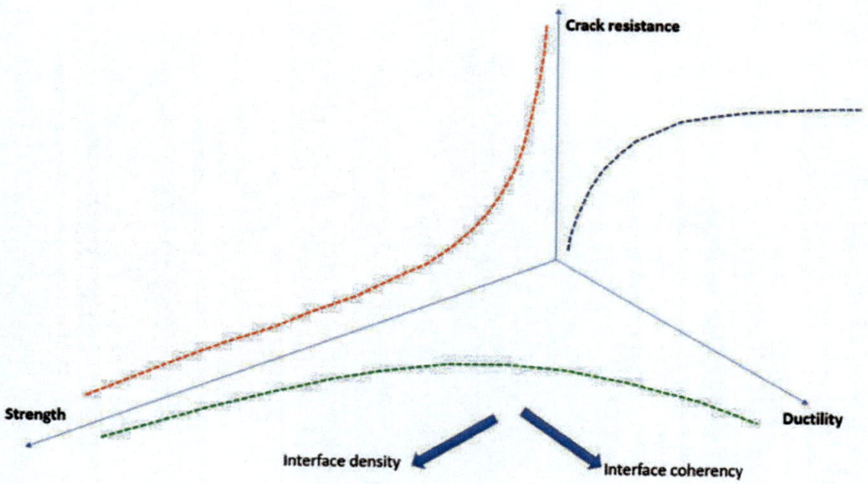

Fig. 3.33 Damage tolerance vs. strength vs. ductility

Cu, while twin boundary energies of the coherent and incoherent twin boundaries are in the range of 30 mJ/m^2 or lower (coherent twin boundary energy is much less than incoherent), the energy of a high-angle grain boundary is around 700 mJ/m^2.

3.4.3 Nanotwinning

The most promising candidates to have actually manifested such a feat are the nanotwinned metals and alloys with a prevalence of ∑3 interfaces, which is the most coherent of all boundaries. Another important feature of NT metals is their improved fatigue resistance and improved microstructural stability under cyclic loads. However, after reaching the critical stress level, crack nucleation is observed as a result of detwinning. It has been observed that a finer twin lamellar thickness and/or spacing is more beneficial in obstructing damage propagation than the coarse-twinned counterparts. An example for nanotwinned pure Cu is shown in Fig. 3.34 (Singh et al. 2011).

A careful study of nanotwinned and traditional fine-grained material behaviors strongly indicates that interface coherency indeed plays a pivotal role in imparting not only high toughness but also improved cracking attributes. Moreover, unlike the conventional nano-sized grains, the nanotwins are found to be quite stable upon thermomechanical treatments. The model of the interaction of the crack with the NC microstructure with nanotwinned structures with coherent ∑3 interfaces is shown in Fig. 3.35.

The crack interaction with the nanotwin is shown in Fig. 3.36.

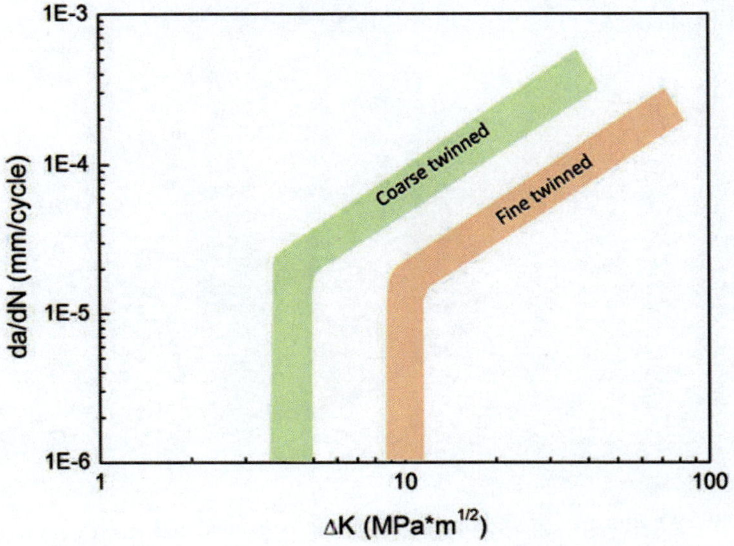

Fig. 3.34 Crack behavior of nanotwinned copper

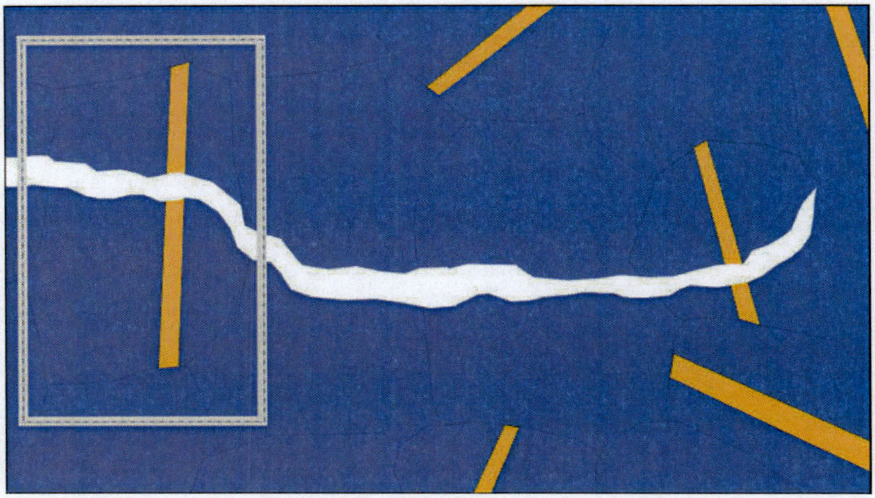

Fig. 3.35 Crack propagation in nanotwinned material

During its propagation the crack encounters twins modifying the crack path and being retarded with increase of the material resistance to damaging.

A well-established model has offered a variant on the Hall-Petch model that allows twins to be given a discounted importance by using an "effective" grain size. This model was shown to perform well compared to the classic Hall-Petch model. Reformulating the Hall-Petch equation entirely in terms of boundary density allows

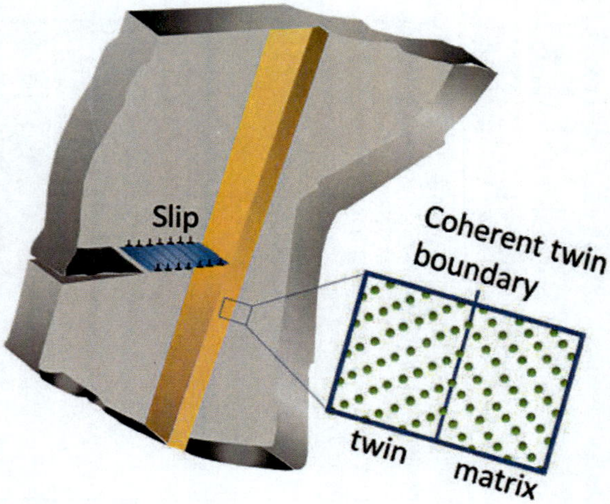

Fig. 3.36 Crack interaction with the nanotwin

an arbitrary number of grain boundary types to be considered, each with individual weighting factors. In this form, the yield strength (σy) can be expressed as

$$\sigma_y = \sigma_0 + \left(\sum_i k_i \rho_i\right)^{\frac{1}{2}}$$

where *ki* are the strengthening coefficients for each type of boundary, which have densities ρi. The exponent is ½, instead of the more familiar −½ because boundary density is inversely related to grain size. The term σ_0 has the same meanings as in the usual Hall-Petch equation and can be thought of as providing the yield strength at infinite grain size, or equivalently zero boundary density. If there is assumed to be a single value for all k_i, and also that mean grain size is inversely proportional to boundary density, then equation reduces to the usual Hall-Petch equation. The assumption that mean grain size is inversely proportional to the boundary density implies a fixed distribution of grain sizes and shapes. Grain boundary density has the added advantage that it can be applied to microstructures where mean grain size is inappropriate. For example, mean grain size is ambiguous in materials with a bimodal grain size or high-aspect-ratio grains, but the interpretation of grain boundary density remains clear.

In the case where grain boundaries are categorized as either twin or high angle, the previous equation can be written more simply as

$$\sigma_y = \sigma_0 + (k_1 \rho_{\mathrm{HA}} + k_2 \rho_{\mathrm{TB}})^{\frac{1}{2}}$$

where ρ_{HA} is the density of high-angle boundaries, ρ_{TB} is the density of twin boundaries, and k_1 and k_2 are their respective strengthening coefficients. This can be rearranged in terms of a relative strengthening effect as

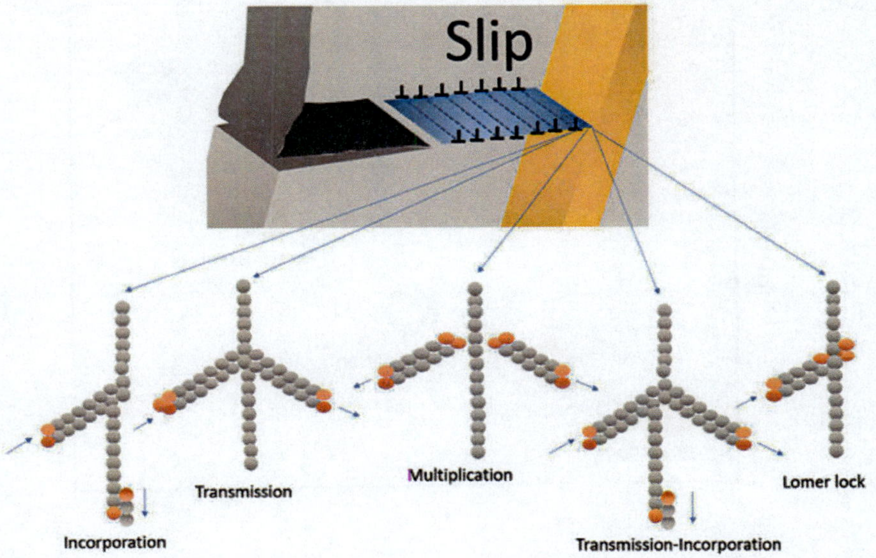

Fig. 3.37 Crack interaction with the nanotwin modes

$$\sigma_y = \sigma_0 + k_1^{\frac{1}{2}}(\rho_{\mathrm{HA}} + m\rho_{\mathrm{TB}})^{\frac{1}{2}}$$

and

$$m = \frac{k_1}{k_2}$$

where m is a unitless parameter that gives the relative strengthening contribution of twin boundaries as compared to high-angle ones. A value of m equal to zero corresponds to twins having no strengthening effect, and $m = 1$ implies equality with other grain boundaries. Armed with a model that can incorporate a continuum of relative boundary strengths, it only remains to determine m.

Now depending on the stress state and on the slip geometry, the crack interaction with twin boundary varies by modifying the crack growth rate. The various modes of interaction are shown in Fig. 3.37.

Each reaction results in a different type of residual slip represented by the resulting Burgers vector:

$$\vec{b}_r = \vec{b}_{\mathrm{incident}} - \vec{b}_{\mathrm{transferred}}$$

Experimental evidences showed the relative efficiency of each mechanism on the shear stress; the qualitative behavior is shown in Fig. 3.38.

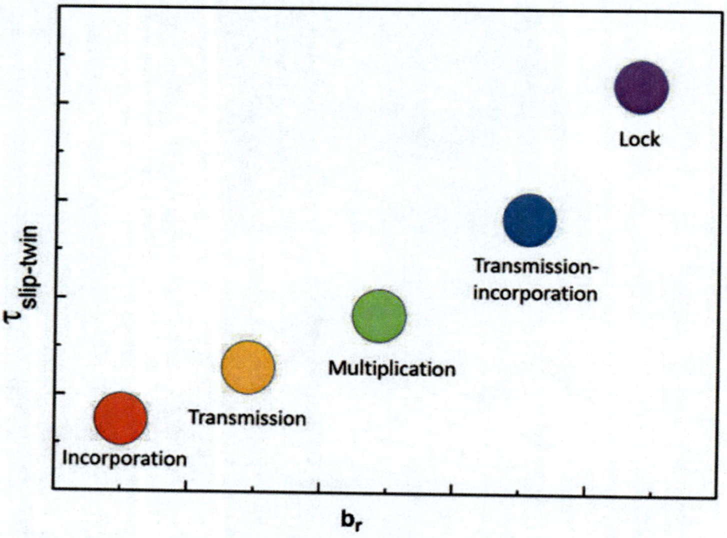

Fig. 3.38 Strength efficiency depending on the slip-twin interaction mechanism

Studies regarding the crack behavior in LSNT and SSNT materials (Fig. 3.39) show a very different growth behavior.

The qualitative crack propagation rate vs. SIF for the different twinned materials is shown in Fig. 3.40.

The twin-free sample has the fastest crack growth rate, the SSNT sample has the slowest rate, and the LSNT sample has a rate between those of the other samples. The interaction between crack and TBs plays an important role in the activation of duplex slip and the formation of a striation. The dislocation density in the SSNT sample is much greater than that in the twin-free and LSNT samples during fatigue. The high density of dislocations in the SSNT samples is mainly attributed to the presence of TBs and the complex interaction with TBs and dislocations. The preexisting nanoscale twins near the crack tip are destroyed and even disappear, and a few new twins are formed around crack tip during fatigue. The annihilation of preexisting twins is mediated by TB migration through the consecutive slip of partial dislocations on TBs. Such detwinning has been reported in the tension–tension fatigue tests of the NT Cu with columnar grains. The formation of new twins manifests the activation of deformation twinning by successive emission of partial dislocations from the GBs on adjacent planes.

As a result of the nanostructured material deformation mechanisms, nanocrystalline systems with $d < 10$ nm should have mechanical properties which resemble those of metallic glasses. This concept was first supported by the observation that nanocrystalline metals exhibit a tension-compression asymmetry of their strength and shear transformation zones operate inside of grain boundary regions during plasticity.

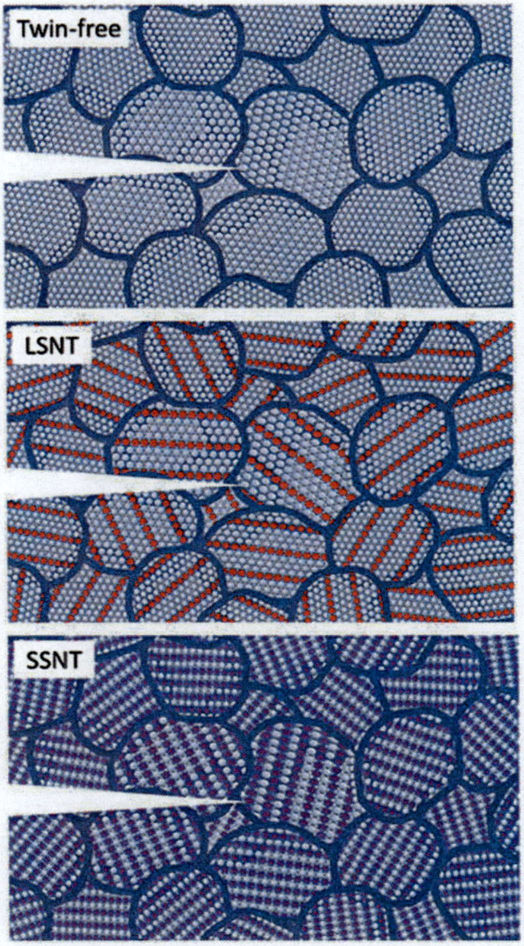

Fig. 3.39 Crack propagating in different twinned density materials

In situ studies performed on nanocrystalline pure palladium under tensile loading allowed to identify twinning, twin boundary migration, and detwinning during the tensile deformation (Kobler et al. 2016). The schematic is shown in Fig. 3.41.

The monitored deformation varied from 0 to 4.7%. The twin activity is accompanied by substantial structural changes in the surrounding grains. In general, the twin activity correlates with local plastic strain manifested in the structural changes. Since Shockley partial dislocations of the FCC lattice have the same Burgers vectors as the intrinsic crystallographic structure elements of twin (Σ3) boundaries, the twin motion indicates the operation of Shockley partial dislocations entering the grain from the boundary or moving along the preexisting twin planes. The twin activity associated with heterogeneous distribution of the plastic strain indicates the necessity of accommodating processes to suppress fracture by local plastic strain incompatibilities.

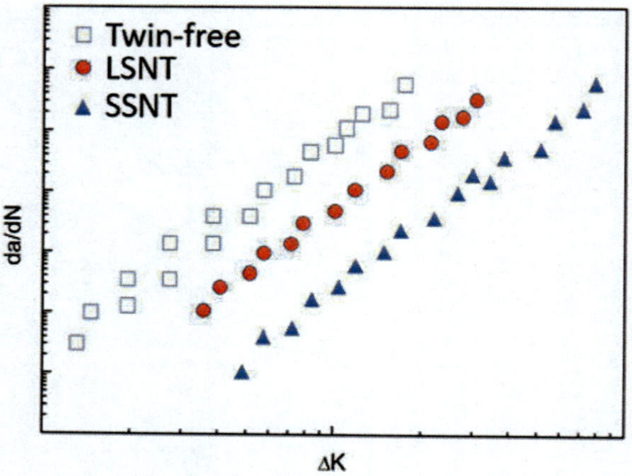

Fig. 3.40 Crack propagating rate in different twinned density materials

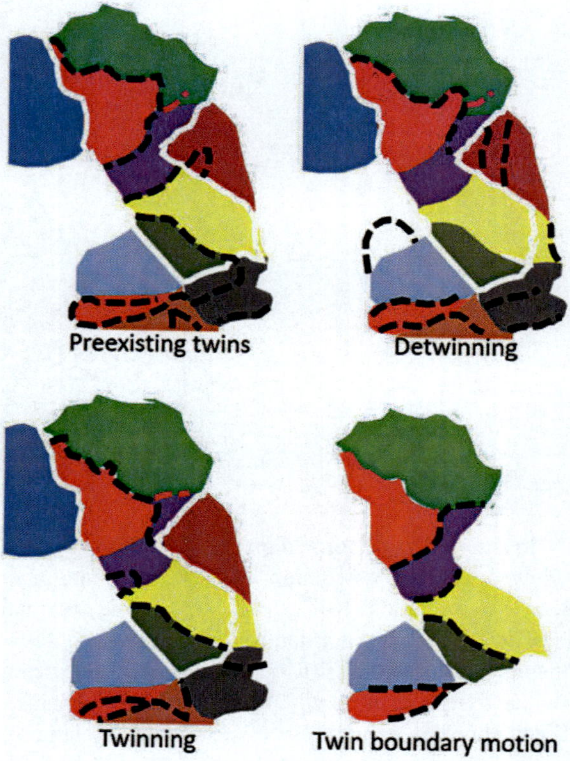

Fig. 3.41 Twinning, twin boundary migration, and detwinning of the center grain (white boundary) during the tensile deformation

The model of slip-twin interaction in the presence of crack is shown in Figs. 3.42 and 3.43 for screw and edge dislocation, respectively.

A full screw dislocation nucleates at the crack tip; it dissociates into leading and trailing Shockley partials. Leading and trailing partials hit CTB and multiply into

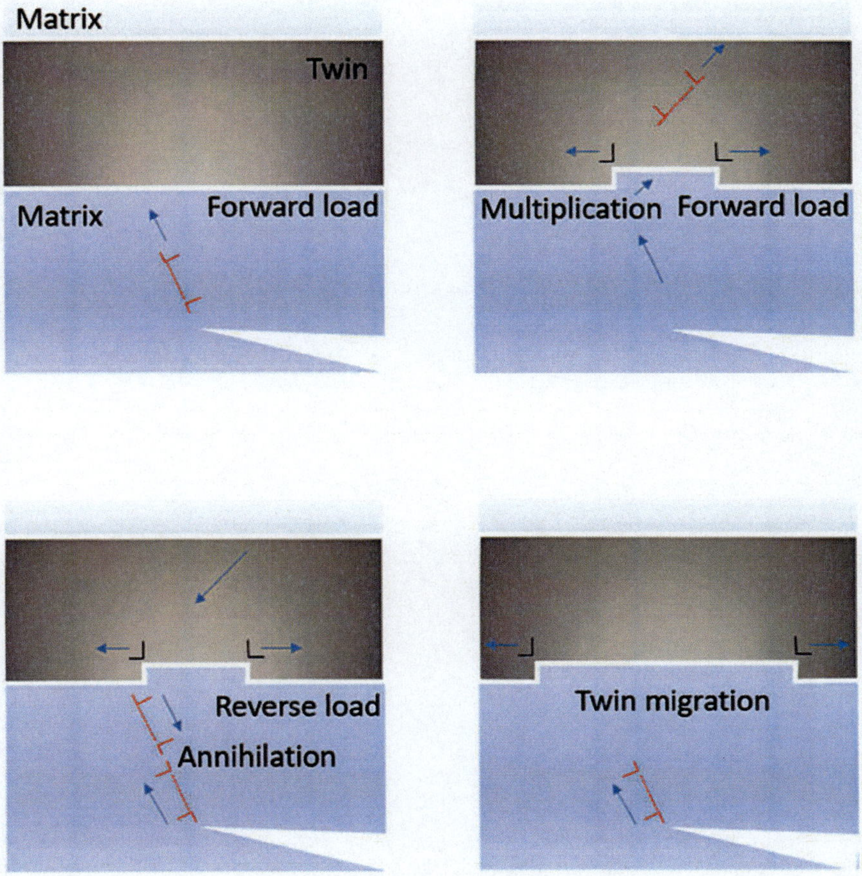

Fig. 3.42 Slip-twin interaction in the case of a screw dislocation

two more full screw dislocations subsequently dissociating into partials on CTB and inside twin. Twinning partials from forward load cause one layer of twin migration. Returning Shockley partials multiply into more twinning partials and matrix Shockley partials which get annihilated by newly nucleating incoming partials of opposite sign. Twinning partials on CTB continue gliding in opposite directions resulting in twin migration again. Another negative pair of Shockley partials nucleate from crack tip.

The sequence is nucleation of pure edge leading and trailing Shockley partials. Incident partials interact with CTB and create unstable sessile residual dislocation and one pair of glissile Shockley partials transmitting inside twin. Returning Shockley partials from inside the twin interact with residual left on CTB and create another set of sessile unstable dislocations. Unstable sessile dislocations dissociate into new stable sessile dislocation, twinning partial and returning glissile Shockley partials

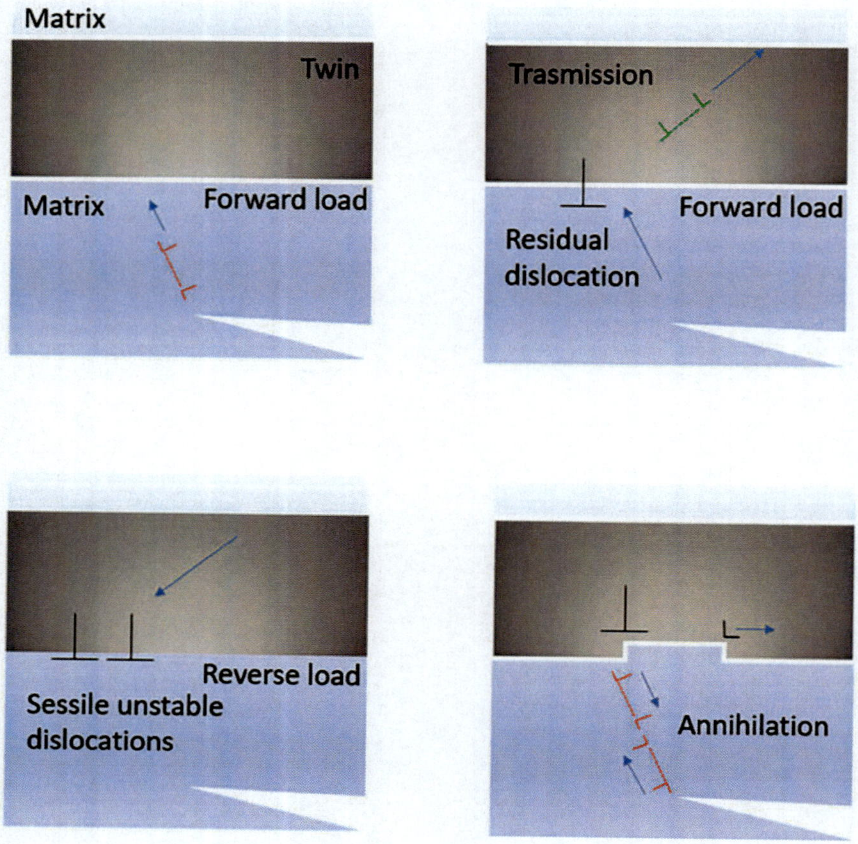

Fig. 3.43 Slip-twin interaction in the case of an edge dislocation

into matrix which subsequently get annihilated with another newly nucleated negative pair of partials from crack tip.

Now, many experimental evidences confirm the beneficial effect of twinning on the fatigue response of nanostructured materials. During crack propagation, its advance is favored by the irreversible glide of dislocations that are emitted by the crack tip as a consequence of stress concentration. As a consequence, the rate of glide irreversibility influences the crack growth rate (Chowdhury et al. 2013). In particular, obstacles such as grain boundaries and coherent twin boundaries act as limiting the slip irreversibility. The rate of irreversibility obviously depends on the nature and structure of the interfaces. The slip emission from a fatigue crack advancing towards a nanotwin is shown in Fig. 3.44.

Basically, with d being the twin spacing, t the thickness, and t_0 the glide strength, the qualitative effect on crack growth behavior is shown in Fig. 3.45.

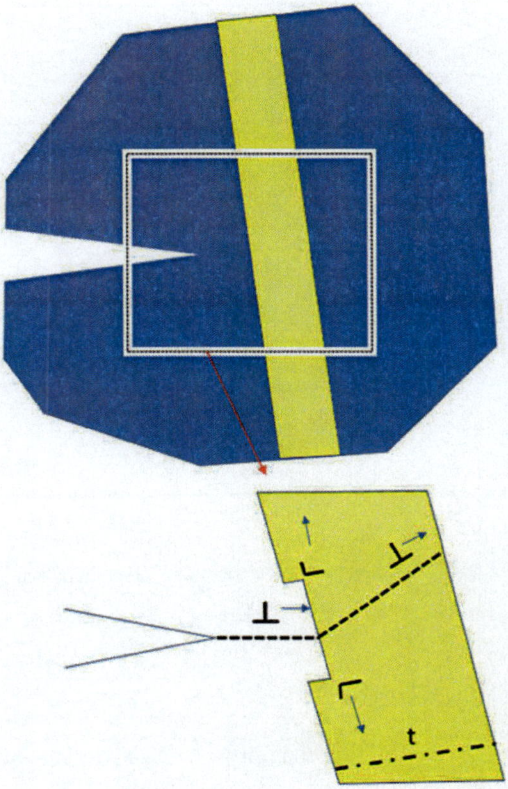

Fig. 3.44 Crack interaction with CTB

When irreversibility is 0, the crack does not propagate. So, the prevalence of twins leads to a direct resistance to FCG. MD simulation studies allowed to model the dislocation-twin interaction effect on the slip irreversibility and, as a consequence, on the FCG. The movement of dislocation glide during cyclic loading is shown in Fig. 3.46.

Cyclical slip-twin interaction involves the incidence of a dissociated full dislocation (screw) on the CTB. The reaction results in the simultaneous incorporation and transmission of extended dislocations. The final locations of these dislocations are two partials c and d and one extended dislocation d. The partials at c and e contribute to migration of the twin by one atomic layer. During reverse flow the extended dislocation at position d is transmitted back into the matrix, again leaving new partials at c′ and e′, which repulse the partials at c and e. The returning crack-bound extended dislocation is annihilated by another incoming dislocation of opposite sign (negative) at location f. The ratio between the irreversible plastic strain and the total plastic shear strain is given by

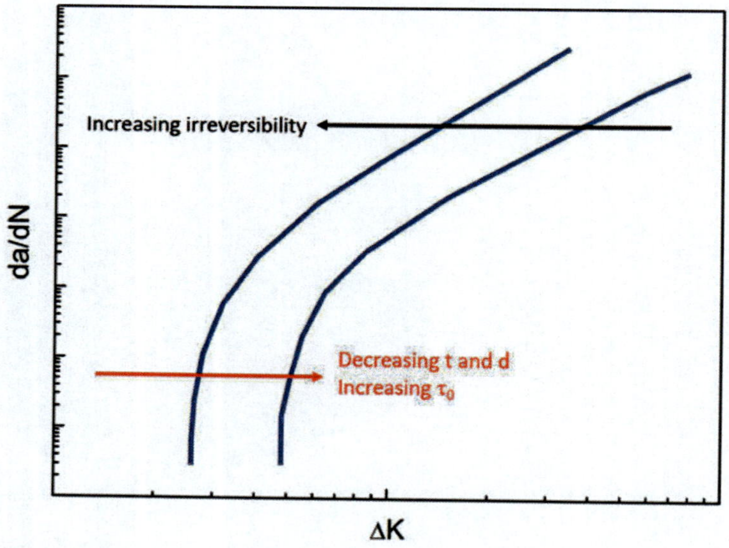

Fig. 3.45 Crack growth behavior as a function of twin properties

$$p = \frac{\gamma_{\text{irr}}}{\gamma_{\text{total}}} = \frac{\gamma_{ce} + \gamma_{c'e'} + \gamma_{af}}{\gamma_{ad} + \gamma_{db} + \gamma_{ce} + \gamma_{c'e'} + \gamma_{bf}}$$

The slip irreversibility is related to the twin lamellae spacing as shown in Fig. 3.47.

Al large twin lamellar spacing, shear starts to become independent on the twin presence. The same behavior is calculated as increasing the crack to twin spacing (Fig. 3.48).

At lower values of *t* and *d* a relatively small number of dislocations are emitted from the source and traverse the entire thickness of the twin. Consequently, a shorter spacing between the source and the twin as well as thinner twins will expedite the return of source-bound positive slip, and at the same time preclude gliding of negative slip sufficiently farther away from the source. Thus the annihilation process occurs in very close proximity to the source. As a matter of fact, calculations performed for pure nanocrystalline Ni show that crack initiation and growth improve by decreasing the twin lamellar spacing (Fig. 3.49).

The same behavior is obtained by calculating the cyclic crack behavior as a function of crack to twin distance (Fig. 3.50).

The model allows to conclude that cracks are retarded as the twin density increases and the twin lamellar distance decreases for a constant material volume.

The microstructure of twinned NC Ni with different twins aspect is shown in Fig. 3.51.

Early experiments indicated that the TBs can strengthen metals by impeding the motion of dislocations as effectively as high-angle GBs. More importantly, ductility

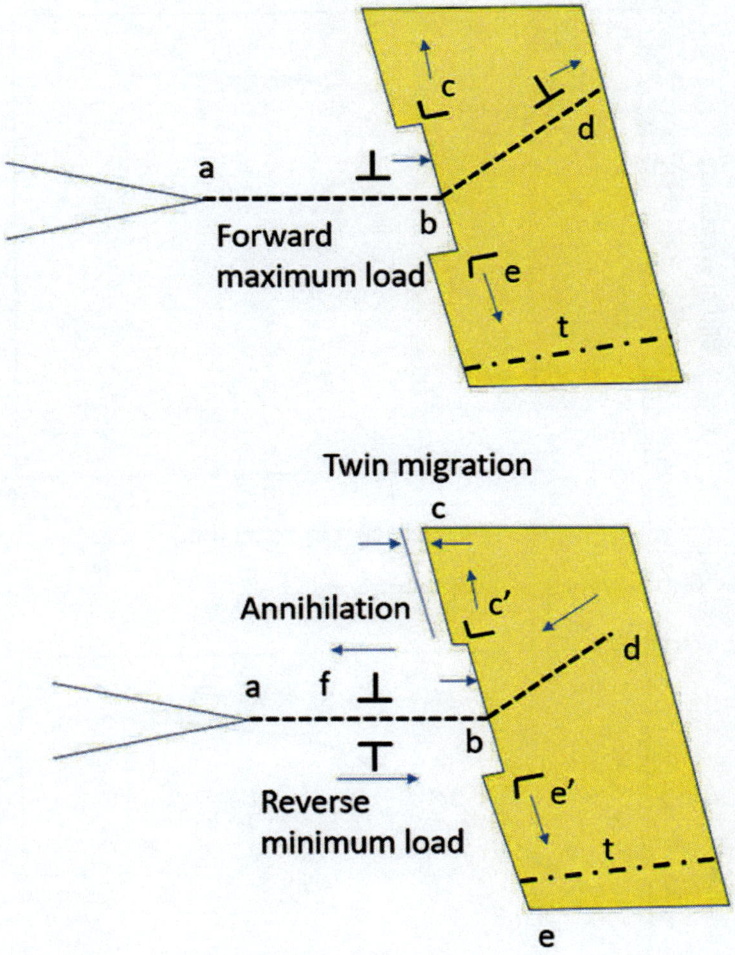

Fig. 3.46 Slip irreversibility schematic

is simultaneously obtained by dislocation sliding along coherent TBs. Therefore, improving strength by TBs is not compromised by a marked reduction of ductility as reported in traditional strengthening approaches (Zhu and Gao 2012). The plastic deformation of metals is modified by the presence of TBs. The dislocation-TB interaction acts in three different ways: First of all, dislocations may either slip transfer across TBs (hard mode I) or glide in between the twin lamellae (hard mode II, by confined-layer slip) as shown in Fig. 3.52.

As previously mentioned, the strengthening effect depends on the twin lamella spacing; the two processes are proportional to $t^{-1/2}$ and t^{-1}, respectively. In the soft mode, dislocations are inclined to initiate from the TB/GB junction and move along the coherent TBs with a very low slip resistance. The activation of any of these three

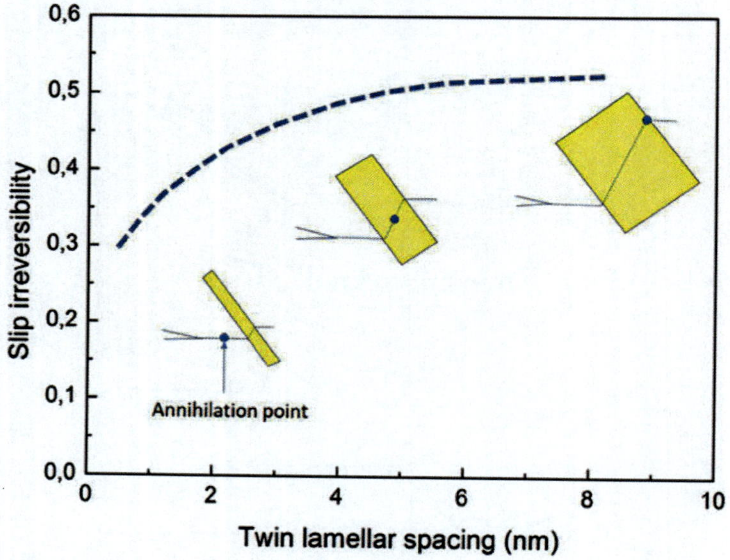

Fig. 3.47 Slip irreversibility vs. twin lamellar spacing

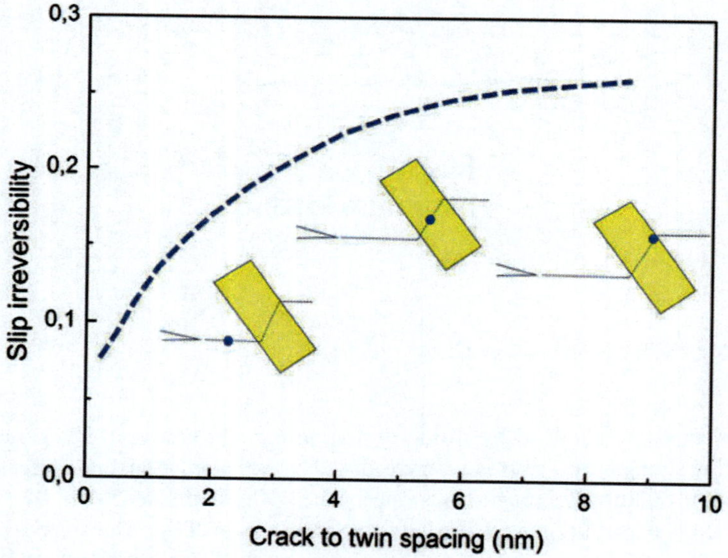

Fig. 3.48 Slip irreversibility vs. crack to twin spacing

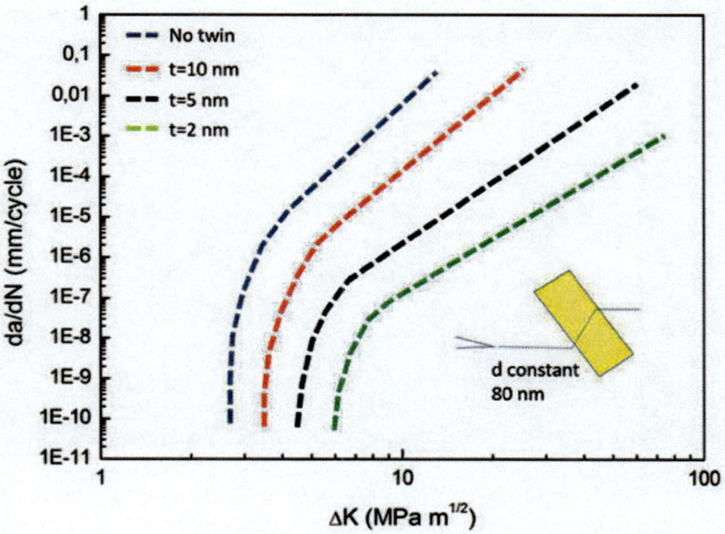

Fig. 3.49 Crack initiation and growth behavior for pure Ni as a function of twin lamellar spacing for a constant crack to twin spacing

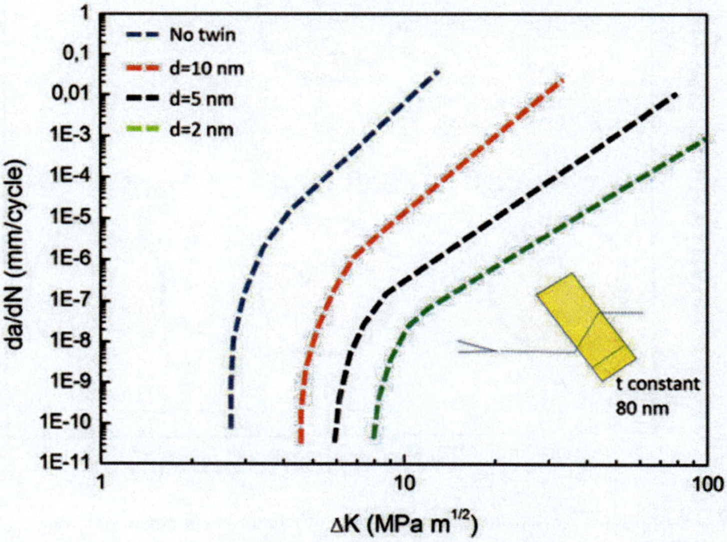

Fig. 3.50 Crack initiation and growth behavior for pure Ni as a function of crack to twin distance for a constant twin lamellar spacing

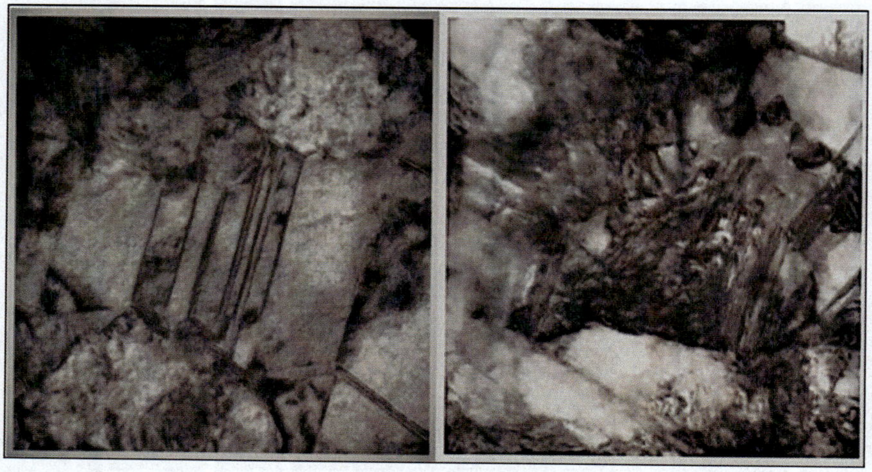

Fig. 3.51 NC Ni with low-density and large-thickness twins (left); high-density and fine-thickness twins (right)

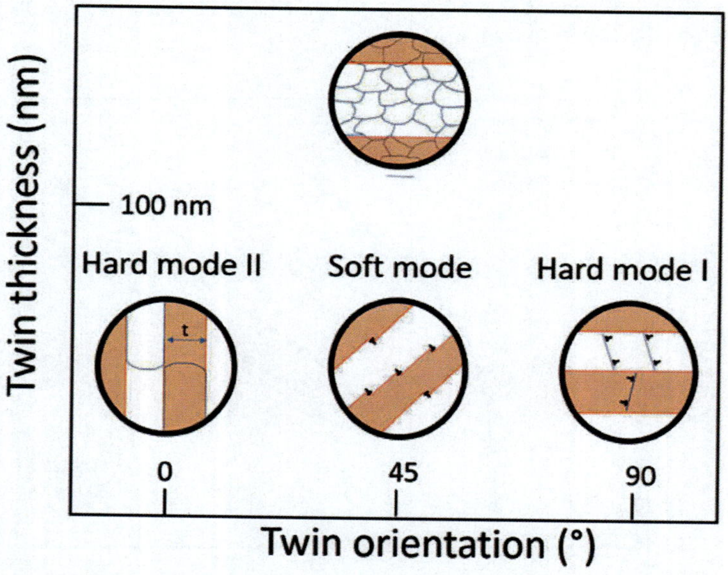

Fig. 3.52 Dislocation-TB interaction schematic with different twin orientations with respect to the loading direction

modes is basically governed by the loading condition, as well as the orientation relationship between operative slip modes and TBs. These unique dislocation dynamics and the presence of nanoscale TBs not only routinely strengthen metals, but also provide sufficient rooms for dislocation nucleation and accumulation, thereby elevating work hardening and tensile ductility (Cao et al. 2018). For the elongation to failure, a pronounced increment is observed upon reducing t, in contrast to the reduction in ductility at smaller grain sizes. It should be mentioned that the boundary energy of coherent TBs is at least one order of magnitude lower than that of GBs. Therefore, the coarsening of nanocrystalline metals (via GB migration) is effectively suppressed by TBs, and NT metals are energetically stable compared to nanograined counterparts under the same grain size and chemical composition (Hu et al. 2017).

3.5 Conclusions

Aside from quantitative improvements in fatigue performance, NC metals also hold the possibility of new insight into the mechanisms responsible for traditional fatigue failure. As a matter of fact, the persistent slip mechanism responsible for conventional fatigue crack initiation may be suppressed when the grain size is below a certain threshold, likely on the order of 100 nm or several hundred nanometers. This length scale threshold is between grain sizes that support collective dislocation activity and grain sizes that support individual dislocation activity. Room-temperature mechanically driven grain growth leads one to suspect that NC metals may evolve such coarse grain structures during fatigue loading, and that the fatigue mechanisms may be influenced more by the evolved grain structure than by the initial structure.

Another important mechanism in nanocrystalline materials is the absorption of dislocations at grain boundaries identified as one of the reasons of the strong increase of strain rate sensitivity as the grain size decreases. NC metals have an increased sensitivity to loading changes with respect to their microcrystalline counterparts. The possibility to obtain the materials' response to strain rate changes is very useful in revealing many deformation mechanisms. The same material is more sensitive to the strain rate variation when the grain size is decreased, demonstrating the strong effect of GBAZ on such mechanical property. A small activation volume of dislocation mobility is responsible for the variation in strain rate sensitivity with decreasing mean grain size of metals. In addition, mechanisms of dislocation generation at grain boundaries coupled with grain rotation and migration are responsible for the whole plastic deformation in NC metals.

The mechanical behavior of materials is governed by crack nucleation and growth. This subject is very crucial for nanostructured materials. In this regard, cracks nucleate preferentially at grains' triple junctions.

One very important factor influencing the growth of the fatigue crack is the interaction between the slip mechanism driving the fatigue crack and potential microstructural barriers that are ahead of the crack tip. The barriers mostly discussed in literature are grain boundaries and twin boundaries. The recent scientific literature demonstrates how nanotwinned NS materials exhibit high damage tolerance.

References

Bobylev SV, Mukherjee AK, Ovidko IA (2009) Emission of partial dislocations from amorphous intergranular boundaries in deformed nanocrystalline ceramics. Scripta Mater 60:36–39. https://doi.org/10.1016/j.scriptamat.2008.08.025

Boyce BL, Padilla HA (2011) Anomalous fatigue behavior and fatigue-induced grain growth in nanocrystalline nickel alloys. Met Trans 42A:1793–1804. https://doi.org/10.1007/s11661-011-0708-x

Cao ZH, Sun W, Yang XB, Zhao JW, Ma YJ, Meng XK (2018) Intersectant coherent twin boundaries governed strong strain hardening behavior in nanocrystalline Cu. Int J Plast 103:81–94. https://doi.org/10.1016/j.ijplas.2018.01.002

Cavaliere P (2008) Strain rate sensitivity of ultra-fine and nanocrystalline metals and alloys. Phisica B 403:569–575. https://doi.org/10.1016/j.physb.2007.09.075

Chen M, Ma E, Hamker KJ, Sheng H, Wang Y, Cheng X (2003) Deformation twinning in nanocrystalline aluminium. Science 300:1275–1277. https://doi.org/10.1126/science.1083727

Cheng S, Spencer JA, Milligan WW (2003) Strength and tension-compression asymmetry in nanostructured and ultrafine-grain materials. Acta Mater 51(15):4505–4518. https://doi.org/10.1016/S1359-6454(03)00286-6

Cheng S, Zhao Y, Wang Y, Li Y, Wang XL, Liaw PK, Lavernia EJ (2010) Structure modulation driven by cyclic deformation in nanocrystalline NiFe. Phys Rev Lett 104:255501. https://doi.org/10.1103/PhysRevLett.104.255501

Chowdhury P, Sehitoglu H, Rateick R, Maier HJ (2013) Modeling fatigue crack growth resistance of nanocrystalline alloys. Acta Mater 61:2531–2547. https://doi.org/10.1016/j.actamat.2013.01.030

Chowdhury P, Sehitoglu H, Rateick R (2016) Recent advances in modeling fatigue cracks at microscale in the presence of high density coherent twin interfaces. Curr Opin Solid Sci Mater Sci 20:140–150. https://doi.org/10.1016/j.cossms.2016.02.003

Da Y, Lu Y, Ni Y (2018) Phase-field simulation of the coupled evolutions of grain and twin boundaries in nanotwinned polycrystals. Appl Math Mech Engl Ed 39:1789–1804. https://doi.org/10.1007/s10483-018-2393-7

Farkas D, Petegem S, Derlet PM, Van Swygenhoven H (2005) Dislocation activity and nano-void formation near crack tips in nanocrystalline Ni. Acta Mater 53:3115–3123. https://doi.org/10.1016/j.actamat.2005.02.012

Fedorov AA, Gutkin MY, Ovidko IA (2003) Transformations of grain boundary dislocations pile-ups in nano- and polycrystalline materials. Acta Mater 51(4):887–898. https://doi.org/10.1016/S1359-6454(02)00433-0

Fensin S, Spearot D, Bieler T et al (2019) Deformation and transitions at interfaces and grain boundaries. JOM 71:1198–1199. https://doi.org/10.1007/s11837-019-03390-2

Furnish TA, Mehta A, Van Campen D et al (2017) The onset and evolution of fatigue-induced abnormal grain growth in nanocrystalline Ni–Fe. J Mater Sci 52:46–59. https://doi.org/10.1007/s10853-016-0437-z

Greer JR, Jang D, Gu XW (2012) Exploring deformation mechanisms in nanostructured materials. JOM 64:1241–1252. https://doi.org/10.1007/s11837-012-0438-6

Gutkin MY, Ovidko IA, Skiba NV (2003) Crossover from grain boundary sliding to rotational deformation in nanocrystalline materials. Acta Mater 51(14):4059–4071. https://doi.org/10.1016/S1359-6454(03)00226-X

Handwerker CA, Pollock TM (2014) Emerging science and research opportunities for metals and metallic nanostructures. JOM 66:1321–1341. https://doi.org/10.1007/s11837-014-1014-z

Hay J, Maier V, Durst K, Göken M (2013) Strain-rate sensitivity (SRS) of nickel by instrumented indentation. In: Shaw G, Prorok B, Starman L (eds) MEMS and nanotechnology, Conference proceedings of the society for experimental mechanics series, vol 6. Springer, New York. https://doi.org/10.1007/978-1-4614-4436-7_8

Hu J, Shi YN, Sauvage X, Sha G, Lu K (2017) Grain boundary stability governs hardening and softening in extremely fine nanograined metals. Science 355:1292–1296. https://doi.org/10.1126/science.aal5166

Jang D (2012) Nano-sized nanocrystalline and nano-twinned metals. In: Bhushan B (ed) Encyclopedia of nanotechnology. Springer, Dordrecht. https://doi.org/10.1007/978-90-481-9751-4

Kobler A, Brandl C, Hahn H, Kubel C (2016) In situ observation of deformation processes in nanocrystalline face-centered cubic metals. J Nanotechnol 7:572–580. https://doi.org/10.3762/bjnano.7.50

Kumar KS, Suresh S, Chisholm MF, Horton JA, Wang P (2003) Deformation of electrodeposited nanocrystalline nickel. Acta Mater 51(2):387–405. https://doi.org/10.1016/S1359-6454(02)00421-4

Li P, Li S, Wang Z et al (2015) Basic criterion for the formation of self-organized dislocation patterns in fatigued FCC pure metals. Sci China Mater 58:921–928. https://doi.org/10.1007/s40843-015-0108-4

Liao XZ, Zhao YH, Zhu YT, Valiev RZ, Gunderov DV (2004) Grain-size effect on the deformation mechanisms of nanostructured copper processed by high-pressure torsion. J Appl Phys 96:636–640. https://doi.org/10.1063/1.1757035

Luo XM, Zhang B, Zhu XF, Zhou YT, Xiao TY, Zhang GP (2016) Local-structure-affected behavior during self-driven grain boundary migration. MRS Commun 6(2):85–91. https://doi.org/10.1557/mrc.2016.10

Ma E (2006) Eight routes to improve the tensile ductility of bulk nanostructured metals and alloys. JOM 58:49–53. https://doi.org/10.1007/s11837-006-0215-5

Maier-Kiener V, Durst K (2017) Advanced nanoindentation testing for studying strain-rate sensitivity and activation volume. JOM 69:2246–2255. https://doi.org/10.1007/s11837-017-2536-y

Morozov NF, Ovidko NF, Sheinerman AG, Aifantis EC (2010) Special rotational deformation as a toughening mechanism in nanocrystalline solids. J Mech Phys Solids 58:1088–1099. https://doi.org/10.1016/j.jmps.2010.04.003

Ovid'ko IA (2007) Review on the fracture processes in nanocrystalline materials. J Mater Sci 42:1694–1708. https://doi.org/10.1007/s10853-006-0968-9

Ovid'ko IA, Reizis AB (2001) Grain-boundary dislocation climb and diffusion in nanocrystalline solids. Phys Solid State 43:35–38. https://doi.org/10.1134/1.1340182

Ovidko IA, Sheinerman AG (2009) Enhanced ductility of nanomaterials through optimization of grain boundary sliding and diffusion process. Acta Mater 57(7):2217–2228. https://doi.org/10.1016/j.actamat.2009.01.030

Ovidko IA, Sheinerman AG (2012a) Nanoscale rotational deformation near crack tips in nanocrystalline solids. J Phys D Appl Phys 45:335301. https://doi.org/10.1088/0022-3727/45/33/335301

Ovidko IA, Sheinerman AG (2012b) Generation and growth of nanocracks near blunt cracks in nanocrystalline solids. Eur J Mech A Solids 33:39–47. https://doi.org/10.1016/j.euromechsol.2011.11.001

Panin VE, Egorushkin VE, Panin AV (2012) Role of local nanostructural states in plastic deformation and fracture of solids. Phys Mesomech 15:1–12. https://doi.org/10.1134/S1029959912010018

Paris PC, Gomez RE, Anderson WE (1961) A rational analytic theory of fatigue. Trend Engng 13 (1):9–14

Peykov D, Martin E, Chromik RR et al (2012) Evaluation of strain rate sensitivity by constant load nanoindentation. J Mater Sci 47:7189–7200. https://doi.org/10.1007/s10853-012-6665-y

Ramesh K (2009) Mechanical failure processes in nanomaterials. In: Nanomaterials. Springer, Boston. https://doi.org/10.1007/978-0-387-09783-1_6

Ramesh KT, Joshi SP, Schuster BE (2007) Instabilities in nanostructured materials. In: Gdoutos EE (ed) Experimental analysis of nano and engineering materials and structures. Springer, Dordrecht. https://doi.org/10.1007/978-1-4020-6239-1_268

Sangid MD, Pataky GJ, Sehitoglu H, Rateick RG, Niendorf T, Maier HJ (2011) Superior fatigue crack growth resistance, irreversibility, and fatigue crack growth–microstructure relationship of nanocrystalline alloys. Acta Mater 59:7340–7355. https://doi.org/10.1016/j.actamat.2011.07.058

Shan Z, Stach EA, Wiezorek JMK, Knapp JA, Follstedt DM, Mao SX (2004) Grain boundary-mediated plasticity in nanocrystalline nickel. Science 305:654–657. https://doi.org/10.1126/science.1098741

Shan Z, Knapp JA, Follstaedt DM, Stach EA, Wierzorek LMK, Mao SX (2008) Inter- and intra-agglomerate fracture in nanocrystalline nickel. Phys Rev Lett 100:105502. https://doi.org/10.1103/PhysRevLett.100.105502

Simokawa T, Nakatani A, Kitagawa H (2005) Grain-size dependence on the relationship between intergranular and intragranular deformation of nanocrystalline Al by molecular dynamics simulation. Phys Rev B71:224110. https://doi.org/10.1103/PhysRevB.71.224110

Singh A, Tang L, Dao M, Lu L, Suresh S (2011) Fracture toughness and fatigue crack growth characteristics of nanotwinned copper. Acta Mater 59:2437–2446. https://doi.org/10.1016/j.actamat.2010.12.043

Stegall DE, Elmustafa AA (2018) The contribution of dislocation density and velocity to the strain rate and size effect using transient indentation methods and activation volume analysis. Metal Mater Trans A 49:4649–4658. https://doi.org/10.1007/s11661-018-4817-7

Su H, Tang Q (2013) MD simulations of loading rate dependence of detwinning deformation in nanocrystalline Ni. Sci China Phys Mech Astron 56:491–497. https://doi.org/10.1007/s11433-013-5010-z

Tian Y (2018) Rise of correlated dislocations in nanotwinned metals against fatigue. Sci China Mater 61:127–128. https://doi.org/10.1007/s40843-017-9167-1

Yu M, Peng XH, Wen PH (2018) Effect of cooperative grain boundary sliding and migration on dislocation emission from interface collinear crack tip in nanocrystalline bi-materials. Acta Mech 229:3901–3913. https://doi.org/10.1007/s00707-018-2196-1

Zheng L, Peng X (2018) Temperature-dependent thermal and chemical stabilities as well as mechanical properties of electrodeposited nanocrystalline Ni. Met Mater Int 24:1293–1302. https://doi.org/10.1007/s12540-018-0146-z

Zhou P, Zhou J, Ye Z, Hong X, Huang H, Xu W (2016) Effect of grain size and misorientation angle on fatigue crack growth of nanocrystalline materials. Mater Sci Eng A663:1–7. https://doi.org/10.1016/j.msea.2016.03.105

Zhu T, Gao HJ (2012) Plastic deformation mechanism in nanotwinned metals: an insight from molecular dynamics and mechanistic modelling. Scr Mater 66:843–848. https://doi.org/10.1016/j.scriptamat.2012.01.031

Chapter 4
Fatigue and Crack Behavior of Nanostructured Metal Alloys and Composites

4.1 Introduction

Grain refinement leads to the strength increase of nanostructured materials in both UFG and NC regimes. In the previous chapters it was shown how grain boundary strengthening allows for the improved plastic deformation before failure in order to avoid catastrophic fracture (Koch 2003; Wang et al. 2002; Sabirov et al. 2008). The plastic deformation of crystalline material is governed by dislocation motion. The interactions between dislocation and barriers like precipitations and grain boundaries (GBs) are extremely important for the plasticity of polycrystalline materials. Among those barriers that interact with dislocations, GB is the most prevalent one. GBs are described as the interfaces between two adjacent crystals with different crystal orientations. The dislocation slip directions are forced to change at GBs and, therefore, a critical stress is required for dislocations to overcome GBs. GBs have a significant effect on the macroscopic mechanical behavior, especially when the volume fraction of GBs is relatively large, e.g., in ultrafine crystalline and nanocrystalline material. GBs exert influences on the mechanical properties of polycrystalline material from different dimensional scales (Ramesh 2009). At the atomic scale, the interfacial properties of GBs, like the interfacial energy and capacity of preventing dislocation, which are the atomic scale characteristics, can be determined by various factors like the misorientation, interfacial energy, external stress field, and electric field. The interfacial properties of GBs attribute to the strength of GBs, resulting in different dislocation-GB interactions, like dislocation accumulation, dislocation transmission, dislocation absorption, and emission. The improved fatigue crack impedance of the nanocrystalline materials originates from the interaction of the crack tip-emitted dislocations with the grain and twin boundaries under the cyclic fatigue loading (Tian and Li 2018).

As a result, the macroscopic mechanical properties can be significantly affected by the interfacial properties of GBs. For example, it has been proposed that GBs with

© Springer Nature Switzerland AG 2021

P. Cavaliere, *Fatigue and Fracture of Nanostructured Materials*,
https://doi.org/10.1007/978-3-030-58088-9_4

low interfacial energy are less susceptible to initiate cracks and damages and can contribute to higher flow stress and ductility (Arutyunyan and Arutyunyan 2018).

In general, it is well known that nanocrystalline materials have a much higher density of grain boundaries and excess free volume than conventional polycrystalline materials. Accordingly, the grain boundary migration and grain growth, which are known to occur normally at high temperature, can do at even room temperature in nanocrystalline materials during high-cycle fatigue. The reduction in average grain size to the nanometer range results in the suppression of traditional intragranular dislocation mechanisms. The higher stress levels which can be accessed and the increased grain boundary (GB) volume fraction in nanocrystalline materials exaggerate the importance of interfaces as plasticity mechanisms begin to depend heavily on the grain boundary network. As average grain size is reduced to the smallest possible nanocrystalline regime ($d < 15$–20 nm), plasticity becomes dominated by phenomena such as GB sliding, grain rotation, and GB migration. A common feature of these new mechanisms is that they all represent collective motion of groups of atoms and should alter the internal grain structure of the material. For example, rotation of two neighboring grains would alter the misorientation between the grains and the character of the GB which separates them. In some earlier experimental studies, it was reported that the amount of solute segregation increases with increasing GB misorientation angle in low-angle GBs (LAGBs). For high-angle GBs (HAGBs), a highly scattered segregation data was measured and no general trend was established in the case of special boundaries. Interestingly, in some special character boundaries, low segregation was detected while a maximum in GB segregation was measured for others. The dependence of solute segregation on GB character has been linked to GB plane orientation in few studies. It was found that GB planes with large Miller indices accommodate larger amounts of solute than coherent twin or symmetrical GBs of similar misorientations (Gupta et al. 2020). The GB character makes the fracture propagation behavior different from that in coarse-grained materials.

The origin of shear localization in nanocrystalline metals has been the subject of much research and is most often attributed to the multitude of grain boundary-mediated deformation mechanisms, such as grain rotation and sliding, operative in nanocrystalline metals (Chan et al. 2011). The plurality of deformation mechanisms has been shown to induce strain-softening behavior and facilitate long-range localization in materials with extremely small grain sizes (<15 nm) via strain partitioning (Mroz et al. 2014). Careful alloying strategies can help mitigate thermal instabilities in pure nanocrystalline metals, as well as provide additional strength. The most common approach is to introduce alloying elements that exhibit a strong propensity for segregation to grain boundaries. The presence of these highly segregating elements at grain boundaries can kinetically pin the boundaries or decrease the grain boundary energy. Classical formulations give grain boundary velocity as proportional to both the thermodynamic driving force (i.e., grain boundary energy) and a kinetic factor; thus both energetic and kinetic efforts to stabilize the grain boundary can effectively reduce the grain boundary velocity under thermal excitation. This alloying strategy has been used quite successfully for promoting thermal stability, enabling the design of nanocrystalline alloys that are resistant to coarsening up to high temperatures (Andrievski 2003, 2014). Solute segregation to grain boundaries has been shown to have a pronounced effect on mechanical properties as well. The presence of both metal and nonmetal dopants at grain boundaries promotes increased strength while mitigating some mechanical instabilities, such as stress-driven coarsening. Not only do alloying additions at grain boundaries affect the properties of nanocrystalline metals, but structural transitions of the grain boundaries due to chemical segregation have pronounced effects on the thermal and mechanical behavior of nanocrystalline metals as well. Nanocrystalline materials with a variety of grain boundary structures, ranging from atomistically sharp with various fractions of special grain boundaries to completely amorphous grain boundaries, have been produced. In particular, the effects of amorphous intergranular films on the mechanical response of nanocrystalline metals can be quite pronounced. Amorphous intergranular films provide mechanical stability by providing sinks for dislocations that propagate across the crystalline regions of the material, as well as strong pinning points as dislocations traverse the grain interior (Feng et al. 2019). In order to design alloys that can exhibit amorphous intergranular films, several key criteria have been proposed: high degree of chemical complexity, large atomic radius mismatch, negative enthalpy of segregation, and a positive enthalpy of mixing. All of these criteria, apart from the negative enthalpy of segregation, echo those proposed for designing fully amorphous alloys. In addition to the spatial extent of the grain boundary, the local atomistic structure—the position of solutes at the grain boundary, structural motifs of the boundary, and disconnection density—plays a large role in both the strength and propensity for shear localization of nanocrystalline metals. These subtle differences in grain boundary structure and energy are often referred to as grain boundary state. The grain boundary state of nanocrystalline metals may be relaxed through annealing at low temperatures, which provides increases in strength with negligible changes in grain size (Balbus et al. 2020).

Prior investigations of nanocrystalline plasticity have shown that strain accommodation mechanisms can occur cooperatively throughout the microstructure, leading to evolution of the interfacial network through large-scale sliding and rotation processes. For example, molecular dynamics simulations allowed to observe the emergence of shear planes in nanocrystalline Ni through the alignment of grain boundary planes via grain rotation, GB sliding, and GB migration, which then resulted in localized shear deformation as well as grain coalescence and growth (Farkas et al. 2005). These observations explain that substantial sliding on the mesoscale can only be accomplished by long-range interconnection of interfaces. In addition, it was showed that strain localization in nanocrystalline metals is highly dependent on the connectivity of the interfacial network. For the finest nanocrystalline grain sizes, the high strain path was formed through only grain rotation and GB sliding, with no significant evidence of dislocation motion. Once neighboring grains align to form a percolating GB network, catastrophic shear deformation occurs suddenly. In contrast to coarse-grained metals whose plastic deformation is governed by intragranular interactions between dislocations, nanocrystalline metals have exhibited structural evolution such as mechanically induced grain growth during loading. The importance of grain structure evolution should be exaggerated during low cycle fatigue loading, where multiple cycles of plastic strain are applied. It has been suggested that cyclic softening or strengthening processes during fatigue are highly dependent on the initial microstructure of the nanocrystalline metal. In order to clarify the details of characteristic features of fatigue behavior and fracture mechanism in nanocrystalline materials, the quantitative evaluation of initial grain boundary microstructure and its change resulting from cyclic stress-induced grain growth is indispensable.

Nanocrystalline Ni with minimal initial defect densities exhibited cyclic hardening during fatigue tests. It was also possible to identify a shift in plasticity from dislocation based to GB mediated during the initial fatigue cycles of nanocrystalline Ni. A strengthening effect during MD simulations of cyclic loading in nanocrystalline Ni is observed, which is ascribed to boundary relaxation and formation of low-energy boundaries (Yasbolaghi and Khoei 2020). Grain growth has also been found to occur during mechanical cycling, as shown by tensile load-unload fatigue. As a matter of fact, grain growth under cyclic stress in nanocrystalline copper specimens ultimately resulted in an increase of grain size by about 30% from the initial average grain size of ca. 20 nm.

4.1.1 Effect of Strain Rate

An interesting model is proposed to validate the effect of grain growth on the strain softening in nanocrystalline pure metals (He et al. 2019). The total strain rate can be decomposed into elastic and plastic parts:

$$\dot{\varepsilon} = \dot{\varepsilon}^e + \dot{\varepsilon}^p$$

The elastic strain rate comes from the stress rate in the linear elastic constitutive relation as

$$\dot{\varepsilon}^e = \boldsymbol{M} : \sigma$$

where $\boldsymbol{M}$ is the elastic compliance tensor. The plastic strain rate is proportional to the deviatoric stress based on the flow theory of plasticity:

$$\dot{\varepsilon}^p = \frac{3\dot{\varepsilon}^p}{2\sigma_e}\sigma'$$

where

$$\sigma'_{ij} = \sigma_{ij} - \frac{\sigma_{kk}\delta_{ij}}{3}$$

and

$$\sigma_e = \sqrt{\frac{3\sigma'_{ij}\sigma'_{ij}}{2}}$$

is the von Mises equivalent stress. The equivalent plastic strain rate is obtained by

$$\dot{\varepsilon}^p = \dot{\varepsilon}\left[\frac{\sigma_e}{\sigma_{\text{flow}}}\right]^m$$

where

$$\dot{\varepsilon} = \sqrt{\frac{2\dot{\varepsilon}'_{ij}\dot{\varepsilon}'_{ij}}{3}}$$

is the equivalent strain rate and

$$\dot{\varepsilon}'_{ij} = \dot{\varepsilon}_{ij} - \frac{\dot{\varepsilon}_{kk}\delta_{ij}}{3}$$

σ_{flow} and m are the flow stress and rate sensitivity exponent, respectively. In nanocrystalline metals, the primary deformation mechanism is boundary mediated instead of intragrain dislocation plasticity in coarse-grained materials. Increased grain boundaries block more lattice dislocations from moving with decreased grain size. The detail of flow stress in NC and MC metals can be found in

$$\sigma_{\text{flow}}^{\text{NC}} = \sigma_0 + M\alpha\mu b\sqrt{\rho_I + \rho_{\text{GB}}}$$

$$\sigma_{\text{flow}}^{\text{MC}} = \sigma_0 + M\alpha\mu b\sqrt{\rho_I} + \sigma_b$$

where α, μ, and M are the Taylor constant, shear modulus, and Taylor factor, respectively; σ_0 and σ_b are the lattice friction stress and back stress, respectively; ρ_I is the dislocation density in the crystal interior of the nanocrystalline; and b is the Burgers constant. ρ_I depends on the balance between accumulation and annihilation of dislocations by dynamic recovery; ρ_{GB} denotes the density of dislocation in the grain boundary dislocation pileup zone. Both ρ_I and ρ_{GB} are functions of the grain size, d_G:

$$\frac{\partial \rho_I}{\partial \varepsilon_p} = M\left(\frac{k}{d_G} + k_1\sqrt{\rho_I} - k_2\rho_I\right)$$

$$\rho_{\mathrm{GB}} = 6d_{\mathrm{GBDPZ}}\frac{\eta^{\mathrm{GB}}}{b\phi^{\mathrm{GB}}}d_G$$

where d_{GBDPZ} is the thickness of GBDPZ and Φ^{GB} is a geometrical factor. The average size of grains is assumed to vary in the following model:

$$d_G = \begin{cases} d_0, \ \varepsilon^p < \varepsilon_0^p \\ d_o + k\left(\varepsilon^p - \varepsilon_0^p\right), \ \varepsilon_0^p < \varepsilon^p < \varepsilon^f \\ d_f, \ \varepsilon^p = \varepsilon^f \end{cases}$$

where

$$k = \frac{d_f - d_0}{\varepsilon^f - \varepsilon_0^p}$$

ε^p is the plastic strain; ε_0^p and ε^f represent the plastic strain values where softening and fracture occur, respectively; d_0 and d_f correspond to the average grain size before and after compressive loading.

Aside from quantitative improvements in fatigue performance, NC metals also hold the possibility of new insight into the mechanisms responsible for traditional fatigue failure. In the recent past, studies on the monotonic strength of NC metals have not only revealed quantitative details regarding the Hall-Petch strengthening effect, but also led to the discovery of mechanistic transitions in dislocation behavior. For example, while individual dislocation slip is still active in NC metals with a grain size in the range of 20–50 nm, there is insufficient space for the collective dislocation interaction mechanisms found in CG metals such as pileups and subgrain formation. Instead, deformation is governed by dislocation nucleation and absorption at grain boundaries. In large-grained metals, persistent slip bands (PSBs) are responsible for the formation of surface extrusions and intrusions, which are the precursor for fatigue crack initiation. This process requires the collective activity of many dislocations within a grain, which as the literature indicates should be on the order of hundreds of nanometers in size. While other researchers have previously noted that crack initiation susceptibility typically decreases with decreasing grain size in metals, it was postulated that the persistent slip mechanism responsible for conventional fatigue crack initiation may be suppressed when the grain size is below a certain threshold, likely on the order of 100 nm or several hundred nanometers. This length scale threshold is between grain sizes that support collective dislocation activity and grain sizes that support individual dislocation activity. Perhaps the explanation for the lack of dramatic improvements in fatigue performance is related to the instability of NC grain structures, which are known to evolve even during storage at room temperature. Observations on room-temperature mechanically driven grain growth lead one to suspect that NC metals may evolve such coarse grain structures during fatigue loading, and that the fatigue mechanisms may be influenced more by the evolved grain structure than by the initial structure. As a matter of fact, modulated grain coarsening is discovered in compressive cyclic deformation of NC Ni-Fe alloy with grain size of 19 nm (Chan et al. 2014). The modulation spacing is about equal to the mean spacing of cracks. This coarsening behavior was mainly driven by crack growth. Where the cracks are absent, the grain coarsening is not observed. Studies suggest that the grain coarsening is generated through the lattice reorientation under the high stress concentration at the crack tip, and this process is assisted with the dislocation accumulation during the cyclic deformation. The dislocations lodged in grain boundaries play an important role, causing rotations across the grain boundaries and, in the larger grains, resulting in a secondary subgrain structure. It is also revealed that twins resulting from the electrodeposition are removed by the cyclic plasticity (Cheng et al. 2010). In addition, there is evidence of reversible fatigue behavior in a nanocrystalline Ni-Fe alloy both in the plastic zone and around the crack tip. In the plastic zone, the deformation is fully recoverable as the crack propagates, and the plastic deformation invokes reversible interactions of dislocation and twinning in the nanograins. But around the crack tip lies a regime with reversible grain lattice reorientation promoted by a change of local stress state. These observations suggest unprecedented fatigue deformation mechanisms in nanostructured systems that are not addressed theoretically (Cheng et al. 2013). The evidences suggest that the coarsened grains are formed through grain lattice reorientation. In some grains, the growth twins disappeared perhaps under the influence of high stress near crack tip, while deformation twinning can be traced in other grains. By contrast, the grain coarsening is not observed where the stress field diminished. In ductile materials, fatigue crack propagation is known to be accommodated by crack blunting and resharpening upon loading and unloading, while the blunting is accomplished through dislocation hardening by irreversible shear on their slip systems. In NC Ni-Fe specimens, although dislocations are active in the PZ, the crack blunting may hardly be fulfilled due to the lack of effective dislocation hardening in nanograins. Conversely, the opposite change of twinning vs. dislocations in the PZ hints that the transient dislocations could interact with twinning during loading cycles, perhaps caused by the emitting dislocations from GBs, but the reversible character may not provide the hardening. Around the crack tip, the reversible grain lattice reorientation represents another deformation mode that challenges the traditional concept of near-tip plasticity. A large grain containing domains with the original nanograins suggests that coarsening should occur through lattice reorientation of the nanograins under the stress field. Consequently, lattice misorientation between nanograins minimizes, forming a united grain. With the stress field set free, the building domains rotate back to their original orientation, reversing the coarsening. The fact that a coarsened grain can transform back to the original nanograins under the influence of stresses is unexpected and hints that the deformation is predominantly carried out at the boundaries of nanograins. This is in contrast not only to CG samples where the deformation is predominated by the dislocations leading to peak width broadening, but also to the irreversible grain growth through the mechanically driven GB migration reported in other NC samples.

Once a crack initially forms in NC metals, it propagates with very little intrinsic toughening mechanisms to offer resistance. Therefore, thorough examination of fatigue crack initiation mechanisms will be paramount in understanding fatigue resistance in NC metals and, arguably more importantly, in devising means to further delay crack initiation to develop metals with truly unparalleled fatigue performance (Furnish et al. 2017). In prior studies on fatigue crack initiation mechanisms in NC alloys, highly localized regions of abnormally large grains (ALGs) [e.g., exceptionally rare, large grains (approx. 1–3 μm) compared to the surrounding matrix (25–115 nm)] were consistently observed directly within crack initiation sites. Evidence of sub-grain formation and deformation-induced texture in the observed ALGs, and their proximity to the initiation sites, suggested that they had undergone some cyclic deformation and it was postulated that the ALGs were responsible for crack initiation. Recent work on notch sensitivity in NC metals, which utilized semicircular notches to initiate fatigue cracks at prescribed locations due to stress concentration at the notch, revealed ALGs directly below the notches in all notched specimens tested under high-cycle conditions (10,000 cycles), corroborating that they were a result of abnormal grain growth (AGG) during fatigue, and that the AGG likely preceded crack initiation. Thus, the AGG did not occur only from the stress gradients imposed by the notch, but required a certain amount of cyclic loading.

Considerable grain growth locally occurred only around fatigue cracks in electrodeposited nanocrystalline Ni-based alloys during high-cycle fatigue (Boyce and Padilla 2011). The observed grain growth in the nanocrystalline Ni-based alloys was due to a non-diffusional and shear stress-driven process. In prior studies on fatigue crack initiation mechanisms in NC alloys, highly localized regions of abnormally large grains (ALGs) [e.g., exceptionally rare, large grains (approx. 1–3 lm) compared to the surrounding matrix (25–115 nm)] were consistently observed directly within crack initiation sites. Evidence of sub-grain formation and deformation-induced texture in the observed ALGs, and their proximity to the initiation sites, suggested that they had undergone some cyclic deformation and it was postulated that the ALGs were responsible for crack initiation. Recent evidence on notch sensitivity in NC metals, which utilized semicircular notches to initiate fatigue cracks at prescribed locations due to stress concentration at the notch, revealed ALGs directly below the notches in all notched specimens tested under high-cycle conditions (10,000 cycles), corroborating that they were a result of abnormal grain growth (AGG) during fatigue, and that the AGG likely preceded crack initiation. It is important to note that in the notched cases, in specimens that failed within 5000 cycles, ALGs were not observed and the material failed by fracture only, without evidence of cyclic damage (similar to fracture during monotonic notched experiments). Thus, the AGG did not occur only from the stress gradients imposed by the notch, but required a certain amount of cyclic loading (Furnish et al. 2016).

An understanding of the deformation behavior of NC metals under the influence of multiaxial loading is important due to several reasons. First, combined stress states are routinely encountered in mechanical components, and effects of stress multiaxiality are fundamental to the understanding of not just ductile failure, but also non-brittle fracture and fatigue mechanisms. Second, several observations in the mechanics of NC metals are not consistent with typical assumption made in classical theories of metal plasticity. The most prominent observation is strength asymmetry under tensile vs. compressive loading (T/C asymmetry). Note that T/C asymmetry is a common feature of deformation in metallic glasses, and in that context, similarities between deformation mechanisms in nanocrystals and glassy phases are also worth exploring. Atomistic simulations have also revealed non-Schmid effects (influence of normal stresses) in dislocation nucleation from GB sources or homogeneous. Atomistic simulation results reveal difference in nucleation stress for dislocations. In NC materials, nucleation stresses under compressive loading are substantially higher by a factor of 3 than under tensile loading. Compression study towards Al nanopillars showed that anisotropy is related to activation of different slip systems. Several recent studies showed plastic anisotropy in Cu with horizontal CTBs in columnar grains under compression, which was rationalized by the hard and soft deformation modes dependent on dislocation activation under various loading directions; meanwhile others found that plastic anisotropy of lamellar TiAl alloys, revealed by micropillar compression, was attributed to the differences in critical resolved shear stress (CRSS) and hardening rate of slip and twinning systems in each constituent phase (Li et al. 2020).

Another unique aspect of NC metals is the nature of the elastic-plastic transition. In CG FCC metals with little or no resistance to depinning of "preexisting" dislocation cores a prominent yield point is indeed absent, and lattice dislocation glide dominates. In NC metals, however, there is both a frictional barrier (in the form of GB sliding resistance) and a dislocation nucleation/propagation barrier to inelastic deformation. Additionally, considerable strain recovery has been observed even after the onset of nonlinear stress-strain curves for NC samples. Interfaces such as grain boundaries (GBs) and/or twin boundaries (TBs) are planar defects present in polycrystalline materials such as CG and NC metals. From a mechanics standpoint, interface acts as a barrier to dislocations and can provide paths for the propagation of intergranular cracks. GBs are also associated with local enhancements in diffusivity, corrosion, and impurity segregation.

Specific deformation mechanisms/modes effectively operate in nanocrystalline and ultrafine-grained bulk materials, ultrathin films, and nanowires, strongly influencing the outstanding mechanical properties of these solids with external and/or internal nanoscale geometries. Lattice dislocation slip in nanocrystalline and ultrafine-grained bulk and thin-film materials shows dramatic behavioral deviations from its conventional counterpart in coarse-grained polycrystals. Also, specific GB deformation modes highly contribute to plastic flow in nanocrystalline materials with finest grains in wide temperature intervals and carry superplasticity in nanomaterials at elevated temperatures. Nanoscale twin deformation effectively operates in nanomaterials with various chemical compositions and structures.

4.1.2 Dislocation-GB Interaction

In contrast to coarse-grained polycrystals where deformation twins are typically generated within grain interiors, in nanomaterials under mechanical load, twins are often generated at GBs. Nanoscale deformation twins are generated at locally distorted GBs, that is, GBs containing local fragments being rich in GB dislocations produced by preceding deformation processes. For instance, such local GB fragments can be formed due to splitting of extrinsic lattice dislocations trapped at GBs (Fig. 4.1).

The splitting of extrinsic dislocations at high-angle GBs is a well-experimentally documented process resulting at its initial stage in the formation of several closely located GB dislocations. These dislocations and preexistent GB dislocations can form a nano-sized wall of GB dislocations located on every slip plane. In a rather typical situation, the extrinsic dislocation that undergoes the splitting transformation at a GB represents a head dislocation of a pileup configuration stopped by the GB. In this case, after the splitting of the head dislocation of the pileup, its second dislocation can reach the GB where this extrinsic dislocation splits into new GB dislocations. Both preexistent GB dislocations and GB dislocations resulted from the splitting transformations of the extrinsic dislocations are capable of forming a nano-sized wall of GB dislocations located on every (or almost every) slip plane. Their transformations followed by emission of partial dislocations into adjacent grain produce nanoscale deformation twins. Another scenario for formation of local deformation-distorted fragments of GBs is related to GB deformation processes (Fig. 4.2).

First, a nanostructured specimen is deformed by GB sliding that produces pileups of GB dislocations stopped by triple junctions of GBs. These dislocations under the applied stress climb along GBs adjacent to triple junctions (Fig. 4.2b–f). Since the rate of GB dislocation slip is much larger than that of diffusion-controlled climb of GB dislocations, the combined slip and climb of GB dislocations typically result in dense, wall-like configurations of climbing GB dislocations that can exist on every (or almost every) slip plane.

The generation of nanotwins at locally deformation-distorted GBs acts through three mechanisms: (1) the cooperative emission of partial dislocations from GBs; (2) the successive events of partial dislocation emission from GBs; and (3) the multiplane nanoscale shear generated at GBs. The former two mechanisms are realized through splitting of the GB dislocations located at local deformation-distorted GB fragments into immobile GB dislocations and mobile partial dislocations. The mobile dislocations move either successively (Fig. 4.3) or cooperatively (Fig. 4.4) on every slip plane within a nanoscale region where a nanoscale twin is thereby generated.

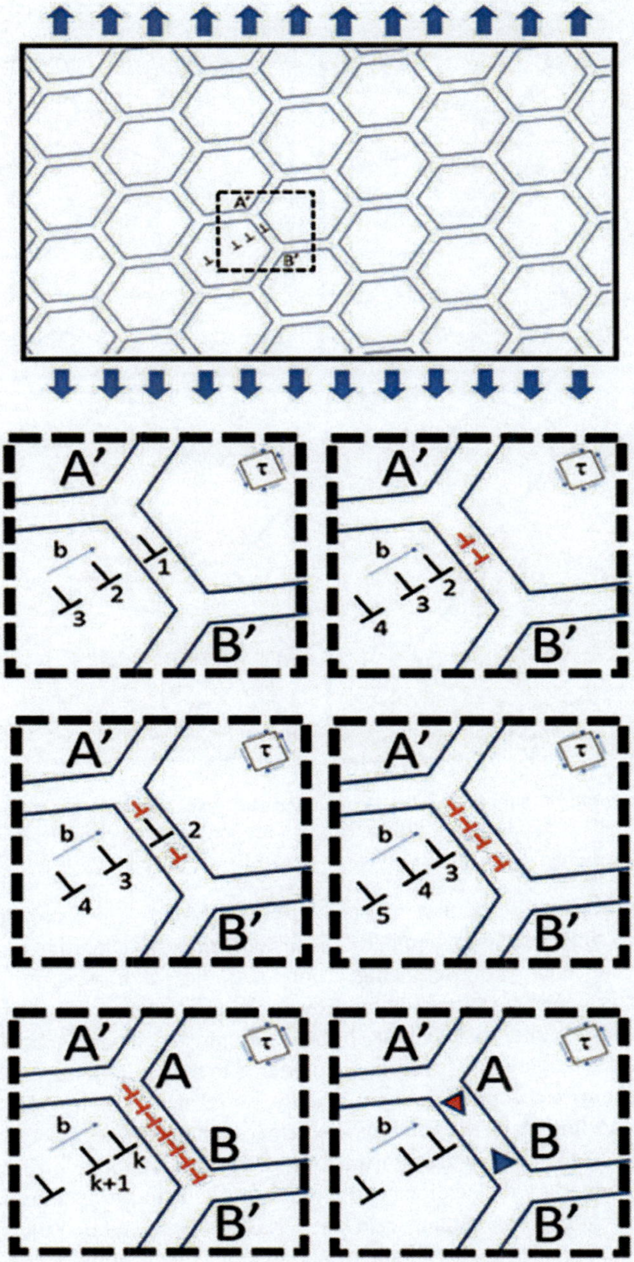

Fig. 4.1 Formation of locally distorted grain boundary due to transformations of lattice dislocation pileup

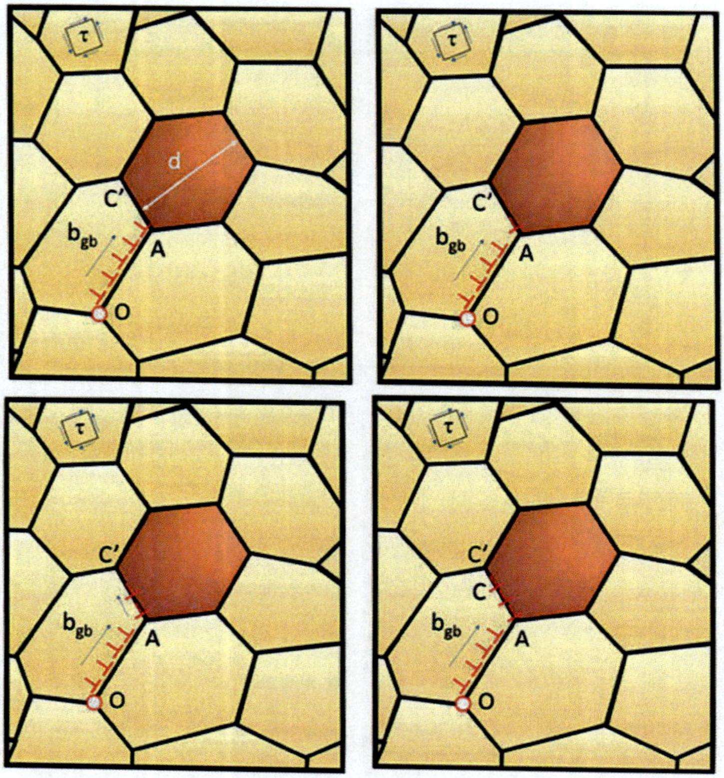

Fig. 4.2 Formation of deformation-distorted fragment of grain boundary due to grain boundary sliding and climb of grain boundary dislocations in a nanocrystalline specimen

The splitting transformation of each GB dislocation at a local deformation-distorted GB fragment into another GB dislocation and a partial dislocation is specified by an energy barrier being around the energy of a partial dislocation. This barrier is lower than that (being around the energy of a full dislocation) for multiplication of partial dislocations. In these circumstances, the splitting transformation is more energetically favored, as compared to the multiplication reaction. For the first mechanism described in Fig. 4.3, in the initial state, a nano-sized wall AC of climbing grain boundary dislocations (red open dislocation signs) exists at grain boundary AC′. Successive transformations of grain boundary dislocations into sessile grain boundary dislocations and mobile partial dislocations move in adjacent grain interior and form a nanoscale twin ACEF. Each such transformation is described as the formation of a dipole of partial lattice dislocations (blue dislocation signs). The nanoscale twin has a wedge-like profile. For the mechanism described in Fig. 4.4, in the initial state, the nano-sized wall AC of climbing grain boundary dislocations (red open dislocation signs) exists at grain boundary AC. The grain boundary dislocations cooperatively transform into sessile grain boundary

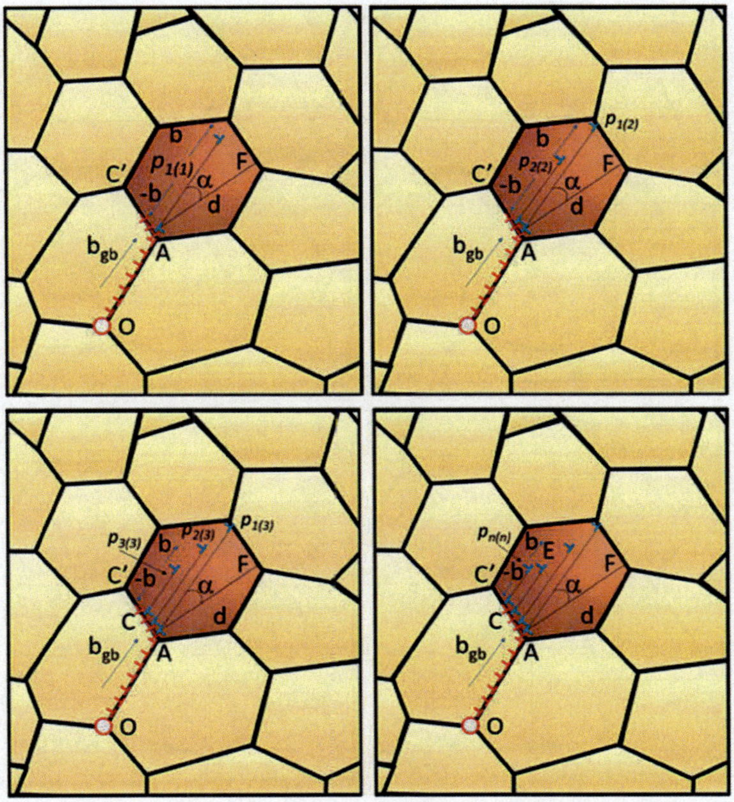

Fig. 4.3 Nanotwin generation through successive processes of dislocation emission from locally distorted grain boundaries in a nanocrystalline specimen at comparatively low external stresses

dislocations and mobile partial dislocations that move in adjacent grain interior. As a result, a nanoscale twin ACEF is formed which grows in parallel with glide of the mobile partial dislocations. Each transformation of a grain boundary dislocation is described as the formation of a dipole of partial lattice dislocations (blue dislocation signs). In doing so, one (immobile) partial dislocation belonging to a dipole configuration is located at the grain boundary AC, whereas another (mobile) partial dislocation of the dipole configuration glides in the adjacent grain interior. The third mechanism for nanotwin formation at a locally distorted GB is multiplane nanoscale shear (Fig. 4.5).

In the initial state, a nano-sized wall AC of climbing grain boundary dislocations (red open dislocation signs) exists at grain boundary AC′. Nanoscale multiplane shear occurs which is characterized by a tiny shear magnitude s. Two walls of non-crystallographic dislocations (blue dislocation signs) with tiny Burgers vectors ±s are formed at grain boundary fragments AC and EF. The nanoscale shear and Burgers vector magnitude s gradually increase. The non-crystallographic

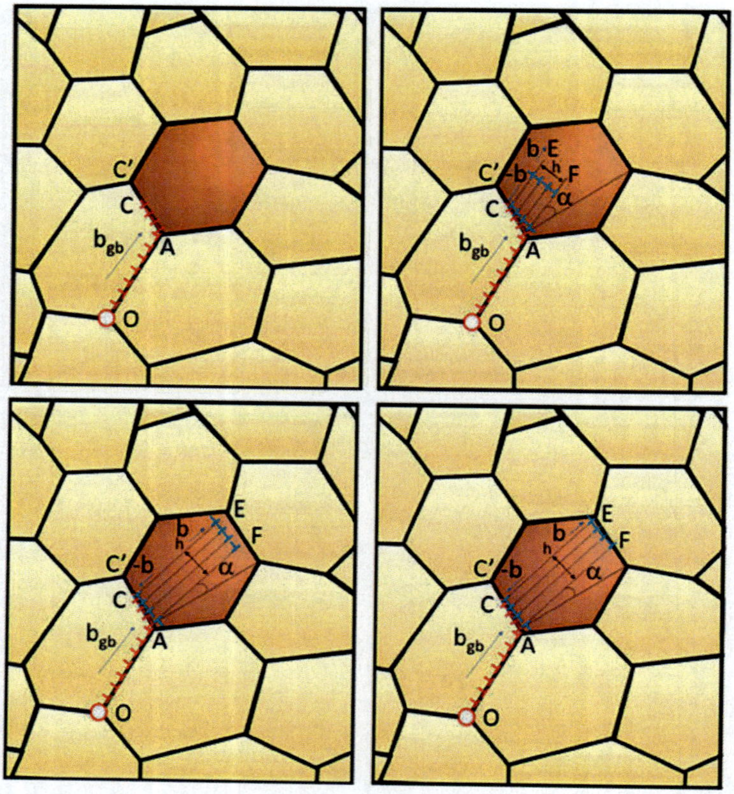

Fig. 4.4 Nanotwin generation through cooperative dislocation emission from a locally distorted grain boundary in a nanocrystalline specimen in the situation where preexistent grain boundary dislocations in the deformation-distorted fragment of the grain boundary are located at every crystallographic plane

dislocations transform into twinning partial dislocations, and the nanotwin is formed. In general, the deformation twinning mechanisms are in competition (Zhou and Zhu 2020). Typically, the most favorable mechanism is that occurring through the cooperative emission of partial dislocations from GBs containing local deformation-distorted fragments (Fig. 4.4), because it is characterized by the lowest critical stress in a wide range of the twin thickness.

In general, nanostructured materials show increased fatigue limit if compared to their microcrystalline counterparts. In fact, nanocrystalline materials demonstrated almost 2–3 times higher fatigue limit than that for ultrafine grain and microcrystalline materials. Anyway, also fracture toughness and fatigue crack growth must be assessed in order to design components for crucial applications. The fatigue ratio in nanocrystalline materials was found to be limited to a much smaller degree than that in conventional microcrystalline materials (Kobayashi et al. 2015). The effects of grain boundary microstructure on fatigue and fracture in nanocrystalline materials

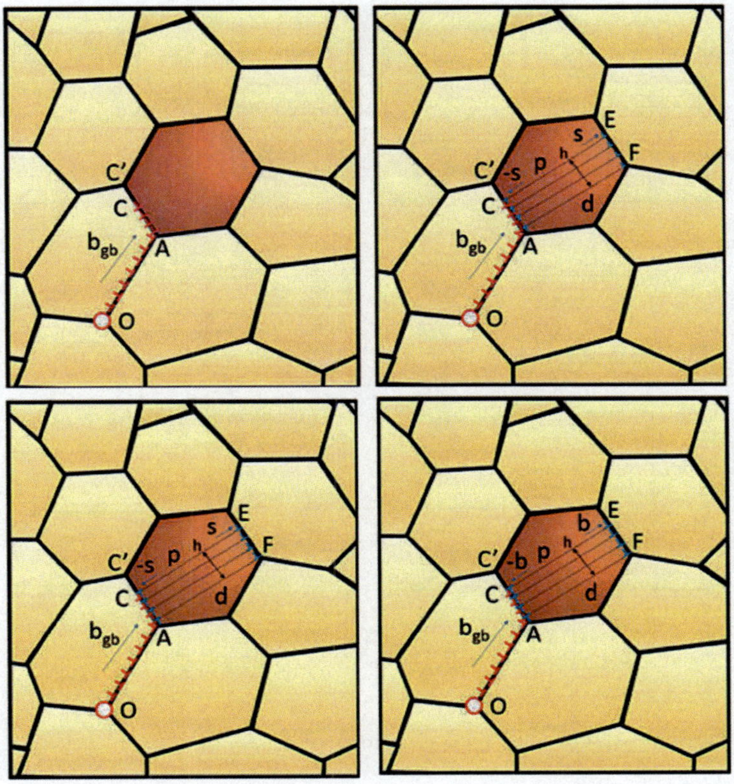

Fig. 4.5 Deformation nanotwinning through nanoscale multiplane shear generated at a locally distorted grain boundary in a nanocrystalline solid

may be more significant than those in conventional microcrystalline materials, because the density of grain boundaries becomes remarkably high in nanocrystalline materials. To date, discussion of the fatigue property of nanocrystalline materials has mostly focused only on the effect of grain refinement. The roles of other microstructural factors, such as texture and grain boundary microstructure, must be taken into account (Kobayashi et al. 2009).

It has been demonstrated that fatigue crack initiation and growth follow very different behavior in UFG and NC regimes (Cavaliere 2009a; Cavaliere and Cabibbo 2008). In general, ductility leads to high plastic deformation before failure leading to a reduction of stress concentrations at the crack tip and at the defect interface by modifying the extension of the plastic and elastic zones (Fig. 4.6); all this allows for the avoiding of catastrophic failure.

The general fracture behavior that can be observed in NC materials is schematically shown in Fig. 4.7.

Materials with low ductile indicator crack as described in the first three pictures, as increasing strain, experience brittle cleavage fracture. The crack moves forward

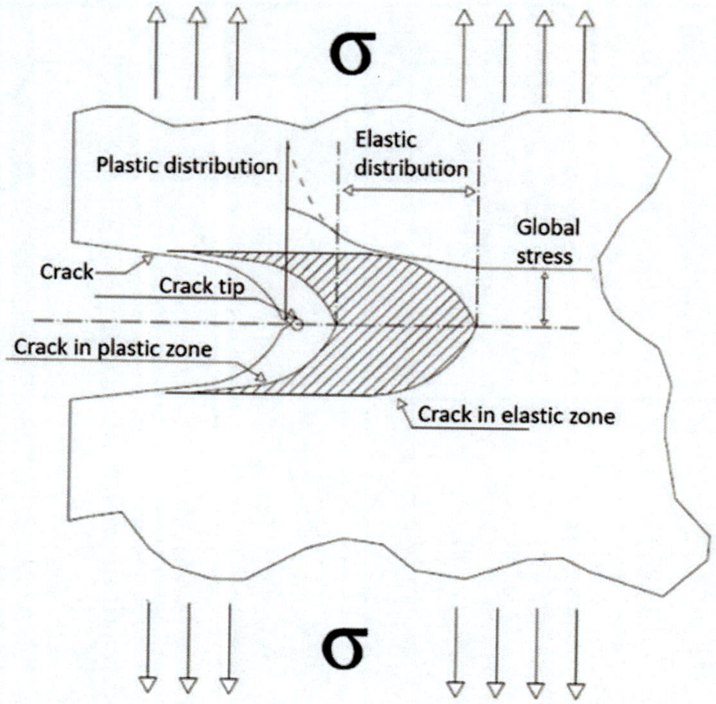

Fig. 4.6 Elastic and plastic zones at the crack tip

with flat surfaces. On the contrary, materials with high ductile indicator own low dislocation energy barrier; dislocation/twinning activities occupy a significant role in the whole crack propagation process, leading to the final material failure by surface necking. The ductility indicator is calculated as $2r_{sur}/r_{usf}$, where r_{usf} is the unstable stacking fault energy and r_{sur} is given by

$$r_{sur\,(ijk)} = \frac{E_{end} - E_{initial}}{2A}$$

which is defined as the change in free energy when a solid is separated into two parts, a large distance along the normal direction of a crystallographic plane. A is the section area perpendicular to the $r_{sur(ijk)}$ plane. $E_{initial}$ is the initial total energy and E_{end} is the amount of energy after displacement along normal direction of the (ijk) plane. MD simulations performed on NC Ni showed three stages of crack propagation: atomic cleavage, void nucleation, and dislocation nucleation (Fig. 4.8).

There were highly concentrated tensile stress and stress triaxiality factors around the crack tip just before the atomic bonds began to break. As the tensile deformation continued the atomic configuration ahead the crack tip showed a crystalline-amorphous state (local structure transformation).

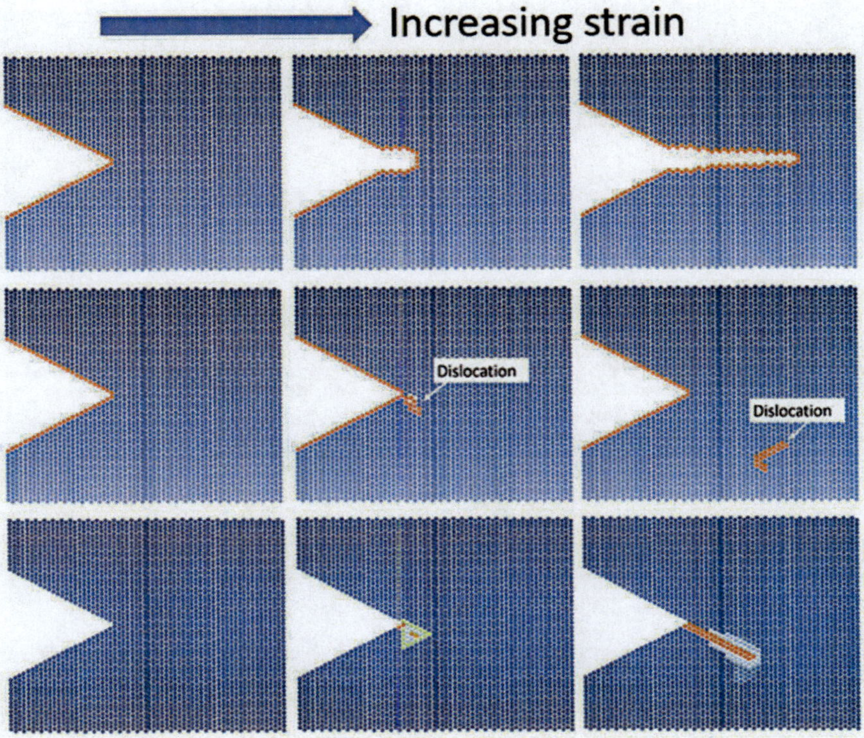

Fig. 4.7 MD simulations of different cracking mechanisms in NC materials

The voids were allowed to nucleate and join the main crack. The surface of the crack was corrugated and inhomogeneous dislocations were further nucleated and emitted from the void. Voids were generated ahead the crack tip, with large tensile stress and stress triaxiality distributions at its local region. The difference was that the concentrated stress was not constrained at a small region in front of the crack tip as occurred at the completed atomic cleavage stage. This means that the voids can still grow under the large local stresses. There was no obvious elastic wave direction due to the disordered atomic debonding around the voids. There were no highly concentrated stress regions ahead the void because the dislocation was being nucleated from the voids. This was experimentally demonstrated for NC Ni-Fe where fatigue fracture was caused by linking of microcracks along grain boundaries owing to joining nanovoids (Yang et al. 2008).

The transition from an intergranular to a transgranular behavior is dependent on the grain boundary. In particular, intergranular behavior is mainly due to the high dislocation density, coupled with the presence of nonequilibrium grain boundaries trapping and accommodating lattice dislocations (Fig. 4.9).

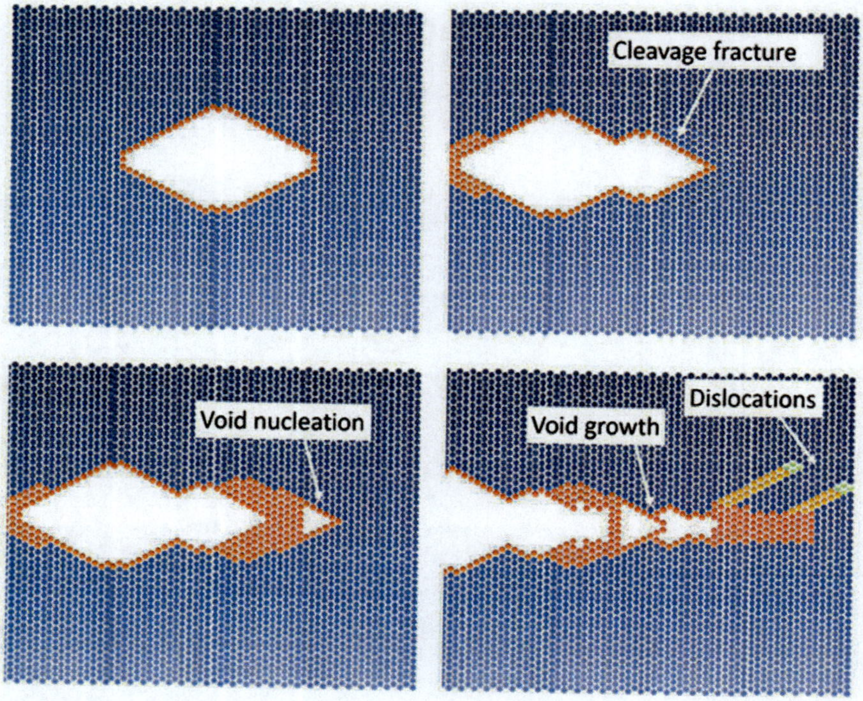

Fig. 4.8 MD simulations of cracking mechanisms in NC Ni

So, it is fundamental for cyclic loading resistance that nanostructured materials exhibit appropriate ductility and grain boundary strength (Hohenwarter and Pippan 2011, 2013, 2015).

As a matter of fact, Fig. 4.10 shows the yield stress vs. ductility of pure Ni with different grain sizes produced by various processing routes (Pineau et al. 2016).

Generally, tensile strength increases with decreasing grain size while ductility follows an inverse behavior; an example for ultrafine-grained and nanocrystalline pure aluminum is shown in Fig. 4.11.

The relationship between UTS and fatigue limit for different nanocrystalline metals and alloys is shown in Fig. 4.12.

While the relationship between ultimate tensile strength and fatigue strength has been strongly emphasized and even touted as a fundamental scaling law, there is a mechanistic disconnect between these properties. By studying the fatigue and tensile behavior in a model Ni-Fe system with grain sizes varying an order of magnitude, studies revealed that tensile strength substantially underestimates the fatigue properties in grain boundary-embrittled nanocrystalline metals; this illustrates the major flaw in these accepted scaling laws. Improved prediction of fatigue properties can be made in this system, however, using the work hardening behavior of the material. For materials that undergo necking in tensile loading, work hardening dictates both

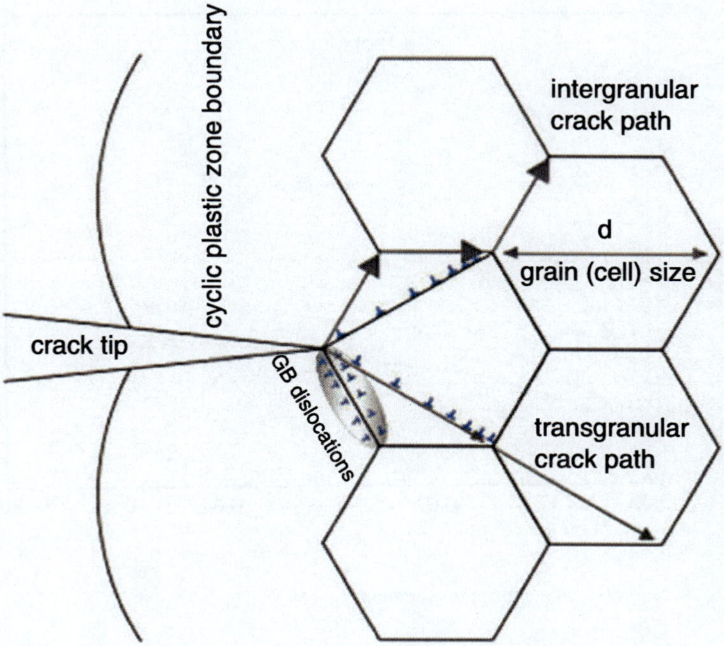

Fig. 4.9 Dislocations slip bands stopped at higher resistance grain boundaries

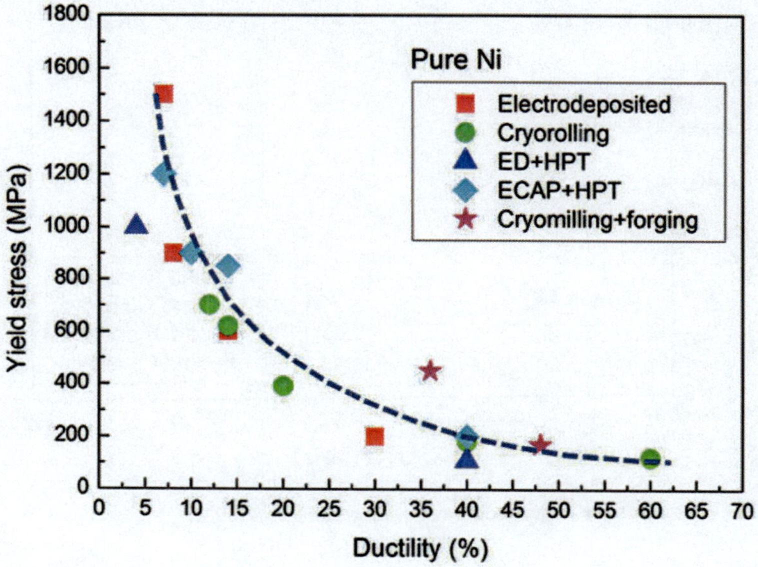

Fig. 4.10 Yield stress vs. ductility for pure Ni

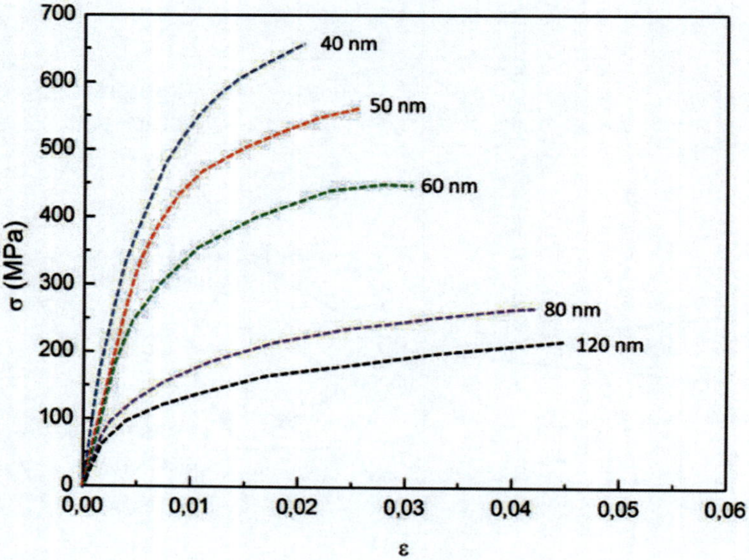

Fig. 4.11 Stress-strain curves for pure aluminum

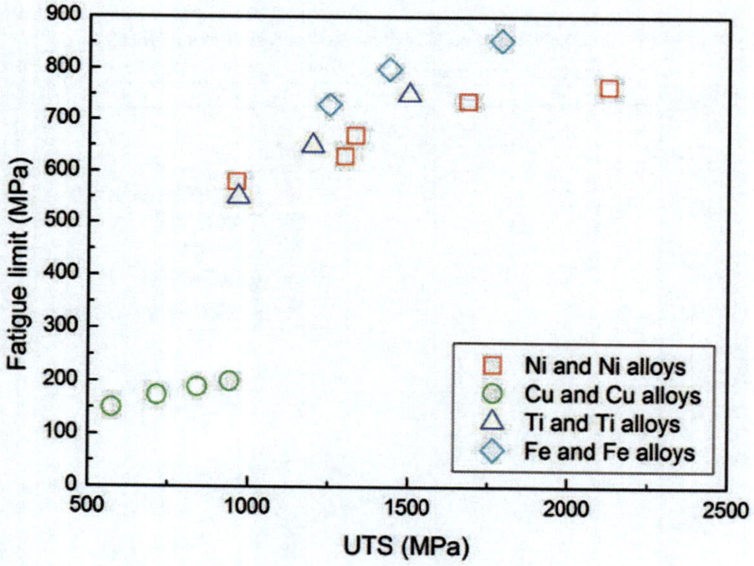

Fig. 4.12 UTS vs. fatigue limit for different nanocrystalline materials

the characteristic flow stress and tensile strength, providing an explanation of why tensile strength often provides reasonable estimates of the fatigue properties in metals. Another key insight drawn from this study is that for the fatigue of nanocrystalline metals, there is an apparent inverse Hall-Petch-type reduction in endurance limit at some grain size (Heckman et al. 2020).

4.2 Fatigue Life of NC Materials

The deep application of NS metals and alloys in the modern industry is related to the increase of the understanding of their damage resistance and of the mechanical mechanisms involved in the deformation, especially under cyclic loading. It has been well established that for conventional metals, the vast majority of the total fatigue life under high-cycle conditions is spent in crack initiation, rather than crack propagation. For coarse-grained metals, initiation is typically attributed to organization of dislocations into localized regions of persistent slip bands (PSBs) within individual grains with cracks initiating at interfaces between intrusions and extrusions caused by the PSBs. In fine-grained materials with grain sizes less than a few micrometers, limited space in the grain interiors inhibits widespread dislocation organization into well-defined PSBs, and new crack initiation processes emerge. For example, in fatigued ultrafine-grained metals with approximately 0.1–1 μm grain size, dislocation substructures are rarely observed and fatigue crack initiation is associated with macroscopic shear bands that span distances much greater than the grain size. Nanocrystalline (NC) metals with grain sizes less than 100 nm present an even more extreme case in which dislocation activity and accumulation within grain interiors are exceedingly low. In the context of fatigue, the severely limited dislocation activity and the deviation from traditional PSB-related mechanisms in NC grains have been suspected to suppress crack initiation under high-cycle conditions. Coupled with larger elastic stresses that can be accommodated due to the higher yield strengths (i.e., cycling in a broader elastic regime with limited or altogether absent plasticity), NC metals possess a potential for superior high-cycle fatigue resistance (Furnish et al. 2017).

4.2.1 Fatigue Endurance in NC Metals and Alloys

As a general behavior, it was observed that the fatigue limit of nanocrystalline metals increases by decreasing the grain size, and the crack initiation susceptibility decreases by increasing the crack growth rate coupled with the grain refinement. The increase of fatigue resistance can be achieved for many materials in stress-controlled tests. Generally, the main damage mechanism has been recognized in the early strain localization and microcrack formation for SPDed materials. In the HCF regime of intermediate to low plastic strain amplitudes, it results in a strong enhancement of the fatigue resistance for materials with grain refinement (Vinogradov and Hashimoto 2003).

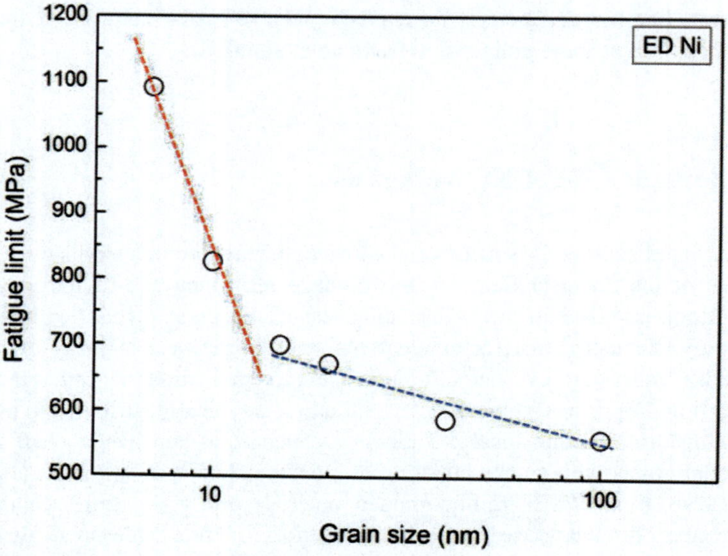

Fig. 4.13 Fatigue limit vs. grain size for pure ED Ni

The variation of fatigue limit with grain size of electrodeposited nickel is shown in Fig. 4.13.

Nanocrystalline Ni has a slightly higher, but reproducible, increase in the tensile stress range needed to achieve a fixed life and in the endurance limit (defined at two million fatigue cycles) compared to the UFG Ni. Both NC and UFG conditions have a significantly higher fatigue endurance limit than the MC metal (Cavaliere 2009a). Even if a grain refinement leads to an increase in the number of cycles to failure at the same stress levels investigated, the data for very close microstructures (20 nm and 40 nm) resulted in a strong function of the ductility; for very high stress levels the material behavior is almost coincident in the nanocrystalline regime. In general, a material that cyclically hardens under a constant range of imposed cyclic loads will exhibit a decreasing strain amplitude with increasing numbers of cycles. Conversely, a material that cyclically softens will show a continual rise in strain amplitude, with increasing fatigue cycles. The material progressively hardens and the hysteresis loop area progressively decreases with increasing cycle number. During plastic deformation dislocations may interact with each other and form locks. This process is irreversible and is seen as the most important contribution to cyclic hardening in conventional metals with grain sizes in the micrometer range and above. Considering the lack of dislocation debris in fatigued NC Ni specimens, this mechanism cannot be seen as the origin of the observed cyclic hardening behavior and an alternate explanation is sought. A variety of nonequilibrium dislocation sources, which can be activated at various stress levels, likely exist at the grain boundaries of as-deposited materials. Such sources are assumed responsible for the reduced hardness of these materials, relative to those with a comparable grain size subjected to an intermediate anneal. Dislocation source exhaustion at a particular stress level would then require an increment in applied stress to activate another set of sources, further sustaining plastic deformation. Assuming, therefore, that dislocation source exhaustion is prevalent, such a phenomenon could account for the observed hardening behavior. In the initial cycles, deformation is accommodated primarily by dislocation motion, while concurrent dislocation source exhaustion accelerates material hardening. In the later stages plastic deformation may rely more heavily on diffusion-based mechanisms such as grain boundary sliding, creep, and grain rotation, reducing the rate of material hardening. Additionally, grain boundary sliding is unlikely to be fully reversible, which may also contribute to the observed hardening behavior. It was largely demonstrated that cyclic hardening occurs over a broad range of loading frequencies in the NC Ni, as the strain amplitude continually decreases throughout the duration of each experiment. In terms of the hardening behavior, there are several potential mechanisms that could act alone, or in concert, to produce the observed effect. They include enhanced dislocation interaction with an increasing number of stacking faults (produced during cycling), environmental effects, and/or strengthening due to a defect source exhaustion mechanism. Interaction between the faulted material and mobile dislocations could result in the observed hardening behavior. Clearly, dislocations can be reabsorbed by the grain boundaries subsequent to fatigue failure; however some level of residual debris would be expected if this mechanism were active. While the interaction between stacking faults and mobile dislocations could partially contribute to the overall hardening behavior, the lack of evidence suggests that it is not the dominant hardening mechanism (Dai et al. 2016). Turning to the fatigue life, it is convenient to consider the total strain range consisting of two components:

$$\Delta\varepsilon_t = \Delta\varepsilon_{\mathrm{el}} + \Delta\varepsilon_{\mathrm{pl}}$$

As a sum of plastic and elastic contributions. The Coffin–Manson relationship relates the total fatigue life (number of cycles to failure N_f) to the plastic strain amplitude as

$$\frac{\Delta\varepsilon_{\mathrm{pl}}}{2} = \varepsilon'_f\left(2N_f\right)^c$$

where ε'_f is the fatigue ductility coefficient (which is often found to be approximately equal to the true fracture ductility in monotonic testing) and c is the fatigue ductility exponent. The elastic strain amplitude is given as

$$\frac{\Delta\varepsilon_{\mathrm{el}}}{2} = \frac{\Delta\sigma}{E} = \frac{\Delta\sigma_a}{E}$$

where E is the Young's modulus. Thus, using the Basquin expression linking the stress amplitude to the total number of reversals to failure $2N_f$,

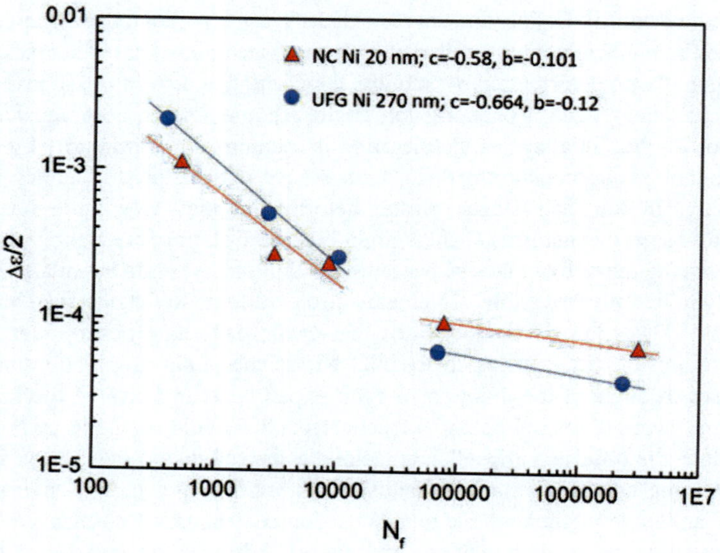

Fig. 4.14 Total strain range as a function of cycles to failure for the electrodeposited NC and UFG Ni

$$\frac{\Delta\sigma}{2} = \sigma_a = \sigma'_f (2N_f)^b$$

where σ'_f is the fatigue strength and b is the Basquin exponent, one obtains

$$\frac{\Delta\varepsilon_{el}}{2} = \sigma'_f (2N_f)$$

The fatigue life under a given total strain is expressed in terms of material constants ε'_f, σ'_f, c, and b:

$$\frac{\Delta\varepsilon_t}{2} = \sigma'_f (2N_f)^b + \varepsilon'_f (2N_f)^c$$

The Coffin–Manson plot for the electrodeposited Ni in NC and UFG conditions is shown in Fig. 4.14.

Kobayashi et al. (2015) quantified the grain growth during fatigue loading of NC Ni-P with starting grain size of 45 nm. The average grain size as a function of the distance from the fracture surface is shown in Fig. 4.15.

The most important finding was that the observed values of the average grain size continuously decrease with increasing distance from the fracture surface, and then ultimately approach to almost constant value of 220 nm at the area 1.3 mm away from the fracture surface. The average grain size in the unfractured specimen was about 220 nm, even after subjected to cyclic stress N > 107, as indicated by the

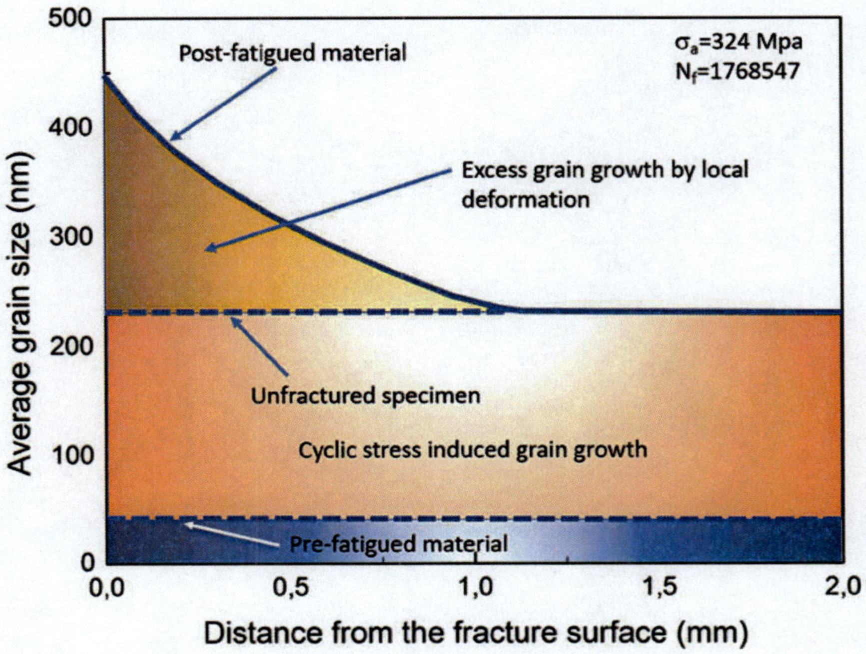

Fig. 4.15 Grain growth as a function of the distance from the fracture surface after fatigue loading of NC Ni-P

broken line. The experimental results may suggest that the grain growth resulting in the average grain size level of 220 nm was caused by the application of cyclic stress. Furthermore, the observed excess and abnormal grain growth around the fracture surface was probably caused by localized plastic deformation which can provide some additional driving force for grain growth. Accordingly, it is very likely that the cyclic stress-induced grain growth in electrodeposited nanocrystalline Ni-P alloy thin sheets may be affected by the level of applied cyclic stress. Also the misorientation angles of grain boundary were changed by fatigue loading demonstrating the rotation-assisted grain growth in NC materials during cyclic loading.

The model of possible mechanisms of cyclic stress-induced grain growth was provided, and of the evolution of "diamond-shaped" grain structure, finally leading to intergranular fatigue fracture caused by a mechanism of shear stress-mediated migration of low-angle boundaries, later coupled by sliding of higher energy random boundaries (Fig. 4.16).

The starting microstructure contained some coarse grains with mean dimensions larger than 100 nm. The microstructure was characterized by a high fraction of low-angle grain boundaries. During the first fatigue cycles, shear bands form at an initial inclination angle of 45° with respect to the loading axis. Since nanograin boundaries can be potential sites for dislocation nucleation the higher density of dislocations can be introduced at the position nearer to shear bands in the nanograin-

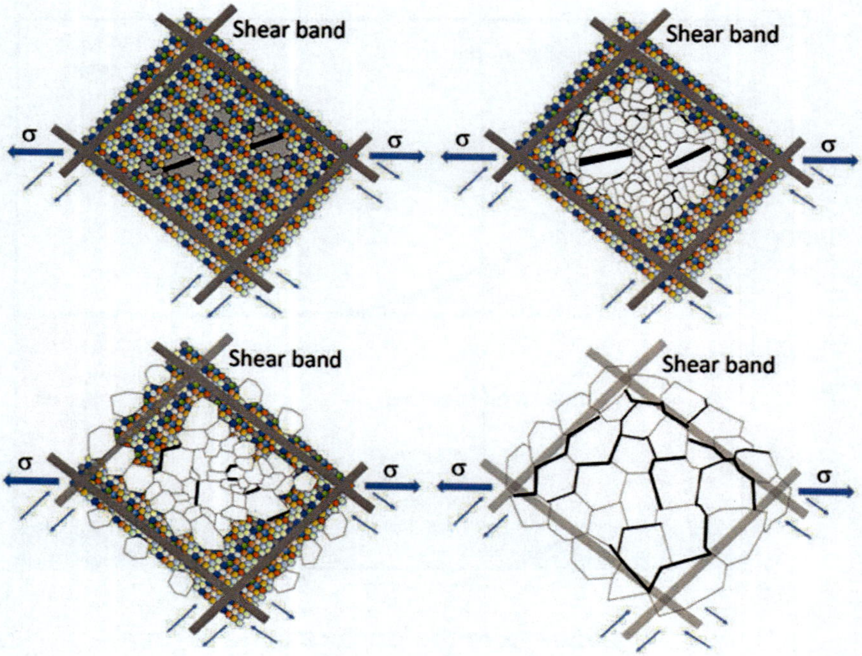

Fig. 4.16 Cyclic stress-induced grain growth and evolution of grain boundary microstructure during high-cycle fatigue

cluster under shear stress. The gradient stored energy-driven migration may dominantly take place towards shear band, resulting in the formation of diamond-shaped grain structure composed of high-energy/random boundaries, i.e., the type of grain boundaries which can slide easily under shear stress. The grain growth caused by the gradient stored energy-driven migration of low-angle boundaries under shear stress takes place in the nanograin cluster. In the interior of nanograin cluster, the misorientation angles of low-angle boundaries may gradually increase then rapidly, due to collective migration of low-angle boundaries by absorbing more dislocations. Some grains seem to grow into the other grains accompanying with shear-coupled migration of low-angle tilt boundaries in the center of the nanograin cluster area surrounded by shear bands. The specific energy of grain boundaries surrounding growing grains tends to increase and become more mobile, finally resulting in the transformation into high-angle–high-energy/random boundaries. Finally, grains grow to submicron size. The high-angle/random boundaries surrounding these oriented grain clusters are laid along the shear bands because of the stored energy gradient, and then the fine-grained structure composed of "diamond-shaped grain cluster" is developed under cyclic deformation. In the case of nanocrystalline Ni-P alloy, the dynamic grain growth occurred during cyclic fatigue deformation even at room temperature, resulting in the changes in grain boundary microstructure and the evolution of the diamond-shaped grain structure, accompanying with the

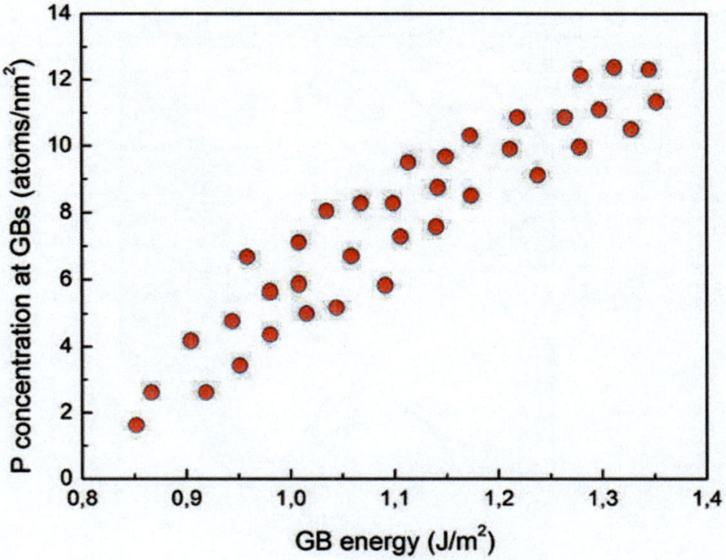

Fig. 4.17 GB energy as a function of P concentration in electrodeposited NC Ni-P alloy

transformation of low-angle boundaries into high-angle random boundaries, and their final positioning along shear bands. Moreover, the observed grain growth may be attributed to a rapid grain boundary migration enhanced by highly enriched P-segregation at high-energy random boundaries. As a matter of fact, the GB energy variation as a function of the P content in Ni-P-electrodeposited alloys is shown in Fig. 4.17.

Grain boundary migration might be enhanced by the presence of P atoms.

In addition, the linkage of a possible mechanism of intergranular fatigue fracture affected by cyclic stress-induced grain growth was demonstrated. Figure 4.18 shows schematic illustrations of the formation of fracture surface through elementary fracture processes of crack nucleation and propagation, leading to intergranular fatigue fracture in the nanocrystalline Ni-P alloy during high-cycle fatigue.

The fatigue fracture occurred along random boundaries whose boundary plane was also the plane subjected to the maximum shear stress as exactly the same as shear bands. Fatigue cracks must nucleate at deformation ledge produced at random boundary by the interaction between sliding grain boundary and PSBs or triple junctions where stress concentration is localized depending on the type of grain boundaries by cyclic deformation. The presence of excess P atoms due to segregation, especially at high-energy random boundaries, can indirectly enhance the characteristic intergranular fatigue fracture through enhanced grain boundary migration and resultant grain growth observed in our electrodeposited Ni-P alloy thin sheets. The model provides also the indication of the aspect that dimples on the fracture surface are larger than the starting grain size. In the case of electrodeposited

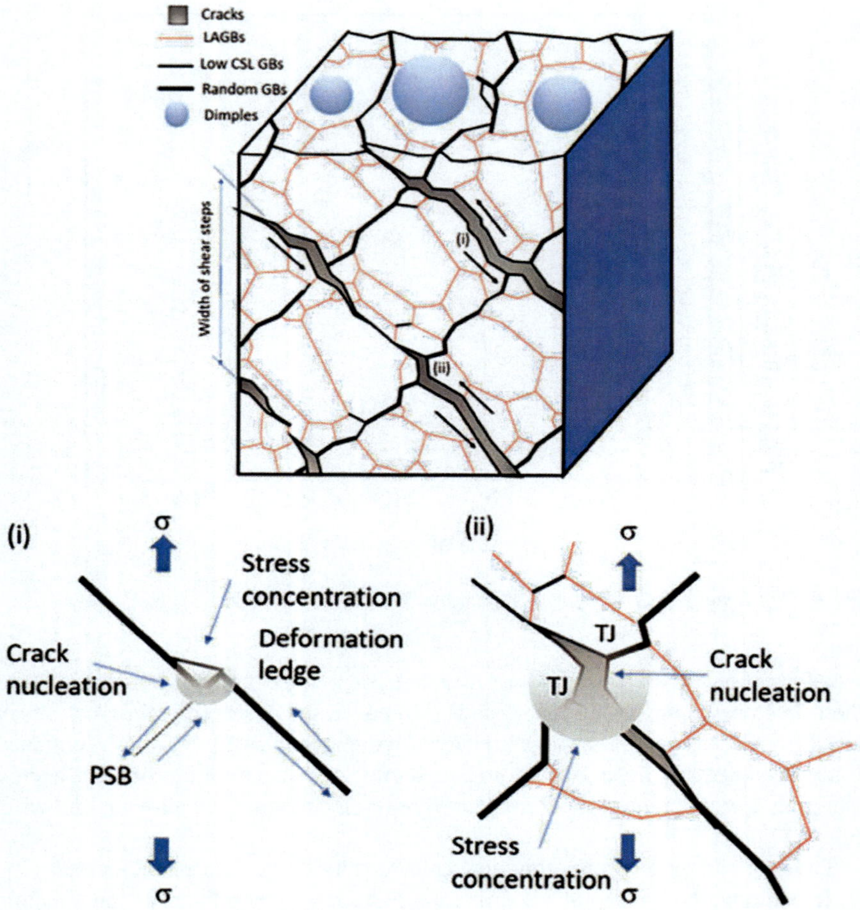

Fig. 4.18 Mechanism of intergranular fatigue fracture at random boundaries and the formation of morphological features of the specimen surface and fracture surface associated with propagation of intergranular fatigue cracks in the nanocrystalline Ni-P alloy specimen during high-cycle fatigue

NC Ni, it has been shown that the dimple size is significantly larger than the average grain size (Fig. 4.19).

The scanning electron microscopy observations of the tensile fractured surfaces revealed a broad population of dimples several times larger than the starting mean grain size for the nanocrystalline nickel, while the mean dimple dimensions for the ultrafine material replicated the original microstructure.

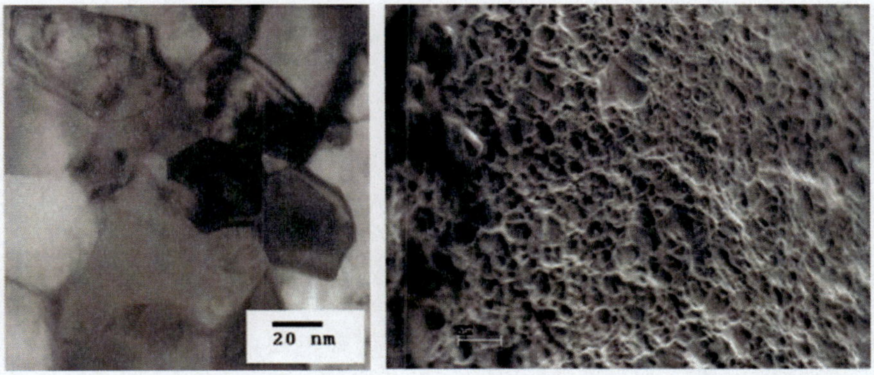

Fig. 4.19 TEM microstructure and fracture surface of NC Ni

4.3 Crack Initiation and Growth in Nanocrystalline Materials

Ductile materials are characterized by their capacity to withstand large deformation and deform permanently. Two basic concepts must be grasped to understand the crack-tip processes, one being dislocation nucleation emission at the crack tip and the other being dislocation mobility. When a crack tip has preexisted in the grain, the crack initially blunted by emitting a limited partial dislocation along the crack, but this was soon taken over by the creation of nanovoids in the surrounding GBs. The consequent result was that the cracks propagated in an intergranular pathway. Plasticity near the crack tip is mediated by partial dislocations. Analyses performed on very fine NC metals (up to 5 nm) revealed that the smaller grain sizes were more ductile and therefore needed higher intensity stresses for cracks to propagate. This summary agreed with the inverse Hall-Petch relationship. With a decreasing grain size, less of the possible generated dislocations are emitted from the crack tip due to the effective suppression of GB. However, the crack can easily grow rather than experience blunting. MD simulations performed on NC Cu indicated that different types of dislocations can be emitted from the crack tip at different temperatures in nanotwinned Cu, leading to a brittle-to-ductile transition (Fig. 4.20).

In Fig. 4.20a, atoms start to be disturbed because of the bond-breaking process. The consequence of these activities is that a twin boundary (TB) is generated (Fig. 4.20b, c). As the strain increases, the newly formed TB extended by an atomic distance (Fig. 4.20d). The newly formed TB shielded the atom at the crack tip and impeded the increase of its stresses, which arrested the bond-breaking process. In turn, the expansion of the newly formed TB no longer continued. The mismatched boundary is a weak link where sliding can easily occur under straining (Fig. 4.20e). Therefore, as the imposed strain increases, the relative movement at the mismatched boundary can trigger the nucleation of a perfect dislocation.

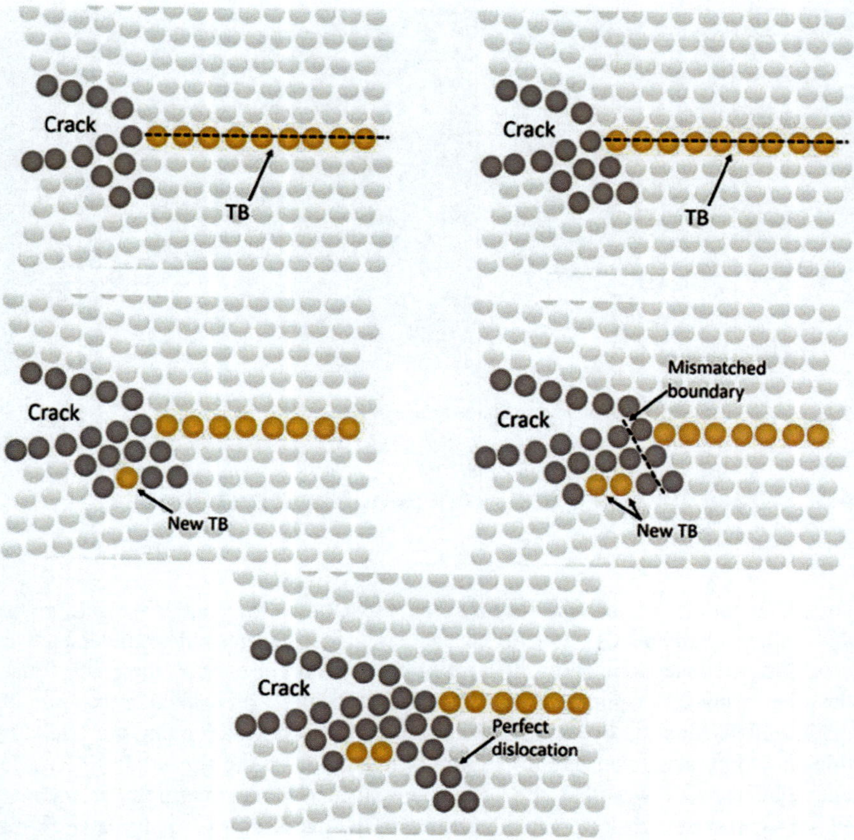

Fig. 4.20 Local crystal structure evolution around the crack tip in NC Cu

4.3.1 Fracture Behavior in NC Materials

In MC materials, a reduction in grain size generally results in an elevation in strength which engenders an increase in fatigue endurance limit during stress-controlled cyclic loading of initial smooth-surfaced laboratory specimens. Since the total fatigue life under such conditions is dominated by crack nucleation and since fatigue cracks generally nucleate at the free surface, grain refinement here is considered to result in improvements in fatigue life as well as endurance limit, with all other structural factors set aside. On the other hand, a coarse grain structure with lower strength and enhanced ductility generally plays a more beneficial role in the strain-controlled fatigue response of metals and alloys. It should be noted, however, that it is often difficult to isolate the sole effects of grain size on fatigue response since other structural factors such as precipitate content, size and spatial distribution, stacking fault energy, attendant equilibrium spacing of partial dislocations, and crystallographic texture are also known to have an important effect on the fatigue characteristics of MC metals.

Intergranular brittle fracture and ductile fracture are experimentally observed in nanocrystalline materials. Plastic flow in such materials occurs at very high stresses close to those needed to induce fracture. In nanocrystalline materials with finest grains, plastic flow is conducted by mostly grain boundary processes. In nanocrystalline materials with intermediate grains, plastic flow is often conducted by both lattice dislocation slip and grain boundary processes. However, if plastic flow and diffusion are not intensive in nanocrystalline materials with intermediate grains and/or these materials contain preexistent nanocracks and pores, brittle fracture tends to occur. They were observed dimpled rupture, dislocation activity at the crack tip, and formation of voids at grain boundaries and triple junctions in the regions ahead of the advancing crack. In the early stages of deformation, dislocations are emitted from grain boundaries under the influence of an applied stress. Triple junction voids and wedge cracks can also result from grain boundary sliding if resulting displacements at the boundary are not accommodated by diffusional or power law creep. These grain boundary and triple junction voids then act as sites for nucleation of the dimples. The deformation and fracture processes are closely related to the coupling of dislocation-mediated plasticity and formation and growth of voids. Whatever the fracture mechanism, it is evident that the fracture will be heavily influenced by microstructural features which have up to now been mostly neglected such as the presence of nanoscale voids or even bubbles and the presence of grown-in twins. The presence of twins has been suggested as an interface control mechanism in coarse-grained metals and may represent a relevant microstructural feature that influences fracture, since many of the NC metals contain grown-in twins. Developing an understanding of the damage tolerance of NC metals and alloys is essential for evaluating their overall usefulness as structural materials or coatings in engineering components. Such understanding should inevitably include comprehensive knowledge of the resistance to fracture initiation and growth under quasi-static and dynamic loading conditions, and of stress- and strain-based total fatigue life and subcritical crack growth under fluctuating loads at different mean stress, cyclic frequency, and environment (Selyutina et al. 2018). It is widely recognized from the wealth of experimental information available in the literature on conventional metals and alloys that grain refinement markedly influences the resistance to fatigue crack initiation and propagation. In the near-threshold regime of fatigue crack growth, where the extent of fatigue crack growth per stress cycle is on the order of a lattice spacing, finer grained materials usually exhibit a relatively faster rate of crack propagation and a lower fatigue threshold stress intensity factor range. This trend is at least partially ascribed to the reduced level of fracture surface roughness seen in the finer grained metal, as the deflections in crack path promoted during crystallographic crack advance are reduced with grain refinement. In addition, any lowering of effective stress intensity factor range due to premature contact between mating crack surface asperities would also be expected to be reduced due to grain refinement. It should be noted, however, that it is often difficult to isolate the sole effects of grain size on fatigue response since other structural factors such as precipitate content, size and spatial distribution, stacking fault energy, attendant equilibrium spacing of partial dislocations, and crystallographic texture are also known to have an important effect on the fatigue characteristics of MC metals (Yang et al. 2008).

When dealing with FCG it is important to distinguish between intrinsic and extrinsic mechanisms (Leitner et al. 2015). Intrinsic mechanisms are responsible for the material separation process when a cracked body is cyclically loaded. The separation process can be caused by mechanisms belonging to three different groups: fracture by accumulated damage, crack advance due to pure deformation, and FCG by cyclic damage. The inherent resistance of the material against these separation processes can be seen as the intrinsic crack growth resistance. However, there also exist so-called extrinsic or shielding mechanisms which can hinder crack growth and reduce FCG rate. These mechanisms reduce the load at the crack tip and thereby the driving force of the FCG process. One possibility is crack deflection, which changes the loading conditions for geometric reasons when the crack deviates from a straight crack path. This can result in an overall reduction of the driving force. So-called contact shielding mechanisms are based on a reduction of the load at the crack tip due to a premature contact of the crack faces. This type of closure can be induced by roughness, oxides, plasticity, or crack bridging. It is important to take these shielding mechanisms into account in order to correctly describe FCG and to give lifetime predictions.

The material mean grain size was also demonstrated to have a strong influence on crack initiation and growth. The fatigue crack rate as a function of ΔK for NC and UFG Ni is shown in Fig. 4.21.

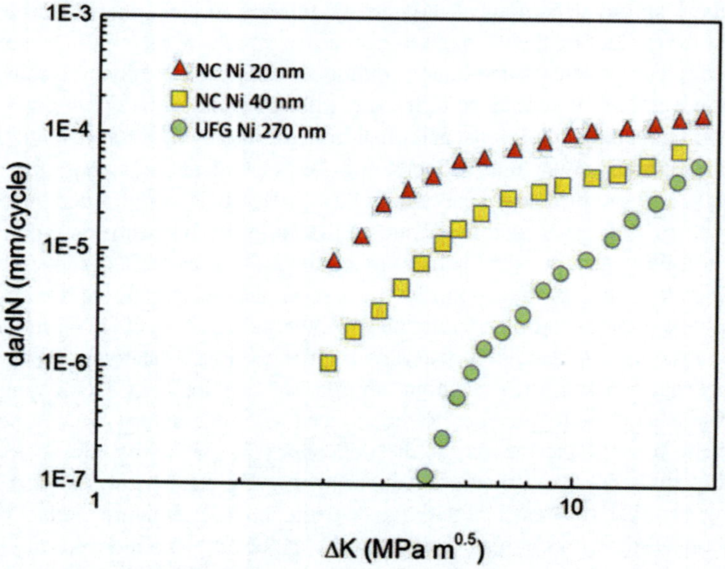

Fig. 4.21 FCB of electrodeposited Ni

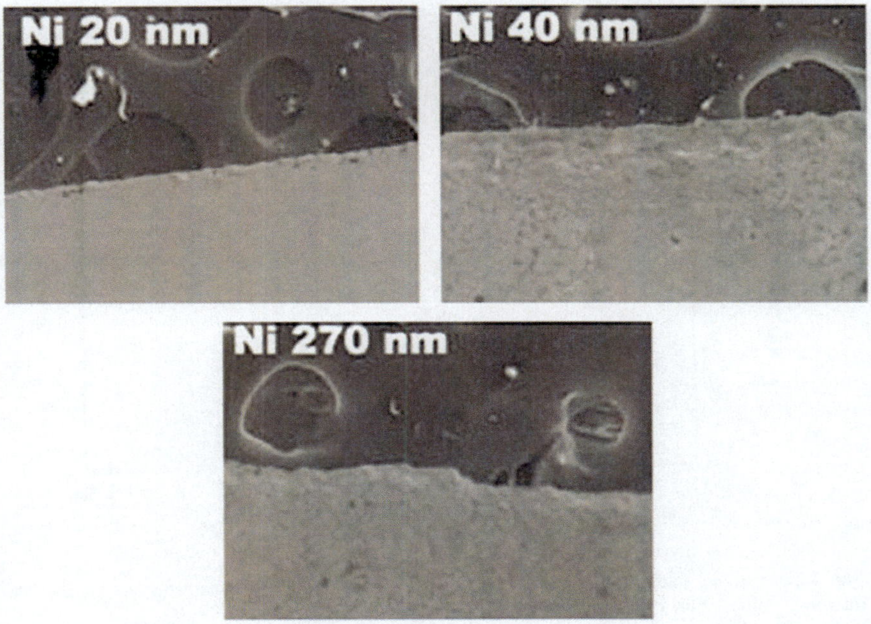

Fig. 4.22 Different crack paths observed for the electrodeposited pure Ni with different mean grain sizes

Overall in terms of stress levels, the material is less sensitive to crack initiation with decreasing mean grain size. On the other hand, the resistance to crack growth decreases with the grain refinement. In order to enforce the assumptions inherent to linear elastic fracture mechanics (LEFM), all data was truncated such that the remaining uncracked ligament was at least 20 times larger than the crack tip plastic zone size. It is evident that the resistance to fatigue crack growth is substantially lower in the NC Ni 20 nm, relative to the UFG Ni with an intermediate behavior for the NC Ni 40 nm. Flat non-tortuous fatigue crack paths are observed in the near-threshold regime. These transition to a combination of dimples and fatigue striations in the mid-ΔK regime and ductile dimples in the high ΔK and overload fracture regimes. The different crack growth rate observed with the different grain sizes is also a function of the different crack paths observed for the different materials (Fig. 4.22).

It is shown how the crack paths become flatter with decreasing mean grain size. As a general behavior, the hardening of the metals produces an increase in ΔK with increasing stress levels of the cyclic loading. The electrodeposited Ni experiences a faster crack growth with decreasing mean grain size. Such a behavior can be explained in terms of dislocation generation and interaction. In electrodeposited pure Ni the structure is dislocation free, and during cyclic deformation causes the generation of partial dislocations at grain boundaries, and the dislocations start to interact as the loading cycles increase forming locks (Cavaliere 2009a). The

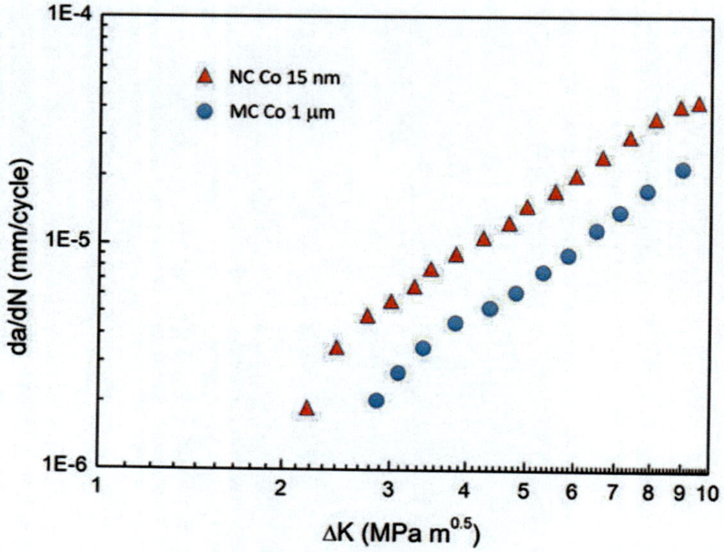

Fig. 4.23 Fatigue crack growth behavior for electrodeposited pure Co with nanocrystalline and microcrystalline grain size

dislocation generation and lock formation are larger with decreasing grain size because of the grain boundary density; in this way the more the structure is fine the more such a phenomenon is pronounced and the hardening increases with decreasing grain size.

Also in NC Co, the crack growth is governed by the crack path (Fig. 4.23).

In the NC Co the path appears to be very flat and governed by the local brittleness of such nanocrystalline metals (Fig. 4.24) while in the case of the microcrystalline material the path appears completely different with localized ductile aspect.

It must be taken into account that the fatigue crack growth depends on the stress range and on the mean stress. Given that cracks experience variable amplitude fatigue cycling, as mentioned before, overloads must be considered. Fatigue overloads can retard the FCG rate by reducing the effective stress intensity factor (SIF) range as well as the mean stress intensity due to the combined effect of various retardation mechanisms. These include plasticity-induced crack closure, residual stress, crack deflection, and surface roughness (Zhang et al. 2020). The fatigue overload has been identified to be due mainly to plasticity-induced crack closure and residual stresses. Overload leads to a modification of the plastic zone shape around the crack tip with the formation of a compressive zone increase during unloading. This reduces the effective stress range leading to crack retardation. Depending on the stress conditions, such a phenomenon can influence the crack closure reducing both the mean stress and the stress range. Roughness-induced crack closure as a secondary effect can be reinforced by plasticity-induced closure, residual stress, and crack deflection, which makes it an important influence factor

Fig. 4.24 Crack path for the NC Co

for crack growth retardation when different degrees of crack tortuosity for materials of various grain sizes are taken into account (e.g., Fig. 4.22). As a matter of fact, in nanocrystalline materials straighter crack with lower surface roughness is observed, which reduces the effects of crack tortuosity and roughness-induced crack closure and hence lessens the fatigue overload retardation effects.

By applying similar loading conditions to electrodeposited Ni from CG to NC regimes, different degrees of crack growth retardation are observed. A smaller grain size can lead to higher strength and hence smaller plastic zone. In other words, the material becomes harder to deform due to the reduced grain size, which reduces the shape misfit between the plastically stretched and the elastically deformed material. Therefore, the primary retardation mechanisms (plasticity-induced crack closure and residual stress) are minimized. Furthermore, grain refinement not only increases the strength of the material but also reduces the tortuosity of the crack path and the roughness of the fractured surface. Crack deflection can reduce the effective SIF and increase the local mode II SIF. The mode II component leads to the relative movement of the crack surfaces along the crack propagation direction and hence strengthens the effect of roughness-induced crack closure at low external loading levels. Consequently, the NC specimen has the smallest grain size and is found to be least affected by the retardation mechanisms.

Most structures undergo low crack growth rates during service life and operate at stress intensity levels in which crack propagation is sensitive to microstructure. Under these circumstances, denoted as the threshold growth regime, the fatigue crack propagation rate exhibits substantial variability and to identify a unique threshold stress intensity level below which the fatigue crack attains a non-propagating state is a formidable task. The microstructural instability of NC metals can lead to catastrophic failure during service and has prevented their use in critical applications (Kalidindi and Schuh 2017). The driving force for grain growth can be eliminated if the energetic preference of grain boundary sites for the solute is strong enough, as governed by the enthalpy of grain boundary segregation (Wang et al. 2020a).

To overcome these limitations and the need for materials with high mechanical/ thermal performance, researchers have proposed several approaches: bimodal grain size distribution, second-phase particle hardening, and introducing twins with nano-scale spacing in grains. Among these approaches nanotwinned (NT) metals stand out with their superior mechanical properties and thermal stability (Singh et al. 2011).

Electrodeposited nanomaterials often show a special behavior concerning the grain growth mechanism. Because of the grain refiners that are used, the grain growth of special grain boundaries is preferred, causing them to grow faster than others. This is called abnormal grain growth and leads to the formation of bimodal microstructures (Boyce et al. 2015), if the temperature that is reduced before the grain size distribution becomes monomodal again. Such bimodal NC/UFG micro-structures are often a good compromise between strength and ductility. The reason for this is that the larger grains can deform much easier and improve the ductility of the material, which can—in the case of fatigue—slow down the crack propagation rate of a crack, while the nanograins provide the high strength of the material (Rathman et al. 2017). There is a good effect on the fatigue performance of bimodal microstructures if the loading conditions are strain controlled, whereas under stress control, this is not the case reported. Other studies reported an improvement of fracture toughness by using bimodal nanocrystalline materials instead of NC materials and that the critical stress intensity factor is larger if the grain size of the large grains in the bimodal microstructure is increased.

Recent studies have revealed that introducing high density of nanotwins into the pure face-centered cubic (FCC) metals can lead to superior strength—much greater than NC metals—and hardness while maintaining thermal stability and ductility. The operating mechanisms of the fatigue crack propagation in NT Cu have been revealed by large-scale atomistic simulations. During fatigue, double striations form at the crack tip in the samples with a high density of nanoscale twins, which is essentially distinct from the fatigue mechanism (i.e., nucleation of nanovoids ahead of the crack tip) in the twin-free counterparts. The formation of double striations is attributed to alternating crack tip blunting and re-sharpening due to dislocation emission and slip. For the samples with a high density of nanoscale twins, the detwinning process dominates the plastic deformation in some grains near the crack tip, and the crack closure becomes pronounced due to the dislocation shielding crack tip. These deformation mechanisms are associated with a high density of preexisting twins

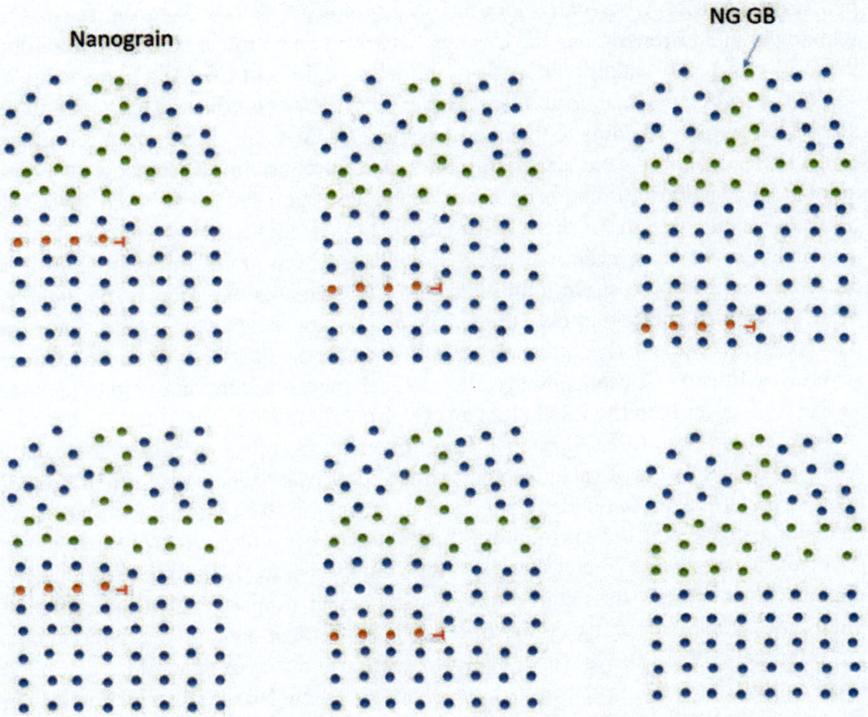

Fig. 4.25 Reverse dislocation motion during fatigue deformation due to the excess volume of nanograin boundaries

and increase the dissipated energy for fatigue. Therefore, introducing nanoscale twins substantially benefits the fatigue resistance of NC metals. The simulation results also show that the fatigue resistance increases as the number of twins per grain increases, which is consistent with the experimental results (Zhou et al. 2015).

A proposed model for the effect of bimodal microstructure on fatigue properties is shown in Fig. 4.25 (Xu et al. 2019).

From MD studies it results that compression causes higher stress level than tension with the same strain level. The reverse behavior of dislocation formation and motion is also due to the reverse change of the excess volume of nanograin boundaries. For example, when the compression strain is sufficiently large during a fatigue cycle, e.g., at the point A on the sinusoidal strain cycle curve, dislocation forms at the nanograin boundaries and starts to move into the micrograin matrix. During this process, the width of the nanograin boundaries is narrowed by approximately one atomic layer. When the strain is reversed, this compression of nanograin boundaries is also recovered accompanying with the reverse motion of dislocations. Finally, when the strain goes back to zero, i.e., the initial state, the excess volume of nanograin boundaries is completely recovered and the fatigue dislocation is also retracted. Different locations of crack nucleation and modes of propagation, though both due to the CG–UFG matrix interface, are symptoms of load. In tension case, CG with lower yield strength than the UFG yields first at much lower stresses. Therefore, the UFG has to carry additional load around weak regions of CG. At a given value of stress, the CG elements deform more, so the UFG elements adjacent to them will be stressed even more, leading to their earlier failure. On the other hand, when subjected to cyclic loads, the plastic strain amplitude is held constant. Macroscopically, the amount of plastic deformation is more dependent upon the UFGs than the CGs. Since the yield strength of the UFG region is higher, they will continue to deform elastically as the CGs reach their yield strength and start to deform plastically. For the same macroscopic strain amplitude, the CG elements experience significantly more plastic deformation in both the tensile and compressive cycles. Also, since the damage models are nearly equivalent in both regions, the fatigue lives of the CGs are severely reduced. Consequentially, the CG elements accumulate more damage sooner and faster than the UFG elements (Nelson et al. 2011).

In threshold regime, FCG rate vs. stress intensity factor range curves are usually segmented as a result of material microstructure effects such as grain/twin boundaries or local dislocation density ahead of the crack tip. The interplay between the advancing crack and the surrounding microstructure stands out as a challenging problem in predicting metal fatigue life. Nanograins were introduced into the engineering materials to impart superior mechanical properties such as ultrahigh yield and fracture strengths on the order of GPas as well as high wear resistance. Unfortunately, many alloys with a microstructure composed of nanograins exhibited low ductility despite high strength. A major advancement for the superior ductility in nanograined materials occurred with the introduction of nanotwins via electrodeposition and deformation processing. The nanotwinned microstructures promote glide motion along the twin lamellae and incorporate dislocations at the interfaces acting as efficient slip barriers. Furthermore, the FCG resistance can be improved with nanosized grains and twins compared to coarse-grained alloys. The improved fatigue crack impedance in this class of materials originates from the interaction of the crack tip-emitted dislocations with the grain and twin boundaries. Admittedly, a comprehensive understanding of the slip transmission mechanisms across a grain/twin boundary is necessary to shed light on the microstructural parameters promoting the FCG resistance of the nanostructured materials.

The schematic of loading and unloading in the presence of nanotwins along the crack path is shown in Fig. 4.26.

In the model a hypothetically linear elastic, continuous medium in which the crack tip-emitted dislocations can glide and interact with the boundary is considered. Meanwhile the glide resistance (or the lattice friction stress) inside a pristine crystal is identical for both forward and reverse glides from the crack tip, which will be denoted as σ_F; the local friction stress varies at the grain boundaries en route the forward and reverse glides owing to the different slip transmission energetics involved at the grain boundaries. Therefore, the glide resistance exerted on the dislocations within the very near proximity of the grain boundaries will be characterized by $\sigma_F^{\text{Forward}}$ and $\sigma_F^{\text{Reverse}}$ for the forward and reverse glides (Fig. 4.27).

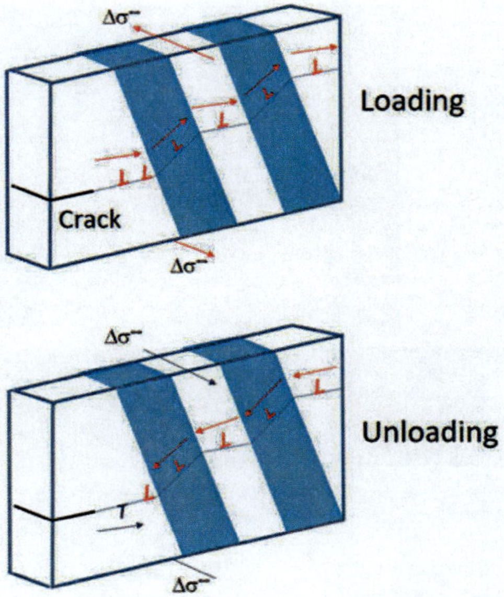

Fig. 4.26 Cross-slip configuration in the presence of nanotwins. Positive dislocations are emitted from the crack tip during loading. The negative dislocations nucleate from the crack tip during unloading and annihilate the positive dislocations if they are within an appropriate distance

The lattice friction stress values, corresponding to the crystal and the boundaries, are evaluated by the modified Peierls-Nabarro (P-N) formulation. Although the forces acting due to the applied stress intensity K_{III} tend to drive the dislocation away from the crack tip, the traction-free surfaces of the crack and the crystal lattice friction apply restoring forces on it. The dislocation interaction force depends on the signs of the dislocations; they are repulsive/attractive for the same/different sign slip vectors.

4.3.2 Cyclic Behavior of Graded Materials

Electrodeposition has been well developed and known for the production of fully dense nanocrystalline materials. Due to its simple and feasible industrial application, this processing technique has been widely used in producing many different nanocrystalline materials (metals, alloys, ceramics, and composites). Figure 4.28 illustrates the various types of electrodeposited coatings.

The properties of the metallic coatings could be easily tailored by changing the experimental variables such as current density, electrolyte solution, electrolyte pH, electrolyte temperature, and additives. Composite coatings consist of a metal or alloy matrix with a dispersion of second-phase particles that could comprise nanotubes, solid lubricants, or hard particles to increase wear and corrosion properties.

Traditionally, the grain boundary network is homogeneously distributed in the polycrystalline material. As a result, the average grain size of a material becomes a representative parameter which is closely related to the mechanical properties. The

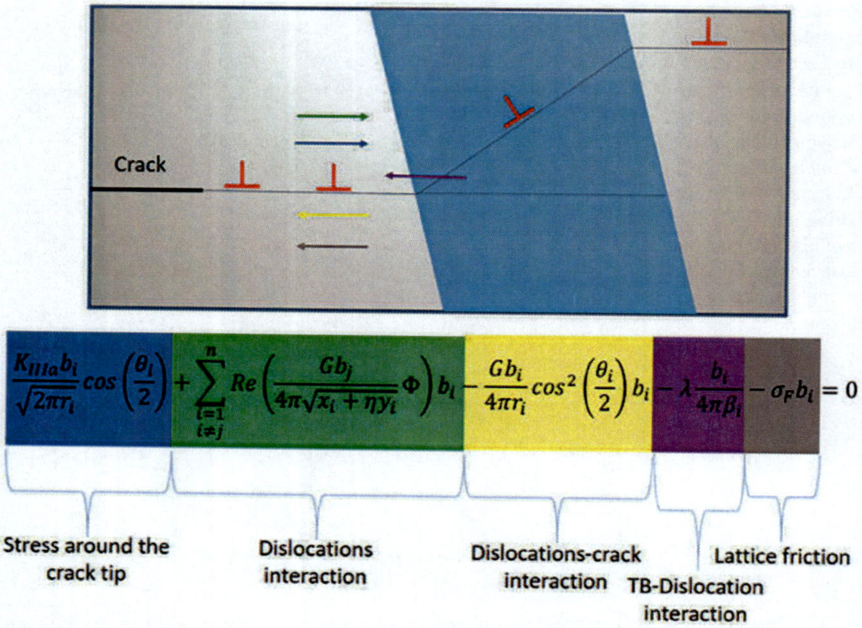

Fig. 4.27 Individual force terms acting on a dislocation. The force terms are due to the applied stress intensity factor, the dislocation-dislocation interactions, the traction-free crack face, the TB-dislocation interaction, and the lattice friction. It should be emphasized that for each K_{III} level applied, each dislocation irrespective of whether it is nucleated by crack tip emission or it is contributing to the initial dislocation density (preexisting slip) should be in force equilibrium as dictated by the right-hand side of the equation

coarse-grained material (grain size is around 10 μm) has a good ductility but a relatively low yield strength. The homogeneous nanocrystalline metals, whose grain sizes are smaller than 100 nm, usually have an ultrahigh yield strength that can be even greater than the counterpart alloyed metals. However, this improved yield strength comes at the expense of ductility. The tensile ductility of a nanocrystalline metal is extremely poor at room temperatures due to its limited strain hardening rate which causes the early necking easily.

One of the promising solutions is introduction of heterogeneous grain boundary network instead of homogeneous grain boundary. These novel materials are usually termed as heterogeneous nanostructured materials (HNMs). It is reported that the HNMs have the potential to ease or even overcome the strength-ductility trade-off. The successful examples of HNMs are bimodal nanostructured material, harmonic nanostructured material, lamellar nanostructured material, gradient nanocrystalline material domain-dispersed nanocrystalline material, and hierarchical nanostructured material. Among those HNMs, the gradient nanocrystalline (GNC) material can be manufactured with the least difficulties. This special material can be obtained by directly processing a coarse-grained material with various surface severe plastic

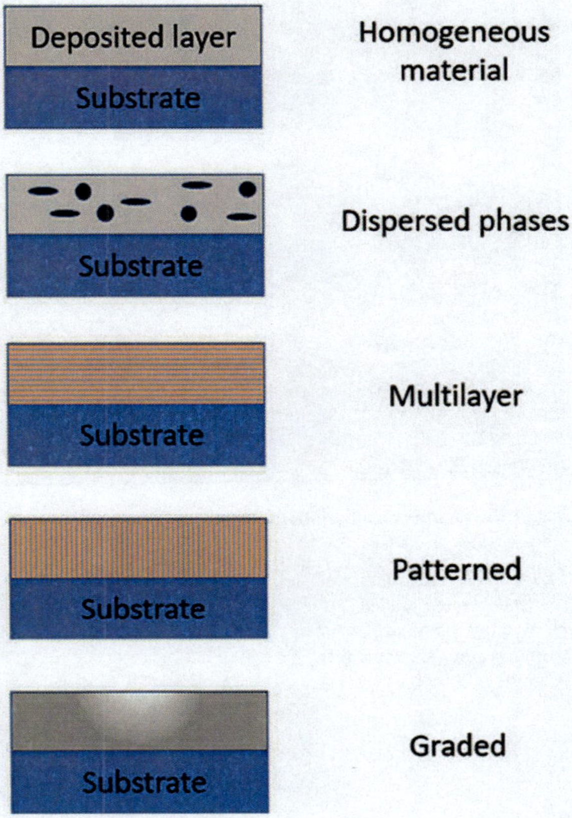

Fig. 4.28 Different types of electrodeposited structures

deformation techniques like shot peening, cryogenic burnishing, surface nano-crystallization, surface mechanical attrition treatment (SMAT), and surface mechanical grinding treatment (Fig. 4.29).

Continuously graded microstructure can be obtained through electrodeposition by varying the processing parameters during deposition (Fig. 4.30).

It has been reported that the yield strength of a GNC material is extremely high.

From the perspective of dislocation activity, high strength is often achieved by impeding the motion of dislocations; yet ductility is associated with their spatial distribution, multiplication, and propagation. As a consequence, an approach to improve the strength or ductility invariably compromises the other, a dilemma known as strength-ductility trade-off (Wu and Fan 2020). This means that these two critical performance indicators of strength and ductility are almost mutually exclusive (Fig. 4.31).

By refining internal grain structures, the yield strength is several times larger than that of coarse-grained (CG) counterparts due to the grain boundary (GB) strengthening, but it comes at the expense of ductility, typically down to a few percentage, owing to the lack of strain hardening capacity. Therefore, the current

Fig. 4.29 Gradient nanocrystalline material

Fig. 4.30 Microstructure of electrodeposited graded Ni-W alloy, grain size variation along the thickness (Cavaliere 2009b)

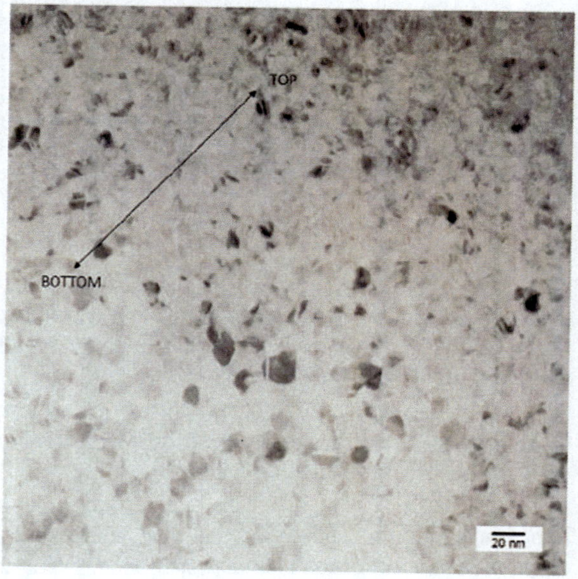

challenge and target are to explore the mechanistic origin and solutions for producing new classes of structural materials with high strength while preserving respectable ductility.

Furthermore, it is observed that the yield strength of GNC materials is significantly higher than the value estimated using the rule of mixture (ROM) which estimates the yield strength by taking the average yield strength of different layers.

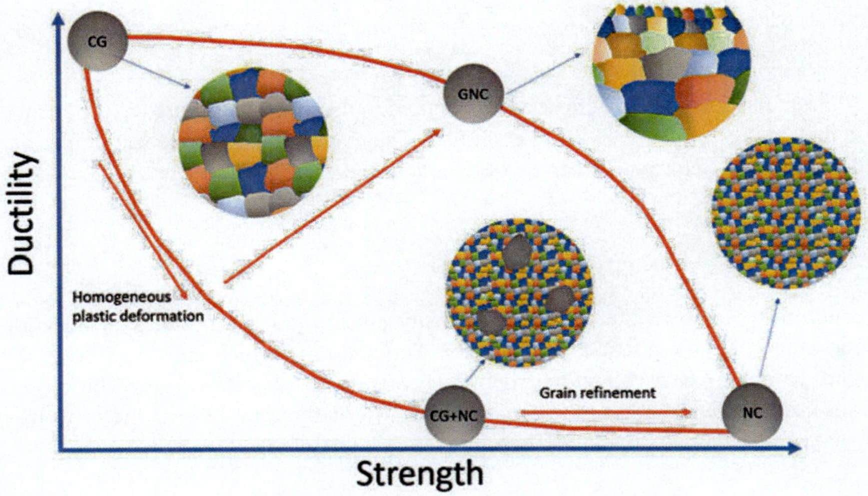

Fig. 4.31 Strength-ductility for the different microstructures

The sharp grain size gradient was argued to be an important reason for the extra high strength. Mechanical incompatibility produced by the grain size gradient will produce a long-range back stress in the GNC material. Dislocation slip in the CG interior is hindered by the back stress, resulting in higher flow stress. The superior mechanical properties of GNC material suggest a unique strengthening mechanism, which is rendered by the gradient GB topography. It has been reported that the gradient structure in a GNC material can produce extra strain hardening (Wang et al. 2020b). This implies that the enhanced dislocation activity in GNC materials could play a critical role in their superior ductility. Another very interesting observation is that during tensile loading, the grain growth in the nanocrystalline layer is extremely severe, and it can even result in the formation of dislocation-free grains near the surface. Observations suggest that grain growth, strain-induced recrystallization, and dislocation annihilation during the deformation of GNC materials might be responsible for their superior mechanical properties.

The flow stress can then be estimated using the simplified Taylor equation:

$$\sigma_f \approx \frac{\alpha b \mu}{M} \sqrt{\rho}$$

If the dislocation density shows a linear relationship with plastic strain ε and is inversely proportional to the average grain size D, then the density can be written as

$$\rho = \rho_0 + \frac{A\varepsilon}{bD}$$

So

$$\sigma_f \approx \frac{\alpha b}{M}\sqrt{\rho_0 + \frac{A\varepsilon}{bD}}$$

From the perspective of dislocation dynamics, this can be attributed to the change in the kinetics of the dislocation evolution. The evolution of the dislocation density during plastic deformation can be described as

$$\frac{d\rho}{d\varepsilon} = \frac{d\rho^+}{d\varepsilon} + \frac{d\rho^-}{d\varepsilon}$$

which is the sum of the dislocation density change per strain due to dislocation nucleation and annihilation, respectively. The first term which is referred to as the work hardening term can be regarded as constant with respect to strain. The second term, usually referred to as the dynamic strain recovery term, follows the first-order kinetics with respect to strain. Therefore

$$\frac{d\rho}{d\varepsilon} = h - rp$$

The integration of the previous equation gives

$$\rho = \rho_0 e^{-r\varepsilon} + \frac{h}{r}(1 - e^{-r\varepsilon})$$

It was reported that both h and r are the functions of the grain size and strain hardening. Based on that understanding, the coefficients h and r should vary with location in a GNC material due to the variation in grain size and dislocation density. In addition, if considering the interaction between each layer due to dislocation migration, an additional dislocation kinetic could rise in GNC material. Therefore, for GNC materials, the dislocation evolution coefficients, h_{GNC} and r_{GNC}, can be described as

$$h_{\mathrm{GNC}} = (1 - f_{\mathrm{GS}})h_{\mathrm{CG}} + f_{\mathrm{GS}}h_{\mathrm{GS}} + \Delta h$$

$$r_{\mathrm{GNC}} = (1 - f_{\mathrm{GS}})r_{\mathrm{CG}} + f_{\mathrm{GS}}r_{\mathrm{GS}} + \Delta r$$

where f_{GS} is the volume fraction of the gradient structure (GS) layer, and h and r are kinetics constants, where the subscript CG indicates coarse-grained layer and GS indicates gradient structure layer, and Δh and Δr are the kinetics constants of the additional effect resulting from the presumed synergistic interaction between layers. The enhanced yield strength of the gradient nanostructure material is the result of the interaction between different layers. In addition to the yield strength, this interaction may also affect the ductility. The additional terms, Δh and Δr, on the total dislocation kinetics of GNC materials, may arise due to the interaction between the interior

CG layer and the surface GS layer. It is demonstrated that the additional kinetics due to the interaction between CG layer and GS layer results in a softening effect due to its negative hardening term, Δh, and positive softening term, Δr. As a matter of fact, the GNC material is even softer than the simple superposition of both layers.

To understand the physics origin for the additional softening effect in GNC materials, the strengthening mechanism in both the CG layer and GS layer should be elucidated. Given the Taylor strengthening equation

$$\frac{\tau}{\mu}\sqrt{D} = \left(\frac{\beta}{\sqrt{D\rho}} + \alpha b\sqrt{D\rho}\right)$$

With different values of $D\rho$, there are two strengthening mechanisms, dislocation starvation and dislocation strengthening, which compete with each other. For dislocation starvation, the dislocation density is low and the dislocation multiplication rate is low as well. Under this condition, the exhaustion of dislocations increases the stress required for plastic deformation. The nanocrystalline material usually has a negative dislocation multiplication rate as manifested from the negative strain hardening rate:

$$\Theta' = \frac{d\rho}{d\varepsilon} = 2\Theta\sigma\left(\frac{M}{\alpha b\mu}\right)^2$$

where Θ' is the dislocation multiplication rate per unit strain and Θ is the strain hardening rate. A negative dislocation multiplication rate implies a dislocation starvation mechanism as increase in dislocation density could lead to the drop in strength. Therefore, the dominant strengthening mechanism for the nanocrystalline part of the GS layer tends to be dislocation starvation, and the dislocation strengthening mechanism should be dominant in the CG layer. However, due to the interaction between the CG and GS layers, the dislocation generated in the CG layer can move to the GS layer due to the high dislocation annihilation rate in the GS layer. Because the GS layer is in the dislocation starvation regime (Fig. 4.32), the increase in dislocation density will produce a softening effect. Overall, this will lead to an additional softening effect in the GNC material.

This phenomenon is known as dislocation-migration-induced strain softening, and its underlying mechanism is schematically illustrated in Fig. 4.33.

The experimental observation supports the proposed mechanism, i.e., dislocations migrate from the CG layer to the GS layer. In the GS layer, the dislocation starvation is the dominated strengthening mechanism and the dislocation density dramatically drops during plastic deformation. Larger plasticity deformation could give a much more refined grain size and drive the strengthening mechanism from dislocation strengthening to dislocation starvation at the GS layer. This can be explained by the dislocation migration which can also induce the transition in strengthening mechanism from dislocation starvation to dislocation strengthening in the nanocrystalline part of GS layer. Dislocation migration from CG layer to GS

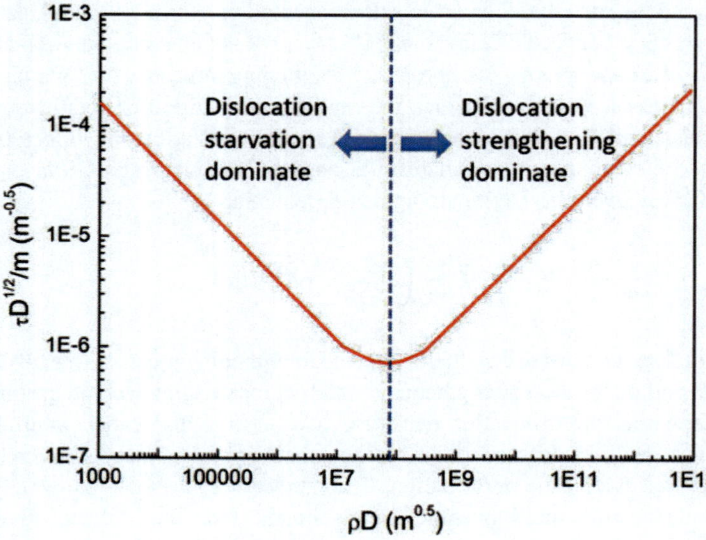

Fig. 4.32 Strengthening mechanisms at different ρD

layer can provide extra dislocations for the GS layer and the strain-induced grain coarsening during plastic deformation can increase the grain size to submicron level. Both effects can increase the value of $D\rho$, thus inducing the transition from dislocation starvation to dislocation strengthening in the GS layer. This transition will first decrease the strain hardening rate and then can provide extra strain hardening rate as the recovered capacity of dislocation accumulation. As a result, the dislocation migration from CG layer to GS layer gives rise to extra ductility to the GS layer.

The imposition of fluctuating mechanical loads on homogeneous crystalline metals and alloys leads to a continuous change in microstructure with increasing fatigue cycles. This change is reflected in progressive accumulation of defects and their reorganization into internal microstructures such as dislocation dipoles and cells, persistent slip bands, and low-angle grain boundaries during different stages of fatigue. Even in initially atomically smooth-surfaced metals and alloys, the significant roughening of surfaces that accompanies cyclic deformation and damage can lead to the formation of intrusions and extrusions. These surface sites serve as micro-notches where fatigue cracks nucleate and advance into the interior during cyclic loading to eventually cause catastrophic failure (Long et al. 2019).

In homogenous metals, refinement of grain size from microcrystalline or coarse-grained to sub-microcrystalline or ultrafine-grained and nanograined scales generally results in an effective increase in strength by suppressing dislocation activities. Both enhanced resistance to fatigue crack initiation and higher fatigue endurance limit develop in UFG and NG metals during stress-controlled high-cycle fatigue. However, grain size refinement in a structural alloy also leads to substantial reduction in ductility, almost without exception, along with deterioration in crack

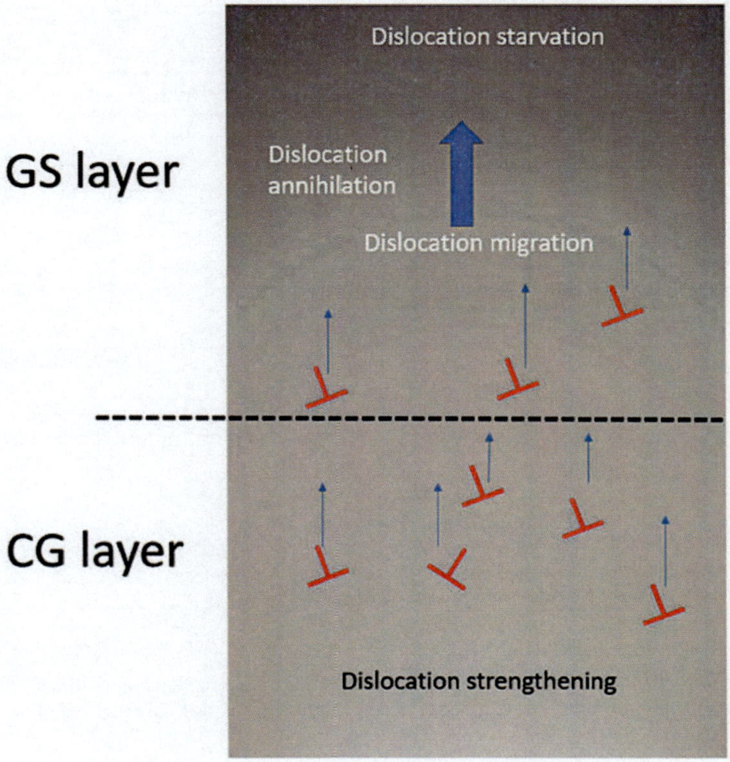

Fig. 4.33 Schematic of the dislocation migration-induced ductility in a GNC material

propagation resistance at low-amplitude cyclic stresses, and exacerbates failure in strain-controlled low-cycle fatigue. Correspondingly, uniform grain size reduction also causes shortened low-cycle fatigue life and pronounced cyclic softening in UFG and NG alloys. These undesirable consequences of fatigue arise from microstructural instability and cyclic deformation-induced local damage accumulation, such as that produced by shear banding and/or abnormal grain coarsening (Xie et al. 2007). These processes, in turn, severely limit the practical utility of grain refinement strategies for homogeneous high-strength alloys in fatigue-critical applications.

Materials in which spatial gradients in structural features are purposely introduced from the surface to the interior are, in some cases, known to exhibit superior mechanical characteristics compared to their homogenous counterparts of appropriately comparable composition.

The advantage to use electrodeposition to produce such kind of materials is represented by the possibility to obtain nanograin materials with a broad grain size ranges with high manufacturing velocity and with a good control of the process obtaining defect-free structures. In addition, electrodeposition allows to obtain structures in which the grain size can be varied to change the consequent mechanical

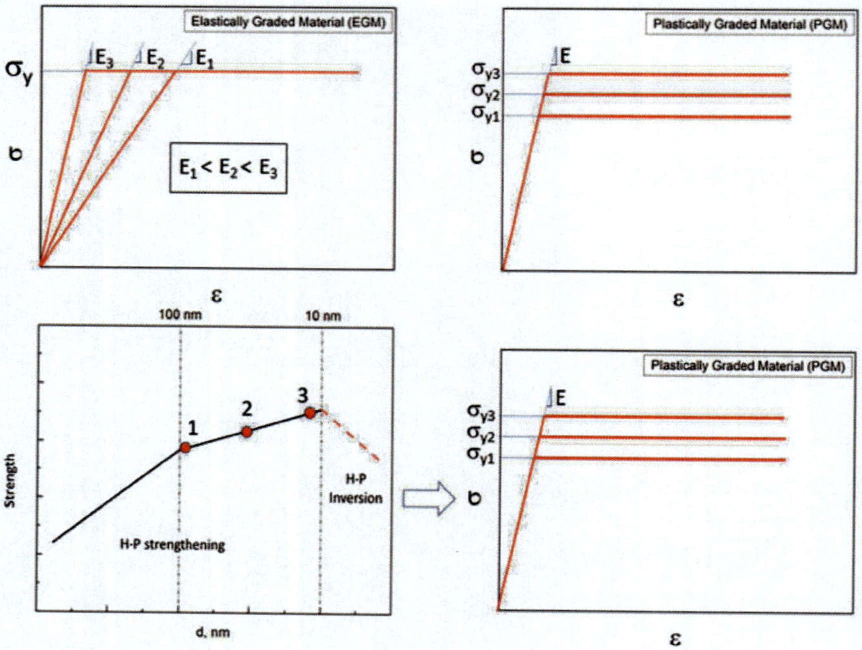

Fig. 4.34 Gradient in elastic and plastic properties of materials as a function of different location and yield strength variation with grain size as described by the Hall-Petch relationship

properties in the space by varying the deposition conditions. Such materials, in which the physical and mechanical properties change continuously or in a discrete way across different interfaces, belong to the so-called functionally graded materials (FGM).

In engineering applications, property gradation offers possibilities to control and optimize material response through redistribution of stresses, either mechanical or thermal as well as relaxation of stress concentration zones, and control of local crack driving force. Such materials can find applications in diverse fields such as optimization of thermomechanical stresses in aircraft parts and space vehicles, damage-resistant surfaces in armored plates and bulletproof vests, and barrier coatings for structural components and industrial tools. The described gradient can be achieved through the variation in the elastic properties (Young's modulus) or in the plastic ones (yield strength) as shown in Fig. 4.34.

In addition, the high strength sensitivity of materials with varying grain size, as described by the well-known Hall-Petch relationship, allows to open a new frontier in materials engineering going to induce plastic-graded mechanical properties by controlling the grain size.

The addition of W to pure Ni has been demonstrated to control the grain size of the final alloy over a very broad range of grain size. In addition, it was demonstrated how the reverse pulsing control allows to obtain a defect-free structure of the deposits if compared with the corresponding conventional cathodic methods. As a consequence, it is possible to change such parameters during the deposition of a single sheet in order to obtain samples in which the grain size varies linearly along the thickness. Such possibility allows that the grade in the microstructure and consequent plastic mechanical properties of nanocrystalline electrodeposited alloys lead to several changes in the driving force for fracture, leading to the possibility of new design philosophy against failure of structural components. We showed how it is possible to obtain a different W content in the alloy by varying the current density and the electrodeposition bath temperature. For a given bath temperature and current density, the grain size of the deposited surface was observed to be a function of tungsten alloying, with the grain size in the range of 140–2 nm achieved by varying tungsten content from 2 to 27 at%. The high solubility of tungsten in nickel resulted in minimum grain boundary segregation. Consequently, a much broader range of grain size was achieved compared to other systems with similar solute content. In addition, nanocrystalline materials exhibit microstructural instability such as grain growth at high stress and deformation levels. The presence of such grain growth can affect the plastic gradient. Multistep indentation experiments can be used to assess such deformation-induced instability for graded materials. The microstructural stability of grain size-graded PGMs was analyzed. For this purpose, multistep indentation was performed quantifying deformation-induced mechanical response of materials.

The solute content of W in the alloy resulted around 20% to obtain a grain size of 20 nm and around 4% for the 100 nm. The produced graded materials were negative-graded configuration in which the grain size varies between 20 and 100 nm from the surface in 50 μm thickness in a linear way (Fig. 4.30) and positive-graded configuration in which the grain size varies between 100 and 20 nm from the surface in 50 μm thickness in a linear way (Fig. 4.35).

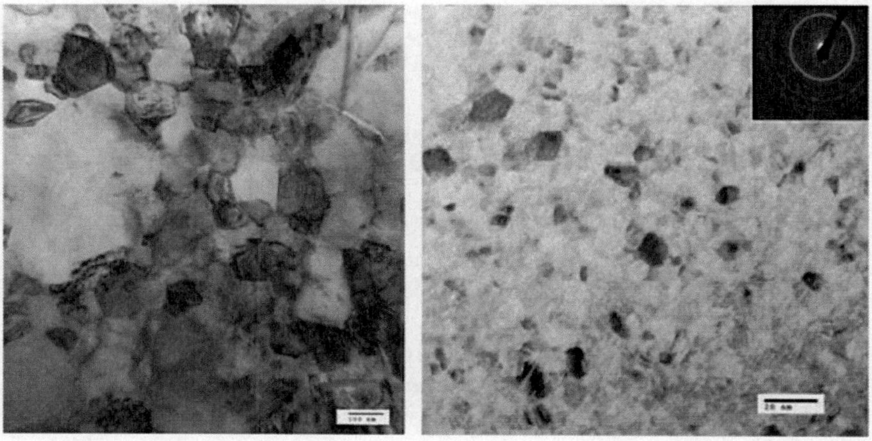

Fig. 4.35 TEM microstructure of the Ni-W alloy with positive-graded grain size; top surface (**a**) and bottom surface (**b**)

4.3.3 Cyclic Indentation of Graded NC Materials

The different grain size and the different microstructure distribution are responsible for the different hardness variation measured for the Ni-W alloys (Fig. 4.36); the indentation curves for the negative-graded Ni-W alloy are shown.

It can be observed that for the Ni-W alloys with constant grain size the difference between the hardness measured in single-step indentations is very close to the values recorded for the multistep ones; such a difference increases for the graded samples with the maximum experienced by the negative one (Fig. 4.37).

Actually it can be seen how the sensitivity to room-temperature hardness variation by changing the loading conditions is proper for nanocrystalline materials. In particular, mechanical instabilities due to unstable grain growth as a consequence of the different loading conditions can lead to loss of strength in critical situations. Such a behavior becomes much more critical in the case of plastically graded materials in which the effect of the microstructural and mechanical grade plays an additional role in the variation of the measured properties as a function of the different loading rates and levels in multistep deformation. In such a case, in particular, the softening–hardening behavior under multistep loading can strongly affect the role of the grade.

The driving force for crack propagation can be controlled by precisely varying the grain size along the interfaces through which the crack is growing. It was widely observed that by decreasing the grain size from microcrystalline to ultrafine and nanocrystalline regime the resistance to crack initiation increases coupled with an increase in crack growth. The obtaining of plastically graded alloys will permit to control both such properties by optimizing the structure distribution against fatigue damage (Cavaliere 2008). The micromechanism involved in the fracture at the interfaces of materials with different mechanical properties is a topic of large scientific and technical interest for the consequent engineering applications. The possibility to induce grading in the microstructure and consequent plastic mechanical properties of nanocrystalline-electrodeposited alloys lead to several changes in the driving force for fracture, leading to the possibility of new design philosophy against failure of structural components. The driving force for crack propagation can be controlled by precisely varying the grain size along the interfaces through which the crack is growing. It was widely observed that by decreasing the grain size from microcrystalline to ultrafine and nanocrystalline regime the resistance to crack initiation increases coupled with an increase in crack growth. The obtaining of plastically graded alloys will permit to control both such properties by optimizing the structure distribution against fatigue damage. As a general behavior, in fact, the spatial variation of material stiffness modifies the stress distribution under a given loading configuration. In addition, the plastic gradient modifies the crack propagation behavior by changing the crack tip toughness. The J-integral calculation is widely considered a powerful approach in the analysis of crack tip elastoplastic state, providing a unique characterization of monotonic, nonlinear fracture in rate-independent materials.

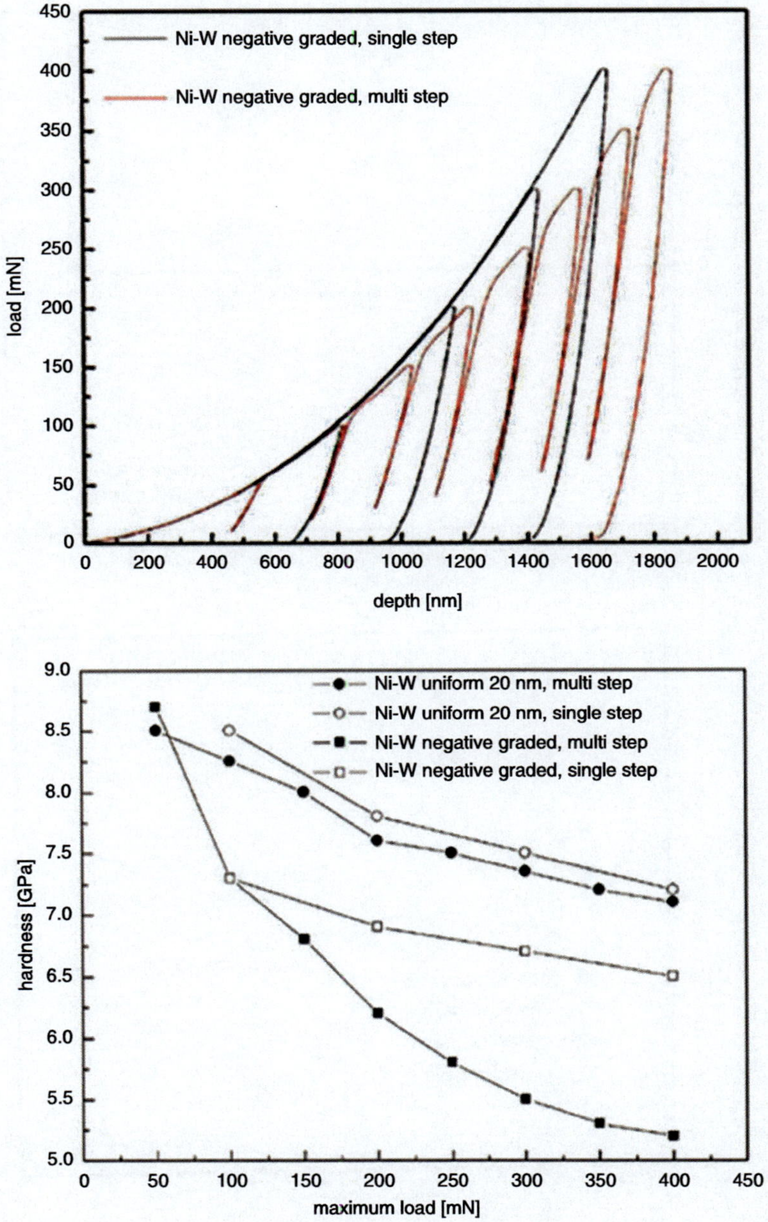

Fig. 4.36 Indentation curves performed in single and multistep by varying the indentation load for the material in negative-graded configuration; Hardness behavior of the different Ni-W alloys with 20 nm constant mean grain size and in negative-graded configuration in single- and multistep conditions

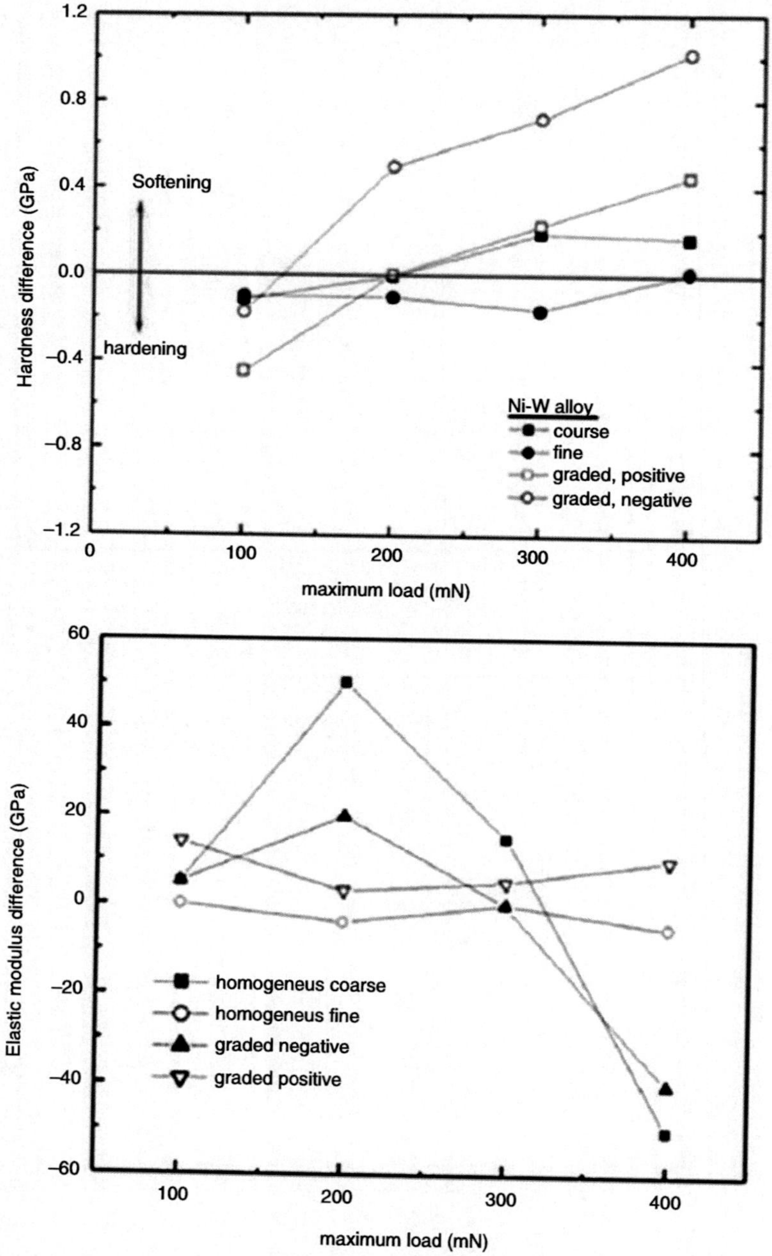

Fig. 4.37 Difference between the hardness (**a**) and elastic modulus (**b**) measured in the single- and multistep indentations as a function of the maximum indentation loads

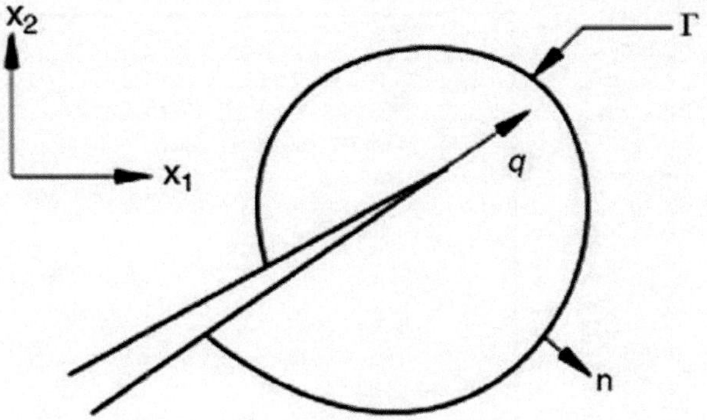

Fig. 4.38 Visualization of Γ contour starting on the bottom surface of the crack and ending at the top surface in anticlockwise direction

Crack propagation occurs when one of several fracture parameters reaches a critical value; the evaluation of effective J-integral is largely recognized as an important method for the analysis of the response of materials in the fracture mechanics problems. It is related to the energy release associated with the crack growth and it gives the measure of the deformation at the crack tip. In the case of linear materials it can be related to the stress intensity factor. In the quasi-statically loaded stationary cracks the J-integral can be defined as

$$J = \lim_{\Gamma \to 0} \int_{\Gamma} \boldsymbol{n} \cdot \boldsymbol{H} \cdot \boldsymbol{q} d\Gamma$$

where Γ is a contour starting on the bottom surface of the crack and ending at the top surface in anticlockwise direction (Fig. 4.38).

The limit $\Gamma \to 0$ indicates that Γ dimension decreases at the crack tip, $\boldsymbol{q}$ is a vector in the direction of the crack growth, and $\boldsymbol{n}$ is the vector perpendicular to the Γ contour. $\boldsymbol{H}$ is described by

$$H = W\boldsymbol{I} - \boldsymbol{\sigma} \cdot \frac{\boldsymbol{du}}{dx}$$

For an elastic behavior of the material W is the strain elastic energy, while for an elastoplastic behavior W is the strain elastic energy plus the plastic dissipation. Now, the J-integral can be expressed as

$$J = \int_{\Gamma} \left(Wdy - T\frac{\partial u}{\partial x} ds \right)$$

where T is the tensile vector perpendicular to Γ, $T_i = \sigma_{ij} n_j$, and u is the displacement in the x-direction. The J-integral is the measure of the released energy and it is $J = 0$ along all the close contours. It was demonstrated that the J-integral defined along a contour surrounding the crack tip is the variation of the potential energy for a virtual extension of the crack:

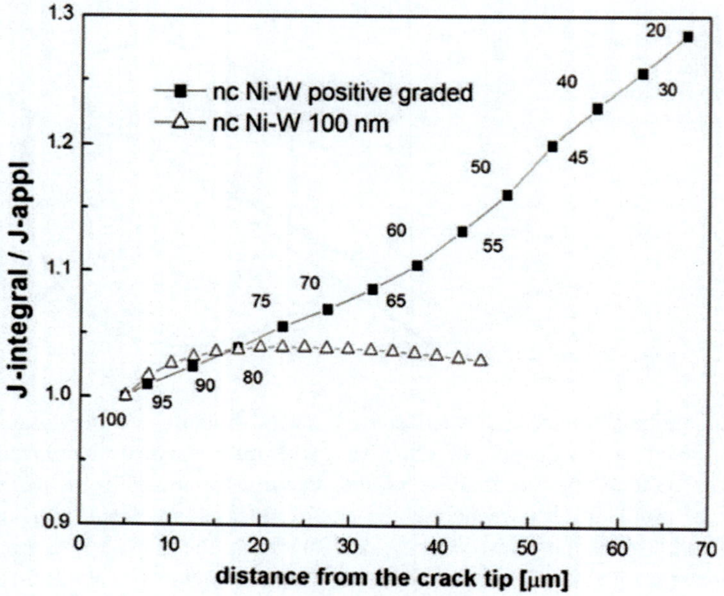

Fig. 4.39 Variation of the *J*-integral as a function of the distance from the crack tip and the comparison with the values of the 100 nm constant grain size Ni-W alloy

$$J = -\frac{\partial V}{\partial a}$$

where *V* is the potential energy.

Actually very interesting results can be obtained by the analysis of the variation of the *J*-integral value for plastically graded materials and from the comparison with the constant grain size one. In fact, in this way it can be shown the potential for such materials to be designed with the correct microstructure variation in order to produce a decrease in the crack propagation and become intrinsically optimal against damage. The variation of the *J*-integral as a function of the distance from the crack tip for the positive-graded Ni-W alloy and the comparison with the values of the 100 nm constant grain size Ni-W alloy are shown in Fig. 4.39.

In this case it can be observed how the *J*-integral increases by passing from the softer to the harder material; this fact demonstrates that the potential energy necessary for the crack propagation decreases in such plastically graded configuration. The principal stress distribution for the negative-graded and 20 nm constant grain size materials is mapped in Fig. 4.40.

The correspondent variation of the J-integral with the distance from the crack tip and the comparison with the positive-graded material can be observed in Fig. 4.41. In such a case the J-integral decreases as passing from the harder to the softer material demonstrating that the potential energy for crack propagation increases in the negative plastic-graded configuration.

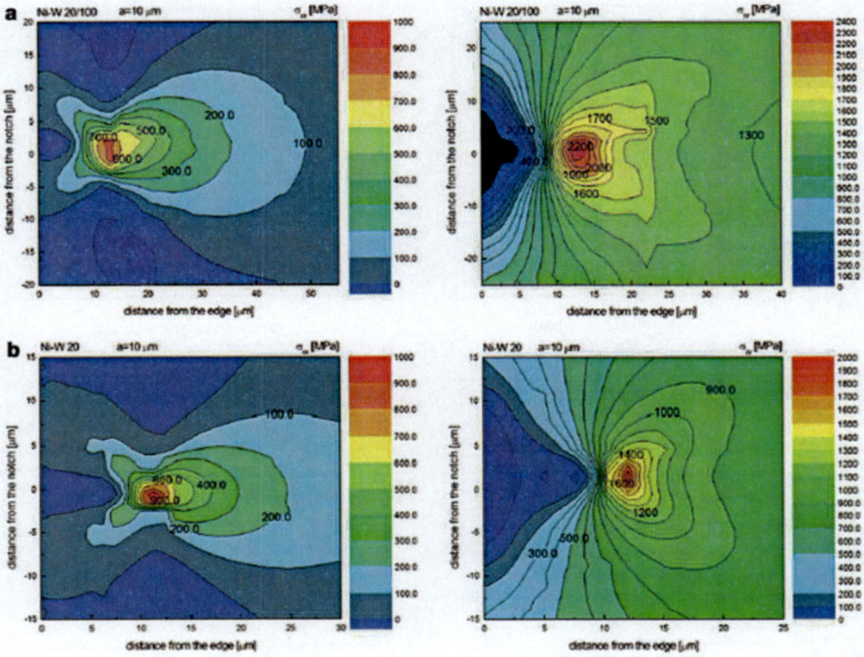

Fig. 4.40 Principal stress maps for the negative-graded (**a**) and 20 nm constant grain size (**b**) Ni-W-electrodeposited nanocrystalline alloys

As a general trend, the driving force for fracture is strongly influenced by whether the crack propagates across interfaces with different mechanical properties. In particular, the energy required for continuing the path increases or decreases linearly in the case of negative or positive plastic-graded properties, respectively.

Through nanoindentation experiments it is possible to calculate mechanical properties such as yield strength, hardness, hardening behavior, and wear characteristics (Fischer-Cripps 2011). The usefulness of instrumented nanoindentation to obtain the fundamental mechanical properties of materials has been widely demonstrated in the past years. Such a technique has much broader application varying from understanding of fundamental materials physics to use as flexible mechanical probe. It is fundamental, in such optics, to investigate the indentation fatigue behavior and relate the resulting data to conventional fatigue and crack evolution tests. In this way it will be possible to elaborate a conventional technique to obtain fatigue properties of materials from direct indentation tests. The last-generation nanoindentation systems are equipped with a numerically controlled loading unit and a high-resolution measurement system for measuring the indentation depth. The loads and the penetrating distances can be controlled. The loading-unloading process can be easily programmed to cyclic loading in different loads and deformation conditions. An extension of this technique is the multistep indentation wherein the sample is loaded-unloaded at the same point with increasing or constant load/depth

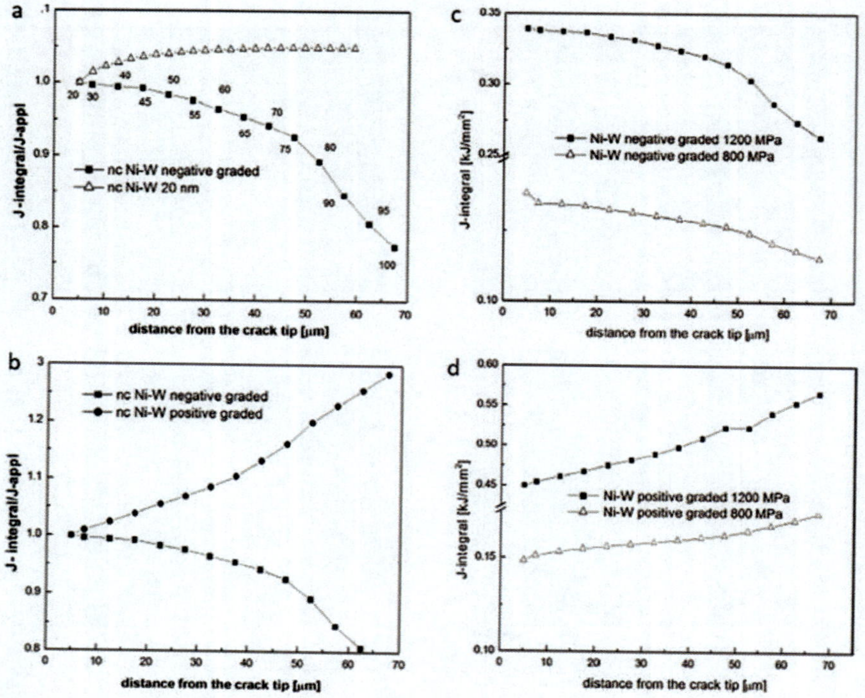

Fig. 4.41 Variation of the J-integral for the plastically graded materials

resulting in more rapid property evaluation. Multistep indentation with increasing load, for example, can provide a rapid hardness measurement of the sample with increasing load along with corresponding microstructure evolution resulting from corresponding increased strains (Cavaliere 2010). It represents a strong independent characterization method since the tested volume of material is scalable with respect to the microstructure. Actually, traditional fatigue tests require a large number of samples to completely characterize the dynamic behavior of metals and alloys (Pan et al. 2006). In addition, it is impossible to observe the microstructural modifications due to cyclic loading variation in a sample-to-sample approach. Cyclic nanoindentation allows to test a material by employing very small samples by obtaining information on fatigue and crack properties by using just one sample.

If the specimen is loaded to a precise maximum load, unloaded, and immediately reloaded a cyclic nanoindentation load curve can be obtained. The loading semi-cycle produces both elastic and plastic deformation in the indentation zone contour, while during unloading a partial recovery of the elastic deformation can be observed. During cyclic loading through a nanoindenter it is possible to observe a steady state in the depth-cycle plot for a certain number of cycles and then an increase in the depth to another step by fixing the maximum indentation load; such a material behavior can be related to fatigue and crack properties. In addition, it is possible to

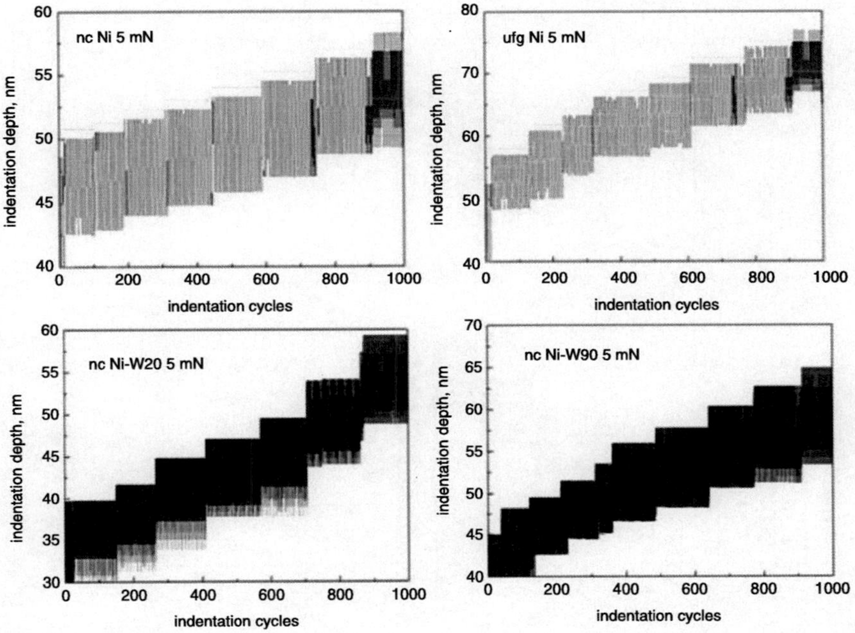

Fig. 4.42 Indentation depth as a function of indentation cycles at 5 mN

provide information about strain hardening, strain rate sensitivity, hardness, and yield properties coupled with fatigue and crack ones by employing the same equipment and then with the same potential error. Especially in nanostructured materials, the dislocation behavior is a strong function of grain structure; with very precise measurements very sensitive dislocation-microstructure interactions can be detected during the unloading-reloading nanoindentation fatigue.

Actually it is well known that in ultrafine grain materials plastic deformation is governed by intracrystalline phenomena such as dislocation slips while in nanocrystalline material the behavior is due to intercrystalline mechanisms such as grain boundary sliding and migration leading to nano-crack formation coupled with dislocation emission from grain boundaries at a certain level of internal stress. For each material such a behavior is related to the hardness-yield relationship giving a precise idea of the grain boundary structure effect on the macroscopic mechanical behavior.

The indentation depth as a function of the indentation number of cycles for a maximum load of 5 (Fig. 4.42) and 10 mN (Fig. 4.43) plots is shown.

The plastic zone propagation can, in this way, be monitored. As a general behavior it is clear the increase in the indentation depth with the number of indentation cycles with differently long plateaus in the depth-cycle steady states. In addition, the bigger the number of cycles during a steady state, the slower the indentation depth propagation; such a behavior is very similar to the one observed in

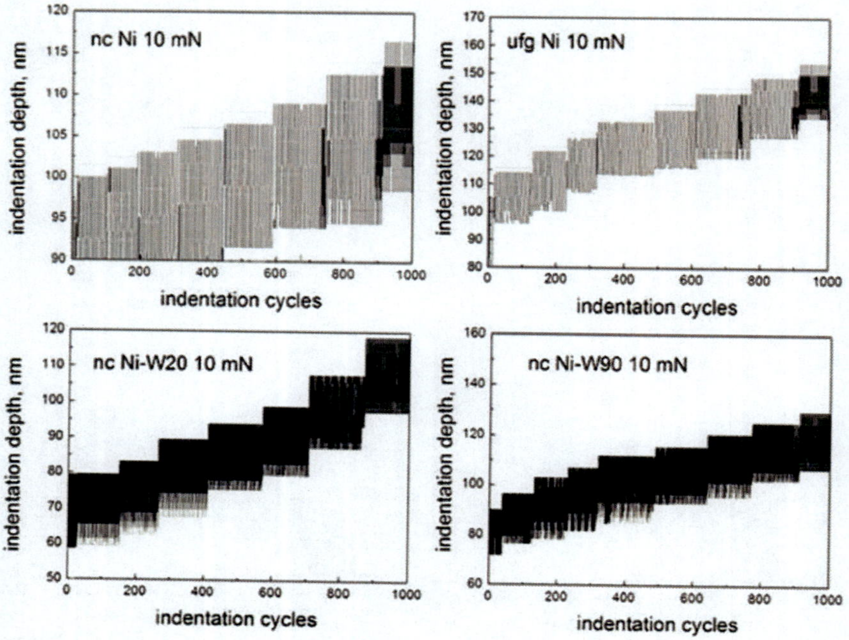

Fig. 4.43 Indentation depth as a function of indentation cycles at 10 mN

conventional fatigue tests being also a good indicator of cyclic softening or hardening of materials. A decrease in the displacement per cycle can be observed with increasing cycle number; such an aspect is due to the increase in the contact area with loading; in this way a power law can be obtained by plotting the total depth as a function of the loading cycles.

By considering the dependence of indentation depth on the number of cycles, it can be observed how there is no substantial modification of indentation depth behavior before the crack occurs. The increase of depth is a power function of the maximum stress depending also on the contact area influenced by the material pileup. During unloading, the partial removal of the stresses leads to dislocation-dislocation interactions in nanograins. This could lead to work hardening of the nanostructured materials upon reloading. It is possible to monitor the crack propagation by comparing the measured change in probe depth between two consecutive maximum loads; so a crack propagation leads to a depth increase between two consecutive cycles. From the plots it can be assumed that the consecutive loading-unloading cycles lead to residual stress accumulation that ends up to crack propagation. The indentation cycle producing the crack is the last one before the depth increase. Such a cycle can also be considered as a failure cycle in a traditional fatigue test. In Fig. 4.44 some examples of indentation steps for different number of cycles at 10 mN can be observed.

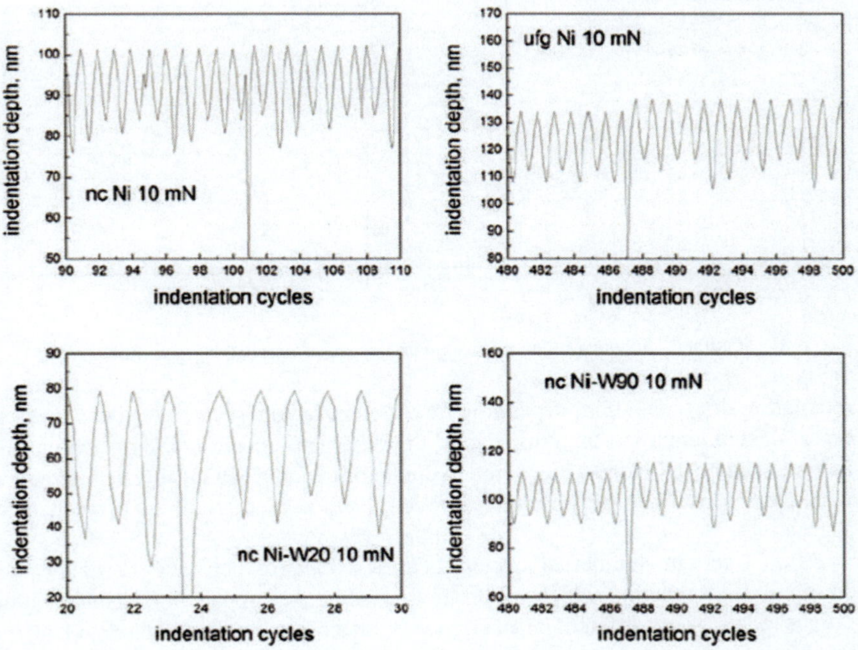

Fig. 4.44 Indentation depth as a function of indentation cycles for electrodeposited Ni and Ni-W alloys

It is possible to extrapolate from the experiments a large number of fatigue data; in Fig. 4.44 the indentation depths are plotted as a function of the number of cycles for the cyclic nanoindentation experiments performed at 5 and 10 mN maximum load; the depths are taken at 50% of the number of cycles in which the materials experienced constant indentation depth. The increase of depth is a power function of the maximum stress. The material behavior can be explained similarly to crack propagation; in static loading the plasticity surrounding the crack tip either blunts the crack or shields the crack tip from the external stress. While the situation is similar in the loading part of the cycle in fatigue loading, however upon unloading, the crack may be sharpened by retracting some dislocations into the crack. Alternatively, the shielding effect of the plastic zone may be reduced. During cyclic loading the crack propagates during the unloading semi-cycle for the modes II and III while for mode I the sharpened crack can propagate in the next loading cycle to compete with dislocation emission, as shown in Fig. 4.44.

During static loading, plastic deformation shields the stress concentration; on the other hand, the unloading semi-cycle of cyclic indentation lets the dislocations to rearrange reducing the internal stresses so the next loading semi-cycle permits the stress concentration to emit more dislocations propagating in the plastic zone. Under such considerations, indentation fatigue has many similarities with crack propagation. In fact, in conventional fatigue, crack propagation is followed by crack

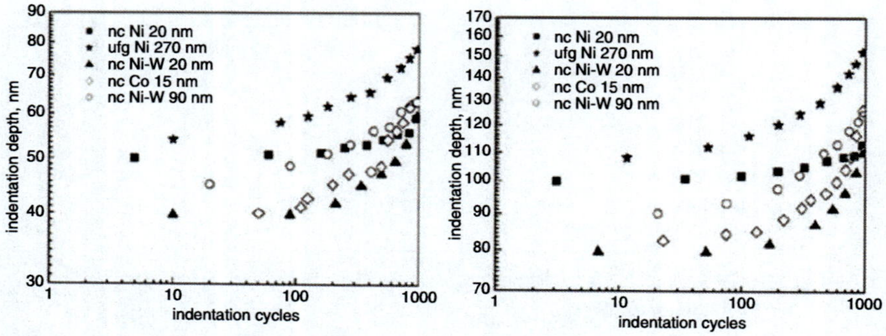

Fig. 4.45 Maximum indentation depth as a function of indentation cycles at 5 and 10 mN

retardation or acceleration, depending on different factors such as residual stresses, crack closure, crack tip blunting, cyclic strain hardening, and crack branching; the indenter can be in this way compared to the tip of the crack and its cyclic interaction with the material can be physically modeled and related to all such mechanical aspects.

After a certain number of cycles there is a balance between the dislocation emission and retraction which results in a steady-state propagation of indentation fatigue depth. The different reached steady states can be directly related to the increase in internal stresses and dislocation generation and movement. The dynamic loading leads to a dynamic process between the effective applied stress and internal stress which is similar to the dislocation generation and annihilation to the crack tip in convention crack propagation tests. The accumulation of plastic deformation during cyclic indentation leads to the nucleation and growth of cracks by increasing the number of cycles; at the same time, this influences the plastic zone proportionally to the maximum load (Fig. 4.45) and the development of plastic zone is directly related to the indentation depth propagation.

Actually the indentation depth, especially at 10 mN maximum load, results over 100 nm with a deformation volume interesting many grains and leading to a bulk plastic zone with respect to the material mean grain size. The cracks propagate with the increase in the number of cycles; there is an increase of cracks close to the indentation zone and such cracks propagate faster as a consequence of the increase in the stress concentration factor.

Actually, in nanoindentation experiments with Berkovich indenter the projected contact area is related to the indentation depth d by the following equation:

$$A = 24.56d^2$$

From the previous equation the equivalent radius a can be obtained; the $K_{\max}$ and the ΔK can be calculated from the following equations:

$$K_{max} = \frac{P_{max}}{2a(\pi a)^{\frac{1}{2}}}$$

$$\Delta K = \frac{\Delta P}{2a(\pi a)^{\frac{1}{2}}}$$

where P_{max} is the maximum indentation load and ΔP is the difference between the maximum and minimum indentation load.

Following such a definition the fracture toughness can be related to the propagation rate in terms of depth variation. In ductile materials the crack growth is dominated by ΔK and the depth-ΔK behavior can be described by the following equation:

$$\frac{da}{dN} = C\Delta K^{n}$$

Very similar to the description of crack length behavior in traditional fatigue crack growth tests:

$$\frac{da}{dN} = C'\Delta K'^{n'}$$

where C′, $\Delta K'$, and n' are the correspondent values for crack growth rate classical curves. The mechanisms related to the evolution of indentation fatigue depth at constant indentation load are comparable to those experienced by the material in fatigue crack growth. In Fig. 4.46 the ΔK vs. depth variation is shown for all the studied materials.

Here the indentation depth behavior is the phenomenological description of the propagation of the plastic zone similar to the crack length variation in the conventional fatigue tests. During cyclic nanoindentation ΔK is the driving force for the plastic zone propagation as a consequence of the stress concentration in the contact zone. There is an increase in ΔK with increasing depth variation rate for all the studied materials and all the maximum load employed during the present nanoindentation experiments. In addition, a decrease can be observed in the nanoindentation fatigue depth propagation rate with increasing number of nanoindentation cycles. For comparison the n and n' values obtained in nanoindentation fatigue and crack growth rate tests on nanostructured metals obtained in previous experiments were related.

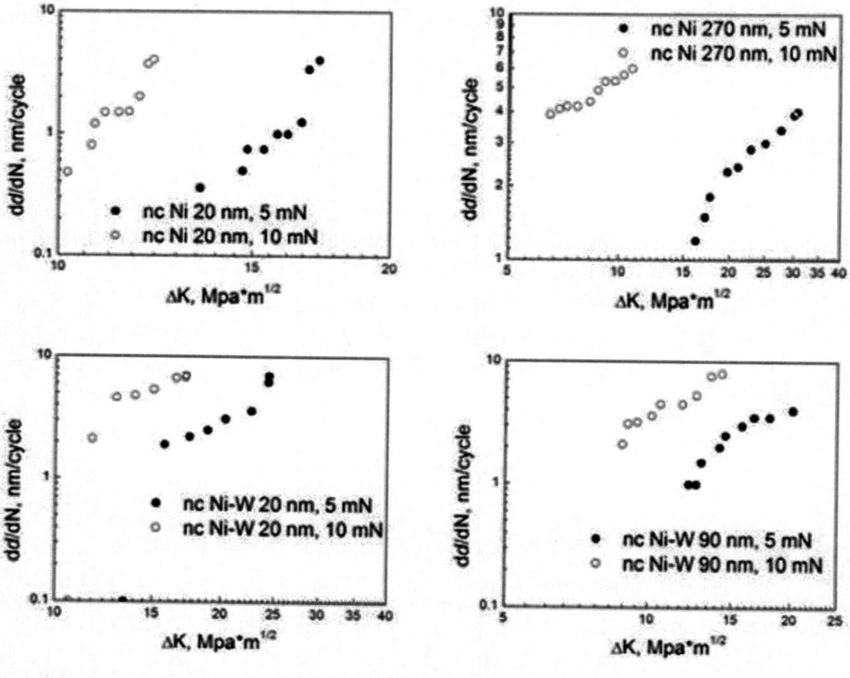

Fig. 4.46 ΔK vs. depth variation for all the studied materials

4.4 Conclusions

Aside from quantitative improvements in fatigue performance, NC metals also hold the possibility of new insight into the mechanisms responsible for traditional fatigue failure. While it was previously noted that crack initiation susceptibility typically decreases with decreasing grain size in metals, it was postulated that the persistent slip mechanism responsible for conventional fatigue crack initiation may be suppressed when the grain size is below a certain threshold, likely on the order of 100 nm or several hundred nanometers. This length scale threshold is between grain sizes that support collective dislocation activity and grain sizes that support individual dislocation activity.

Once a crack initially forms in NC metals, it propagates with very little intrinsic toughening mechanisms to offer resistance. Therefore, thorough examination of fatigue crack initiation mechanisms will be paramount in understanding fatigue resistance in NC metals and, arguably more important, to devise means to further delay crack initiation to develop metals with truly unparalleled fatigue performance.

It has been demonstrated that fatigue crack initiation and growth follow very different behavior in UFG and NC regimes. In general, ductility leads to high plastic deformation before failure leading to a reduction of stress concentrations at the crack tip and at the defect interface by modifying the extension of the plastic and elastic zones; all this allows for the avoiding of catastrophic failure. Materials with low ductile indicator crack as increasing strain experience brittle cleavage fracture. The crack moves forward with flat surfaces. On the contrary, materials with high ductile indicator own low dislocation energy barrier; dislocation/twinning activities occupy a significant role in the whole crack propagation process, leading to the final material failure by surface necking.

As a general behavior, it was observed that the fatigue limit of nanocrystalline metals increases with decreasing grain size, and the crack initiation susceptibility decreases with increasing crack growth rate coupled with the grain refinement. The increase of fatigue resistance can be achieved for many materials in stress-controlled tests. Generally, the main damage mechanism has been recognized in the early strain localization and microcrack formation for SPDed materials. In the HCF regime of intermediate to low plastic strain amplitudes, there results a strong enhancement of the fatigue resistance for materials with grain refinement.

Introducing high density of nanotwins into the pure face-centered cubic (FCC) metals can lead to superior strength—much greater than NC metals—and hardness while maintaining thermal stability and ductility.

Materials in which spatial gradients in structural features are purposely introduced from the surface to the interior are, in some cases, known to exhibit superior mechanical characteristics compared to their homogenous counterparts of appropriately comparable composition.

The advantage to use electrodeposition to produce such kind of materials is represented by the possibility to obtain nanograin materials with a broad grain size ranges with high manufacturing velocity and with a good control of the process obtaining defect-free structures. In addition, electrodeposition allows to obtain structures in which the grain size can be varied to change the consequent mechanical properties in the space by varying the deposition conditions. Such materials, in which the physical and mechanical properties change continuously or in a discrete way across different interfaces, belong to the so-called functionally graded materials (FGM). As a general trend, the driving force for fracture is strongly influenced by whether the crack propagates across interfaces with different mechanical properties. In particular, the energy required for continuing the path increases or decreases linearly in the case of negative or positive plastic-graded properties, respectively.

References

Andrievski RA (2003) Review stability of nanostructured materials. J Mater Sci 38:1367–1375. https://doi.org/10.1023/A:1022988706296

Andrievski RA (2014) Review of thermal stability of nanomaterials. J Mater Sci 49:1449–1460. https://doi.org/10.1007/s10853-013-7836-1

Arutyunyan AR, Arutyunyan RA (2018) Application of the Griffith energy concept to the formulation of the strength criteria for nonlinear-elastic medium with a crack. Mech Solids 53:349–353. https://doi.org/10.3103/S0025654418070130

Balbus GH, Wang F, Gianola DS (2020) Suppression of shear localization in nanocrystalline Al-Ni-Ce via segregation engineering. Acta Mater 188:63–78. https://doi.org/10.1016/j.actamat.2020.01.041

Boyce BL, Padilla HA (2011) Anomalous fatigue behavior and fatigue-induced grain growth in nanocrystalline nickel alloys. Metal Mater Trans A42:1793–1804. https://doi.org/10.1007/s11661-011-0708-x

Boyce BL, Furnish TA, Padilla HA et al (2015) Detecting rare, abnormally large grains by x-ray diffraction. J Mater Sci 50:6719–6729. https://doi.org/10.1007/s10853-015-9226-3

Cavaliere P (2008) Crack tip plasticity in plastically graded Ni–W electrodeposited nanocrystalline alloys. Comput Mater Sci 41:440–449. https://doi.org/10.1016/j.commatsci.2007.05.007

Cavaliere P (2009a) Fatigue properties and crack behavior of ultra-fine and nanocrystalline pure metals. Int J Fatigue 31:1476–1489. https://doi.org/10.1016/j.ijfatigue.2009.05.004

Cavaliere P (2009b) Mechanical properties of nanocrystalline metals and alloys studied via multi-step nanoindentation and finite element calculations. Mater Sci Eng A 512:1–9. https://doi.org/10.1016/j.msea.2009.03.008

Cavaliere P (2010) Cyclic deformation of ultra-fine and nanocrystalline metals through nanoindentation: similarities with crack propagation. Procedia Eng 2:213–222. https://doi.org/10.1016/j.proeng.2010.03.023

Cavaliere P, Cabibbo M (2008) Effect of Sc and Zr additions on the microstructure and fatigue properties of AA6106 produced by equal-channel-angular-pressing. Mater Charact 59:197–203. https://doi.org/10.1016/j.matchar.2006.12.013

Chan T, Backman D, Bos R, Sears T, Brooks I, Erb U (2011) In situ heat generation and strain localization of polycrystalline and nanocrystalline nickel. In: Thermomechanics and infra-red imaging, Conference proceedings of the society for experimental mechanics series, vol 7. Springer, New York. https://doi.org/10.1007/978-1-4614-0207-7_3

Chan T, Zhou Y, Brooks I et al (2014) Localized strain and heat generation during plastic deformation in nanocrystalline Ni and Ni–Fe. J Mater Sci 49:3847–3859. https://doi.org/10.1007/s10853-014-8099-1

Cheng S, Zhao Y, Wang Y, Li Y, Wang X-L, Liaw PK, Lavernia EJ (2010) Structure modulation driven by cyclic deformation in nanocrystalline NiFe. Phys Rev Lett 104:255501. https://doi.org/10.1103/PhysRevLett.104.255501

Cheng S, Lee SY, Li L, Lei C, Wang X-L, Ungar T, Wang Y, Liaw PK (2013) Uncommon deformation mechanisms during fatigue-crack propagation in nanocrystalline alloys. Phys Rev Lett 110:135501. https://doi.org/10.1103/PhysRevLett.110.135501

Dai PQ, Zhang C, Wen JC et al (2016) Tensile properties of electrodeposited nanocrystalline Ni-Cu alloys. J Mater Eng Perform 25:594–600. https://doi.org/10.1007/s11665-016-1881-2

Farkas D, Willemann M, Hyde B (2005) Atomistic mechanisms of fatigue in nanocrystalline metals. Phys Rev Lett 94:165502. https://doi.org/10.1103/PhysRevLett.94.165502

Feng H, Tang J, Peng J, Wu H (2019) Nanoscale amorphization effect on dislocation emission from an elliptical blunt crack tip in deformed nanocrystalline and ultrafine-grained materials. Mech Mater 134:98–105. https://doi.org/10.1016/j.mechmat.2019.04.019

Fischer-Cripps AC (2011) Applications of nanoindentation. In: Nanoindentation, Mechanical engineering series. Springer, New York. https://doi.org/10.1007/978-1-4419-9872-9_12

Furnish TA, Boyce BL, Sharon JE, O'Brien CJ, Clark BG, Arrington CL, Pillars JL (2016) Fatigue stress concentration and notch sensitivity in nanocrystalline metals. J Mater Res 31(6):740–752. https://doi.org/10.1557/jmr.2016.66

Furnish TA, Mehta A, Van Campen D, Bufford DC, Hattar K, Boyce BL (2017) The onset and evolution of fatigue-induced abnormal grain growth in nanocrystalline Ni–Fe. J Mater Sci 52:46–59. https://doi.org/10.1007/s10853-016-0437-z

Gupta A, Zhou X, Thomson GB, Tucker GJ (2020) Role of grain boundary character and its evolution on interfacial solute segregation behavior in nanocrystalline Ni-P. Acta Mater 190:113–123. https://doi.org/10.1016/j.actamat.2020.03.012

He X, Zhu L, Liu J, An L (2019) Grain growth-induced strain softening in nanocrystalline magnesium: experiments and modelling. Mater Res Express 6:108002. https://doi.org/10.1088/2053-1591/ab3e8e

Heckman NM, Padilla HA, Michael JR, Barr CM, Clark BG, Hattar K, Boyce BL (2020) Rethinking scaling laws in the high-cycle fatigue response of nanostructured and coarse-grained metals. Int J Fatigue 134:105472. https://doi.org/10.1016/j.ijfatigue.2020.105472

Hohenwarter A, Pippan R (2011) Fracture toughness evaluation of ultrafine-grained nickel. Scr Mater 64:982–985. https://doi.org/10.1016/j.scriptamat.2011.02.007

Hohenwarter A, Pippan R (2013) Fracture of ECAP-deformed iron and the role of extrinsic toughening mechanisms. Acta Mater 61:2973–2983. https://doi.org/10.1016/j.actamat.2013.01.057

Hohenwarter A, Pippan R (2015) Fracture and fracture toughness of nanopolycrystalline metals produced by severe plastic deformation. Philos Trans A Math Phys Eng Sci 373:20140366. https://doi.org/10.1098/rsta.2014.0366

Kalidindi AR, Schuh CA (2017) Stability criteria for nanocrystalline alloys. Acta Mater 132:128–137. https://doi.org/10.1016/j.actamat.2017.03.029

Kobayashi S, Kamata A, Watanabe T (2009) Roles of grain boundary microstructure in high-cycle fatigue of electrodeposited nanocrystalline Ni–P alloy. Scripta Mater 61:1032–1035. https://doi.org/10.1016/j.scriptamat.2009.08.021

Kobayashi S, Kamata A, Watanabe T (2015) A mechanism of grain growth-assisted intergranular fatigue fracture in electrodeposited nanocrystalline nickel–phosphorus alloy. Acta Mater 91:70–82. https://doi.org/10.1016/j.actamat.2015.03.028

Koch C (2003) Optimization of strength and ductility in nanocrystalline and ultrafine grained metals. Scr Mater 49:657–662. https://doi.org/10.1016/S1359-6462(03)00394-4

Leitner T, Hohenwarter A, Pippan R (2015) Revisiting fatigue crack growth in various grain size regimes of Ni. Mater Sci Eng A 646:294–305. https://doi.org/10.1016/j.msea.2015.08.071

Li Q, Xue S, Zhang Y, Sun X, Wang H, Zhang X (2020) Plastic anisotropy and tension-compression asymmetry in nanotwinned Al–Fe alloys: an in-situ micromechanical investigation. Int J Plasticity 132:102760. https://doi.org/10.1016/j.ijplas.2020.102760

Long J, Pan Q, Tao N, Dao M, Suresh S, Lu L (2019) Improved fatigue resistance of gradient nanograined Cu. Acta Mater 166:56–66. https://doi.org/10.1016/j.actamat.2018.12.018

Mroz KP, Bigos A, Kucharski S et al (2014) Ni-W electrodeposited coatings on low carbon steel substrate: fatigue observations. J Mater Eng Perform 23:3459–3466. https://doi.org/10.1007/s11665-014-1149-7

Nelson S, Ladani L, Topping T, Lavernia E (2011) Fatigue and monotonic loading crack nucleation and propagation in bimodal grain size aluminum alloy. Acta Mater 59:3550–3570. https://doi.org/10.1016/j.actamat.2011.02.029

Pan D, Nieh TG, Chen MW (2006) Strengthening and softening of nanocrystalline nickel during multistep nanoindentation. Appl Phys Lett 88:161922. https://doi.org/10.1063/1.2197289

Pineau A, Benzerga AA, Pardoen T (2016) Failure of metals III: fracture and fatigue of nanostructured metallic materials. Acta Mater 107:507–544. https://doi.org/10.1016/j.actamat.2015.07.049

Ramesh K (2009) Plastic deformation of nanomaterials. In: Nanomaterials. Springer, Boston. https://doi.org/10.1007/978-0-387-09783-1_5

Rathman D, Marx M, Motz C (2017) Crack propagation and mechanical properties of electrodeposited nickel with bimodal microstructures in the nanocrystalline and ultrafine grained regime. J Mater Res 32(24):4573–4582. https://doi.org/10.1557/jmr.2017.353

Sabirov I, Estrin Y, Barnett MR, Timokhina I, Hodgson PD (2008) Tensile deformation of an ultrafine-grained aluminium alloy: micro shear banding and grain boundary sliding. Acta Mater 56:2223–2230. https://doi.org/10.1016/j.actamat.2008.01.020

Selyutina NS, Borodin EN, Petrov YV (2018) Structural–temporal peculiarities of dynamic deformation of nanostructured and nanoscaled metals. Phys Solid State 60:1813–1820. https://doi.org/10.1134/S1063783418090275

Singh A, Tang L, Dao M, Lu L, Suresh S (2011) Fracture toughness and fatigue crack growth characteristics of nanotwinned copper. Acta Mater 59:2437–2446. https://doi.org/10.1016/j.actamat.2010.12.043

Tian L, Li L (2018) A review on the strengthening of nanostructured materials. Int J Curr Eng Technol 8(2):236–249. https://doi.org/10.14741/ijcet/v.8.2.7

Vinogradov A, Hashimoto S (2003) Fatigue of severely deformed metals. Adv Eng Mater 5:351–358. https://doi.org/10.1002/adem.200310078

Wang Y, Chen M, Zhou F, Ma E (2002) High tensile ductility in a nanostructured metal. Nature 419:912–915. https://doi.org/10.1038/nature01133

Wang J, Han L, Huang Y, Liu Y, Wang Z (2020a) Temperature-dependent evolution of strength of nanocrystalline Ni(Mo) alloys at the Mo solubility limit. Mater Sci Eng A 786:139326. https://doi.org/10.1016/j.msea.2020.139326

Wang L, Li M, Tan H et al (2020b) Enhanced mechanical properties of a gradient nanostructured medium manganese steel and its grain refinement mechanism. J Mater Eng Perform 29:3812. https://doi.org/10.1007/s11665-020-04903-w

Wu H, Fan G (2020) An overview of tailoring strain delocalization for strength-ductility synergy. Prog Mater Sci 113:100675. https://doi.org/10.1016/j.pmatsci.2020.100675

Xie J, Wu X, Hong Y (2007) Shear bands at the fatigue crack tip of nanocrystalline nickel. Scripta Mater 57:5–8. https://doi.org/10.1016/j.scriptamat.2007.03.027

Xu W, Ramirez K, Gomez S, Lee R, Hasan S (2019) A bimodal microstructure for fatigue resistant metals by molecular dynamics simulations. Comput Mater Sci 160:352–359. https://doi.org/10.1016/j.commatsci.2019.01.026

Yang Y, Imasogie B, Fan GJ, Liaw PK, Soboyejo WO (2008) Fatigue and fracture of a bulk nanocrystalline NiFe alloy. Metal Mater Trans 39A:1145–1156. https://doi.org/10.1007/s11661-008-9487-4

Yasbolaghi R, Khoei AR (2020) Micro-structural aspects of fatigue crack propagation in atomistic scale via the molecular dynamics analysis. Eng Fract Mech 226:106848. https://doi.org/10.1016/j.engfracmech.2019.106848

Zhang W, Simpson CA, Lopez-Crespo P, Mokhtarishirazabad M, Buslaps T, Pippan R, Whiters PJ (2020) The effect of grain size on the fatigue overload behaviour of nickel. Mater Des 189:108526. https://doi.org/10.1016/j.matdes.2020.108526

Zhou H, Zhu P (2020) Correlated necklace dislocations in highly oriented nanotwinned metals. J Zhejiang Univ Sci A 21:294–303. https://doi.org/10.1631/jzus.A1900637

Zhou X, Li X, Chen C (2015) Atomistic mechanisms of fatigue in nanotwinned metals. Acta Mater 99:77–86. https://doi.org/10.1016/j.actamat.2015.07.045

Chapter 5
Fatigue and Crack Behavior of Bulk Nanostructured Metal Alloys and Composites

5.1 Introduction

It has been largely shown how severe plastic deformation allows for the strong increase of strength in metals and alloys. The advent of nanocrystalline and ultrafine grain metals has raised new questions regarding the validity of existing strengthening theories (Koch et al. 2017). In the strengthening mechanisms for microcrystalline alloys, dislocations are considered to move only within the matrix. No dislocation-dislocation or dislocation-grain boundary interactions are considered. Now, any polycrystalline metal can be broadly classified into two distinct regions: (a) matrix and (b) grain boundary. The hypothesis of noninteracting dislocations, though valid for microcrystalline metals, is a questionable assumption for ultrafine grain microstructures. To understand the issue further, consider the typical length scale of dislocation movement at room temperatures. At room-temperature conditions this is in the range of 1–5 μm (as evidenced from cell size and PSB dimensions). This observed length scale of movement is in fact larger than the grain size in ultrafine grain (<1 μm). Consequently, dislocation interaction with grain boundaries and associated strengthening mechanism changes becomes important. This change in strengthening mechanism is amply illustrated in the shift of strength-ductility combinations for UFG materials compared to their microcrystalline counterparts.

By taking into account the influence of two types of boundaries in UFG materials, i.e., HAGB and LAGB, the strengthening contribution to the yield stress can be given by

$$\sigma_Y = \sigma_0 + \sigma_{GB} + \sigma_{SGB} = \sigma_{P\text{-}N} + \sigma_{SSS} + \sigma_{GB} + \sigma_{SGB}$$

where σ_0 takes into consideration the strengthening contributions of Peierls-Nabarro stress, $\sigma_{P\text{-}N}$; and σ_{SSS} accounts for the solid solution strengthening; and σ_{GB} and σ_{SGB} are the strengthening contributions from HAGB and LAGB, respectively (Munoz et al. 2017).

© Springer Nature Switzerland AG 2021

P. Cavaliere, *Fatigue and Fracture of Nanostructured Materials*,
https://doi.org/10.1007/978-3-030-58088-9_5

The Peierls-Nabarro stress can be calculated with

$$\sigma_{\text{P-N}} = \frac{G}{1-\upsilon} e^{-\frac{2\pi a}{b(1-\upsilon)}} s$$

a is the interplanar spacing of the primary slip, G is the shear modulus, and b is the Burgers vector. EBSD analysis also enables an evaluation of LAGB contribution to the increase of strength. This can be treated as dislocation strengthening (Taylor strengthening):

$$\sigma_{\text{SGB}} = \alpha GMB\sqrt{\rho_0 + \rho_{\text{LAGB}}}$$

ρ_{LAGB} is the dislocation density stored in LAGBs (subgrains):

$$\rho_{\text{LAGB}} = \frac{3(1-f)\overline{\theta}_{\text{LAGB}}}{bL}$$

where f is the HAGB fraction, θ_{LAGB} the average misorientation of the LAGB, and L the average boundary spacing of subgrains. In addition, the strengthening contribution from the HAGB follows the Hall-Petch relationship as

$$\sigma_{\text{GB}} = k\left(\frac{f}{d}\right)^2$$

The major contribution to the material strength comes from the HAGB, for the processed material. However, an increase in the amount of the LAGB contribution for the severely deformed material can be explained by the reduction in the grain size (as a consequence of the great deformation introduced during the ECAP process due to the fragmentation of the initial coarse grains by the generation of new subgrains which will evolve into grains with the increase in the deformation). As a matter of fact, studies performed on pure iron processed via ECAP gave the results summarized in Fig. 5.1.

Here, anisotropy due to the different processing routes can strongly influence mechanical properties such as fatigue and crack propagation (Bruder et al. 2018).

A disadvantage of bulk UFG metals is their highly nonequilibrium structure with large internal stresses leading to their low thermal stability and limited plasticity due to crack nucleation in GB regions (Goto et al. 2007).

The enhancement of mechanical properties of NC metals and alloys through the finding of methods to prevent premature crack formation in their structures is a sophisticated task of materials science. This is because the crack nucleation and propagation mechanisms depend on many factors such as the loading mode, temperature, chemical, phase composition, stacking fault energy, grain size, texture, and impurities (Waitz et al. 2014). The nature of SPD techniques inevitably empowers the SFE to have a fundamental impact on the formation of nanostructures and their

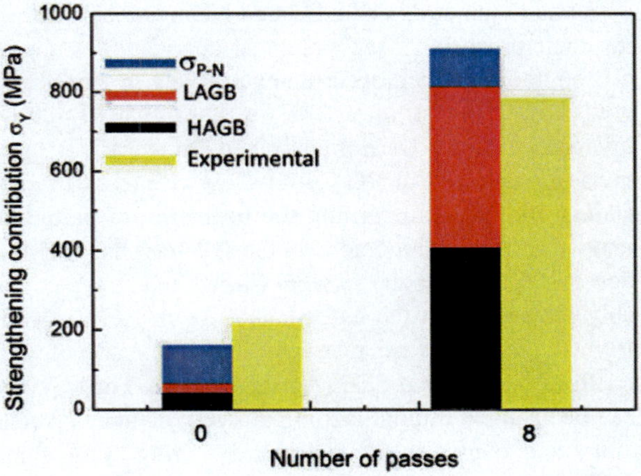

Fig. 5.1 Strengthening mechanism during ECAP of pure iron

corresponding properties. The comprehensive exploration of the importance of SFE during SPD is vital both scientifically for understanding microstructure evolution and deformation mechanisms and technologically concerning its applications in guiding material preparation (An et al. 2019). In materials with high and medium SFE, geometrically necessary dislocations and statistically stored dislocations usually develop at the onset of plastic deformation. Geometrically necessary dislocations function to maintain the compatibility of deformation and accommodate the shear strain gradients in various parts of a material when suffering inhomogeneous deformation, while statistically stored dislocations are formed by trapping each other in a random way under uniform deformation. Owing to the active motions of dislocations facilitated by cross-slip in these materials, they are inclined to be assembled into boundaries with strain, and the low-energy dislocation structures are achieved via their mutual trapping, rearrangement, and annihilation. Consequently, geometrically necessary boundaries (GNBs) and incidental dislocation boundaries (IDBs) are derived from the geometrically necessary dislocations and the statistically stored dislocations, respectively, to bear the large heterogeneous or homogeneous plastic strain. When SPD techniques are applied at room temperature and a low strain rate, the choice of the grain refinement mechanism is mainly regulated by the SFE of the material. In materials with high and medium SFE, complicated dislocation activities including their multiplication, accumulation, interaction, and spatial rearrangement manage the formation of nanostructures during SPD. The overall fatigue properties in both the LCF region and HCF regime can be increased significantly with lowering of the SFE in NC alloys in comparison with UFG Cu. Also, it should be mentioned that better LCF properties might be obtained through low-temperature annealing to relax the nonequilibrium GB structures in these NC alloys with low SFE. Unveiling the roles of SFE in cyclic deformation responses of UFG and NC materials is crucial, since it provides an excellent opportunity to thoroughly comprehend the fatigue mechanisms and extend the understanding of their structure-property relation, while it also furnishes an efficient compass to the design of UFG and NC materials with superior properties, broadening their prospective technological application. Fatigue-induced grain growth via GB migration, which occurred extensively in UFG Cu, was improved significantly with lowering of the SFE in NC Cu-Al alloys. The essential fatigue micromechanisms were transformed from the GB migration in UFG Cu to other local GB activities in NC Cu-Al alloy with low SFE, such as atom shuffling or GB sliding and rotation, causing the formation of partial/full amorphous structures or nano-voids and nano-cracks in the GB regions. Highly localized shear bands were developed substantially in UFG Cu, while this type of surface damage was remarkably lessened with the reduction of SF in NC Cu-Al alloys. This might originate from the enhanced microstructural stability and the change in GB character as a result of the lowered SFE (An et al. 2014). For wavy-slip materials, the slip mode can be changed through adding alloying elements (such as adding Zn or Al to Cu), and this is confirmed to be primarily owing to the change in stacking fault energy (SFE) and short-range order (SRO). It has been fully reported that the change of the SFE and slip mode may affect many material behaviors during plastic deformation. For example, by adding elements to reduce SFE, the microstructure evolution and refinement mechanism will be changed and the strength-ductility cooperation will be improved during the SPD process; the dislocation evolution and plastic deformation mechanism will be transformed; and even the fatigue cracking modes of twin boundaries will be affected significantly (Zhang et al. 2013).

So, a possible way to overcome these drawbacks is to utilize the GB segregation hardening. So far, attention has been given mainly to adverse GB segregation that leads to an intergranular embrittlement and/or decrease in corrosion resistance of coarse-grained polycrystalline metallic materials. Meanwhile, it is believed that the GB segregation hardening is a new approach that could help to avoid premature crack formation and enhance ductility and thermal stability of the NC metals and alloys having, unlike their coarse-grained polycrystalline counterparts, a dense GB network (Murashkin et al. 2015). An example of the difference between pure Al and Al-Co with GB segregation is shown in Fig. 5.2.

At the same shear strain level (20%), pure aluminum shows void formation and high stress concentration at triple junctions. On the contrary, the UFG Al-Co does not show the presence of voids.

A model taking into account the strengthening contributions for SPD alloys is shown by Chembarisova et al. (2012). The following strengthening mechanisms are considered to be operating in the alloys: work hardening τ_p, solution hardening τ_c, precipitate hardening τ_f, lattice friction τ_0 (the power of Peierls-Nabarro). These mechanisms interact with each other in a certain way. Their combination may be simple as well as complicated, when each mechanism has an effect on others.

UFG materials in general exhibit a combination of high yield/tensile strength and a low uniform ductility. Thus, although the strengths of UFG materials are high their ductility is low compared to their coarse grain counterparts. As a matter of fact, SPD pure titanium has tensile and fatigue properties comparable to those belonging to Ti6Al4V for both ECAP and HPT (Semenova et al. 2009; Medvedev et al. 2015; Plekhov et al. 2009; Naseri et al. 2018).

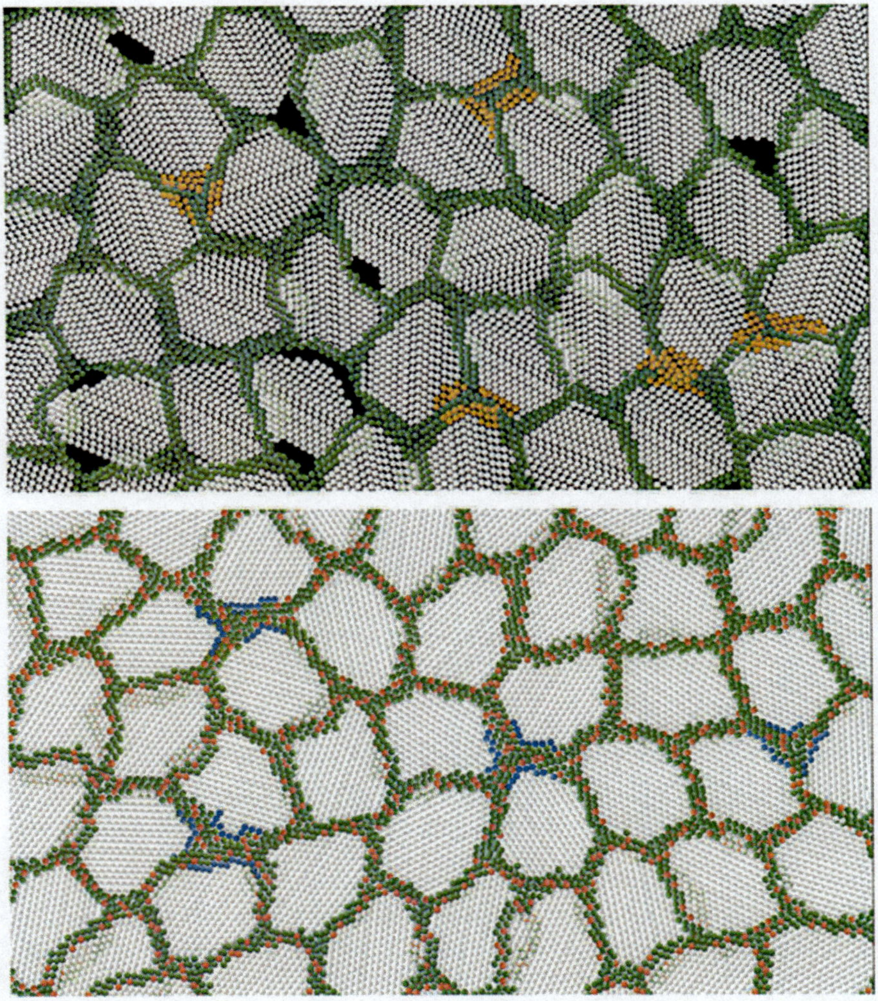

Fig. 5.2 MD simulations of shear straining (20%) of EFG Al and Al-Co

Now, ductility of any metal is determined by its dislocation storage ability. Thus, as long as dislocation storage is homogeneous, a uniform deformation is achieved (e.g., uniform elongation region in a tensile test). Heterogeneous dislocation accumulation on the other hand causes internal notch formation with consequent onset of failure (e.g., nonuniform elongation region in a tensile test). At steady-state stress, the ductility is therefore determined by the following condition:

$$\dot{\rho}_{\text{nucleation}} - \dot{\rho}_{\text{annihilation}} = \dot{\rho}_{\text{storage}}$$

The low ductility of UFG materials as compared to their coarse grain counterparts therefore indicates a lower dislocation storage rate that can be detrimental of fatigue life especially at very high plastic strain amplitudes (Saitova et al. 2009) for those high-strength materials. It is shown for different aluminum alloys where fatigue strength increases after one ECAP pass and then it decreases by increasing the number of passes (Esmaeili et al. 2018; Namdar and Jahromi 2015). Although the obtained improvement by one pass of ECAP disappeared in LCF regime with increasing number of passes, this improvement sustained in HCF regime.

The cyclic deformation responses of SPDed metals as available in the related literature are mixed, ranging from cyclic hardening to cyclic softening depending on various factors. These factors include the purity of the metal, the level of the imposed cyclic stress/strain, and the processing parameters. The effect from the above factors has been categorized into three fundamental influencing factors in this review.

The primary influencing factor is the microstructure stability of the initial microstructure. This includes the effects of the purity of the metals, the level of development of a UF-grained microstructure, and the grain geometry of the final microstructure. The latter two are affected by the processing technique and the parameters used (Valiev 2007). In short, the above influences the cyclic deformation response by changing the driving force for the grain coarsening phenomenon. The second influencing factor is the cyclic life span, which provides, or not, the satisfaction of the required kinetics for grain coarsening phenomenon. There are reports in which the thermodynamic driving force is in abundance but the short cycling time available limits the extent of grain coarsening. Lastly, the applied cyclic plastic strain level is also an influencing factor. Such an influence is based on changing the level of energy input for the grain coarsening phenomenon. In the case of high-purity metals, this effect is often overshadowed by the limitation by the cyclic life span of the sample and is only apparent at low-imposed plastic strain amplitudes (Kwan and Wang 2012).

A softening effect induced by cyclic loading is observed after EACP in SS316L especially with increasing strain amplitude (Kaneko et al. 2013). However, it is less prominent than that in pure FCC metals manufactured by SPD, which is attributed to the high stability of twin boundaries. Nanostructure formation by twinning of low-SFE materials through SPD processing is a promising technology to tailor novel microstructures with desired multifunctional properties (Ueno et al. 2011; Korn et al. 2011).

In addition, the inapplicability of Hart's stability criterion has been shown for UFG materials because of the observation of micro-shear appearance. Another notable phenomenon was the increase in uniform ductility with a reduction in the strain rate of testing. This increase in ductility was accompanied by a decrease in flow stress. It may be mentioned here that for UFG materials a slight increase in the strain rate sensitivity has been observed.

With this increase of strength accompanied with a reduction in ductility, many strategies have been developed in order to obtain acceptable levels in order to increase also the fatigue life and the damage tolerance. The first observation is that mechanical properties in UFG materials are direction dependent as a consequence of grain elongation and textures (Rathmayr et al. 2013). The fatigue properties in severely deformed UFG materials fabricated by ECAP depend on the specific shear texture induced during the fabrication which promotes orientation of the easy glide crystallographic planes to be aligned parallel to the shear plane in ECAP (Higuera-Cobos et al. 2014). The high-cyclic fatigue life in severely deformed UFG materials fabricated by ECAP can be improved through the controlled degradation of the ECAP-induced texture allowing effective rotation of the texture components corresponding to the easy dislocation glide plane from the ideal simple shear position (Orlov and Vinogradov 2011).

5.1.1 Anisotropy in SPDed Materials

Fracture toughness is strongly dependent on the grain size and shape as well as on the orientation of the specimen (Kazemi and Seifi 2018). In detail, the crack growth resistance is low for cracks parallel to the shear plane acting on a direction parallel to the longer axis of the elongated grains with respect to a perpendicular direction. So, the axial direction (AD), radial direction (RD), and tangential directions (TD) must be considered for HPT materials (Fig. 5.3).

And for ECAPed materials the radial (R) and extrusion (E) directions must be considered (Fig. 5.4).

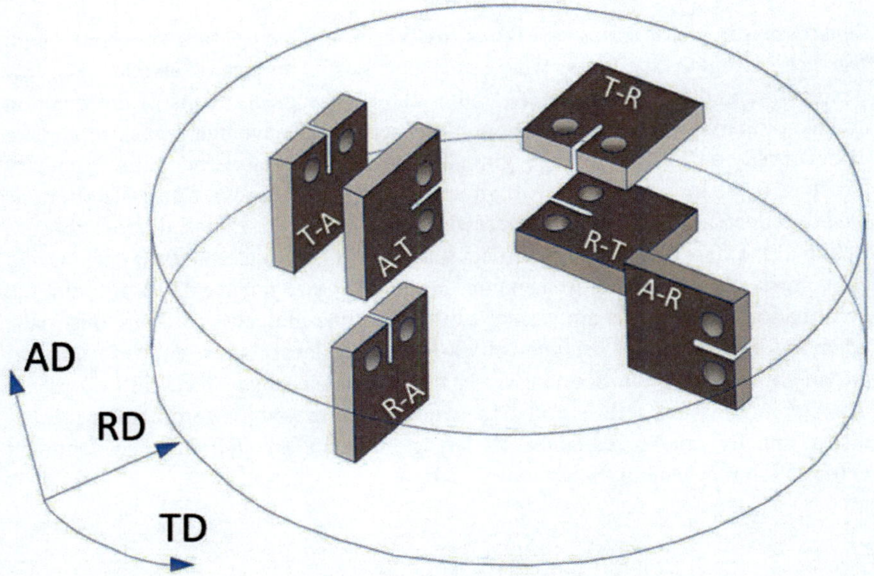

Fig. 5.3 Fracture toughness sample orientation for HPTed materials

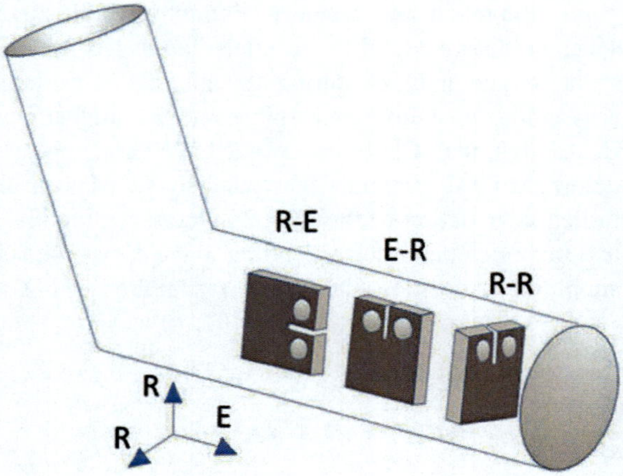

Fig. 5.4 Fracture toughness sample orientation for ECAPed materials

The most employed tests performed on UFG materials produced via severe plastic deformation are mode I type on pre-cracked samples (Fig. 5.5).

As a general behavior, in ultrafine-grained materials more grains lie in the plastic zone with respect to their microcrystalline counterparts. So, the potential interaction of the crack with grain boundaries can retard the FCG (Holzapfel et al. 2007).

5.2 Fatigue Life of UFG Materials

Severe plastic deformation (SPD) is useful in producing bulk materials with enhanced strength, hardness, wear, and superplastic properties at relatively low temperature and high strain rates (Valiev 2000). The mechanisms of deformation and the properties of the material not only depend on the average grain size, but are also strongly influenced by the grain size distribution and the grain boundary structure (e.g., low-angle versus high-angle grain boundaries). Small-sized grains lead to reduced flaw sizes and increased difficulties for the imposed stress concentration at the flaw to exceed the critical toughness of the material, thus suppressing early crack nucleation and propagation. In the grain size range of 100–500 nm, the deformation mechanisms are similar to those in fine-grained traditional materials, whereas for grain sizes in the range of 50–100 nm, dislocations are emitted from, and are annihilated at, grain boundaries; in the grain size range of 10–50 nm, partial dislocation emission and deformation twinning are the major deformation mechanisms; and for grain sizes below 10 nm, grain boundary sliding is the dominant deformation mechanism (Sabirov et al. 2015).

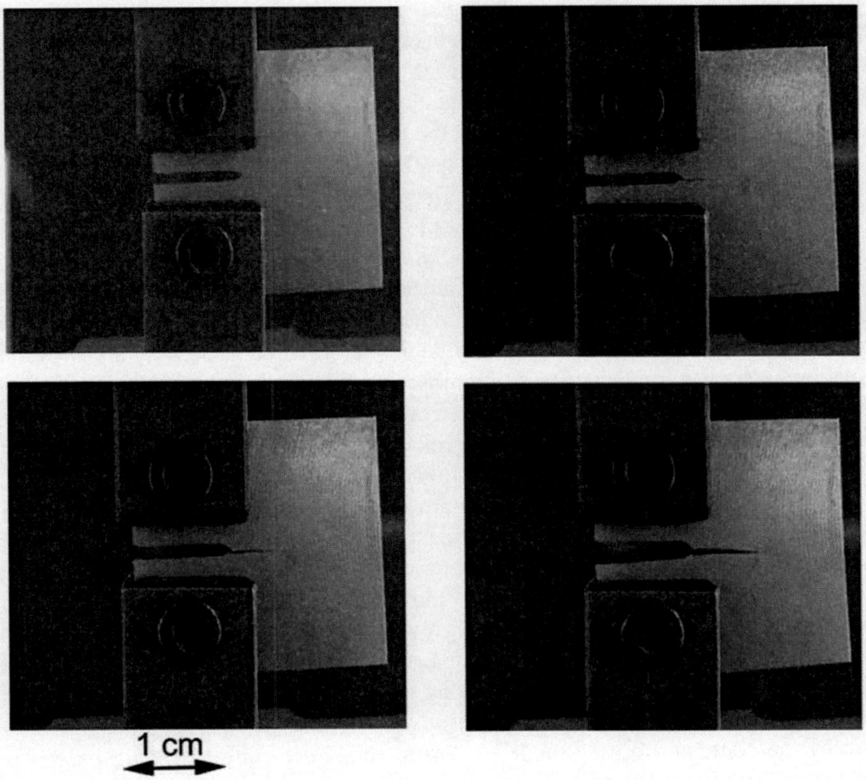

Fig. 5.5 Mode I crack propagation tests on ECAPed samples

5.2.1 Deformation Mechanisms in UFGed Materials

In UFG regime, the fundamental mechanisms for failure change with respect to the microcrystalline regime. This is due first of all to the aspect that the submicron grain sizes of UFG metals are similar or smaller in length scale than the cell size observed in single-crystal or microcrystalline alloys. In addition, UFG grain sizes have dimensions comparable to the typical PSB width and wall thickness combined together. Consequently, significant difference in the damage evolution of a UFG metal (compared to microcrystalline metals) is expected. The UFG material reveals substructural features—subgrains, dislocation cells, and X-ray coherent diffraction regions. The structure bears traces of the ECAP process. The cells are often elongated in the shear direction along the last ECAP plane. The details of the structure are dependent on the ECAP route; the Bc route yields the structure of highest equiaxiality. Further, the scatter of the cell size is often substantial.

Quantitative determination of grain or cell size of materials processed by ECAP is complicated by the fact that the size is varying broadly between hundreds of nanometers and some microns and by not-well-defined boundaries in transmission electron microscopy (TEM) or electron back scattering diffraction (EBSD) images. The mutual orientation of structural units cannot be satisfactorily described as homogeneous high-angle random orientation (Soliman et al. 2012). That is why the term grain size is often replaced by the term cell size to point out that there are areas containing crystallites with very similar orientation. There are regions where the low-angle boundaries are frequent. These regions can be described as "zones of nearby oriented grains" and they seem to play an important role in the localization of cyclic plasticity, crack initiation, and early crack propagation (Kunz et al. 2014).

The deep application of ultrafine-grained nanocrystalline metals in the modern industry is related to the increase of the understanding of their damage resistance and of the mechanical mechanisms involved in the deformation, especially under cyclic loading. As a general behavior, it was observed that the fatigue limit of nanocrystalline metals increases with decreasing grain size, and the crack initiation susceptibility decreases with increasing crack growth rate coupled with the grain refinement (Hohenwarter and Leitner 2019). It is shown how the increase of fatigue resistance can be achieved for many materials in stress-controlled tests. Generally, the main damage mechanism has been recognized in the early strain localization and microcrack formation for the ECAP materials. The localization of cyclic plasticity is very pronounced and starts within the individual grains. Long slip bands develop in regions of nearby oriented grains by growth and concatenating of individual slip bands. The fatigue damage within the slip bands consists of formation of cavities and elongated voids, which later on link along the planes with the highest cyclic shear stress. Sufficiently large fatigue cracks initiated by this mechanism propagate later on by the common opening mode. A plastic zone in fracture mechanical sense is generated at the crack tip due to cyclic stress concentration. The plastic zone manifests itself by creation of secondary slip bands, which further predetermines the crack path on the microscopic level.

The small grain size of these materials makes them quite different from the bulk in a number of ways. For example, at high and moderate strain rates, the nanomaterials exhibit very low room-temperature ductility. This is related to extremely low work hardening rate according to the Considère criterion. However, at very low strain rates (less than $10^{-4}\ s^{-1}$) very large strains are usually observed in FCC nanomaterials. This is due to an extremely high strain rate sensitivity exponent m which for these conditions is analogous to that observed in high-temperature superplasticity of micro-grained materials. Consequently, the deformation mechanism at room temperature and low strain rate resembles that for high-temperature superplasticity. Indeed, Coble creep (enabling both the grain boundary sliding and grain rotation) can operate according to the Ashby–Verrall model (Antolovich and Armstrong 2014).

5.2.2 *Fatigue Life of UFGed Materials*

In general, a strong decrease in fatigue properties was shown for SPD materials in the LCF regime of intermediate to high plastic strain amplitudes. On the contrary, in the HCF regime of intermediate to low plastic strain amplitudes, there results a strong enhancement of the fatigue resistance for materials with grain refinement. In addition, it was observed that an annealing treatment, subsequent to the ECAP process, produces an enhancement in the LCF properties due to the increase in ductility. Such a behavior is obtained by recovering partially the grain boundary region, which has been most heavily distorted during processing. The oriented distribution of defects along the shear plane in the last ECAP processing is one of the major mechanisms of SB formation. Two kinds of shear bands (SBs), i.e., PSB-like SB and step-like SB, were found in cyclically deformed ultrafine-grained copper processed by ECAP. The formation of PSB-like SBs and step-like SBs may be attributed to the local reversible and irreversible deformation, respectively. Two mechanisms of cyclic plastic deformation were proposed: the temperature-independent bulk dislocation mechanism taking place in the entire loaded volume and the temperature-dependent localized mechanism consisting of cooperative grain boundary (GB) sliding along the shear plane of the last ECAP pass taking place in the surface layer and leading to formation of surface fatigue markings. The temperature-dependent grain coarsening and microstructural changes in the coarsened grains are believed to be the governing reasons for the decrease in fatigue life. In high-strength materials a very different behavior was observed; in such a case the severe plastic deformation appears as a very strong tool in the improving of mechanical performances (Chukin et al. 2016).

It was shown that every production route entails specific microstructural features. Differences exist between materials produced by different techniques, e.g., regarding the level of impurities or the type of grain boundaries, with a higher purity and a dominance of high-angle grain boundaries typical for SPD materials. Thus, for a valid comparison of materials from different production routes it must be ensured that the effect of variations of microstructural features other than the grain size is negligible for the measured mechanical properties (Leitner et al. 2015). Given the increase of strength with decreasing material grain size in the UFG regime, a summary plot of Al, Cu, Fe, and Ti alloys processed via ECAP has been drawn (Fig. 5.6).

It shows the variation of the rate fatigue limit/UTS with respect to the material grain size.

A summary of fatigue limit of ECAP metals and alloys has been obtained by comparing the fatigue limit of the corresponding coarse-grained materials (Fig. 5.7). The summary was obtained by the available scientific literature for those experiences where the optimization of the ECAP process is devoted to the maximum increase of fatigue properties of the SPD materials (Djavanroodi et al. 2010).

The presence of minor concentrations of Sc and Zr produces, in Al alloys, a strong increase in strength, ductility, and fatigue life. This is due to the formation of

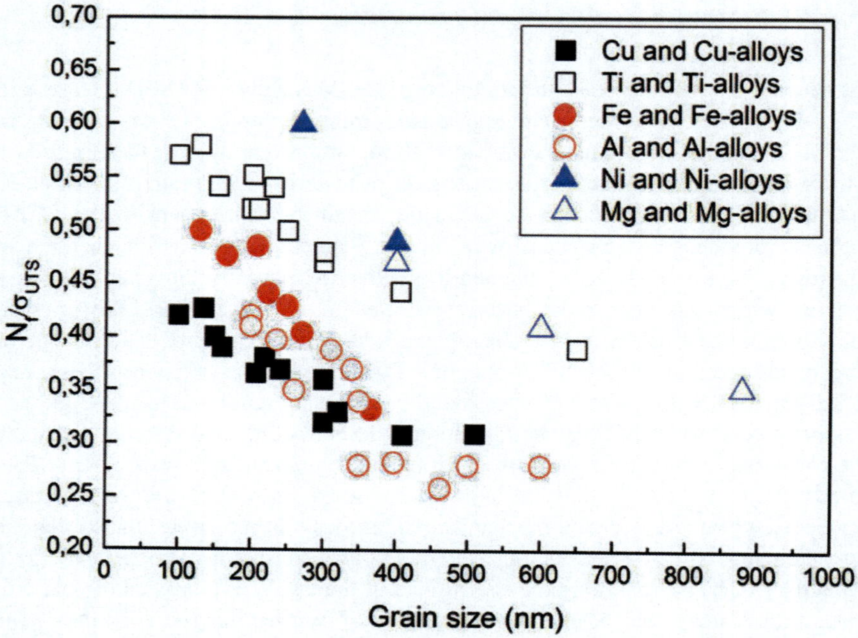

Fig. 5.6 Fatigue endurance limit/ultimate tensile strength vs. grain size for various metals and alloys after SPD

very finely dispersed Al_3 (Sc, Zr) particles acting as grain boundary pinners via a Zener effect and then as grain structure stabilizers up to very high temperatures (Kendig and Miracle 2002). The strengthening mechanism is due to the coherent thermodynamically stable particles with ordered $L1_2$ structure. The presence of the very fine Al_3Zr and Al_3 (Sc, Zr) precipitates is capable of producing a very fine microstructure which is stable at high temperatures (Vinogradov et al. 2003).

The enhanced fatigue limit can be attributed to a shift of the onset of microplasticity to higher stress levels. However, the fatigue behavior of UFG metals depends not only on the grain size-related strength but also on other characteristics of cyclic deformation and damage processes such as deformation-induced grain coarsening and fatigue crack-promoted grain growth. It has been suggested that a certain fatigue time and cumulative strain are required for the release of stored strain energy to trigger dislocation motion, recrystallization and associated grain growth, as well as crack initiation. As a matter of fact, it is shown that for pure Cu, grain coarsening occurs only at growth rates below 10^{-10} m/cycle and is more pronounced at lower crack growth rates. The occurrence of grain coarsening can now have two effects: on the one hand it seems reasonable that the dissipated energy during cycling is transmitted into plastic deformation and movement of grain boundaries instead of creation of new surface area and crack advancement (Kozlov et al. 2015). On the other hand, grain coarsening takes place in front of the crack tip leading to a totally

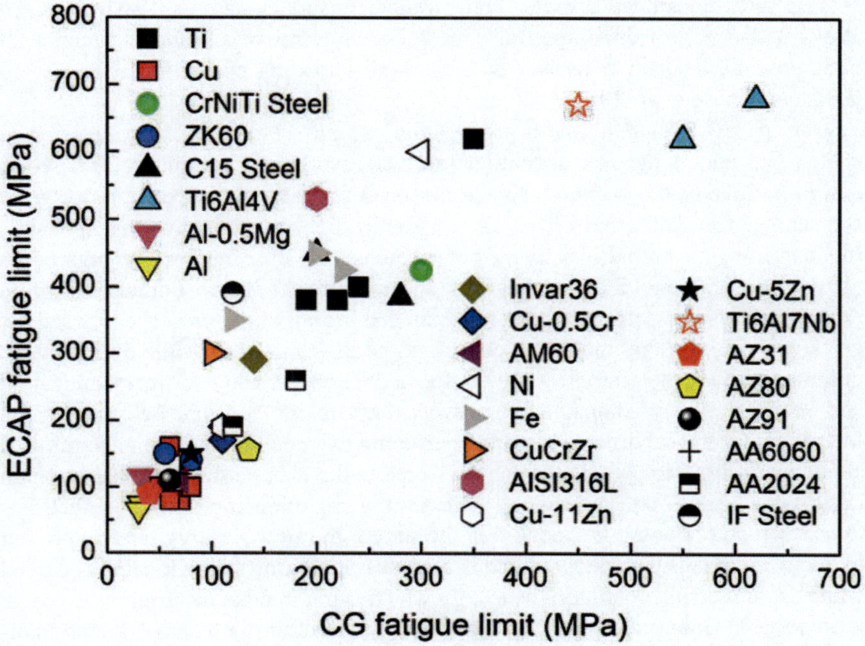

Fig. 5.7 Fatigue endurance limit for ECAP metals and alloys vs. the fatigue limit of the corresponding coarse-grained materials

different surrounding area for crack growth (Horky et al. 2013). In this case crack propagation does not occur in an ultrafine-grained material with high strength and low ductility but in a coarse-grained material which is apparently softer and more ductile. Grain coarsening and related cyclic softening behavior in nano- or ultrafine-grained materials were frequently observed during cyclic deformation in the high-cycle fatigue regime by several groups using different sample geometries and testing frequencies. It was associated with the low stability of UFG or nanostructured materials especially those exhibiting high purity and recently also with the mode of testing, i.e., stress- or strain-controlled cyclic loading. Grain coarsening was identified as a thermally activated process which depends also on the plastic strain amplitude. It is a dynamic recrystallization process, meaning nucleation and growth of grains, but it also involves subsequent migration of large-angle grain boundaries (Glezer 2008). The boundary migration seems to be of particular importance because an exclusive nucleation and growth mechanism would lead to smaller grain sizes due to the large number of nucleation sites in the heavily deformed SPD materials. With crack propagation, grain coarsening was found in the vicinity of a crack irrespective of the material's purity, but during cyclic deformation, coarsening was only observed in high-purity material. Cyclic deformation in the HCF regime showed that grain coarsening is mainly time controlled; that is, it is more intense at higher number of loading cycles and therefore larger accumulated plastic strain.

This is demonstrated also for high-strength materials such as Ti-Nb ECAPed alloys where cyclic softening effects are found in terms of a gradual reduction of hardness and dislocation density with increasing loading cycles from 1e6 to 1e9 (Delshadmanesh et al. 2017).

During crack growth a smaller propagation rate and therefore increased dissipated energy per crack length expansion lead to increased grain coarsening. So it seems that in both cases the accumulation of plastic deformation is responsible for grain coarsening. The processes of dynamic recrystallization and grain growth responsible for coarsening are both driven by the defect content or rather the stored energy of the material but the annihilation of defects and migration of grain boundaries (grain growth) need some activation energy. Grain coarsening in the presence of a fatigue crack depends more on the stored energy of the SPD materials while during cyclic deformation it is more associated with the activation enthalpy for recrystallization and grain growth phenomena. A possible explanation for this behavior is the following: for the occurrence of grain coarsening two conditions have to be fulfilled simultaneously. Namely, (1) the energy stored in the material has to reach a certain critical value and (2) the activation barrier for the migration/annealing of defects has to be overcome. For the first condition the sum of the stored energy already present in the material plus the energy which is accumulated during cyclic loading is crucial while for the second condition the energy which is put into the material in one cycle is important. Grain coarsening occurs locally and within the coarser grains traditional persistent slip processes can take place leading to the initiation of a fatigue crack. During cycling in the HCF regime most part of the lifetime is devoted to crack initiation, which is much easier in larger grains where surface extrusions and intrusions can form. If the material coarsens and larger grains form, crack initiation takes place much earlier and the overall HCF life is markedly reduced. If the applied stress is not too high and coarsening occurs also during propagation of the initiated crack, crack growth in nanomaterials can be retarded and the instability of the microstructure which favors crack initiation also is able to decrease crack growth rates.

The presence of pinning precipitates inhibits grain growth by impeding grain boundary mobility. The presence of these trialuminides increases the recrystallization temperature, resulting in a very stable microstructure under heat treatment. Zr-containing and (Zr + Sc)-containing dispersoids were observed to be distributed along the extrusion directions both within the grains and on some grain boundaries (Fig. 5.8).

ECAP processing provides an important procedure for obtaining an ultrafine microstructure in the materials studied. The ECAP process proved very effective in reducing grain size to the sub-micrometer range in both alloys. The mean grain size of the 6106 Zr and 6106 Sc alloys was very close to 0.2 μm. The mean grain size dimension was 200 nm in the 6106 Zr alloy while it was 170 nm in 6106 Sc (Fig. 5.9).

The dispersoids remain stable also after solution treatment and aging; the corresponding fatigue curves are shown in Fig. 5.10.

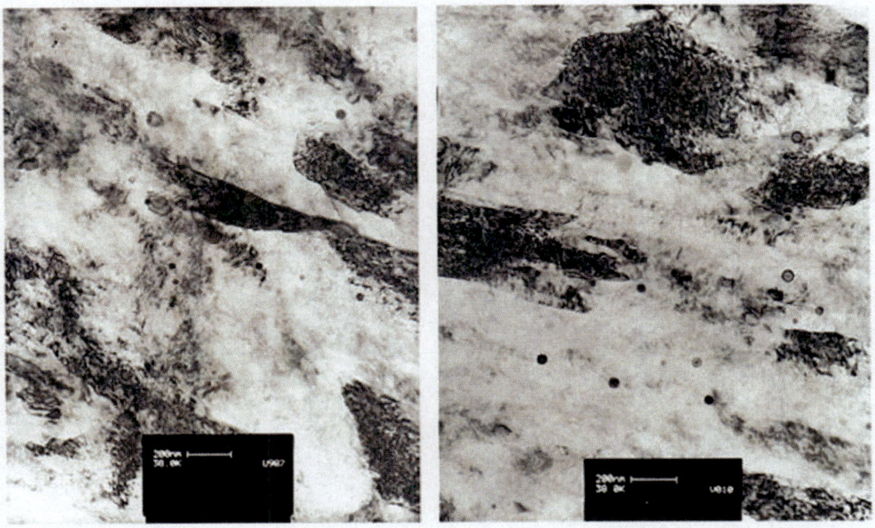

Fig. 5.8 Al_3Zr (**a**) and $Al_3(Zr1-x,Scx)$ (**b**) dispersoids in extruded 6106 aluminum alloy

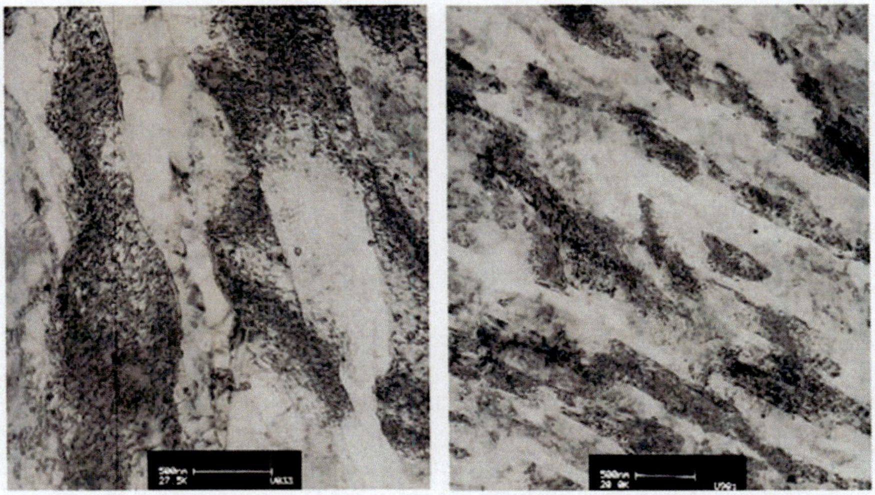

Fig. 5.9 Al_3Zr (**a**) and $Al_3(Zr1-x,Scx)$ (**b**) dispersoids in ECAPed 6106 aluminum alloy (four passes, B_C route)

A strong increase in fatigue properties compared to the as-heat-treated condition was observed for both alloys after ECAP and aging; this was particularly true for the Zr-Sc-modified material. The heat treatment and ECAP process prove very effective in increasing the fatigue resistance of the materials, compared to the as-extruded condition (Cavaliere and Cabibbo 2009). The increase in fatigue life after ECAP has been explained as the result of decreased damage nucleation thanks to the reduction

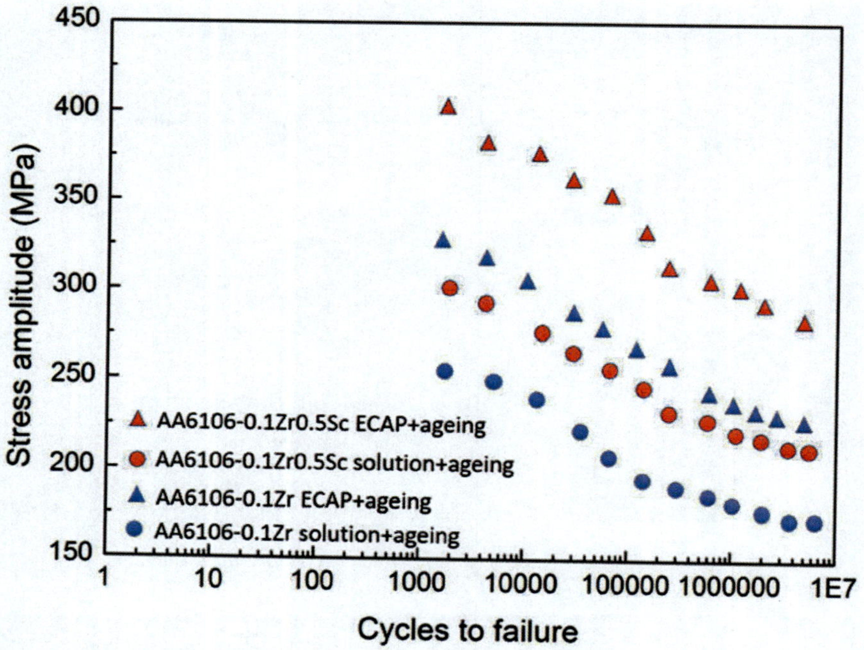

Fig. 5.10 Fatigue properties of the 6106 Zr alloy in the solution treated + aged, and solution treated + ECAP + aged conditions, and the 6106 ZrSc alloy in the same conditions

in grain size and a reduction in the number of dislocations in slip bands (Hoppel et al. 2011). This may be due to a decrease of the damage depth of surface intrusions or extrusions, or due to an increase of the yield stress that prevents macroscopic plastic deformation at the beginning of load-controlled cycling (Chung et al. 2002). The different fatigue resistance between heat-treated and ECAP materials can be attributed to the higher cyclic softening that coarse grain materials exhibit compared to sub-micrometer grain materials. ECAP appears to play a fundamental role in increasing monotonic and dynamic mechanical properties of such dilute Zr- and Sc-modified aluminum alloys.

For UFG materials, three mechanisms have been found to contribute to the observed cyclic softening, which are (1) shear band formation, (2) coarsening of the fine-scaled grain structure, as well as (3) a reduction of the defect density, in especially dislocation density, or a combination of them. These mechanisms can lead to early strain localization, fatigue damage, and failure of the samples. The small grain size and the resulting large grain boundary fraction are suspected to be responsible for these microstructural instabilities as it offers large driving forces for grain growth (Kaneko et al. 2014; Kapp et al. 2017). The reason for enhanced fatigue limits or HCF performance can be explained by a shift of the onset of microplasticity, necessary to induce fatigue damage, to higher stress levels than in the CG condition. Unfortunately, the performance of such materials in the low-cycle-fatigue (LCF) regime, where higher plastic strain amplitudes are present, is deteriorated and cyclic softening is promoted (Limooei et al. 2018). For UFG materials, which will be in the focus of the current work, cyclic softening has been found to occur not only under strain-controlled conditions (decreasing stress amplitude) for purity levels above 99.9%, but also in stress-controlled experiments above a certain stress amplitude, where the softening is reflected in a continuously increasing strain amplitude. For UFG materials, three mechanisms have been found to contribute to the observed cyclic softening, which are (1) shear band formation, (2) coarsening of the fine-scaled grain structure, as well as (3) a reduction of the defect density, in especially dislocation density, or a combination of them. These mechanisms can lead to early strain localization, fatigue damage, and failure of the samples. The small grain size and the resulting large grain boundary fraction were suspected to be responsible for these microstructural instabilities as it offers large driving forces for grain growth (Kapp et al. 2017). Under low-cycle-fatigue (LCF) conditions, i.e., large plastic strain amplitudes, nanocrystalline (NC) or ultrafine-grained (UFG) structures become unstable due to their large amount of grain boundary area. In AZ31 Mg alloy processed via ECAP a pronounced cyclic softening is observed as a consequence of marked global and local grain coarsening/recrystallization (Zuberova et al. 2007). Local grain coarsening leads to the formation of dense shear localization. The characteristic distribution of defects introduced by equal-channel angular pressing was believed to be related to grain coarsening, which induced plastic instability and cyclic softening (Zhu et al. 2011). In the present case the following grain coarsening process during cyclic deformation is proposed. The shear plane of ECAP is not a crystallization plane. Its influence can be only reflected by means of oriented distribution of defects, resulting in deformation taking place along a family of planes approximately parallel to the shear plane of the last pressing. As such, the distribution of defects, such as dislocation, is inhomogeneous, and is about 45° with respect to the loading axis. It provides a prerequisite for the recrystallized nucleus within the inhomogeneous regions. In the case of reversible deformation, the annihilation of dislocation increased the dislocation density gradient over which local inhomogeneous dislocation structures distribute. When the dislocation density gradient accumulated to a certain level, the recrystallized nucleus formed in the lower dislocation density regions, and the recrystallized boundaries formed in the higher dislocation density regions. Once a recrystallization nucleus forms, it grows at the expense of the defects within the region of microstructural inhomogeneity. As such, the direction of larger axis of recrystallized grains is nearly 45° with respect to the loading axis. In cyclic deformation, shear bands are the primary plastic instability of a concern since they often nucleate cracks. For ultrafine-grained copper, shear bands and coarse grains with substructures were observed (Mughrabi and Hoppel 2000, 2010). Concerning the recrystallization and grain coarsening during cyclic deformation of ultrafine-grained copper, two possible processes were proposed. One is that local grain coarsening takes place first and shear band forms later. The other is that shear localization occurs first and a new coarsened grain appears subsequently within the shear band. Cyclic stress-strain response of UFG Cu under stress-controlled loading changes from cyclic softening to cyclic hardening with stress amplitude decreasing; the UFG samples under strain-controlled loading show rapid cyclic softening, which occurs under all applied plastic strain amplitudes in early stage of testing; the softening is faster under higher plastic strain amplitude. In other evidences, it was shown that the presence of small quantities of impurities is very effective in reducing coarsening and recrystallization with positive effect on the fatigue resistance also at very high cycle number (Khatibi et al. 2010). For low-purity copper, the following conclusions can be made. The formation of SBs is not directly related to the cyclic softening. SBs and cracks initiated within them were observed in both cyclic softening and cyclic hardening. No detectable grain coarsening was witnessed near the SBs observed by EBSD and TEM, which means there is no inevitable relationship between formation of SBs and grain coarsening. All of these should be related to the impurities in copper matrix. During cyclic deformation, the oriented distribution of defects along the shear plane of the last ECAP processing is one of the major mechanisms of SB formation (Xu et al. 2008). These features are confirmed for ECAPed copper with minor alloying (Xu et al. 2007).

The nonlinear stress-strain behavior of UFG materials suggests that when such materials are strained, there are larger internal back stresses than in their CG counterparts. There are several independent studies into the cyclic deformation behavior of ultrafine-grained materials produced via SPD that have shown evidence of cyclic softening effects.

The cyclic softening ratio is defined by

$$1 - \left[\frac{(\Delta\varepsilon_{\text{pl}}/2)_{N\tilde{a}}}{(\Delta\varepsilon_{\text{pl}}/2)_{Nf/2}}\right]$$

where

$$(\Delta\varepsilon_{\text{pl}}/2)_{N\tilde{a}}$$

is the plastic strain amplitude at a number of cycles approximately equal to a, $a = 1$ for strain-controlled testing and $a = 1000$ for stress-controlled testing, and

$$(\Delta\varepsilon_{\text{pl}}/2)_{Nf/2}$$

is the plastic strain amplitude at half of the total fatigue life. The cyclic softening ratio is a quantitative measure of the degree of softening that a material undergoes during a fatigue test. The material cyclic softening is mainly due to grain coarsening. Grain coarsening was found to proceed during fatigue testing at low strain rates, low plastic strain amplitudes, and also higher temperatures. These conditions are the same as those that would lead to cyclic softening, and hence these two phenomena are likely to be related. This means that the coarse grains that have developed from fatigue testing result in changes to the mechanical behavior of the material we observe as cyclic softening. It was suggested that this fatigue-induced grain coarsening is due to a dynamic recrystallization process. As a matter of fact, fatigue life of UFG pure Al increases with increasing strain rate. Based on the strain rate sensitivity, the hardening in the beginning of fatigue experiments, and the reduction of dislocation density during fatigue, dislocation movement combined with thermally activated annihilation of dislocations at the grain boundaries has been suggested as the dominating deformation mechanism for UFG Al at room temperature (May et al. 2008).

The mechanism by which the heavily deformed ultrafine-grained microstructure changes into a coarse microstructure is unknown, whether it occurs by nucleation and grain growth processes or migration of high-angle grain boundaries. What is known is that this dynamic recrystallization process is time, temperature, and plastic strain amplitude dependent and also thermally activated. Evidence of slip banding was found in ECAP materials (e.g., Cu) in the form of parallel microcracks and extrusions. The shear bands were found to be 45° to the tension and compression axis aligning with the plane of maximum resolved shear stress. These slip bands developed over large areas of the sample and spanned for millimeters despite the material having a grain size in the UFG regime. It was also found during fatigue testing that crack initiation began at the shear bands except for tests involving high strain amplitudes. It was suggested that microcracks nucleate along these slip bands and propagate until a critical length. It was found after ECAP processing that the microstructure was dominated by bands of elongated grains that coincide with the shear plane of the ECAP die. Such a microstructure results with decreased HCF performance as it is likely to result in macroscopic shear banding and easier crack initiation. It may also lead to more pronounced cyclic softening effects which again contribute to decreased HCF performance. Their occurrence was observed not only above, but also below a certain cyclic stress (strain) amplitude value—persistent slip band threshold. Density of the slip markings increases with increasing stress (strain) amplitude and number of cycles. Nevertheless, failure does not occur even if slip markings cover the whole surface (Navratilova et al. 2013).

This softening in HCF properties is also due to grain growth of UFG materials. Grain growth is often associated with thermally activated grain boundary migration. The grain growth is primarily stress driven, not diffusion driven (Zhao et al. 2019). Here the transition from LAGBs to HAGBs suggests that continuous dynamic recrystallization acts as the grain growth mechanism during room-temperature HCF fatigue of UFG cp-Ti. Because the fatigue-induced grain growth was very commonly reported to occur locally near the crack surfaces the large strains in a cyclic plastic zone around the crack tip should be responsible for this phenomenon. Formation of nonequilibrium grain boundaries containing numerous dislocations within grain boundaries (i.e., sub-grains) in UFG metals is an immediate consequence of severe plastic straining. The nonequilibrium grain boundaries are manifested by the existence of high energy, excess defect volume, and long-range stress field. Accordingly, the rate of atomic transport in UFG metals is enhanced by many orders of magnitude compared to that in coarse-grained metals because of the large number of diffusion paths. Recently, all available data were summarized on

GB diffusion in UFG Cu and Ni obtained by severe plastic deformation (SPD) methods and, employing the well-known correlation between the GB energy and diffusivity, arrived at the conclusion that the GB energy in UFG materials obtained by SPD is by about 30% higher than the average GB energy in well-annealed coarse-grained polycrystals. It is interesting that anomalously fast GB diffusivity in UFG Ni obtained by high-pressure torsion (HPT) survived nearly full recrystallization prior to or during annealing, indicating that the nonequilibrium state is partially "inherited" by the GBs formed during recrystallization from their counterparts in the as-processed material (Zimmermann et al. 2020).

This could explain why the grain growth already occurred at a relatively low homologous temperature (i.e., room temperature). The dynamically recrystallized grains are expected to have a lower dislocation density compared with the severe plastic deformation-induced UFG microstructure. This means that one should expect to see a much lower value of geometrically necessary dislocations (GND) in post-fatigued material. The GND density reduction is an important evidence to support that continuous dynamic recrystallization is responsible for the room-temperature HCF fatigue-induced grain growth.

Any model attempting to account for the cyclic behavior of UFG materials has to describe the hysteresis loop shape and the cyclic stress-strain curve with rapid hardening at low plastic strain amplitudes and nearly saturated (or slowly increasing) response at large strains (Vinogradov et al. 2001). In pure ECAPed metals the dislocation density is relatively low in the grain body and is high near the grain boundary. Assuming that the mobile dislocations pass through the grain and disappear in the grain boundary region (the grain boundaries act in this sense as an effective sink for dislocations), the kinetic equation for dislocation density can be written in a simple form:

$$\frac{d\rho}{d\gamma} = \frac{2}{\boldsymbol{b}\boldsymbol{L}} - \frac{2}{\boldsymbol{b}} y\rho$$

where L is the slip path of dislocations with the Burgers vector b and y is the so-called annihilation length, which serves as a phenomenological parameter of the model. The first term on the right-hand side describes the rate of dislocation multiplication with the strain increase while the second term accounts for the strain-induced decrease of dislocation density (dislocation dynamic recovery and annihilation of a general kind). The shear flow stress is controlled by the average total dislocation density as

$$\tau = \alpha\mu \boldsymbol{b}\sqrt{\rho}$$

where α is a geometrical factor of the order of 0.5 and μ is the shear modulus. By combining the previous equations

$$\frac{d\tau}{d\gamma} = \frac{(\alpha\mu\boldsymbol{b})^2}{2\tau}\left[\frac{2}{\boldsymbol{b}\boldsymbol{L}} - \frac{2y}{\boldsymbol{b}}\frac{\tau^2}{(\alpha\mu\boldsymbol{b})^2}\right]$$

If dislocation accumulation takes place uniformly in the grain body, the L value is usually reciprocal to the square root of the forest dislocation density. For the cell structure, the slip path is related to the cell diameter. In UFG materials, the grain size (of 0.2–0.3 μm) is smaller than the typical cell size (of 0.5 μm) and the grain boundaries form the main barriers for dislocation motion. It is, therefore, plausible to relate the dislocation mean free path to the average grain size if the grain size is small enough (the case of UFG materials). By integrating the previous equation

$$\tau = \alpha\mu\boldsymbol{b}\sqrt{\frac{1 - \exp\left(-\frac{2y\gamma}{\boldsymbol{b}}\right)}{Ly}}$$

At large enough strains, the saturation stress is attained due to equilibrium between the dislocation multiplication and annihilation:

$$\tau_s = \frac{\alpha\mu\boldsymbol{b}}{\sqrt{Ly}}$$

By employing this equation one can calculate peak stresses, σ_s, at different plastic strain amplitudes, giving a crude approximation of the cyclic stress-strain curve. Furthermore, in the low strain limit

$$\tau = \alpha\mu\boldsymbol{b}\sqrt{\frac{2\gamma}{\boldsymbol{b}\boldsymbol{L}}} = 2\alpha\mu(\gamma\boldsymbol{b})^{\frac{1}{2}} \cdot d^{-\frac{1}{2}}$$

corresponding to the Ashby's expression of the Hall-Petch work-hardening stress. This analysis can be useful because it provides a vivid picture of the cyclic behavior of UFG materials, showing that most characteristic features of their cyclic response can be understood from the standpoint of ordinary dislocation kinetics constrained by grain boundaries.

Alloying is very effective in reducing coarsening also for the testing at high temperatures because of the precipitation hardening. Anyway, coarsening effect overcomes the precipitation hardening and stabilization as the temperature is increased above a certain level (Canadinc et al. 2008).

5.3 Damage Tolerance of UFG Materials

In MC materials, a reduction in grain size generally results in an elevation in strength which engenders an increase in fatigue endurance limit during stress-controlled cyclic loading of initially smooth-surfaced laboratory specimens. Since the total fatigue life under such conditions is dominated by crack nucleation and since fatigue cracks generally nucleate at the free surface, grain refinement here is considered to result in improvements in fatigue life as well as endurance limit, with all other structural factors set aside. On the other hand, a coarse grain structure with lower strength and enhanced ductility generally plays a more beneficial role in the strain-controlled fatigue response of metals and alloys. It should be noted, however, that it is often difficult to isolate the sole effects of grain size on fatigue response since other structural factors such as precipitate content, size and spatial distribution, stacking fault energy, attendant equilibrium spacing of partial dislocations, and crystallographic texture are also known to have an important effect on the fatigue characteristics of MC metals. The presented data supported the notion that fatigue notch sensitivity increases with decreasing grain size. Intergranular brittle fracture and ductile fracture are experimentally observed in nanocrystalline materials. Plastic flow in such materials occurs at very high stresses close to those needed to induce fracture (San Marchi et al. 2006). In nanocrystalline materials with finest grains, plastic flow is conducted by mostly grain boundary processes. In nanocrystalline materials with intermediate grains, plastic flow is often conducted by both lattice dislocation slip and grain boundary processes. However, if plastic flow and diffusion are not intensive in nanocrystalline materials with intermediate grains and/or these materials contain preexistent nano-cracks and pores, brittle fracture tends to occur.

In general, hardening or softening with ΔK variation plays a crucial role in the material behavior (Fig. 5.11).

Anisotropic microstructures lead to orientation-dependent mechanical properties, especially concerning the resistance against static and cyclic crack growth. In principle, the anisotropy of the mechanical properties is not a drawback and, as in nature, can be used as a toughening strategy. For example, fracture toughness was found to be relatively low for cracks introduced parallel to elongated grains. However, the existence of this weak crack path leads to a significantly enhanced crack growth resistance for crack propagation perpendicular to the elongated grains (Leitner et al. 2017). So, a comparison should be possible if the same material with different grain sizes is analyzed.

5.3.1 *Crack Initiation and Growth in UFGed Materials*

As a matter of fact, pure electrolytic iron subjected to eight passes in the ECAP route Bc was employed for comparison with its microcrystalline counterpart. The MC iron measured 5 and 20 μm of mean grain size while the ECAPed iron had 350 nm mean

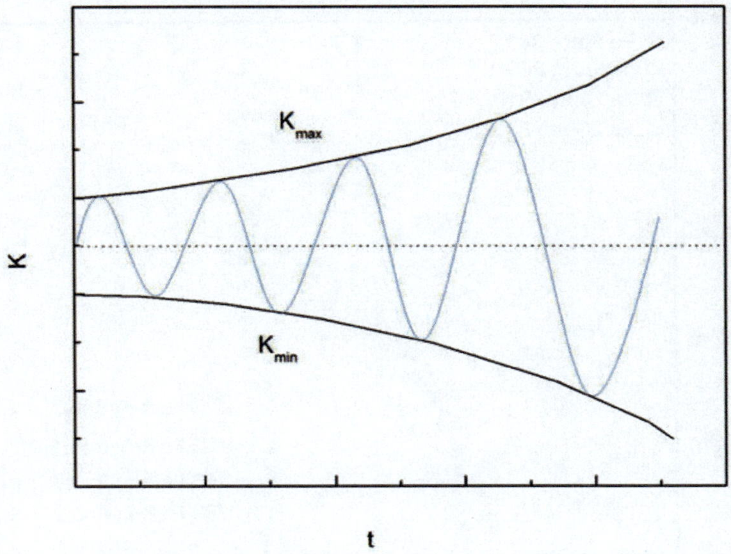

Fig. 5.11 Example of K variation with time during cyclic loading

grain size. The curves of crack propagation as a function of the applied ΔK are shown in Fig. 5.12.

The influence of grain refinement on the crack initiation is negligible. The highest difference is underlined for the crack growth behavior. In particular, faster crack growth is recorded for R-R samples processed via ECAP. This is mainly due to the texture evolution during ECAP with the grain size equiaxed because of the Bc processing after eight passes.

It was found that the fracture toughness of UFG Ni produced by severe plastic deformation (SPD) shows a remarkable combination of high fracture toughness and strength. Especially in materials produced with SPD techniques anisotropy in mechanical properties was observed. NC and UFG metals are also promising materials for the use in applications subjected to cyclic loading, as grain refinement is accompanied by an increase in the resistance against crack initiation. However, also the fatigue crack growth (FCG) behavior has to be considered for an evaluation of the entire fatigue lifetime or the defect tolerance. The analysis of severe plastic deformation on the crack growth behavior has been extended to UFG Ni. The microcrystalline Ni measured a mean grain size of 40 μm, while the UFG Ni measured 200 nm after eight passes of ECAP in Bc route. The curves belonging to the MC and UFG Ni are shown in Fig. 5.13.

When dealing with FCG it is important to distinguish between intrinsic and extrinsic mechanisms. Intrinsic mechanisms are responsible for the material separation process when a cracked body is cyclically loaded. The separation process can be caused by mechanisms belonging to three different groups: fracture by accumulated damage, crack advance due to pure deformation, and FCG by cyclic damage. The

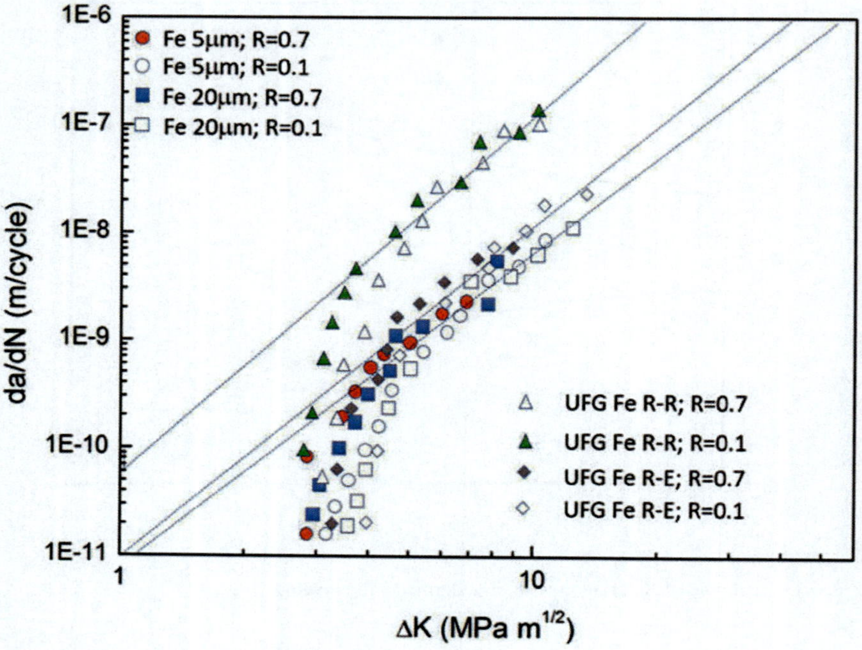

Fig. 5.12 Crack behavior of Fe after ECAP

inherent resistance of the material against these separation processes can be seen as the intrinsic crack growth resistance. However, there also exist so-called extrinsic or shielding mechanisms which can hinder crack growth and reduce the FCG rate. These mechanisms reduce the load at the crack tip and thereby the driving force of the FCG process. One possibility is crack deflection, which changes the loading conditions for geometric reasons when the crack deviates from a straight crack path. This can result in an overall reduction of the driving force. So-called contact shielding mechanisms are based on a reduction of the load at the crack tip due to a premature contact of the crack faces. This type of closure can be induced by roughness, oxides, plasticity, or crack bridging. It is important to take these shielding mechanisms into account in order to correctly describe FCG and to give lifetime predictions. In the case of UFG Ni, crack initiation is faster for all the UFG materials in all the tested conditions. In addition, crack growth is faster for the R-R samples. In general, larger crack propagation rates and lower fatigue threshold stress intensity factors are observed in the near-threshold regime of fatigue crack growth in materials with refined grain size. Refined microstructures are effective in suppressing crack initiation due to high strength, whereas they are deleterious to crack propagation due to lack of ductility. In addition, microstructure instabilities during cyclic loading, such as grain growth and macroscopic shear banding, might also decrease the fatigue lifetime.

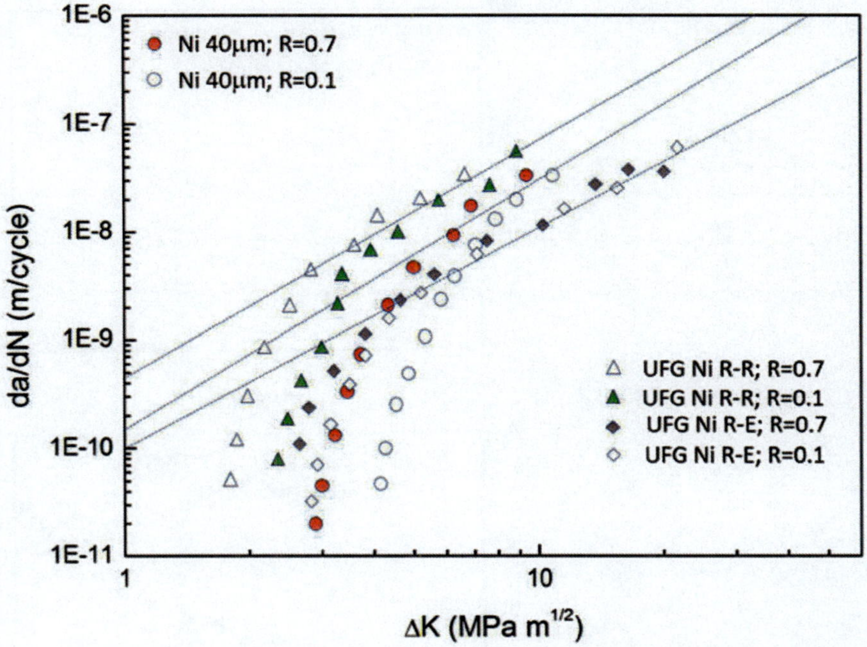

Fig. 5.13 Crack behavior of Ni after ECAP

An increase in the SPD temperature in UFG metals leads to an increase in the activation energy of grain boundary diffusion. The dependence of the activation energy of grain boundary diffusion on the SPD temperature is found to be determined by the kinetics of the competing processes of defect accumulation at grain boundaries and the diffusion accommodation of defects (Chuvil'deev et al. 2017). Pure Ti grade 2 is produced by equal-channel angular pressing with eight passes in route Bc by employing a 90° corner channel at the temperature of 425 °C; the extruded billets measured 45 mm in diameter and 100 mm in length. Pure Al was produced via ECAP with 12 passes in route Bc by employing a 90° corner channel at room temperature. Tests performed on ECAPed Ti and Al at $R = 0.25$ showed the behavior described by the curves in Fig. 5.14. For the tested materials the value of yield stress is indicated.

For such metals it can be observed that the grain refinement, due to the severe plastic deformation, produces a decrease in ΔK threshold and an increase in crack propagation rate. The primary mechanism responsible for the accelerated fatigue crack growth rate observed with decreasing grain size is the reduction in crack path deflection with grain refinement. Microstructural size scales can play a dominant role in crack morphology as well as the mode of fracture, especially near the threshold regime. Periodic deflections in the fatigue crack at grain boundaries during crystallographic fracture can lead to a relatively tortuous crack path in coarser grain materials. ECAP UFG Ti exhibits a comparatively straight crack path, relative to

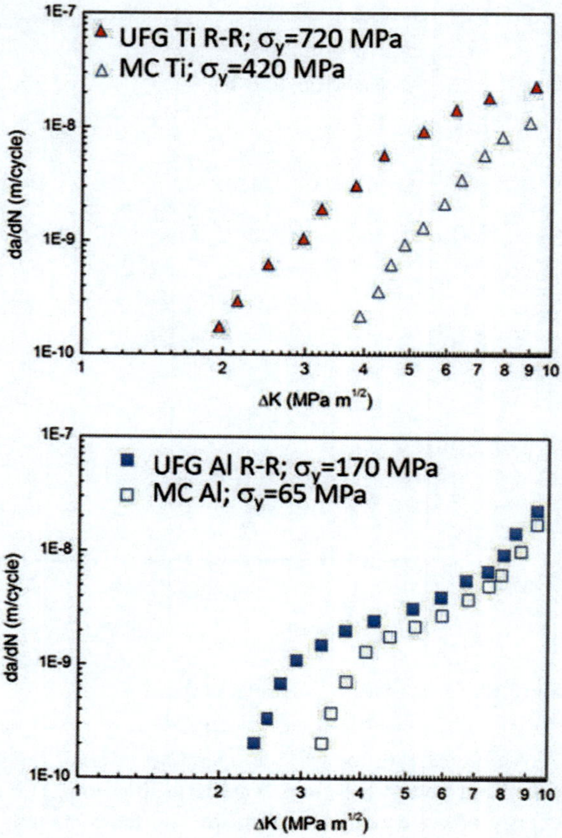

Fig. 5.14 Crack behavior of Ti and Al after ECAP

its MC counterpart. The reduction in effective driving force for crack propagation in the MC material is again assumed to result from a combination of crack deflection and asperity contact. This can also be substantiated by the sensitivity-to-load ratio of the two materials (Cavaliere 2009).

As mentioned before, one of the main aspects of nanostructuring through SPD is the material anisotropy. Anisotropic microstructures lead to orientation-dependent mechanical properties, especially concerning the resistance against static and cyclic crack growth (Ritchie 2011). The analyses of anisotropy on pre-cracked fatigue samples of pure iron produced via HPT showed very different crack paths for the different sample orientations (Fig. 5.15).

In the T-A sample the crack is very flat with an orientation of 75° (Leitner et al. 2017). After a small angle of the crack inclination in R-A sample, again a flat path with an orientation of 75° can be observed. The T-R sample shows the most tortuous path during crack propagation. This behavior is demonstrated by the crack propagation curves shown in Fig. 5.16, where the resistance to crack growth is enhanced for the T-R sample with respect to the A-T sample with a very flat behavior.

Fig. 5.15 Crack aspect of NS pure iron produced via HPT; the samples were obtained from different orientations with respect to the torsion one

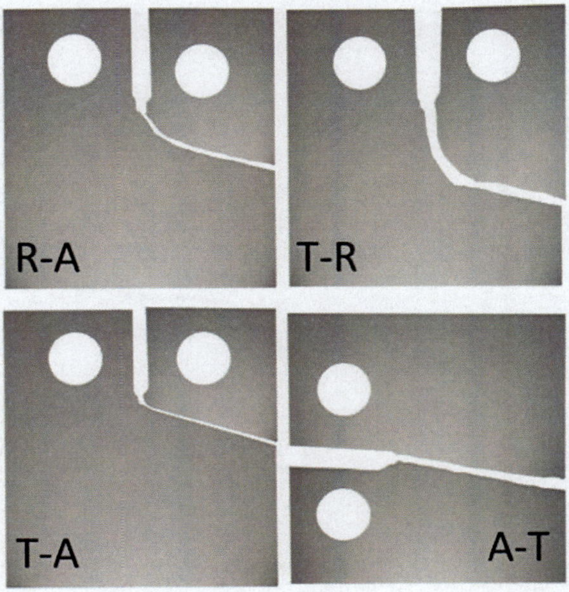

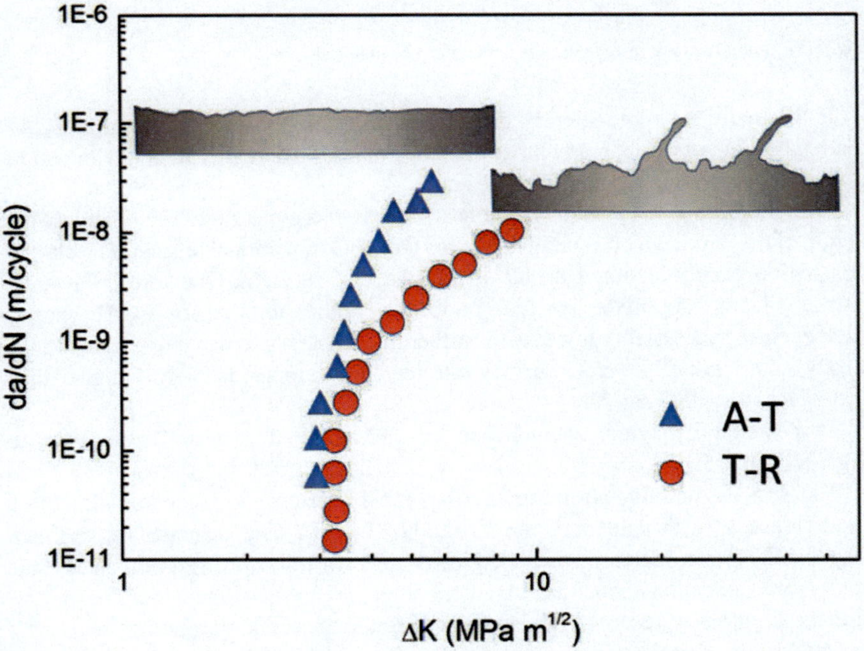

Fig. 5.16 Crack initiation and growth behavior of HPTed A-T and T-R samples

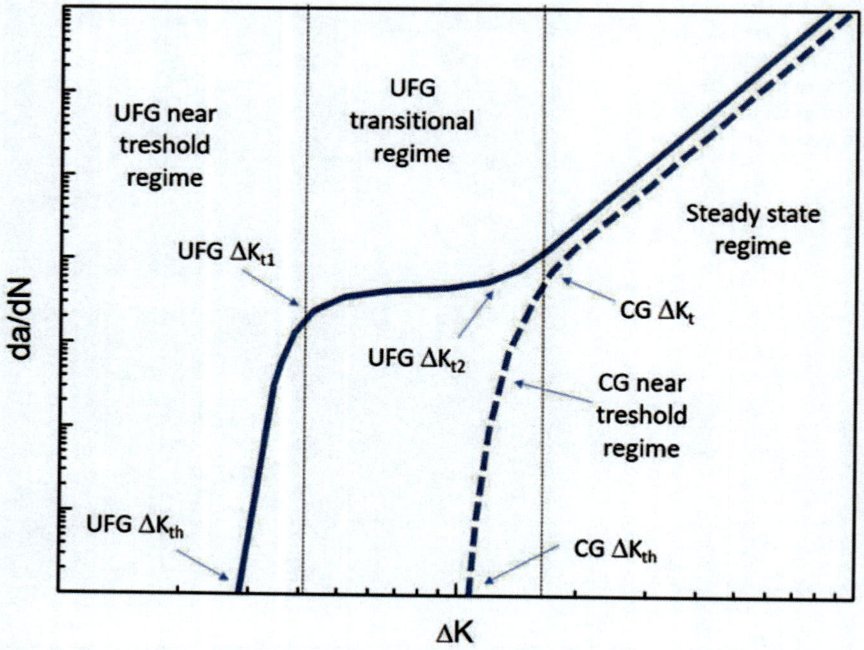

Fig. 5.17 Fatigue crack propagation in UFG and CG materials

In the figure the zoomed crack path of both the samples is also shown. In ECAP materials, it is noted that many times the crack initiation and growth behavior can be illustrated as the schematic in Fig. 5.17.

With increasing ΔK just above the near-threshold regime, there is a relatively flat region of the curve which is defined here as the UFG transitional regime. The start of this region can be denoted as ΔK_{t1}, and the end as ΔK_{t2}. The threshold stress intensity factor amplitude, ΔK_{th}, is noticeably higher for the non-ECAP sample and becomes successively lower with further ECAP passes. It can be seen that in the steady-state regime, the crack growth rate for UFG samples is slightly higher than that of the non-UFG sample.

This model behavior is confirmed in the ECAPed materials belonging to Figs. 5.10 and 5.18.

The Sc + Zr-modified alloy shows comparable resistance to crack initiation but a faster crack growth rate than the Zr-modified alloy. The principal mechanism responsible for an increase in crack growth rate with decreasing grain size is the crack path deflection reduction which accompanies grain refinement. The microstructural regime represents a fundamental aspect in the crack morphology and in the fracture mode, especially near the threshold regime. The deflection of cracks due to grain boundaries in the coarse-grained materials creates a complex and tortuous crack path, while intergranular fracture becomes predominant as the nanocrystalline regime is approached. In addition, the predominance of intergranular fracture,

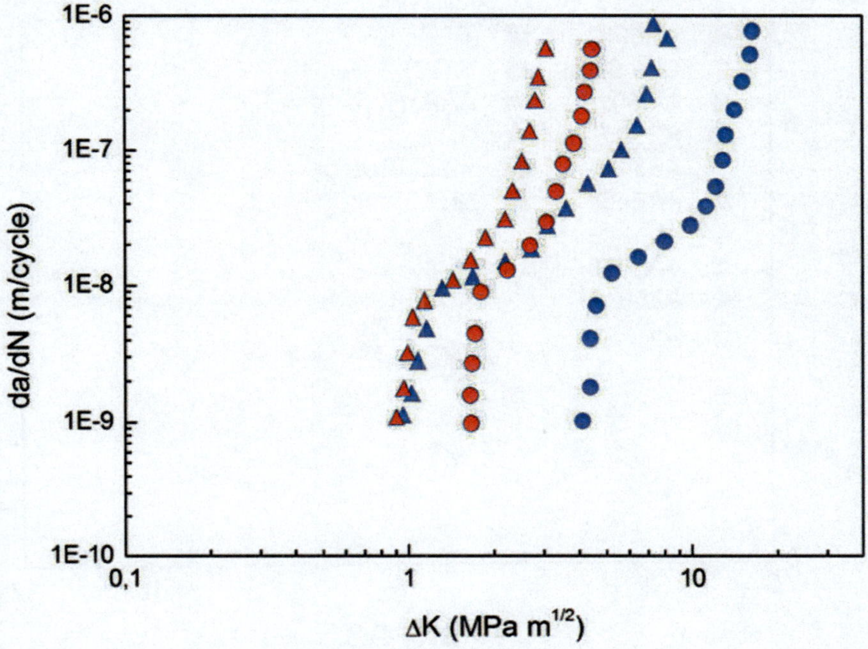

Fig. 5.18 Crack growth behavior in AA6006-Zr-Sc after ECAP and heat treatment

expected in the UFG and NC regimes, is accompanied by a large number of undeflected grain boundary paths oriented normally to the loading direction. The crack path tortuosity increases with increasing grain size, and so the reduction of the effective stress intensity factor range leads to the lowering of crack growth rate in the coarse grain alloys (Meyer et al. 2008).

This behavior is confirmed for many other SPD materials; a summary of the threshold stress of metals and alloys after SPD vs. the threshold stress measured for their microcrystalline counterpart is shown in Fig. 5.19.

All the summarized data show that UFG materials have a crack initiation faster with respect to their microcrystalline counterpart.

Small surface cracks are strongly influenced by microstructural features and hence it is important to relate the reverse plastic zone size at the crack tip to the fracture surface morphology. By calculating the reverse plastic zone size

$$r_{cp} = \frac{1}{2\sqrt{2}\pi}\left(\frac{\Delta K_{eff}}{2\sigma'_{0.2c}}\right)^2$$

where ΔK_{eff} is the effective stress intensity factor range and $\sigma'_{0.2c}$ is the cyclic 0.2% proof stress. The ultrafine-grained alloy is seen to have higher crack growth rates throughout the stress intensity factor ranges. In addition to this, the threshold stress

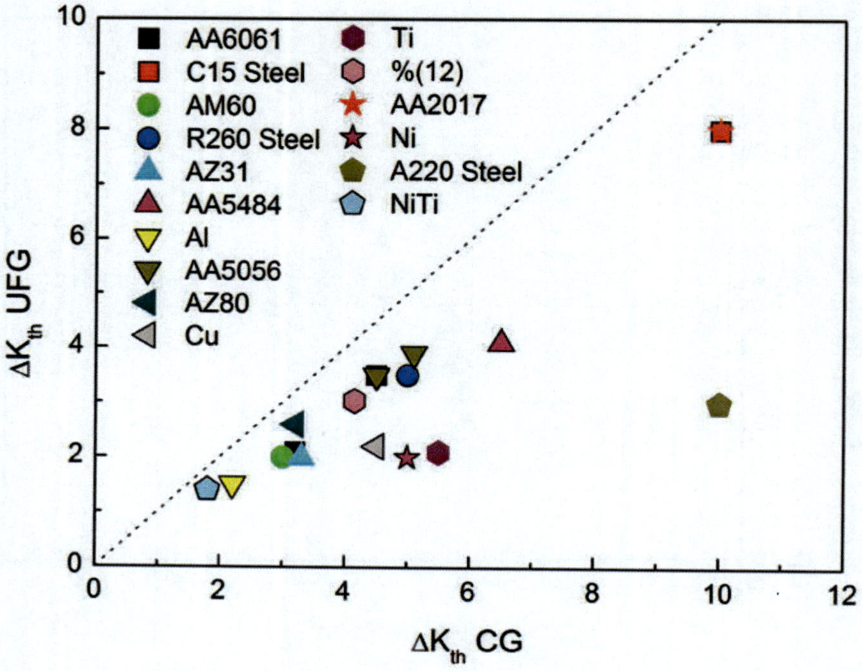

Fig. 5.19 ΔK_{th} comparison of UFG and MC materials

intensity factor range is lower in the ultrafine-grained alloy and also the stress intensity factor range at which catastrophic failure occurs is much smaller for the ultrafine-grained alloy.

The transition occurs when the crack tip cyclic plastic zone is approximately equal to the grain size. However, in ultrafine-grained materials, the crack tip plastic zone will always be significantly greater than the grain size even at threshold stress intensity factor ranges. Crack tip plasticity will always involve many grains for ultrafine-grained materials:

$$\Delta K_t \tilde{\sigma}_y' \sqrt{d^*}$$

where ΔK_t is the transition point stress intensity factor, σ_y' is the cyclic yield stress, and d^* is the characteristic microstructural dimension responsible for fracture. By assuming that the threshold stress intensity factor range approximately equals the transition stress intensity factor range, the equation can be written as

$$\Delta K_{\mathrm{th}} \tilde{A} + B\sqrt{d^*}$$

where ΔK_{th} is the threshold stress intensity factor amplitude, and A and B are material constants. These equations qualitatively explain the decrease in threshold stress intensity factor ranges with ultrafine-grained materials. The presence of slip markings that were found tens of microns from the crack indicates that the assumption that the transition between the near-threshold and steady-state regimes occurs when the reversible plastic zone (RPZ) is equal to the grain size may be an oversimplification (Goto et al. 2010, 2013a). It seems that near-threshold fatigue crack growth behavior in UFG materials is not a straightforward process and may depend on a variety of microstructural parameters in addition to grain size. A decrease in the grain size increases the yield stress and the ultimate tensile strength, and also improves the fatigue strength, especially of materials with low stacking fault energy, making crack nucleation difficult. On the other hand, a small grain size may have a negative effect on the fatigue crack growth resistance, as cracks can propagate along the grain boundaries (Dobatkin et al. 2018).

5.3.2 *Microstructural Behavior of Deformed UFGed Materials*

Another approach to explain fatigue threshold variations takes into account that cell formation occurs during cyclic deformation for ductile BCC and FCC metals with a high stacking fault energy.

As these cells are in the vicinity of the crack tip, they are the critical microstructural features that affect fatigue crack propagation behavior. An equation was derived by taking into account the crack tip plastic strain, crack tip plastic zone size, and average glide distance of dislocations:

$$\Delta K_{\text{th}} = 2\left[\frac{\varnothing \rho b E d}{3\sigma_y}\right]^{\frac{1+n_b'}{2}} \sigma_y \sqrt{\pi d}$$

where b is the magnitude of the Burgers vector, E is the Young's modulus, n_b' is the cyclic strain hardening exponent, ρ is the average mobile dislocation density, and Ø is the constant for plastic flow in a polycrystalline material. This equation takes into account other factors such as yield stress, mobile dislocation density, and susceptibility for cyclic hardening as well as grain size.

There are several distinct types of fracture surfaces that mark different stages and indicate different mechanisms of crack growth. Small surface cracks are strongly influenced by microstructural features and hence it is important to relate the reverse plastic zone size at the crack tip to the fracture surface morphology. By considering the previously defined reverse plastic zone size, it was found that a planar fracture surface occurs when

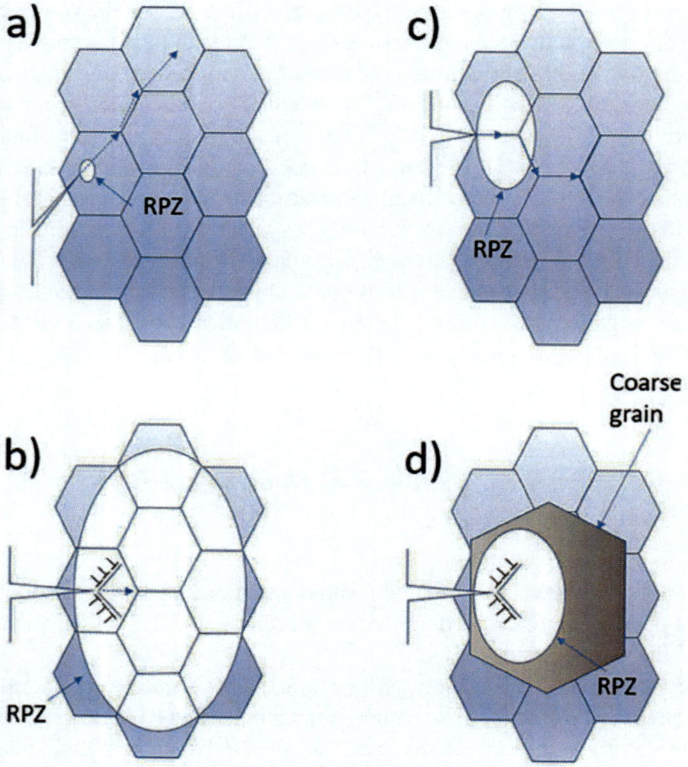

Fig. 5.20 Crack growth mechanisms in UFG materials (Goto et al. 2013b)

$$\frac{r_{cp}}{d} < 1$$

A granulated surface occurs when

$$1 < \frac{r_{cp}}{d} < 2$$

and a striated surface occurs when

$$\frac{r_{cp}}{d} < 3$$

By considering Fig. 5.20, a planar fracture surface occurs when $r_{cp}/d < 1$ (a), a granulated surface occurs when $r_{cp}/d = 1$–2 (b), a striated surface occurs when $r_{cp}/d < 3$ (c), and a striated plateau formation occurs in a coarsened grain (d). When r_{cp} is smaller than d, crack growth occurs by a combination of grain boundary sliding and deformation within grains with the crack propagating in the shear direction.

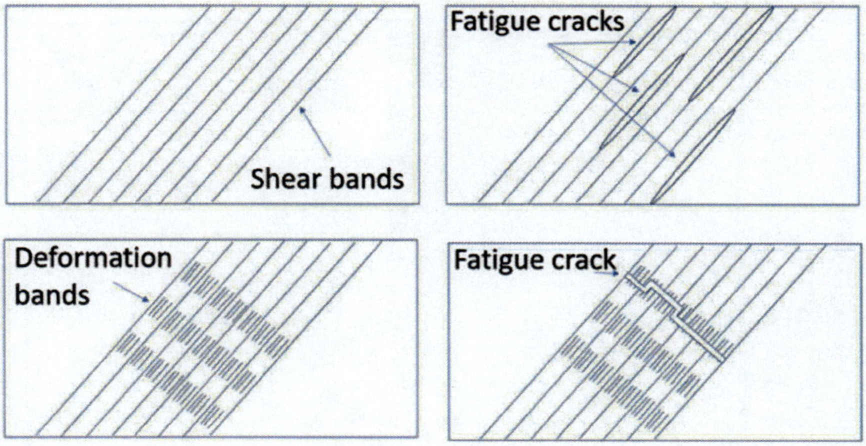

Fig. 5.21 Formation of shear bands, deformation bands, and cracks during fatigue loading

Studies performed on the grain boundaries of UFG copper processed by SPD have shown that these grain boundaries contain high energy, are nonequilibrium, and contain excess volume and long-range stress fields. This leads to accelerated diffusion possibly making grain boundary sliding relatively easy even at room temperatures during fatigue. When r_{cp} is about 1–2 times the size of d, in certain grain boundaries, there is a concentrated incompatibility of plastic deformation in adjacent grains, and an intergranular crack path results. The resulting granular fracture surface may lead to lower crack growth rates. When r_{cp} is more than three times the size of d, crack growth propagates through the mechanism of duplex slip and plastic blunting. Striations result and become evident in the resulting fracture surface. If grain coarsening occurs, the fracture surface morphology is that of striated plateaus.

Normally, fatigue cracking can either take place along the shear bands at low or medium strain range or occur along the coarse deformation bands at high strain range, which has not been reported previously in UFG materials (Zhang et al. 2005). It is suggested that fatigue cracking depends on the applied strain range or the degree of strain localization in the UFG materials. At high strain amplitude, the strain localization induced by pure shear deformation is insufficient to carry out the applied plastic strain. Consequently, a conjugated strain localization, approximately perpendicular to the shear bands, appeared in the form of the coarse deformation bands, which seemed to carry out more strain than shear bands (Fig. 5.21).

Therefore, fatigue cracking is prone to nucleate along the coarse deformation bands rather than along the shear bands at high strain range. Meanwhile, the coarse deformation bands are basically parallel to the maximum shear stress plane of the specimen, which also fulfils the requirement in mechanics.

Studies performed on ECAP IF steel with different routes underlined a very different mechanical and microstructural behavior between the two conditions (Fig. 5.22).

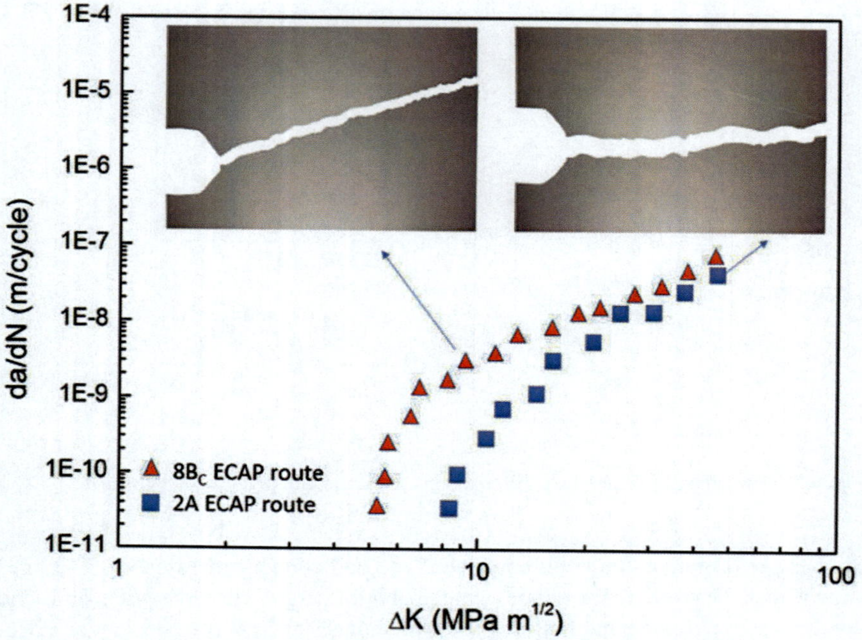

Fig. 5.22 RR sample crack propagation in ECAP IF steel produced via different routes

The ECAP processing route significantly influences the crack advance. Specifically, a significant deviation from the expected direction of crack growth normal to the loading direction is notable, where the deviation from the normal direction is a function of the ECAP processing route (Niendorf et al. 2010). As a matter of fact, the route 4C-processed UFG IF steel, characterized by cyclic softening, possesses a high volume fraction of LAGBs, whereas the microstructure of the route 4E-processed material, which displays a stable CSS response, is dominated by HAGBs. Under cyclic loading, the two types of grain boundaries (LAGBs and HAGBs) impose different fatigue responses. The HAGBs remain stable, preventing any significant fatigue-induced changes in the microstructure of the route 4E-processed IF steel, thus giving way to stable CSS response. In contrast, the LAGBs in the route 4C-processed material are rearranged continuously throughout the cyclic loading, eventually leading to cyclic softening (Niedorf et al. 2007).

In ECAP low-carbon steel, the intergranular based fracture surface was dominant in the ECAP steel, whereas a transgranular fracture surface was obtained in the as-received material (Okayasu et al. 2008). The intergranular fracture characteristics in the ECAP steel might have occurred through grain boundary sliding. This sliding is associated with the high dislocation density or pileup of dislocations along the grain boundary (Fig. 5.23).

The GBS can be related to the dislocation movement or pileup of dislocations along the grain boundary. The slip on the primary glide plane leads to a buildup of

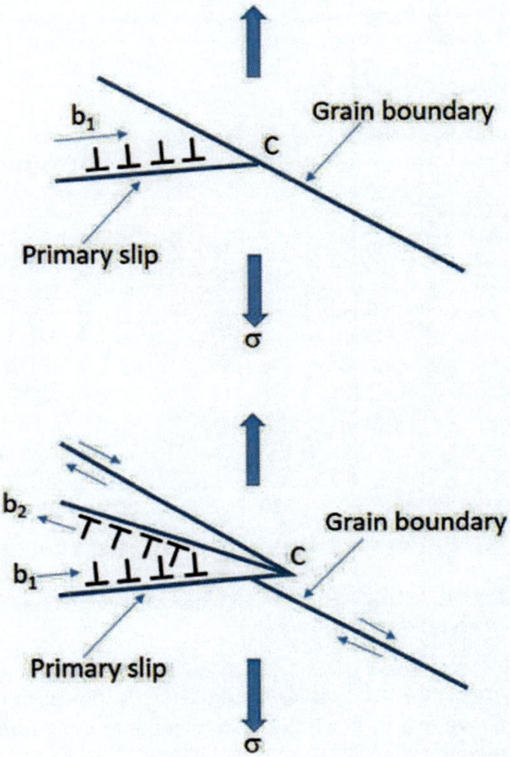

Fig. 5.23 Model of grain boundary sliding: (**a**) primary slip; (**b**) secondary slip

stress at its intersection with the grain boundary at C (Fig. 5.23a) and secondary grain sliding occurs in the absence of relaxation of this stress concentration (Fig. 5.23b). Microstructures dominated by HAGBs are stable under cyclic loading, whereas large fractions of LAGBs result in cyclic instability (Niedorf et al. 2013). The cyclic stability related to the fraction of HAGBs is confirmed also for SPDed 316L SS (Renk et al. 2012). Low-carbon steel with martensite microstructure becomes a difficult-to-work material due to its high tensile strength. If it is solely treated by ECAP, the processing will be inefficient and not practicable for the extensive engineering application, too. This indicates that the combination of grain refinement and heat treatment will be an efficient method to enhance its strength and ductility (Pang et al. 2012), even fatigue strength (Fig. 5.24).

5.4 Conclusions

The cyclic deformation responses of SPDed metals as available in the related literature are mixed, ranging from cyclic hardening to cyclic softening depending on various factors. These factors include the purity of the metal, the level of the

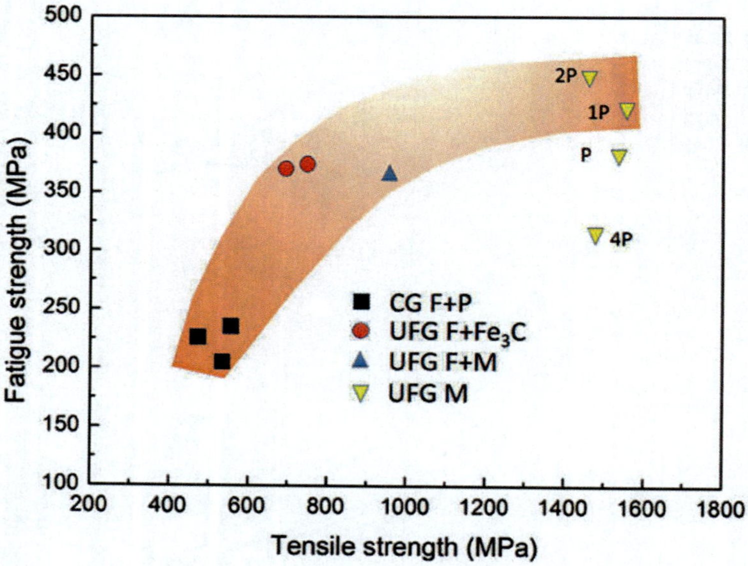

Fig. 5.24 Fatigue properties of coarse-grained (CG) and ultrafine-grained (UFG) low-carbon steels as a function of microstructure

imposed cyclic stress/strain, and the processing parameters. The effect from the above factors has been categorized into three fundamental influencing factors in this review.

The primary influencing factor is the microstructure stability of the initial microstructure. This includes the effects of the purity of the metals, the level of development of a UF-grained microstructure, and the grain geometry of the final microstructure. The latter two are affected by the processing technique and the parameters used. In short, the above influences the cyclic deformation response by changing the driving force for the grain coarsening phenomenon. The second influencing factor is the cyclic life span, which provides, or not, the satisfaction of the required kinetics for grain coarsening phenomenon.

With this increase of strength accompanied with a reduction in ductility, many strategies have been developed in order to obtain acceptable levels in order to increase also the fatigue life and the damage tolerance. The first observation is that mechanical properties in UFG materials are direction dependent as a consequence of grain elongation and textures. The fatigue properties in severely deformed UFG materials fabricated by ECAP depend on the specific shear texture induced during the fabrication which promotes orientation of the easy glide crystallographic planes to be aligned parallel to the shear plane in ECAP. The high-cyclic fatigue life in severely deformed UFG materials fabricated by ECAP can be improved through the controlled degradation of the ECAP-induced texture allowing effective rotation of the texture components corresponding to the easy dislocation glide plane from the ideal simple shear position. Fracture toughness is strongly dependent on the grain size and shape as well as on the orientation of the specimen. In detail, the crack growth resistance is low for cracks parallel to the shear plane acting on a direction parallel to the longer axis of the elongated grains with respect to a perpendicular direction. As a general behavior, in ultrafine-grained materials more grains lie in the plastic zone with respect to their microcrystalline counterparts. So, the potential interaction of the crack with grain boundaries can retard the FCG.

In general, a strong decrease in fatigue properties was shown for SPD materials in the LCF regime of intermediate to high plastic strain amplitudes. On the contrary, in the HCF regime of intermediate to low plastic strain amplitudes, it results in a strong enhancement of the fatigue resistance for materials with grain refinement. In addition, it was observed that an annealing treatment, subsequent to the ECAP process, produces an enhancement in the LCF properties due to the increase in ductility. Such a behavior is obtained by recovering partially the grain boundary region, which has been most heavily distorted during processing. The oriented distribution of defects along the shear plane in the last ECAP processing is one of the major mechanisms of SB formation. For UFG materials, three mechanisms have been found to contribute to the observed cyclic softening, which are (i) shear band formation, (ii) coarsening of the fine-scaled grain structure, as well as (iii) a reduction of the defect density, in especially dislocation density, or a combination of them. These mechanisms can lead to early strain localization, fatigue damage, and failure of the samples. The small grain size and the resulting large grain boundary fraction are suspected to be responsible for these microstructural instabilities as it offers large driving forces for grain growth.

Normally, fatigue cracking can either take place along the shear bands at low or medium strain range or occur along the coarse deformation bands at high strain range. It is suggested that fatigue cracking depends on the applied strain range or the degree of strain localization in the UFG materials. At high strain amplitude, the strain localization induced by pure shear deformation is insufficient to carry out the applied plastic strain. Consequently, a conjugated strain localization, approximately perpendicular to the shear bands, appeared in the form of the coarse deformation bands, which seemed to carry out more strain than shear bands. Therefore, fatigue cracking is prone to nucleate along the coarse deformation bands rather than along the shear bands at high strain range. Meanwhile, the coarse deformation bands are basically parallel to the maximum shear stress plane of the specimen, which also fulfils the requirement in mechanics.

References

An XH, Wu SD, Wang ZG, Zhang ZF (2014) Enhanced cyclic deformation responses of ultrafine-grained Cu and nanocrystalline Cu–Al alloys. Acta Mater 74:200–214. https://doi.org/10.1016/j.actamat.2014.04.053

An XH, Wu SD, Wang ZG, Zhang ZF (2019) Significance of stacking fault energy in bulk nanostructured materials: insights from Cu and its binary alloys as model systems. Prog Mater Sci 101:1–45. https://doi.org/10.1016/j.pmatsci.2018.11.001

Antolovich SD, Armstrong RW (2014) Plastic strain localization in metals: origins and consequences. Prog Mater Sci 59:1–160. https://doi.org/10.1016/j.pmatsci.2013.06.001

Bruder E, Gangaraju C, Lapovok R (2018) Influence of equal channel angular pressing on high cycle fatigue behaviour of ultrafine-grained iron: role of anisotropy. Mater Sci Eng A 711:650–658. https://doi.org/10.1016/j.msea.2017.11.075

Canadinc D, Maier HJ, Gabor P, May J (2008) On the cyclic deformation response of ultrafine-grained Al–Mg alloys at elevated temperatures. Mater Sci Eng A 496:114–120. https://doi.org/10.1016/j.msea.2008.04.071

Cavaliere P (2009) Fatigue properties and crack behavior of ultra-fine and nanocrystalline pure metals. Int J Fatigue 31:1476–1489. https://doi.org/10.1016/j.ijfatigue.2009.05.004

Cavaliere P, Cabibbo M (2009) Effect of Sc and Zr additions on the microstructure and fatigue properties of AA6106 produced by equal-channel-angular-pressing. Mater Charact 59:197–203. https://doi.org/10.1016/j.matchar.2006.12.013

Chembarisova RG, Latypov MI, Alexandrov IV (2012) Kinetic modelling of flow stress of metals and alloys processed by severe plastic deformation. Rev Adv Mater Sci 31:100–108

Chukin MV, Emaleeva DG, Polyakova MA et al (2016) Status and prospects of deformational methods for refining the microstructure of bulk materials. Metallurgist 60:299–305. https://doi.org/10.1007/s11015-016-0290-y

Chung CS, Kim JK, Kim HK, Kim WJ (2002) Improvement of high-cycle fatigue life in a 6061 Al alloy produced by equal channel angular pressing. Mater Sci Eng A 337:39–44. https://doi.org/10.1016/S0921-5093(02)00010-2

Chuvil'deev VN, Myshlyaev MM, Nokhrin AV et al (2017) Effect of the severe plastic deformation temperature on the diffusion properties of the grain boundaries in ultrafine-grained metals. Russ Metall 2017:413–425. https://doi.org/10.1134/S0036029517050044

Delshadmanesh M, Khatibi G, Zare Ghomsheh M, Lederer M, Zehetbauer M, Danninger H (2017) Influence of microstructure on fatigue of biocompatible β-phase Ti-45Nb. Mater Sci Eng A 706:83–94. https://doi.org/10.1016/j.msea.2017.08.098

Djavanroodi F, Ebrahimi M, Rajabifar B, Akramizadeh S (2010) Fatigue design factors for ECAPed materials. Mater Sci Eng A 528:745–750. https://doi.org/10.1016/j.msea.2010.09.080

Dobatkin SV, Skrotzki W, Rybalchenko OV, Terentev VF, Belyakov AN, Prosvirnin DV, Raab GI, Zolotarev EV (2018) Structural changes in metastable austenitic steel during equal channel angular pressing and subsequent cyclic deformation. Mater Sci Eng A 723:141–147. https://doi.org/10.1016/j.msea.2018.03.018

Esmaeili A, Shaeri MH, Noghani MT, Razaghian A (2018) Fatigue behavior of AA7075 aluminium alloy severely deformed by equal channel angular pressing. J Alloys Compd 757:324–332. https://doi.org/10.1016/j.jallcom.2018.05.085

Glezer AM (2008) A new approach to a description of structural-phase transformations under a very severe plastic deformation. Russ Phys J 51:480–491. https://doi.org/10.1007/s11182-008-9067-5

Goto M, Han SZ, Kim SS, Kawagoishi N, Lim CY (2007) Significance of non-equilibrium grain boundaries in surface damage formation of ultrafine-grained copper in high-cycle fatigue. Scripta Mater 57:293–296. https://doi.org/10.1016/j.scriptamat.2007.04.040

Goto M, Ando Y, Han SZ, Kim SS, Kawagoishi N, Euh K (2010) The effect of microstructural inhomogeneity on the growth paths of surface-cracks in copper processed by equal channel angular pressing. Eng Fract Mech 77:1914–1925. https://doi.org/10.1016/j.engfracmech.2010.01.011

Goto M, Teshima N, Han SZ, Euh K, Yakushiji T, Kim SS, Lee J (2013a) High-cycle fatigue strength and small-crack growth behavior of ultrafine-grained copper with post-ECAP annealing. Eng Fract Mech 110:218–232. https://doi.org/10.1016/j.engfracmech.2013.07.018

Goto M, Kamil K, Han SZ, Euh K, Kim SS, Yokoho Y (2013b) Effects of grain refinement due to severe plastic deformation on the growth behavior of small cracks in copper. Int J Fatigue 50:63–71. https://doi.org/10.1016/j.ijfatigue.2012.02.020

Higuera-Cobos OF, Berríos-Ortiz JA, Cabrera JM (2014) Texture and fatigue behaviour of ultrafine grained copper produced by ECAP. Mater Sci Eng A 609:273–282. https://doi.org/10.1016/j.msea.2014.05.011

Hohenwarter A, Leitner T (2019) Fatigue crack-growth properties of SPD-metals. In: Gdoutos E (ed) Proceedings of the first international conference on theoretical, applied and experimental mechanics. ICTAEM 2018, Structural integrity, vol 5. Springer, Cham. https://doi.org/10.1007/978-3-319-91989-8_79

Holzapfel C, Schäf W, Marx M, Vehoff H, Mücklich F (2007) Interaction of cracks with precipitates and grain boundaries: understanding crack growth mechanisms through focused ion beam tomography. Scr Mater 56:697–700. https://doi.org/10.1016/j.scriptamat.2006.12.025

Hoppel HW, May I, Prell M, Goken M (2011) Influence of grain size and precipitation state on the fatigue lives and deformation mechanisms of CP aluminium and AA6082 in the VHCF-regime. Int J Fatigue 33:10–18. https://doi.org/10.1016/j.ijfatigue.2010.04.013

Horky J, Khatibi G, Setman D, Weiss B, Zehetbauer MJ (2013) Effect of microstructural stability on fatigue crack growth behaviour of nanostructured Cu. Mech Mater 67:38–45. https://doi.org/10.1016/j.mechmat.2013.07.008

Kaneko Y, Hayashi S, Vinogradov A (2013) Cyclic response of SUS316L stainless steel processed by ECAP. Mater Trans 54(9):1612–1618. https://doi.org/10.2320/matertrans.MH201308

Kaneko Y, Tomita R, Vinogradov A (2014) Low-cycle fatigue of Fe-20%Cr alloy processed by equalchannel angular pressing. IOP Conf Series Mater Sci Eng 63:012160. https://doi.org/10.1088/1757-899X/63/1/012160

Kapp MW, Kremmer T, Motz C, Yang B, Pippan R (2017) Structural instabilities during cyclic loading of ultrafine-grained copper studied with micro bending experiments. Acta Mater 125:351–358. https://doi.org/10.1016/j.actamat.2016.11.040

Kazemi MA, Seifi R (2018) Effects of crack orientation on the fatigue crack growth rate and fracture toughness of AA6063 alloy deformed by ECAP. Mater Sci Eng A 733:71–79. https://doi.org/10.1016/j.msea.2018.07.042

Kendig KL, Miracle DB (2002) Strengthening mechanisms of an Al–Mg–Sc–Zr alloy. Acta Mater 50:4165–4175. https://doi.org/10.1016/S1359-6454(02)00258-6

Khatibi G, Horky J, Weiss B, Zehetbauer MJ (2010) High cycle fatigue behaviour of copper deformed by high pressure torsion. Int J Fatigue 32:269–278. https://doi.org/10.1016/j.ijfatigue.2009.06.017

Koch CC, Langdon TG, Lavernia EJ (2017) Bulk nanostructured materials. Metal Mater Trans A 48:5181–5199. https://doi.org/10.1007/s11661-017-4298-0

Korn M, Lapovok R, Bohner A, Hoppel HW, Mughrabi H (2011) Bimodal grain size distributions in UFG materials produced by SPD—their evolution and effect on the fatigue and monotonic strength properties. Kokove Mater 49:51–63

Kozlov EV, Popova NA, Koneva NA (2015) Plastic deformation of ultrafine-grained metallic materials. Russ Metall 2015:264–268. https://doi.org/10.1134/S0036029515040072

Kunz L, Lukas P, Navratilova L (2014) Strain localization and fatigue crack initiation in ultrafine-grained copper in high- and giga-cycle region. Int J Fatigue 58:202–208. https://doi.org/10.1016/j.ijfatigue.2013.04.027

Kwan CCF, Wang Z (2012) The cyclic deformation behavior of severe plastic deformation (SPD) metals and the influential factors. Metals 2:41–55. https://doi.org/10.3390/met2010041

Leitner T, Hohenwarter A, Pippan R (2015) Revisiting fatigue crack growth in various grain size regimes of Ni. Mater Sci Eng A 646:294–305. https://doi.org/10.1016/j.msea.2015.08.071

Leitner T, Hohenwarter A, Ochensberger W, Pippan R (2017) Fatigue crack growth anisotropy in ultrafine-grained iron. Acta Mater 126:154–165. https://doi.org/10.1016/j.actamat.2016.12.059

Limooei MB, Zandrahimi M, Ebrahimi R (2018) Low-cycle fatigue behavior of Al1070 severely deformed by equal channel angular pressing process. Iran J Sci Technol Trans Mech Eng 42:99–105. https://doi.org/10.1007/s40997-017-0085-1

May J, Amberger D, Dinkel M, Hoppel HW, Goken M (2008) Monotonic and cyclic deformation behavior of ultrafine-grained aluminium. Mater Sci Eng A 483-484:481–484. https://doi.org/10.1016/j.msea.2006.12.177

Medvedev A, Ng HP, Lapokov R, Estrin Y, Lowe TC, Anumalasetty VN (2015) Comparison of laboratory-scale and industrial-scale equal channel angular pressing of commercial purity titanium. Mater Lett 145:308–311. https://doi.org/10.1016/j.matlet.2015.01.051

Meyer LW, Sommer K, Halle T, Hockauf M (2008) Crack growth in ultrafine-grained AA6063 produced by equal-channel angular pressing. J Mater Sci 43:7426–7431. https://doi.org/10.1007/s10853-008-2725-8

Mughrabi H, Hoppel HW (2000) Proceedings of the symposium B on structure and mechanical properties of nanophase materials—theory and computer simulations. MRS Fall Meeting, Boston, 634:B2.1.1–B2.1.12

Mughrabi H, Hoppel HW (2010) Cyclic deformation and fatigue properties of very fine-grained metals and alloys. Int J Fatigue 32:1413–1427. https://doi.org/10.1016/j.ijfatigue.2009.10.007

Munoz JA, Higuera OF, Cabrera JM (2017) High cycle fatigue of ARMCO iron severely deformed by ECAP. Mater Sci Eng A 681:85–96. https://doi.org/10.1016/j.msea.2016.10.010

Murashkin M, Sabirov I, Prosvirnin D, Ovidko I, Terentiev V, Valiev R, Dobatkin S (2015) Fatigue behavior of an ultrafine-grained Al-Mg-Si alloy processed by high-pressure torsion. Metals 5:578–590. https://doi.org/10.3390/met5020578

Namdar M, Jahromi SAJ (2015) Influence of ECAP on the fatigue behavior of age-hardenable 2xxx aluminum alloy. Int J Miner Metall Mater 22:285–291. https://doi.org/10.1007/s12613-015-1072-4

Naseri R, Hiradfar H, Shariati M et al (2018) A comparison of axial fatigue strength of coarse and ultrafine grain commercially pure titanium produced by ECAP. Archiv Civil Mech Eng 18:755–767. https://doi.org/10.1016/j.acme.2017.12.005

Navratilova L, Kunz L, Novy F, Mintach R (2013) Development of cyclic slip bands in UFG copper in gigacycle fatigue. Acta Met Slovac 19(2):88–93. https://doi.org/10.12776/ams.v19i2.92

Niedorf T, Canadinc D, Maier HJ, Karaman I (2007) On the microstructural stability of ultrafine-grained interstitial-free steel under cyclic loading. Met Trans 38A:1946–1955. https://doi.org/10.1007/s11661-007-9154-1

Niedorf T, Bohner A, Hoppel HW, Goken M, Valiev RZ, Maier HJ (2013) Comparison of the monotonic and cyclic mechanical properties of ultrafine-grained low carbon steels processed by continuous and conventional equal channel angular pressing. Mater Des 47:138–142. https://doi.org/10.1016/j.matdes.2012.12.025

Niendorf T, Rubitschek F, Maier HJ, Canadinc D, Karaman I (2010) On the fatigue crack growth–microstructure relationship in ultrafine-grained interstitial-free steel. J Mater Sci 45:4813–4821. https://doi.org/10.1007/s10853-010-4511-7

Okayasu M, Sato K, Mizuno M, Hwang DY, Shin DH (2008) Fatigue properties of ultra-fine grained dual phase ferrite/martensite low carbon steel. Int J Fatigue 30:1358–1365. https://doi.org/10.1016/j.ijfatigue.2007.10.011

Orlov D, Vinogradov A (2011) The control of texture to improve high-cyclic fatigue performance in copper after equal channel angular pressing. Mater Sci Eng A 530:174–182. https://doi.org/10.1016/j.msea.2011.09.069

Pang JC, Yang MX, Yang G, Wu SD, Li SX, Zhang ZF (2012) Tensile and fatigue properties of ultrafine-grained low-carbon steel processed by equal channel angular pressing. Mater Sci Eng A 553:157–163. https://doi.org/10.1016/j.msea.2012.06.005

Plekhov O, Naimark O, Valiev R, Semenova I (2009) Investigation of energy balance in nanocrystalline titanium under cyclic loading. In: Boukharouba T, Elboujdaini M, Pluvinage G (eds) Damage and fracture mechanics. Springer, Dordrecht. https://doi.org/10.1007/978-90-481-2669-9_21

Rathmayr GB, Hohenwarter A, Pippan R (2013) Influence of grain shape and orientation on the mechanical properties of high pressure torsion deformed nickel. Mater Sci Eng A 560:224–231. https://doi.org/10.1016/j.msea.2012.09.061

Renk O, Hohenwarter A, Pippan R (2012) Cyclic deformation behavior of a 316L austenitic stainless steel processed by high pressure torsion. Adv Eng Mater 14(11):948–954. https://doi.org/10.1002/adem.201200015

Ritchie RO (2011) The conflicts between strength and toughness. Nat Mater 10:817–822. https://doi.org/10.1038/nmat3115

Sabirov I, Enikeev NA, Murashkin MY, Valiev RZ (2015) Multifunctional properties of bulk nanostructured metallic materials. In: Bulk nanostructured materials with multifunctional properties, Springerbriefs in materials. Springer, Cham. https://doi.org/10.1007/978-3-319-19599-5_3

Saitova LR, Hoppel HW, Goken M, Semenova IP, Valiev RZ (2009) Cyclic deformation behavior and fatigue lives of ultrafine-grained Ti-6AL-4V ELI alloy for medical use. Int J Fatigue 31:322–331. https://doi.org/10.1016/j.ijfatigue.2008.08.007

San Marchi C, Robinson SL, Yang NYC, Lavernia EJ (2006) Fracture of nanocrystalline aluminum. In: Gdoutos EE (ed) Fracture of nano and engineering materials and structures. Springer, Dordrecht. https://doi.org/10.1007/1-4020-4972-2_331

Semenova IP, Alimgareeva GK, Latysh VV, Lowe T, Valiev RZ (2009) Enhanced fatigue strength of commercially pure Ti processed by severe plastic deformation. Mater Sci Eng A 503:92–95. https://doi.org/10.1016/j.msea.2008.07.075

Soliman MS, El-Danaf EA, Almajid AA (2012) Enhancement of static and fatigue strength of 1050 Al processed by equal-channel angular pressing using two routes. Mater Sci Eng A 532:120–129. https://doi.org/10.1016/j.msea.2011.10.072

Ueno H, Kakihata K, Kaneko Y, Hashimoto S, Vinogradov A (2011) Enhanced fatigue properties of nanostructured austenitic SUS 316L stainless steel. Acta Mater 59:7060–7069. https://doi.org/10.1016/j.actamat.2011.07.061

Valiev RZ (2000) SPD processing and enhanced properties in metallic materials. In: Lowe TC, Valiev RZ (eds) Investigations and applications of severe plastic deformation, NATO science series (Series 3. High technology), vol 80. Springer, Dordrecht. https://doi.org/10.1007/978-94-011-4062-1_29

Valiev RZ (2007) The new trends in fabrication of bulk nanostructured materials by SPD processing. J Mater Sci 42:1483–1490. https://doi.org/10.1007/s10853-006-1281-3

Vinogradov AY, Stolyarov VV, Hashimoto S, Valiev RZ (2001) Cyclic behavior of ultrafine-grain titanium produced by severe plastic deformation. Mater Sci Eng A 318:163–173. https://doi.org/10.1016/S0921-5093(01)01262-X

Vinogradov A, Washikita A, Kitagawa K, Kopylov VI (2003) Fatigue life of fine-grain Al-Mg-/Sc alloys produced by equal-channel angular pressing. Mater Sci Eng A 349:318–326. https://doi.org/10.1016/S0921-5093(02)00813-4

Waitz T, Schranz W, Tröster A (2014) Nanoscale phase transformations in functional materials. In: Saxena A, Planes A (eds) Mesoscopic phenomena in multifunctional materials, Springer series in materials science, vol 198. Springer, Berlin. https://doi.org/10.1007/978-3-642-55375-2_2

Xu CZ, Wang QJ, Zheng MS, Zhu JW, Li JD, Huang MQ, Jia QM, Du ZZ (2007) Microstructure and properties of ultra-fine grain Cu–Cr alloy prepared by equal-channel angular pressing. Mater Sci Eng A 459:303–308. https://doi.org/10.1016/j.msea.2007.01.105

Xu C, Wang Q, Zheng M, Li J, Huang M, Jia Q, Zhu J, Kunz L, Buksa M (2008) Fatigue behavior and damage characteristic of ultra-fine grain low-purity copper processed by equal-channel angular pressing (ECAP). Mater Sci Eng A 475:249–256. https://doi.org/10.1016/j.msea.2007.04.074

Zhang ZF, Wu SD, Li YJ, Liu SM, Wang ZG (2005) Cyclic deformation and fatigue properties of Al–0.7 wt.% Cu alloy produced by equal channel angular pressing. Mater Sci Eng A 412:279–286. https://doi.org/10.1016/j.msea.2005.08.221

Zhang ZJ, An XH, Zhang P, Yang MX, Yang G, Wu SD, Zhang ZF (2013) Effects of dislocation slip mode on high-cycle fatigue behaviors of ultrafine-grained Cu–Zn alloy processed by equal-channel angular pressing. Scripta Mater 68:389–392. https://doi.org/10.1016/j.scriptamat.2012.10.036

Zhao P, Chen B, Kelleher J, Yuan G, Guan B, Zhang X, Tu S (2019) High-cycle-fatigue induced continuous grain growth in ultrafine grained titanium. Acta Mater 174:29–42. https://doi.org/10.1016/j.actamat.2019.05.038

Zhu R, Wu YJ, Ji WQ, Wang JT (2011) Cyclic softening of ultrafine-grained AZ31 magnesium alloy processed by equal-channel angular pressing. Mater Lett 65:3593–3596. https://doi.org/10.1016/j.matlet.2011.07.111

Zimmermann J, Sharma A, Divinski SV, Rabkin E (2020) Relative grain boundary energies in ultrafine grain Ni obtained by high pressure torsion. Scripta Mater 182:90–93. https://doi.org/10.1016/j.scriptamat.2020.03.008

Zuberova Z, Kunz L, Lamark TT, Estrin Y, Jnacek M (2007) Fatigue and tensile behavior of cast, hot-rolled, and severely plastically deformed AZ31 magnesium alloy. Met Trans 38A:1924–1940. https://doi.org/10.1007/s11661-007-9109-6

Chapter 6
Creep in Nanostructured Materials

6.1 Introduction

As mentioned in the previous chapters, one of the most remarkable phenomena that nanostructured materials exhibit is the time-dependent deformation, or creep, at room temperature. Nanostructured materials exhibit creep at lower temperatures with respect to their microcrystalline counterparts (Murty 2017). There are two important parameters, grain size and diffusivity, which affect strongly the creep behavior of a polycrystalline material. By increasing diffusivity and/or reducing the grain size, the creep rate of a polycrystalline material can be enhanced (Gollapudi et al. 2010a). For example, in the case of the Coble model and grain boundary sliding model, the creep rate depends on the third power of the mean grain size. Going from a grain size of 1 μm to 10 nm should increase creep rate by 10^6 or more if the diffusivity of grain boundary is significantly larger for NC materials.

6.1.1 Creep Mechanisms

With the volume fraction being increasingly fundamental with decreasing grain size, diffusion creep is considered to be significant (Yamakov et al. 2002; Nie et al. 2017). In addition, with high diffusivity and nanometer grain size, NC materials are expected to creep at room temperature. The high-temperature deformation of crystalline materials is given by the following (Bird–Dorn–Mukherjee) equation:

$$\dot{\varepsilon} = \frac{ADGb}{kT}\left(\frac{b}{d}\right)^{p}\left(\frac{\sigma}{G}\right)^{n}$$

© Springer Nature Switzerland AG 2021

P. Cavaliere, *Fatigue and Fracture of Nanostructured Materials*,
https://doi.org/10.1007/978-3-030-58088-9_6

A is a dimensionless constant, G is the shear modulus, b is the magnitude of the Burgers vector, k is Boltzmann constant, T is the absolute temperature, p is the inverse grain size exponent, and n is the stress exponent.

Among the established diffusional creep mechanisms in coarse-grained materials are the Nabarro–Herring creep that involves vacancy flow through the lattice and Coble creep that involves vacancy flow along the grain boundaries (Chokshi 2009). The associated equations are

$$\dot{\varepsilon}_{\mathrm{NH}} = \frac{A_{\mathrm{NH}} D_L G b}{kT}\left(\frac{b}{d}\right)^2\left(\frac{\sigma}{G}\right)$$

where D_L is the lattice diffusion coefficient; the exponents $p = 2$ and $n = 1$; and the dimensionless constant $A_{\mathrm{NH}} = 28$. On the other hand, Coble creep involves vacancy flow along grain boundaries, and the related equation is

$$\dot{\varepsilon}_{\mathrm{CO}} = \frac{A_{\mathrm{CO}} D_{\mathrm{gb}} G b}{kT}\left(\frac{b}{d}\right)^3\left(\frac{\sigma}{G}\right)$$

where D_{gb} is the grain boundary diffusion coefficient; the exponents $p = 3$ and $n = 1$; and the dimensionless constant $A_{\mathrm{CO}} = 33$.

As described in Chap. 1, grain boundary volume fraction increases with decreasing grain size and triple junction volume fraction displays greater grain size dependence than grain boundary (Voyiadjis and Deliktas 2010). So, the standard diffusion creep equation to accommodate for diffusion along triple lines was modified and this leads to the following expression for triple-line diffusion creep:

$$\dot{\varepsilon}_{\mathrm{TL}} = \frac{K_{\mathrm{TL}} D_{\mathrm{TL}} \Omega \delta^2 \sigma}{kTd^4}$$

where K_{TL} is a constant depending on the geometry and boundary conditions, D_{TL} is the triple-line diffusion coefficient, Ω is the atomic volume, and δ is the grain boundary width.

In addition to grain boundary sliding, diffusion along grain boundaries and along triple junctions has been proposed to account for the plastic deformation without dislocation motion below the critical grain size (Park et al. 2006). Diffusion is particularly relevant to the deformation of nanocrystalline metals because typical dislocation mechanisms are suppressed and grain boundary-controlled deformation processes become dominant (Mishra et al. 1999, Diffusion in nanocrystalline materials 2007). The creep rate due to grain boundary diffusion (Coble creep) is given by

$$\dot{\varepsilon} = A' \frac{\sigma \Omega_D}{kT} \frac{\delta_D D_{\mathrm{gb}}}{d^3}$$

where δ_D is the effective diffusional width of the grain boundary, Ω_D is the activation or atomic volume, $\sigma\Omega_D$ is the work performed by the stress during an elementary diffusion jump in the grain boundary, D_{gb} is the grain boundary diffusion coefficient, and A' is a constant.

For conditions under which transition takes from one diffusion creep mechanism to the other and also from diffusion-controlled mechanism to intergranular dislocation power law creep, the following form of the Bird–Dorn–Mukherjee equation has been proposed:

$$\dot{\varepsilon}_{PL} = \frac{AD_L Gb}{kT}\left(\frac{\sigma}{G}\right)^n$$

where $n \geq 3, p = 0$, and $D = D_L$ in the Bird–Dorn–Mukherjee equation.

When the grain diameter becomes of the order of the grain boundary width, the grain size dependence of the strain rate changes from d^{-3} to d^{-2}. This implies that the grain boundary width, δ, is non-negligible compared with the grain size (d) in the nanocrystalline metals (Kawasaki and Langdon 2007). Taking grain boundary width into consideration, the Coble creep rate equation is modified to

$$\dot{\varepsilon} = 4K\frac{\sigma\Omega_D D_{gb}}{kT}\left(\frac{2\delta}{d^3} - \frac{\delta^2}{d^4}\right)$$

where K is a material constant. In the limit of large grain size, i.e., $d \geq \delta$, the second term inside the bracket of the equation is negligible compared to the first one. In the limit of small grain size, $d \approx 2\delta$, the second term is comparable to the first one, and the resulting creep rate becomes

$$\dot{\varepsilon} = 3K\frac{\sigma\Omega_D D_{gb}}{kT}\frac{1}{d^2}$$

It is evident that the grain size dependence of grain boundary diffusion creep becomes identical to the Nabarro–Herring creep model when the grain boundary width becomes comparable to the grain size (Keblinski et al. 1998; Kolobov et al. 2004).

Therefore, these extend the validity of the Coble creep equation from coarse-grained polycrystals to nanocrystalline materials, provided that the creep mechanism remains under the control of grain boundary diffusion process (Kolobov et al. 2002). This implies that grain boundary diffusion and grain boundary sliding form an integral process of creep deformation (Yang et al. 2006). Grain boundary sliding is needed to accommodate the homogeneous grain elongation induced by Coble creep, with the entire deformation process being controlled by grain boundary diffusion (Korla and Chokshi 2014). In NC Ni, creep deformation at room temperature appears to be dominated by a Coble creep mechanism, in which enhanced diffusion along intercrystalline components including grain boundaries, triple junctions, and quadruple nodes controls the mass transport process. The calculated creep rate at room temperature, based on the Coble creep mechanism, is in agreement with the experimental results. Nevertheless, the creep deformation mechanism might change from a diffusion-controlled process at room temperature to multiple mechanisms at 373 K. Above ambient temperature, lattice dislocation gliding and grain boundary sliding may play an important role in deformation (Yin et al. 2001).

It has been widely reported that creep deformation in NC metals is mainly driven by three fundamental processes: grain boundary (GB) diffusion, GB sliding and migration, and dislocation nucleation from GBs with increasing applied stress (Wang et al. 2013). These thermally activated deformation mechanisms lead to a temperature dependence of strength in NC metals (Hu et al. 2012). It is well known that a rate-controlling process can be characterized by

$$\nu = \nu_0 N \exp\left[\left(-\frac{\Delta G(T,\sigma)}{k_b T}\right)\right]$$

via the temperature T- and stress tensor σ-dependent activation Gibbs free energy ΔG, where ν is the resultant frequency, ν_0 is the trial frequency, N is the equivalent source number, and k_b is the Boltzmann constant. The steady-state creep rate tensor of NC metals can be expressed as

$$\dot{\varepsilon} = A\left(\frac{1}{d}\right)^p \exp\left(\frac{\Delta S}{k_p}\right) \exp\left(-\frac{\Delta Q - \sigma : \Omega}{k_b T}\right)$$

where A is a dimensionless pre-factor tensor related to the source number N and the trial frequency ν_0, d is the grain size, and the exponent p arises from the fact that each successful activated event contributes to a unit strain; the strains, along with N, are usually functions of d. Here, the activation free energy is

$$\Delta G = \Delta Q - \sigma : \Omega - T\Delta S$$

where ΔQ is the activation energy at $T = 0$ K and $\sigma = 0$.

$$\Omega = \frac{\partial \Delta G}{\partial \sigma}$$

is the activation volume tensor. ΔS is the activation entropy. This implies dependence of the creep rate on the activation entropy, etc.; the rate is proportional to

$$\exp\left(\frac{\Delta S}{k_b}\right)$$

By employing the uniaxial tensile stress instead of the stress tensor, the steady-state creep rate tensor can be rewritten as

$$\dot{\varepsilon} = \dot{\varepsilon}_0 \exp\left(-\frac{\Delta H - T\Delta S}{k_b T}\right)$$

with

$$\Delta H = \Delta Q - \sigma\Omega$$

That can also be expressed as

$$\Delta H = -\frac{\partial \ln \dot{\varepsilon}}{\partial\left(\frac{1}{k_b T}\right)}$$

This is actually from the Gibbs–Helmholtz equation:

$$\Delta H = \left[\frac{\partial\left(\frac{\Delta G}{T}\right)}{\partial\left(\frac{1}{T}\right)}\right]_\sigma$$

ΔS can be given by

$$\Delta S = k_b \ln \dot{\varepsilon} + \frac{\Delta H}{T} - k_b \ln \dot{\varepsilon}_0$$

The established constitutive equation for Coble creep is expressed as

$$\dot{\varepsilon}_C = \frac{\delta b^2 v_D}{d^3} \exp\left(-\frac{\Delta G_C}{k_b T}\right)$$

where δ is the GB width and v_D is the trial frequency of migration of a point defect. The displacive creep that is accommodated by GB sliding and GB dislocation nucleation are described as

$$\dot{\varepsilon}_{\mathrm{GBS}} = \varepsilon^T \frac{3\delta}{d} v_{\mathrm{GBS}} \exp\left(-\frac{\Delta G_{\mathrm{GBS}}}{k_b T}\right)$$

and

$$\dot{\varepsilon}_{\mathrm{disl}} = \frac{b}{d} v_{\mathrm{disl}} \exp\left(-\frac{\Delta G_{\mathrm{disl}}}{k_b T}\right)$$

ε^{T} is the unit local strain caused by each activation of GB sliding and b/d is the unit strain contributed by each dislocation nucleation. v_{GBS} and v_{disl} are the trial frequency of GB sliding and dislocation nucleation from GB, respectively. The activation volume is an important parameter for distinguishing deformation mechanisms, which provides a link between modeling and experiment and is given by

$$\Omega = \sqrt{3} k_b T \frac{\partial \ln \dot{\varepsilon}}{\partial \sigma}$$

The activation volume is a function of temperature. The activation volume of dislocation-nucleation-dominant creep decreases almost linearly with increasing temperature. Two features of the activation volume for diffusional creep are found: (1) the activation volume is greater than one atomic volume, and (2) it increases with increasing T. The conventional opinion is that Coble creep is involved in vacancy diffusion within GBs. It seems to be one atom's decision, and usually presents an activation volume of the order of unit b^3. The large activation volume for diffusion-dominant creep indicates that such a deformation mechanism involves a collective decision of multiple atoms. The activation volume of such a mechanism becomes larger at higher temperatures, along with a larger activation enthalpy and activation entropy (Sergueeva et al. 2006).

In most cases, creep diffusion is enhanced along the grain boundaries and triple junctions. It has long been suggested that GBs and triple junction points are preferred paths for self-diffusion, as opposed to the grain interior, with diffusivity values order of magnitude more than the latter. Thus, as the volume of GBs and triple junction points increases for nanocrystalline metals, the contribution of diffusional transport along these pathways leading to creep increases significantly.

With decreasing grain size in nanocrystalline materials to 10 nm or lower, the effects of triple junctions on the deformation become significant because the volume fraction of triple junctions is very high (Yin and Whang 2005). The volume fraction of triple junction approaches that of grain boundaries when the grain size is decreased to ~3 nm. Triple junctions of grain boundaries play the role of enhanced diffusivity pipes, nuclei of second-phase segregation, sources of lattice dislocations during plastic deformation, drag centers of grain boundary migration upon grain growth, nucleation site for voids, etc. Grain boundary triple junctions could act as softening regions during deformation in sharp contrast to grain boundaries, which are well-known hardeners (Mohamed 2007). The triple junctions play the role of softeners possibly due to their low density in which diffusion transport is fast (Tjong and Chen 2004).

6.1.2 Creep Characterization of NC Materials

In general, consolidated nanocrystalline samples are too small to be used in conventional, large-scale mechanical tests. In this regard, indentation method appears to be an effective route to study the creep deformation of nanocrystalline materials (Zhang et al. 2004). Nano/microindentation is a fast and effective technique to characterize the mechanical properties of materials at the nanometer/micrometer scales. Due to its unique feature of small size-scaled characterization, the nano/microindentation has been widely utilized in the mechanical characterization of various materials, especially nanomaterials, nanostructured materials, biomaterials, and thin films. Within such small detected volume in nano/microindentation tests, the strain magnitude, stress level, and strain rate could be considerably high, which are hardly achieved in conventional tensile tests. In normal uniaxial tensile (compression) test, the cross-head movement rate of a testing machine is preset and the load cell and the extension meter record, as functions of time, the load and sample displacement, respectively. Then, based on the geometry of a tested sample, load and displacement are correspondingly converted to stress and strain, and then stress rate and strain rate can be calculated. The strain rate is linked to the cross-head movement rate and a preset cross-head movement rate corresponds generally to a constant strain rate. In this sense, normal uniaxial tensile (compression) test is carried out under a sustained strain rate, during which the stress varies with the deformation characterized by strain. In contrast, a sustained stress is applied on a sample in uniaxial tensile (compression) creep test and creep deformation strain is recorded with creep time. Based on the creep deformation behavior, creep deformation is usually divided into three stages of the primary (transient) stage, the secondary steady stage, and the tertiary (unstable) stage. In the primary stage, the strain rate decreases with time and reaches a steady value in the secondary stage, while in the tertiary stage, the strain rate increases until failure of the crept sample. The creep strain rate in the secondary steady stage is approximately constant independent of time, which is the most critical creep strain rate. This is because the period of steady-state creep spends, in most cases, the largest portion of the creep life, and the higher the steady creep rate is, the shorter the survival life of the crept material will be. In indentation creep tests there is a challenging issue of which parameters should be used to represent and describe the loading level and the degree of deformation. This is because the stress distribution is completely inhomogeneous inside the tested sample under an indentation load and the inhomogeneous stress distribution causes inhomogeneous deformation. It is suggested that hardness might be the appropriate parameter to describe the loading level, since hardness in nanoindentation tests is defined as the averaged pressure on the projected contact area. Indentation depth might be the appropriate parameter to describe the deformation, because indentation depth is the only measurable parameter in nanoindentation creep under a given indentation load (Choi and Jang 2020). Indentation creep is actually the inhomogeneous deformation varying with time under a given load, in which the indentation depth increases and the hardness decreases monotonically with time. Commonly, the creep deformation will eventually saturate under a given compressive indentation load.

Indentation creep, in which the extent of deformation is measured during both loading and unloading, has been studied in nanocrystalline materials by monitoring the changes in the indentation size occurring beneath the indenter as a function of time. In the process, the indenter is pushed into the surface at a fixed rate of indentation until a predetermined load or penetration depth is reached, and then the load is held fixed while the indenter continues to creep into the material (Fig. 6.1).

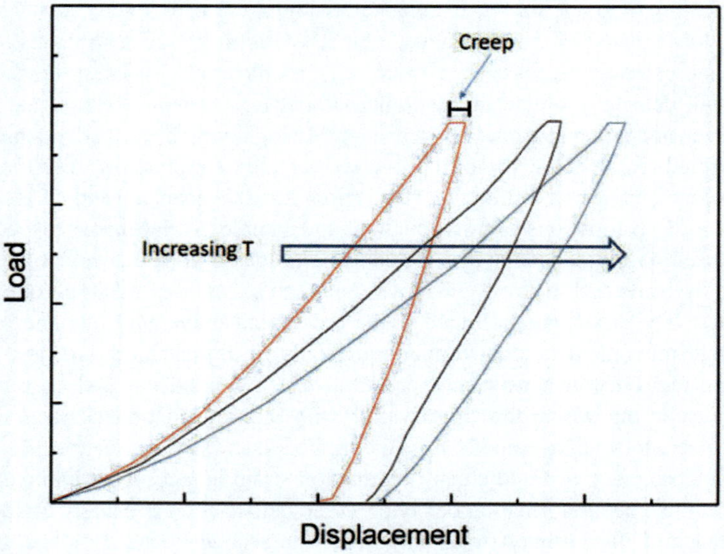

Fig. 6.1 Creep tests through nanoindentation

Indentation creep technique can be used to determine the power law stress exponent and activation energy. The results obtained from indentation creep translate readily into those determined from tensile loading. Nanoindentation creep and uniaxial creep are expected to have a similar stress exponent because the underpinning deformation mechanism is the same (Liu et al. 2012). However, the creep rate under nanoindentation is typically much faster than that under uniaxial creep. The faster creep rate is caused by the facts that the stress state under nanoindentation is more complex and severe than the uniaxial condition (Wang et al. 2010).

Creep displacement, h, is a function of time, t, and follows the empirical relation:

$$h(t) = h_0 + a(t - t_0)^p + kt$$

where h_0 and t_0 are the initial depth and time during creep period and a, p, and k are fitting constants.

By employing Berkovich indenter, indentation strain rate and hardness are given by

$$\dot{\varepsilon} = \frac{1}{h}\frac{dh}{dt}$$

$$H = \frac{P}{24.5h_c^2}$$

The creep stress exponent can be easily calculated from the slope of ln of strain rate versus H curves. In fact, the activation volume is given by

$$V^* = \frac{9kT}{\partial H/\partial \ln \dot{\varepsilon}_{\text{eff}}}$$

And the strain rate m (1/n) is

$$m = \frac{\partial H}{\partial \ln \dot{\varepsilon}_{\text{eff}}}$$

Take into account the effective strain rate in the plastic zone beneath the indenter.

The first reports on Ni-based NC alloys with various grain sizes support either Coble or Nabarro–Herring creep leading to a first conclusion that while grain boundary diffusion is the operating mechanism in nanocrystalline creep, a combined mechanism involving dislocation creep and grain boundary diffusion governs deformation in coarse-grained materials. Other evidences about NC palladium showed that creep rates were two to four orders of magnitude smaller than the values predicted by the equation for Coble creep. Calculated creep curves, assuming grain boundary diffusion for different grain sizes (notice increase in strain rate by six orders of magnitude when grain size is decreased from 1 μm to 10 nm), led to the conclusion that prevalence of low-energy grain boundaries together with inhibition of dislocation activity caused by small grain sizes is responsible for low strain rates and higher-than-expected creep resistance.

The contribution from diffusion creep mechanisms through intercrystalline regions can be significant for smaller grain sizes (Kawasaki and Langdon 2017). It was suggested that dynamic creep should be taken into account when analyzing the stress-strain curves of nanocrystalline materials.

Experiments performed on NC Cu (30 nm) revealed that the steady-state creep is proportional to an effective stress:

$$\sigma_e = \sigma - \sigma_0$$

where σ is the applied stress, and σ_0 is the threshold stress. Using σ_e, the creep rates were found to be of the same order of magnitude as those calculated from the equation for Coble creep. The existence of a threshold stress implied that the grain boundaries do not act as perfect sources and sinks of atoms (or vacancies). Hence the rate of grain boundary diffusion is limited by the emission and absorption of atoms (or vacancies). The energy needed for the emission and absorption of atoms (or vacancies) results in the threshold stress (Cai et al. 2000). The mechanism of creep was identified as “interface-controlled diffusional creep” (Meyers et al. 2004).

In UFG materials, the development of meso- and macrobands of localized deformation during deformation of fine-grained materials has been postulated. This leads to an apparent creep activation energy lower than the microcrystalline counterpart. This difference is attributed to the significant contributions to the overall deformation of grain boundary sliding controlled by grain boundary diffusion (Kolobov et al. 2008).

6.1.3 *Microstructural Features*

A model for GB sliding has been developed by Langdon (2006) in which sliding occurs by the movement of dislocations along, or adjacent to, the boundary by a combination of climb and glide leading to increased creep rates at smaller grain sizes. The first type of GB sliding refers to the relative displacement of adjacent grains where the grains essentially retain their original shape but they become visibly displaced with respect to each other. This type of GB sliding is designated as Rachinger sliding (Fig. 6.2).

It occurs in a polycrystalline matrix under creep conditions where there is a net increase in the number of grains lying within the gauge length along the direction of the tensile stress as the diffusion along the GBs causes the grains in the transverse direction to migrate to axial direction under shear (Zhang et al. 2013).

The second type of GB sliding, known as Lifshitz sliding (Fig. 6.3), occurs exclusively in Nabarro–Herring and Coble diffusion creep processes and it is related to boundary offsets that develop as a direct consequence of stress-directed diffusion of vacancies. The GB sliding that is expected in nanocrystalline metals is a combination of these two types of GB sliding; it should be understood that although it originates from the diffusional flow along GBs, the strain accumulated during this process needs to be accommodated via other deformation mechanisms (Panin et al. 2017). These strain accommodation mechanisms can be purely elastic at low loads while at high loads sliding can be accommodated by dislocation-mediated

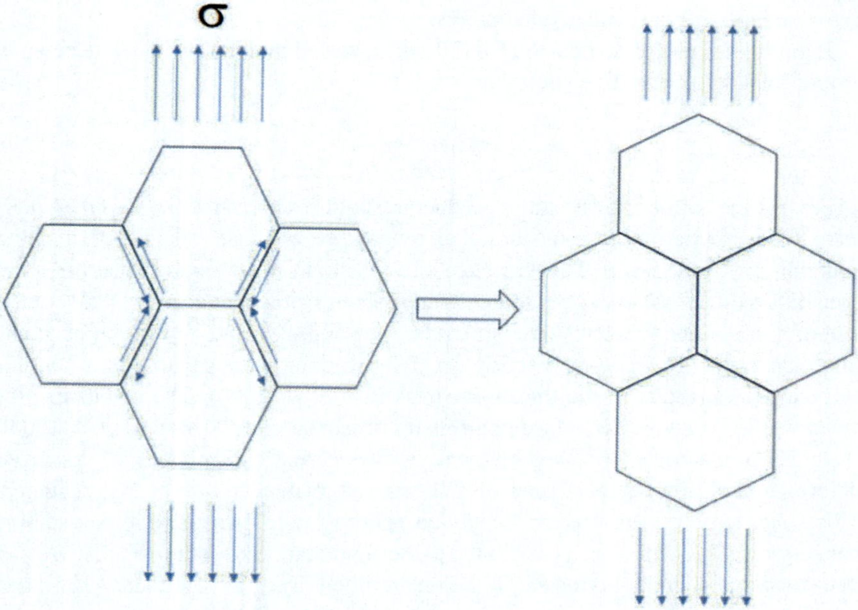

Fig. 6.2 Schematic of Rachinger sliding

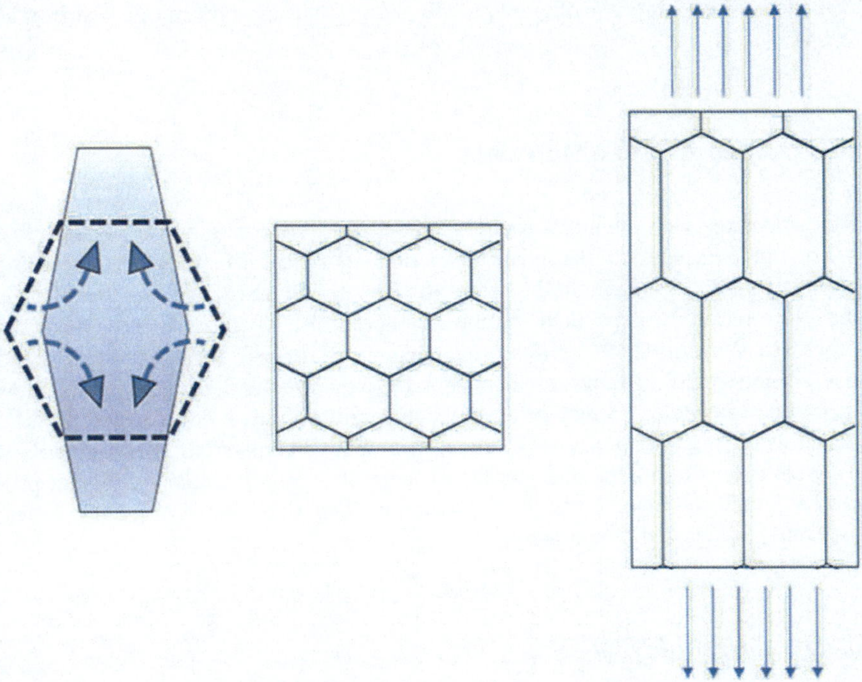

Fig. 6.3 Schematic of Lifshitz sliding

deformation mechanisms originating at GBs. Thus, GB sliding results in much higher creep rates than purely diffusional mechanisms such as Coble creep. At high temperatures, the aforementioned creep mechanisms in nanocrystalline metals would cross over to dislocation-mediated creep including glide and climb of dislocations as immobile dislocation structures at GBs at room temperature can become mobile at elevated temperatures.

MD simulations show that prior to the onset of significant grain growth, the deformation was shown to proceed via the mechanism of Coble creep accompanied by grain boundary sliding. While grain growth is generally known to decrease the creep rate due to the increase of the average grain size, the results revealed an enhanced creep rate at the onset of the grain growth, when rapid boundary migration occurred. The enhanced creep rate was shown to arise from topological changes during the initial growth phases, which enhanced both the stress-induced grain boundary diffusion fluxes and grain boundary sliding. Dislocations generated as a result of grain rotation-induced grain coalescence and grain boundary decomposition in the vicinity of certain triple junctions were also shown to contribute to the deformation.

Grain growth-induced vacancy effects have a strong influence on diffusion-controlled creep of nanocrystalline materials. The obvious decrease of the Nabarro–Herring and Coble creep rate due to the usual increase in grain size is partly compensated by inhibition of grain growth owing to vacancy generation (Estrin et al. 2004).

6.2 Creep in UFG Materials

By comparison with the initial coarse-grained state, processing by ECAP leads to a considerable change in the creep properties in bulk UFG materials including increases and/or decreases in the creep life. It is important to note that these trends may be noticeably dependent on the number of ECAP passes. UFG materials processed by ECAP differ quantitatively from their coarse-grained (CG) counterparts in terms of their characteristic structural parameters and thus their creep behavior cannot be easily compared with that documented for CG materials. It is important to note in this respect that UFG materials are characterized by great extension of internal interfaces; therefore, grain boundary diffusion processes have to be involved in the formation of their structure-sensitive properties, especially at elevated temperature.

6.2.1 Effect of SPD

The first studies of creep in UFG materials were performed on pure metals in order to isolate the basic mechanisms related to grain refinement and grain boundary structure. In pure aluminum, under same testing conditions of stress and temperature the measured minimum creep rates in the ECAP material with ultrafine grain size were slower and creep lifetimes markedly longer than in the same material in a coarse-grained unpressed state (Saxl et al. 2009).

In pure Cu (after four ECAP passes with route Bc), the minimum creep rate for the ECAP material is about one to two orders of magnitude less than that of coarse-grained copper. Another important aspect is represented by the effect on the number of passes on the creep life. As mentioned in the previous chapters, the high-angle grain boundary fraction increases with increasing number of passes. Here, creep in aluminum and copper showed that after the first pass, the boundaries were predominantly low-angle boundaries in character, and the population of high-angle grain boundaries had considerably increased in materials during repetitive pressings. This indicates that high-angle grain boundaries have a lower strengthening effect under creep than low-angle ones. Further, the refinement of the dislocation substructure during ECAP increases the creep resistance and leads to a decrease in the minimum creep rate (Dvorak et al. 2010).

Experiments performed on pure Cu and pure Al with routes A, Bc, and C show the general trend that the minimum creep rate sharply increases as a function of the number of passes (for all the routes) up to four passes, and then the saturation leads to small increases with the number of passes. An inverse behavior is observed for the time to fracture. Another observation is that there are not very significant differences in the creep behavior of samples prepared by the various ECAP processing routes. This can be mainly due to the grain growth of the unstable pure metals deformed at high temperature.

6.2.2 Creep Rate in UFGed Materials

Now, for elevated and high-temperature creep, the minimum creep rate generally varies with the applied stress through a power law relationship:

$$\dot{\varepsilon}_{\mathrm{min}} \tilde{A} \sigma^{n}$$

And the stress exponent is

$$n = \left(\frac{\partial \ln \dot{\varepsilon}_{\mathrm{min}}}{\partial \ln \sigma} \right)_T$$

And the stress exponent *m* is

$$m = -\left(\frac{\partial \ln t_f}{\partial \ln \sigma} \right)_T$$

With t_f being the time to fracture. For the pure metals, higher values of the times to fracture t_f for the pressed samples may reflect also a considerable increase of creep plasticity (creep fracture strain) in the pressed samples in comparison to the unpressed ones.

The activation energy for creep is given by

$$Q_C = \left[\frac{\partial \ln \dot{\varepsilon}_{\mathrm{min}}}{\partial \left(-\frac{1}{kT} \right)} \right]_{\sigma}$$

For pure metals, the high fraction of nonequilibrium high-angle grain boundaries in the pressed material reduces activation energy of deformation to values in the direction of those of thermally activated grain boundary processes like the grain boundary diffusion.

The stress dependence of the grain boundary sliding rate can be described by

$$\dot{\varepsilon}_{\mathrm{gb}} \cong A^{*} \sigma^{n^{*}}$$

where $n^{*} \approx 0.8n$. Thus, the slightly lower values of *n* for the ECAP pure metals may reflect the effect of more intensive grain boundary sliding in the creep of the ECAP material in comparison to its coarse-grained counterpart. The mechanism which most probably plays the dominant role in the power law creep of ECAPed pure metals is dislocation climb controlled by lattice self-diffusion. Supposing that grain boundary sliding is controlled by grain boundary diffusion, which usually is assumed to be about 0.7 times that for lattice self-diffusion, results give support to the assumption that grain boundary sliding is increasingly important in the ECAPed pure metals. Further, the contribution of grain boundary sliding to the total strain increases at the same temperature with a decrease in the applied stress. Finally, since high-angle grain boundaries are necessary in order to achieve GBS, an increase in the fraction of high-angle boundaries with increasing number of ECAP passes will essentially lead to an increasing contribution of GBS to the total creep strain. This is confirmed by the increase of n with increasing number of passes.

It is important to note that high-angle grain boundaries have lower strengthening effect under creep conditions than low-angle ones (sub-boundaries). This softening by high-angle boundaries may be explained by their influence on the evolution of the dislocation structure, namely by modifying the rates of generation and annihilation of dislocations during creep. Further, since high-angle grain boundaries are necessary in order to achieve grain boundary sliding, an increase in the fraction of high-angle boundaries with increasing number of ECAP passes will essentially lead to an increasing contribution of sliding to the total creep strain as it was observed experimentally (Saxl et al. 2009). Consequently, it could be proposed that the boundary structure with highly prevailing sub-boundaries is responsible for the high creep resistance after a low number of passes.

Kawasaki et al. (2011) confirmed that the ultrafine grains produced by ECAP in high-purity aluminum are not sufficiently stable when testing in high-temperature creep so that the dominant creep process is an intragranular dislocation mechanism as in metals with large grain sizes (Fig. 6.4).

There is a potential for using pressed materials to obtain new flow processes in high-temperature creep provided that the ultrafine grains are reasonably stable at elevated temperatures. So, scandium additions of 0.2% to the Al-3% Mg alloy or pure Al are sufficient to retain a small grain size at high temperatures and this leads to creep data where the stress exponent is ~2 and flow occurs by grain boundary sliding as in superplastic deformation.

A knowledge of the operating creep deformation mechanisms in UFG materials assists in understanding creep ductility in these materials. The total creep strain generally consists of several contributions:

$$\varepsilon = \varepsilon_g + \varepsilon_{\text{gbs}} + \varepsilon_d + \varepsilon_v$$

where ε_g, ε_{gbs}, ε_d, and ε_v are the strains caused by intragranular dislocation processes including dislocation glide and climb, grain boundary sliding (GBS), stress-directed diffusion of vacancies, and intergranular void nucleation and growth, respectively (Sklenicka et al. 2005, 2012). The first fundamental observation is that not all these processes are independent of each other. By observing the creep behavior of pure Al

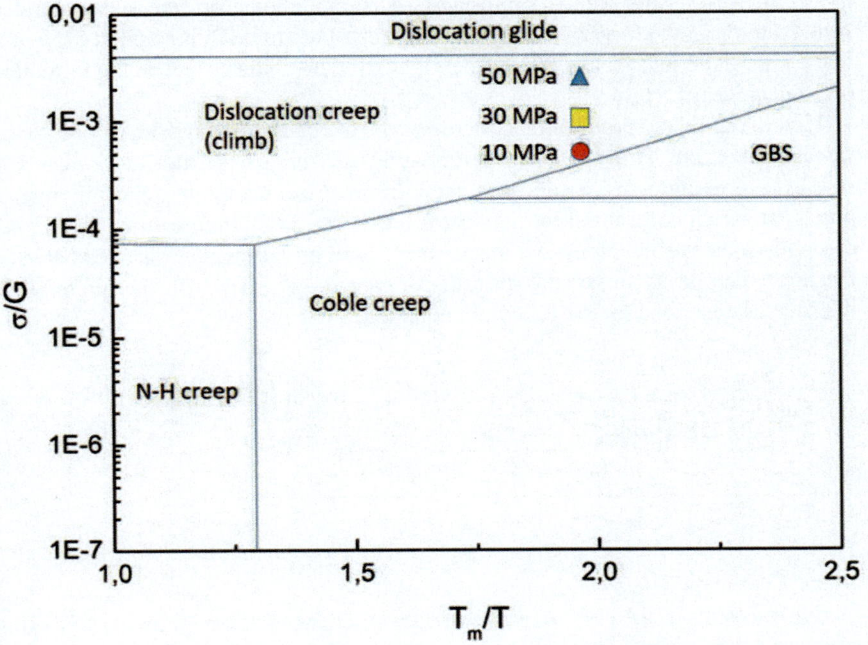

Fig. 6.4 Deformation mechanisms for pure aluminum with the indication of the results for ECAPed material tested at different stress levels

and Cu in comparison with alloys where Zr and Sc are added as grain stabilizers, it was shown that the main creep behavior shifts from intergranular dislocation mechanism (for pure metals) to GBS (in the Zr- and Sc-modified alloys). Pinning improves the kinetic stability of the grains by increasing the energy barrier for grain boundary sliding and rotation and by inhibiting grain coarsening, under extremely long-term creep conditions (Mathaudhu and Boyce 2015). So, by adding grain stabilizers, the pinning effect does not allow grain growth leading to creep deformation governed by intragranular dislocation mechanism leading to GBS and intergranular cavitation.

It appears that the ductility enhancement after higher numbers of ECAP passes is associated with the increase in the fraction of high-angle grain boundaries and with a consequent change in the flow process due to the increasing significance of grain boundary sliding. It is now well established that grain boundary sliding occurs in UFG materials due to the presence of an excess of extrinsic dislocations in the nonequilibrium grain boundaries.

Another important issue is represented by the textures and potential preferential grain orientation after ECAP, all properties influencing the creep behavior of SPD metals and alloys (Caturla and Nieh 2006). Experiments performed on Al-Si-Na alloy reveal a uniform grain size (250 nm) after eight passes by employing route A (Orozco-Caballero et al. 2014). The deformation imposed upon the material during the ECAP processing causes strain-induced precipitation from the supersaturated solid solution in the form of homogeneously distributed fine Si precipitates (50 nm) in the aluminum matrix. The presence of these fine precipitates aids in the formation of a fine microstructure.

First of all, ECAP produces and increases the material ductility with respect to the as-cast condition. The operating deformation mechanism in the Al-Si materials should be constant substructure slip creep. In this case, the evolution of the fine Si precipitates will control the mechanical properties at high temperature. The higher creep resistance of the processed materials is based on the smaller distance between the fine precipitates. In general, the stress exponent n varies with the interparticle distance as described by the following equation:

$$n(\lambda) = \left.\frac{\partial \ln \dot{\varepsilon}}{\partial \ln \left(\frac{\sigma}{E}\right)}\right|_T = \left.\frac{\partial \ln \left[\left(\frac{\sigma}{E}\right)^8 \cdot (\lambda/b)^3\right]}{\partial \ln \left(\frac{\sigma}{E}\right)}\right|_T = \left.\frac{\partial \ln \left(\frac{\sigma}{E}\right)^8}{\partial \ln \left(\frac{\sigma}{E}\right)}\right|_T + \left.\frac{\partial \ln \left(\frac{\lambda}{b}\right)^3}{\partial \ln (\sigma/E)}\right|_T$$
$$= 8 + 3\left.\frac{\partial \ln \lambda}{\partial \ln \sigma}\right|_T$$

where λ is the particle spacing.

Evidences on HPTed P92 steel showed that the UFG state exhibits a lower value of the stress exponent. The decrease in the stress exponent, *n*, in UFG state can be associated with the grain size reduction. It is generally accepted that in UFG materials, the grain boundary-mediated processes play an important role. Among the creep processes significantly influenced by grain size are processes such as grain boundary sliding and migration, enhanced recovery of dislocations at HABGs, and diffusion creep. The material was characterized by strong texture; at slow strain rates, GBS is the main mechanism. Creep data for high stresses can be modeled by dynamic recovery of dislocations at HAGBs (Kral et al. 2018). HAGBs accelerate creep strain rate through enhanced recovery of dislocations via grain boundary diffusion.

6.3 Creep in NC Materials

Nanocrystalline metals, with a mean grain size of less than 100 nanometers, have greater room-temperature strength than their coarse-grained equivalents, in part owing to a large reduction in grain size. However, this high strength generally comes with substantial losses in other mechanical properties, such as creep resistance, which limits their practical utility (Sklenicka et al. 2013). The degradation of creep resistance in nanocrystalline materials is in part due to an increase in the volume fraction of grain boundaries, which lack long-range crystalline order and lead to processes such as diffusional creep, sliding, and rotation (Darling et al. 2016).

6.3.1 Creep Mechanisms in NC Materials

The mechanisms involved during creep in NC metals and alloys are different with respect to those acting in CG materials. As described in the previous chapters, the well-known generation and multiplication mechanisms of dislocations from intragranular sources in CG metals are significantly suppressed in nanocrystalline (NC) metals due to the limit of intragranular space. Dislocation activity in NC metals involves nucleation or emission of partial/full dislocations from grain boundary (GB) sources, their propagation across the grains on limited slip planes, and final absorption in the opposite GBs. In addition, dislocations in NC metals are highly movable and the dislocation structure generated during loading can be completely recovered upon unloading and no permanent dislocation network is accumulated during a loading-unloading cycling (Pal and Meraj 2019). A reduction or removal of the applied stress can make propagating dislocations glide back and get absorbed in the GBs where they originated. This reverse dislocation glide and absorption processes have been connected with the negative creep observed in strain dip experiments on NC pure metals. The recovery of dislocation structure not only can occur in the unloading regime, but also takes place concurrently with the generation of dislocations in the loading regime (Ghazi and Kysar 2016). A dynamic equilibrium process of the dislocation generation and recovery would be expected after a short initial strain interval upon loading at given applied stress and strain rate, which leads to inferior "strain-hardening ability" or a plain and even dropping flow curve and hence makes NC metals possess low tensile ductility in spite of their very high (yield) strength caused by high critical stress required for nucleation or emission of dislocations from GBs (Hu et al. 2015).

Therefore, it has been a critical issue to stabilize the dislocation structure and increase the dislocation storage rate so as to achieve a good combination of both strength and ductility. In fact, many attempts have been made to stabilize dislocation structure and elevate the strain-hardening ability of NC metals and alloys by forming nanotwined structures or introducing twin boundaries (TBs) and/or stacking faults (SFs) through reducing their stacking fault energy (SFE). In metals and alloys, SFE is an important material parameter influencing the nucleation and movement of dislocations and the type of dislocation patterns. Generally, low SFE makes it easy for a full dislocation to split into two partials leaving behind a SF and/or form the deformation twins, and, in contrast, for metals and alloys with high SFE, the width of SFs between the two dissociated partial dislocations decreases, and the SFs and twins are less likely to form. As the effective barriers to dislocation movement and also effective sources/sinks of dislocations, the TBs and SFs can increase the dislocation storage rate and elevate the strain-hardening ability through their interactions with dislocations and hence lead to significant improvements in both strength and tensile ductility. Indeed, recent studies on the tensile mechanical properties of NC metals and alloys have found that reducing SFE through alloying can effectively elevate the strain-hardening ability and hence lead to a great improvement in strength and ductility. In addition to the mechanical behavior in the loading regime, the creep behavior after loading would also be affected by SFE through its influence on the dislocation structure generated in the loading regime and its recovery process during holding (Blum and Li 2007). On the other hand, due to high volume fraction of GBs and junctures in NC metals, GB-mediated processes such as GB sliding and GB diffusion can also take place concurrently with the dislocation activity. These two mechanisms compete mutually as loading strain rate changes due to their different thermal activation natures and also produces an important effect on the dislocation structure generated in the loading regime.

While stress exponent values in the range of 1–2 are indicative of deformation controlled by grain boundaries, specifically diffusion creep or grain boundary sliding, n values larger than 4 are usually attributed to lattice dislocation-based deformation processes (Gollapudi et al. 2010b).

6.3.2 Strain Rate Effect

It has been widely accepted that the Coble creep and the GB sliding accommodated by thermally activated GB diffusion are two possible mechanisms responsible for the creep deformation of NC metals and alloys even at room temperature due to their high volume fractions of GBs and triple junctions. The creep strain rates produced by these two mechanisms can be calculated by the following two equations, respectively:

$$\dot{\varepsilon}_{\mathrm{CO}} = 148\frac{\delta\Omega}{\pi d^3 kT} D_{\mathrm{gb0}} \exp\left(-\frac{Q_{\mathrm{gb}}}{kT}\right)\sigma_H$$

$$\dot{\varepsilon}_{\mathrm{GBS}} = 2\cdot 10^5 D_{\mathrm{gb0}} \exp\left(-\frac{Q_{\mathrm{gb}}}{kT}\right)\left(-\frac{Gb}{kT}\right)\left(\frac{b}{d}\right)^3\left(\frac{\sigma_H}{G}\right)^2$$

For the dislocation-mediated mechanism, a high critical shear stress is required for partial/full dislocations to nucleate and emit from GBs. This stress condition cannot be, however, well satisfied at low strain rate because of the stress relaxation role caused by the GB-mediated mechanisms such as the GB sliding accommodated by thermally activated GB diffusion (Cao et al. 2012). The creep and stress relaxation behavior of metals are a thermally activated plastic flow, where the shear strain rate in face-centered cubic metals is expressed as

$$\dot{\gamma} = \dot{\gamma}_0 \exp\left(-\frac{\Delta G(\tau_{\mathrm{eff}})}{kT}\right)$$

where k is the Boltzmann constant, γ_0 is a pre-exponential constant, and ΔG is the stress-assisted activation energy and is a reduced function of the effective shear stress τ_{eff}:

$$\Delta G = \Delta F - \tau_{\text{eff}} \Delta V^*$$

where ΔF and ΔV^* are the Helmholtz free energy for overcoming barriers to dislocation motion and activation volume, respectively. τ_{eff} accounts for the stress needed to overcome the short-range barrier responsible for the temperature and strain rate dependence, which is defined by

$$\tau_{\text{eff}} = \tau_a - \tau_u$$

where τ_a is the applied shear stress, and τ_u is the athermal internal stress derived from long-range obstacles. High-energy GB relaxation without grain growth occurs through decreasing GB dislocation density and local disorder, thereby leading to an increase in H and a decrease in plasticity.

The GB sliding would preferentially initiate at low strain rate due to its low threshold stress and become the dominant mechanism accommodating the applied plastic strain. As strain rate increases, this GB sliding would be suppressed due to the increased difficulty for GB diffusion. A much higher stress is thus needed for the GB sliding to proceed until a much higher strain rate at which the dislocation mechanism is activated (Mohamed and Li 2001). Such a conclusion is supported by the results of the m values measured in the loading regime.

The emitted dislocations from GBs would propagate across the grains and some of them would be eventually absorbed in the opposite GBs. The dislocation propagation would be hindered by GB pining points such as GB edges and also by some intragranular barriers such as twins and SFs as well as other dislocations, which must be overcome by the assistance of thermal activation. Furthermore, the dislocation absorption would involve the rearrangement of dislocation structure, which also needs the assistance of atomic diffusion. So the dislocation propagation and absorption would also be strain rate dependent. By considering atomic jump or migration between dislocation core and GB, a theoretical expression for the probability of a dislocation absorbed by GB (P) was derived:

$$P = \left[1 - (1-p)^I\right]^J = \left\{1 - [1 - \exp(X)]^I\right\}^J$$

where $p = \exp(X)$ is the probability value of a dislocation being successfully absorbed by GB:

$$X = -\frac{\Delta G + \tau^* b^3}{kT}$$

It is the number of a dislocation attempting to jump GB and being absorbed during a success absorption, which is related to strain rate, d; J is the total number of atoms in a dislocation participating in the absorption process or the total number of dislocations emitted from GB, which depends on d and the grain boundary structure; ΔG is the activation energy for atomic migration from dislocation core to GB; and τ^* is the effective shear stress acted on the dislocations. According to this expression, a low strain rate can result in a large P and, in contrast, a high strain rate can promote a small P and hence an effective dislocation accumulation or storage or a high dislocation density due to the enhanced dislocation nucleation and the increased difficulties for the dislocation absorption. As a result, a high internal (back) stress would thus develop inside the grain due to the inhomogeneous distribution of the stored dislocations inside the grain and/or the possible formation of the dislocation pileups against the opposite GBs. Such a dislocation structure developed in the loading regime can provide a reasonable explanation for the increased hardness with increasing strain rate. Since the stress level and stress state persist in the initial holding regime, the dislocation activities do not cease and the stored dislocations will still propagate forward and produce plastic strain, i.e., the creep strain, when they are absorbed in the opposite GBs. Furthermore, since the stress condition for the dislocation nucleation can be still satisfied well in the initial holding regime, new dislocations are able to continue to nucleate from GBs and produce the creep strain in the same way as the stored dislocations. As the number of the stored dislocations reduces and the stress level drops, the rapid creep process mediated by the rapid dislocation absorption would slow down and would be gradually replaced by the slow creep process mediated by the thermally activated GB sliding; a steady-state creep process thus occurs in the subsequent holding period. At low strain rate since fewer dislocations can be stored in the loading regime and low stress level, the GB sliding becomes the dominant mechanism. The dislocation structure generated in the loading regime and their relaxation in the holding regime will be different as a function of SFE. For low SFE, it is much easier for partials to emit successively from GBs or for full dislocations emitted from GBs to split into partial dislocations, which will lead to the formations of SFs and deformation twins within the grains. When the glide partial or full dislocations meet with the twins, the dislocation-TB interactions will occur. The easy glide of dislocations parallel to the TBs can contribute primarily to plasticity, while these inclined dislocations to TBs would be blocked and can cross the TBs only under much higher stress, which contributes primarily to strength. The TBs are also the effective sources of dislocations. Some glide partial or full dislocations would be accumulated near TBs and would trigger the activation of secondary twins by partial dislocation emission from the other side of the TBs. As a result, a high dislocation storage rate or a high density of dislocations would be expected in the same loading condition. On the other hand, the SFs and twins would severely impede full dislocation to cross-slip or climb and thus suppress the dynamic recovery or hinder the annihilation of these full dislocations, which are also in favor of a large storage rate for dislocations, although such a cross-slip or climb process is believed to be very difficult in NC metals due to the limited intragranular space. The above effects of SFE agree with the experimental phenomenon that the NC metals and alloys with lower SFE exhibit always the higher strain-hardening ability than that with higher SFE. In addition, lower SFE can reduce the critical stress for the dislocation nucleation or emission from GBs, which is also a possible factor increasing the density of the stored dislocations due to the earlier initiation of the dislocation emission process.

The resistance of coherent twin boundaries (CTBs) against dislocation motion is much more moderated, thereby optimizing both the strength and ductility of metals. CTBs are not fast diffusion channels in comparison with GBs, resulting in more thermal stability of nanotwinned (NT) structures compared with the twin-free NG counterparts.

The indentation creep behaviors depend on the experimental conditions of load, creep time, and temperature, and the micro/nanostructure of a crept material. The micro/nanostructure under the experimental conditions gives the material responses and hence yields the load-, time-, and temperature-dependent deformation, which can be characterized by the creep parameters of strain rate sensitivity, activation energy, activation volume, and stress exponent. The four parameters describe the material intrinsic creep behaviors and thus are micro/nanostructure related. It is desirable to let the creep intrinsic parameters be independent of the external loading conditions at least within certain ranges of the loading conditions. To achieve this goal, the multi-temperature approach was developed based on the hypothesis that the temperature dependency of deformation is expressed by the Arrhenius equation, and the activation energy is independent of temperature within the temperature range and depends on applied load only (Yang et al. 2018).

In the multi-temperature approach, the plot of plastic strain rate under a certain stress level versus reciprocal absolute temperature gives a value of pre-factor and a value of the apparent activation energy. Then, conducting the multi-temperature approach under various stress levels leads to many values of the pre-factor and the apparent activation energy. From the values of apparent activation energy, one is able to determine the intrinsic activation energy and the activation volume through the previously described equations. The stress dependency is expressed by the following power law:

$$\dot{\varepsilon}_p(\sigma, T) = \dot{\varepsilon}_{00}\left(\frac{\sigma}{\sigma_0}\right)^{n'} \exp\left(-\frac{\Delta G_0}{kT} + \frac{\sigma V}{\sqrt{3}kT}\right)$$

where σ_0 is a reference stress, which magnitude is one and has the same unit as σ. Here, the "athermal stress exponent n" is redefined to be the exponent for the pre-factor in the Arrhenius equation, rather than the traditional "stress exponent n" for the creep strain rate in the conventional creep studies:

$$\frac{1}{m} = n' + \frac{\sigma V}{\sqrt{3}kT}$$

The previous equation shows the relationship among the strain rate sensitivity m, athermal stress exponent n', and activation volume V. Only when the ratio of mechanical work over the thermal energy approaches zero

$$\frac{1}{m} = n'$$

In the case of nanoindentation creep tests the stress dependency is evaluated through the indentation hardness:

$$\dot{\varepsilon}(H,T) = \dot{\varepsilon}_0(H)\exp\left(-\frac{\Delta G}{kT}\right)$$

where $\dot{\varepsilon}_0(H)$ and ΔG are the pre-factor and the apparent (nominal) activation energy. The slope of the linear fitting of the experimental data under a given hardness gives the value of the apparent activation energy ΔG, while the fitted intercept determines the logarithmic temperature-independent creep rate $\ln(\dot{\varepsilon}_0(H))$. The apparent activation energy denotes real energy barrier, reduced by applied stress (hardness), to against the dislocation motion, GB and TB movement, and/or diffusion, etc. The activation volume, as a thermodynamic conjugate of flow stress, represents the generalized change in volume caused by flow stress and the mechanical work, e.g., the product of activation volume and flow stress, can reduce the activation energy. In dislocation-mediated deformation, the activation volume represents the generalized area swept over by mobile dislocations with an averaged value of Burgers vector. In indentation creep, hardness represents the averaged applied compressive stress on the projected contact area. Thus, a common and simple relationship between the apparent activation energy, hardness, and activation volume can be expressed by

$$\Delta G = \Delta G_0 - \frac{H}{2}V$$

where ΔG_0 is a hardness-independent intrinsic activation energy or activation energy at zero hardness. So, the Arrhenius equation for the indentation creep can be rewritten as

$$\dot{\varepsilon}(H,T) = \dot{\varepsilon}_{00}\left(\frac{H}{H_0}\right)^{n'}\exp\left(-\frac{\Delta G_0}{kT} + \frac{HV}{2kT}\right)$$

6.4 Creep in Thin Films

Competing deformation mechanisms in nanocrystalline metals result in increased rate sensitivity which may not be uniform across all time scales: diffusion-controlled processes are important at slow strain rates and may dominate for large fractions of GBs, while intragranular mechanisms are dominant at high strain rates. The competition between the two processes is expected to control the strain rate sensitivity and activation volumes in nanocrystalline metals.

At nominally elastic stresses, the mechanical stability of nanocrystalline metal films is strongly influenced by creep strain. The mechanical behavior in the slow rate regime is strongly influenced by diffusional creep. Dislocation-based mechanisms are dominant at higher strain rates and are different from those for coarse-grained metals as GB regions play an important role in dislocation activity.

Although GB sliding has been suggested as a likely deformation mechanism for creep in nanocrystalline metals, it must be accommodated by accumulation of strain by other elastic and plastic deformation mechanisms. At low stresses, GB sliding can be accommodated via elastic and diffusional mechanisms, while the contribution of dislocation-based mechanisms increases significantly at higher stresses. GB sliding can manifest itself in the form of its accommodation mechanisms such as dislocation activity originating at GBs.

The kinetics of thermal activation suggest that an exponential relationship could capture the effect of the applied stress, σ, and the energy barriers a_p and a_{ss} for thermally activated primary and steady-state creep:

$$\eta_1 = \exp\left(a_p - a_1\sigma\right)$$

$$\eta_2 = \exp\left(a_{ss} - a_2\sigma\right)$$

The creep strain ε_c for the total creep response of the material can follow an equivalent nonlinear Burgers viscoelastic model as

$$\varepsilon_c = \frac{\sigma}{E_h}\left[\frac{E_h}{\eta_2}t + \left(1 - \exp\left(-\frac{E_h}{\eta_1}t\right)\right)\right]$$

where E_h corresponds to the delayed elasticity component quantifying the duration of primary creep and is a direct consequence of competition between strain hardening and recovery in the material. The stress nonlinearity introduced in the exponential relationships is a direct consequence of the kinetics of thermal activation for creep deformation and provides an adequate representation of the overall creep behavior of nanocrystalline films.

The strain rate sensitivity and creep of nanocrystalline metal films at room temperatures are further accentuated at elevated temperatures which are common in thin-film applications. Plastic deformation of nanocrystalline metals is heavily influenced by the applied strain rate due to competing mechanisms of dislocation plasticity at higher temperature-compensated strain rates and creep mechanisms at lower temperature-compensated strain rates. The creep rates for nanocrystalline FCC metals are expected to increase with temperature; thus their strain rate sensitivity and the activation volume should greatly depend on temperature. As a matter of fact the higher creep resistance of NC Ni-W shows that addition of W dominates the creep response over reduction in grain size. The creep behavior of NC Ni-W can be explained by dislocation interactions with grain boundaries (Singh et al. 2020).

At temperatures higher than room temperature, the GB-mediated diffusional creep rate is surpassed by dislocation glide and climb-driven creep rates (Zhang et al. 2019). The dominant thermally activated mechanism was suggested to originate in three processes, all involving interactions of mobile dislocations with GBs. On the other hand, creep mechanisms, such as Coble creep and other diffusional processes, were ruled out on activation volume grounds. In contrast to FCC materials, the strain rate sensitivity of BCC materials decreases markedly as the grain size is reduced to ultrafine and nanocrystalline dimensions. For BCC metals, the observed creep rates exhibited a linear stress dependence for each grain size, and a grain size scaling factor that deviates from d^{-3} to d^{-2} as the grain size is increased. This, together with a positive curvature in the Arrhenius diagram for the creep rates, provides evidence for a combination of both Coble and Nabarro–Herring creep operating in parallel. Visual analysis reveals that extensive diffusion occurred within the GBs during the high-temperature simulations. In addition, lattice diffusion is observed within the grain interiors as a result of vacancies that are being emitted from the GBs and subsequently populate the entire grain (Millett et al. 2008).

6.4.1 Size Effect

Experimental evidences on Au thin films show that failure at large strains (10–30%) was characterized by voiding at all temperatures and strain rates. Nucleation and growth of voids in nanocrystalline FCC metals have been reported to be a dominant damage mechanism. The void-based damage in Au thin films was more distributed at rates prone to creep while the damage at higher rates was more localized near the fracture surface and was due to shear-driven void nucleation and growth. As expected, void growth and density increased with increasing temperature and reduced strain rate.

At the lower strain rates, 10^{-5} to 10^{-2} s^{-1}, deformation mechanisms such as GB diffusion, dislocation-based creep, and GB sliding play an important role in plastic deformation. Coble creep ($v \sim 1$ b^3) with strain rates 10^{-7} to 10^{-9} s^{-1} has been suggested to operate in nanocrystalline metals. Similarly, GB sliding may occur even at low homologous temperatures and low stresses in nanocrystalline metals as GBs have been shown to undergo sliding due to diffusional processes. Although GB sliding can initially be accommodated elastically, dislocation activity near GBs is eventually necessary. Thus, GB sliding involves material volumes that are larger than purely diffusional mechanisms, resulting in v ~ 10 b^3. Finally, at elevated temperatures, dislocation-based creep dominates with much higher activation volumes. These GB processes occur together with other deformation mechanisms that ensure geometrical compatibility. In other words, a coupling of no less than two deformation mechanisms is necessary. Among possible processes are GB sliding, diffusion, dislocation glide, intragranular dislocation climb, and intragranular elastic deformation. Since most of these processes are thermally activated, their contribution would vary with strain rate and, perhaps more importantly, temperature. Thus, the activation volume for GB diffusional processes would be higher than the expected true activation volume of 1 b^3 for a purely diffusional process. At low homologous temperatures, diffusional creep involves conventional lattice dislocation slip to accommodate GB sliding. In this mechanism, lattice dislocations are supplied from GBs as the motion of GB dislocations along the GBs leads to generation of lattice dislocations in order to accommodate the incompatible sliding process. Such dislocation-based accommodation processes for GB sliding would also result in an increase in activation volume as compared to purely diffusional processes. During the process of GB sliding, GB dislocations slide along the GBs and would pile up at the triple points. The resulting stress concentration generated at triple points can lead to dissociation of GB dislocations into crystal dislocations that result in slip in the grain interior, and GB dislocations that can glide and climb along the adjacent GBs resulting in grain rotation. Elevated temperatures would result in an increased contribution of such a thermally activated dislocation climb-based mechanism due to pileup of GB dislocations at the triple points and would result in increased activation volume.

At strain rates between 10^{-3} and 10 s^{-1} (regime II), the rate-controlling deformation process can be mediated by dislocations at GBs, analogously to forest hardening in coarse-grained FCC metals, but the amount of material participating in the rate-limiting process is restricted by GBs, thus resulting in lower activation volumes (Van Petegem et al. 2009). In regime II, GB defects, such as GB ledges and sessile dislocations, act as dislocation sources due to high local stresses. Due to small grain size, the time for dislocation propagation may not be the rate-limiting process anymore, and dislocation nucleation due to GB defects assumes this role in nanocrystalline metals. The apparent activation volumes calculated in regime II are close to those reported before for dislocation-mediated plasticity. A dislocation, or a partial, nucleated at a GB can move across a grain with its ends moving along the GB. During this process, the dislocation ends can be pinned at an obstacle at the GB. The depinning of such a dislocation is a thermally activated event and can be the rate-limiting process in plastic deformation. These dislocation-mediated deformation mechanisms involve GBs and hence result in much smaller activation volumes compared to coarse grain measurements. An expression for the activation volume for dislocation depinning was obtained as follows:

$$v^* = r^2 b \sin^{-1}\left(\frac{L+w^*}{2r}\right) - \left(\frac{L+w^*}{2}\right) b \sqrt{r^2 - \left(\frac{L+w^*}{2}\right)^2} - \left[r^2 b \sin^{-1}\left(\frac{L}{2r}\right) - \left(\frac{L}{2}\right) b \sqrt{r^2 - \left(\frac{L}{2}\right)^2}\right]$$

where L is the distance between the pinning points, w^* is the activation distance, b is the Burgers vector, and r is the radius of the dislocation loop. The radius of the dislocation loop can be defined in terms of the dislocation line tension T_L, effective shear stress τ^*, and the Burgers vector b. Substitution of empirical definitions for T_L and τ^* leads to the final expression for the radius of dislocation loop:

$$r = \frac{T_L}{\tau^* b} = \frac{L}{2}\left[1 - \left(\frac{kT \ln\left(\frac{\dot{\gamma}_0}{\dot{\gamma}}\right)}{G_0}\right)^{\frac{1}{q}}\right]^{\frac{1}{p}}$$

where p and q are empirical constants in the mechanical threshold stress model, k is the Boltzmann constant, T is the temperature, $\dot{\gamma}$ is the applied strain rate, and $\dot{\gamma}_0$ is the reference material strain rate. Thus, the unexpected decrease in the activation volume with increasing temperature levels can be explained effectively by assuming thermally activated depinning of dislocations, and therefore this mechanism could be dominant in regime II for nanocrystalline films.

Regime III belongs to the strong hardening acting in NC materials. The flow stress of FCC metals is known to be a function of temperature, strain rate, and state variables. Assuming a single-state variable, S, the expression for flow stress is given by

$$\sigma = f(T, \dot{\varepsilon}, S)$$

In the form of partial differentials

$$\partial\sigma = \left(\frac{\partial\sigma}{\partial\dot{\varepsilon}}\right)_{T,S} \partial\dot{\varepsilon} + \left(\frac{\partial\sigma}{\partial T}\right)_{\dot{\varepsilon},S} \partial T + \left(\frac{\partial\sigma}{\partial S}\right)_{\dot{\varepsilon},T} \partial S$$

For a tensile test at constant temperature and strain rate, the first two terms become zero. In this case the evolution of flow stress depends on the evolution of the state variable, S, with increasing strain. The state variable, S, is generally related to the dislocation density, ρ, in coarse-grained metals and the evolution of such a variable has been explained in terms of two terms, namely athermal hardening and dynamic recovery of dislocations. The athermal hardening term is related to the rate of dislocation accumulation and, thus, the average dislocation spacing, while the dynamic recovery term is temperature and strain rate dependent. The evolution of the state variable with strain accumulation can, therefore, be described by a linear hardening law as

$$\frac{\partial S}{\partial \varepsilon} = \theta_0 \left(1 - \frac{\sigma}{\sigma_S}\right)$$

where θ_0 is a parameter related to the athermal hardening and σ_s is the saturation stress which is also temperature and strain rate dependent and can follow an Arrhenius relationship of the form

$$\frac{\sigma_S}{\sigma_{S0}} = 1 - \frac{kT}{\Delta G_{SS}} \ln \left(\frac{\dot{\varepsilon}_0}{\dot{\varepsilon}}\right)$$

where σ_{s0} is the saturation stress at 0 K, i.e., the athermal component of saturation stress; ΔG_{ss} is the activation energy associated with the saturation stress; and $\dot{\varepsilon}_0$ is a reference strain rate ($\dot{\varepsilon}_0 \sim 10^7 s^{-1}$) for FCC metals. A similar relationship can be assumed to describe the temperature- and rate-dependent behavior using the energy term ΔG_f.

$$\frac{\sigma}{S} = 1 - \frac{kT}{\Delta G_f} \ln \left(\frac{\dot{\varepsilon}_0}{\dot{\varepsilon}}\right)$$

By employing these equations, *n* expression for the tangent modulus, θ, can be obtained as

$$\theta = \frac{\partial \sigma}{\partial \varepsilon_p} = \theta_0 \left(1 - \frac{kT}{\Delta G_f} \ln \left(\frac{\dot{\varepsilon}_0}{\dot{\varepsilon}}\right)\right) \cdot \left(1 - \frac{\sigma}{\sigma_S}\right)$$

or

$$\frac{\sigma}{\sigma_S} = \left(1 - C_1 \frac{\partial \sigma}{\partial \varepsilon_p}\right)$$

The parameter C_1 is itself a function of strain rate and temperature:

$$\frac{1}{C_1} = \theta_0 \left(1 - \frac{kT}{\Delta G_f} \ln \left(\frac{\dot{\varepsilon}_0}{\dot{\varepsilon}}\right)\right)$$

Thus, if $1/C_1$ is plotted as a function of

$$\varepsilon_T = T \ln \frac{\dot{\varepsilon}_0}{\dot{\varepsilon}}$$

the constants θ_0 and ΔG_f can then be obtained and the hardening behavior of nanocrystalline thin film can be described over the broad range of temperatures and strain rates.

Experiments showed that the primary creep of nanocrystalline films has significant duration, thus accumulating large strains, which presents long-term reliability concerns. Contrary to the creep response at low stresses, $\sigma < 100$ MPa for which diffusional transport along GBs can be described by a linear viscoelasticity model, the creep response at higher stresses requires nonlinear viscoelasticity models, which are only valid for stresses smaller than the elastic limit, above which creep associated with dislocation glide and climb may be described by a power law relationship (François et al. 2012). Such a nonlinear viscoelasticity model describes the material behavior in the transition regime between purely diffusional transport along GBs at low stresses which is described by a linear model and dislocation glide and climb which are described by a power law relationship at stresses above the elastic limit.

Nanocrystalline Au thin films with 850 nm thickness and average grain size of 64 nm were tested in tension in the strain rate range of 10^{-5} to 10 s^{-1} and at temperatures 25–110 °C (Karanjgaokar and Chasiotis 2016). The nanoscale grain size resulted in high values for the elastic limit, yield strength, and ultimate tensile strength at room temperature, which increased with strain rate. The apparent activation volumes calculated from the yield strength values as a function of temperature and strain rate followed bilinear trends, which indicated that, in a given combination of temperature and strain rate, the rate-limiting mechanism could fall in one of the two regimes: regime I that is dominated by creep-influenced plastic deformation, and regime II that is dominated by dislocation-mediated plasticity. At room temperature, the activation volume was ~6.4 b^3 which was an indication of a limiting GB diffusive creep mechanism. This activation volume increased with temperature to ~29.5 b^3 at 110 °C, which indicates a shift in the dominant creep mechanism to dislocation climb creep, as an accommodation mechanism for GB sliding. On the contrary, the apparent activation volumes in regime II decreased with temperature. Thermally activated depinning of dislocations was identified as the possible rate-controlling mechanism for this unexpected trend in activation volumes in regime II, which was supported by a literature model for dislocation depinning in nanocrystalline metals. A temperature increase from room to 110 °C leads to an order of magnitude increase in the initial primary creep rate and an order of magnitude increase in the steady-state creep rate. Results suggest a strong influence of the initial (defect) state of the material on creep response. The creep strain was partially recovered during the zero stress component of each creep cycle, including the experiments at room temperature. The percent of recovery was rather independent of temperature, which supports the argument that back stresses developed in the material during creep strain accumulation due to the grain heterogeneity drove strain recovery. The calculated creep exponents suggest that the rate-controlling mechanism at 110 °C was predominantly dislocation-mediated creep, while diffusional creep processes controlled the creep response at room temperature. The contribution of dislocation-mediated creep increased with temperature to dominate at 80 °C and higher, as suggested by the high steady-state creep rates.

6.5 Conclusions

Nanostructured materials exhibit creep at lower temperatures with respect to their microcrystalline counterparts. With the volume fraction being increasingly fundamental with decreasing grain size, diffusion creep is considered to be significant. In addition, with high diffusivity and nanometer grain size, NC materials are expected to creep at room temperature. In addition to grain boundary sliding, diffusion along grain boundaries and along triple junctions has been proposed to account for the plastic deformation without dislocation motion below the critical grain size. Diffusion is particularly relevant to the deformation of nanocrystalline metals because typical dislocation mechanisms are suppressed and grain boundary-controlled deformation processes become dominant. It has been widely reported that creep deformation in NC metals is mainly driven by three fundamental processes: grain boundary (GB) diffusion, GB sliding and migration, and dislocation nucleation from GBs with increasing applied stress. These thermally activated deformation mechanisms lead to a temperature dependence of strength in NC metals.

In UFG materials, the development of meso- and macrobands of localized deformation during deformation of fine-grained materials has been postulated. This leads to an apparent creep activation energy lower than the microcrystalline counterpart. This difference is attributed to the significant contributions to the overall deformation of grain boundary sliding controlled by grain boundary diffusion. High-angle grain boundaries have a lower strengthening effect under creep than low-angle ones. Further, the refinement of the dislocation substructure during ECAP increases the creep resistance and leads to a decrease in the minimum creep rate. For pure metals, the high fraction of nonequilibrium high-angle grain boundaries in the pressed material reduces activation energy of deformation to values in the direction of those of thermally activated grain boundary processes like the grain boundary diffusion. It appears that the ductility enhancement after higher numbers of ECAP passes is associated with the increase in the fraction of high-angle grain boundaries and with a consequent change in the flow process due to the increasing significance of grain boundary sliding. It is now well established that grain boundary sliding occurs in UFG materials due to the presence of an excess of extrinsic dislocations in the nonequilibrium grain boundaries.

The degradation of creep resistance in nanocrystalline materials is in part due to an increase in the volume fraction of grain boundaries, which lack long-range crystalline order and lead to processes such as diffusional creep, sliding, and rotation. It has been widely accepted that the Coble creep and the GB sliding accommodated by thermally activated GB diffusion are two possible mechanisms responsible for the creep deformation of NC metals and alloys even at room temperature due to their high volume fractions of GBs and triple junctions.

Competing deformation mechanisms in nanocrystalline metals result in increased rate sensitivity which may not be uniform across all time scales: diffusion-controlled processes are important at slow strain rates and may dominate for large fractions of GBs, while intragranular mechanisms are dominant at high strain rates. The competition between the two processes is expected to control the strain rate sensitivity and activation volumes in nanocrystalline metals.

The strain rate sensitivity and creep of nanocrystalline metal films at room temperatures are further accentuated at elevated temperatures which are common in thin-film applications.

References

Blum W, Li YJ (2007) Flow stress and creep rate of nanocrystalline Ni. Scripta Mater 57:429–431. https://doi.org/10.1016/j.scriptamat.2007.04.041

Cai B, Kong QP, Lu L, Lu K (2000) Low temperature creep of nanocrystalline pure copper. Mater Sci Eng A 286:188–192. https://doi.org/10.1016/S0921-5093(00)00633-X

Cao ZH, Wang L, Hu K, Huang YL, Meng XK (2012) Microstructural evolution and its influence on creep and stress relaxation in nanocrystalline Ni. Acta Mater 60:6742–6754. https://doi.org/10.1016/j.actamat.2012.08.047

Caturla MJ, Nieh TG (2006) Modeling the effect of texture on the deformation mechanisms of nanocrystalline materials at the atomistic scale. In: Chuang TJ, Anderson PM, Wu MK, Hsieh S (eds) Nanomechanics of materials and structures. Springer, Dordrecht. https://doi.org/10.1007/1-4020-3951-4_28

Choi I-C, Jang J-I (2020) A survey of nanoindentation studies on HPT-processed materials. Adv Eng Mater 22:1900648. https://doi.org/10.1002/adem.201900648

Chokshi AH (2009) Unusual stress and grain size dependence for creep in nanocrystalline materials. Scripta Mater 61:96–99. https://doi.org/10.1016/j.scriptamat.2009.03.009

Darling KA, Rajagopalan M, Komarasamy M, Bhatia MA, Hornbuckle BC, Mishra RS, Solanki KN (2016) Extreme creep resistance in a microstructurally stable nanocrystalline alloy. Nature 537:378–381. https://doi.org/10.1038/nature19313

Diffusion in nanocrystalline materials. In: Diffusion in solids. Springer series in solid-state sciences, vol. 155. Springer, Berlin; 2007. https://doi.org/10.1007/978-3-540-71488-0_34

Dvorak J, Sklenicka V, Kral P, Svoboda M, Saxl I (2010) Characterization of creep behavior and microstructure changes in pure copper processed by equal-channel angular pressing. Rev Adv Mater Sci 25:225–232

Estrin Y, Gottstein G, Shvindlerman LS (2004) Diffusion controlled creep in nanocrystalline materials under grain growth. Scripta Mater 50:993–997. https://doi.org/10.1016/j.scriptamat.2004.01.002

François D, Pineau A, Zaoui A (2012) Viscoelasticity. In: Mechanical behaviour of materials, Solid mechanics and its applications, vol 180. Springer, Dordrecht. https://doi.org/10.1007/978-94-007-2546-1_5

Ghazi N, Kysar JW (2016) Experimental investigation of plastic strain recovery and creep in nanocrystalline copper thin films. Exp Mech 56:1351–1362. https://doi.org/10.1007/s11340-016-0169-7

Gollapudi S, Rajulapati KV, Charit I et al (2010a) Understanding creep in nanocrystalline materials. Trans Indian Inst Metals 63:373–378. https://doi.org/10.1007/s12666-010-0050-9

Gollapudi S, Rajulapati KV, Charit I, Koch CC, Scattergood RO, Murty KL (2010b) Creep in nanocrystalline materials: role of stress assisted grain growth. Mater Sci Eng A 527:5773–5781. https://doi.org/10.1016/j.msea.2010.05.048

Hu L, Huo R, Zhou J et al (2012) The effects of the finest grains on the mechanical behaviours of nanocrystalline materials. J Nanopart Res 14:677. https://doi.org/10.1007/s11051-011-0677-4

Hu J, Sun G, Zhang X, Wang G, Jiang Z, Han S, Zhang J, Lian J (2015) Effects of loading strain rate and stacking fault energy on nanoindentation creep behaviors of nanocrystalline Cu, Ni-20 wt.% Fe and Ni. J Alloys Compd 647:670–680. https://doi.org/10.1016/j.jallcom.2015.06.094

Karanjgaokar N, Chasiotis I (2016) Creep behavior of nanocrystalline Au films as a function of temperature. J Mater Sci 51:3701–3714. https://doi.org/10.1007/s10853-015-9687-4

Kawasaki M, Langdon TG (2007) Principles of superplasticity in ultrafine-grained materials. J Mater Sci 42:1782–1796. https://doi.org/10.1007/s10853-006-0954-2

Kawasaki M, Langdon TG (2017) Applying conventional creep mechanisms to ultrafine-grained materials. In: Charit I, Zhu Y, Maloy S, Liaw P (eds) Mechanical and creep behavior of advanced materials, The minerals, metals & materials series. Springer, Cham. https://doi.org/10.1007/978-3-319-51097-2_10

Kawasaki M, Sklenicka V, Langdon TG (2011) Creep behavior of metals processed by equal-channel angular pressing. Kovove Mater 49:75–83. https://doi.org/10.4149/km.2011.1.75

Keblinski P, Wolf D, Gleiter H (1998) Molecular-dynamics simulation of grain-boundary diffusion creep. Interface Sci 6:205–212. https://doi.org/10.1023/A:1008664218857

Kolobov Y, Grabovetskaya G, Ivanov K et al (2002) Grain boundary diffusion and mechanisms of creep of nanostructured metals. Interface Sci 10:31–36. https://doi.org/10.1023/A:1015128928158

Kolobov YR, Ratochka IV, Ivanov KV et al (2004) Characteristic features of diffusion-controlled processes in ordinary and ultrafine-grained polycrystalline metals. Russ Phys J 47:840–856. https://doi.org/10.1007/s11182-005-0004-6

Kolobov YR, Lipnitskii AG, Nelasov IV et al (2008) Investigations and computer simulations of the intergrain diffusion in submicro- and nanocrystalline metals. Russ Phys J 51:385. https://doi.org/10.1007/s11182-008-9062-x

Korla R, Chokshi AHA (2014) Constitutive equation for grain boundary sliding: an experimental approach. Metal Mater Trans A 45:698–708. https://doi.org/10.1007/s11661-013-2017-z

Kral P, Dvorak J, Sklenicka V, Masuda T, Horita Z, Kucharova K, Kvapilova M, Svobodova M (2018) The effect of ultrafine-grained microstructure on creep behaviour of 9% Cr steel. Materials 11:787. https://doi.org/10.3390/ma11050787

Langdon TG (2006) Grain boundary sliding revisited: developments in sliding over four decades. J Mater Sci 41:597–609. https://doi.org/10.1007/s10853-006-6476-0

Liu Y, Huang C, Bei H, Hu W (2012) Room temperature nanoindentation creep of nanocrystalline Cu and Cu alloys. Mater Lett 70:26–29. https://doi.org/10.1016/j.matlet.2011.11.119

Mathaudhu SN, Boyce BL (2015) Thermal stability: the next frontier for nanocrystalline materials. JOM 67:2785–2787. https://doi.org/10.1007/s11837-015-1708-x

Meyers MA, Mishra A, Benson DJ (2004) Mechanical properties of nanocrystalline materials. Prog Mater Sci 51:427–556. https://doi.org/10.1016/j.pmatsci.2005.08.003

Millett PC, Deasai T, Yamakov V, Wolf D (2008) Atomistic simulations of diffusional creep in a nanocrystalline body-centered cubic material. Acta Mater 56:3688–3698. https://doi.org/10.1016/j.actamat.2008.04.004

Mishra RS, McFadden SX, Valiev RZ et al (1999) Deformation mechanisms and tensile superplasticity in nanocrystalline materials. JOM 51:37–40. https://doi.org/10.1007/s11837-999-0010-1

Mohamed FA (2007) Interpretation of nanoscale softening in terms of dislocation-accommodated boundary sliding. Metal Mater Trans A 38:340–347. https://doi.org/10.1007/s11661-006-9057-6

Mohamed FA, Li Y (2001) Creep and superplasticity in nanocrystalline materials: current understanding and future prospects. Mater Sci Eng A 298:1–15. https://doi.org/10.1016/S0928-4931(00)00190-9

Murty KL (2017) Creep, deformation and fracture studies of materials for various technologies in the Nuclear Materials Research Group at NC state. In: Charit I, Zhu Y, Maloy S, Liaw P (eds) Mechanical and creep behavior of advanced materials, The minerals, metals & materials series. Springer, Cham. https://doi.org/10.1007/978-3-319-51097-2_1

Nie K, Wu W, Zhang X et al (2017) Molecular dynamics study on the grain size, temperature, and stress dependence of creep behavior in nanocrystalline nickel. J Mater Sci 52:2180–2191. https://doi.org/10.1007/s10853-016-0506-3

Orozco-Caballero A, Menon SK, Cepeda-Jimenez CM, Hidalgo-Manrique P, McNelley TR, Ruano OA, Carreno F (2014) Influence of microstructural stability on the creep mechanism of Al–7 wt % Si alloy processed by equal channel angular pressing. Mater Sci Eng A 612:162–171. https://doi.org/10.1016/j.msea.2014.06.017

Pal S, Meraj M (2019) Investigation of reorganization of a nanocrystalline grain boundary network during biaxial creep deformation of nanocrystalline Ni using molecular dynamics simulation. J Mol Model 25:282. https://doi.org/10.1007/s00894-019-4177-2

Panin VE, Panin, AV, Pochivalov YI (2017) et al. Scale invariance of structural transformations in plastically deformed nanostructured solids. Phys Mesomech 20:55–68. https://doi.org/10.1134/S1029959917010052

Park K, Lee CS, Shin DH et al (2006) Reappraisal of grain boundary diffusion creep equations for nanocrystalline materials. Met Mater Int 12:107–113. https://doi.org/10.1007/BF03027465

Saxl I, Sklenička V, Ilucová L, Svoboda M, Dvořák J, Král P (2009) The link between microstructure and creep in aluminum processed by equal-channel angular pressing. Mater Sci Eng A 503:82–85. https://doi.org/10.1016/j.msea.2008.03.056

Sergueeva AV, Mara NA, Valiev RZ, Mukherjee AK (2006) Elevated temperature deformation characteristics of nanocrystalline materials. In: Zhu YT, Varyukhin V (eds) Nanostructured materials by high-pressure severe plastic deformation, NATO science series (II: Mathematics, physics and chemistry), vol 212. Springer, Dordrecht. https://doi.org/10.1007/1-4020-3923-9_39

Singh PS, Ling Z, Pharr GM, de Boer MP (2020) Creep of a thermally stable nanocrystalline nickel tungsten alloy as measured by high temperature nanoindentation. Mater Sci Eng A 784:139309. https://doi.org/10.1016/j.msea.2020.139309

Sklenicka V, Dvorak J, Kral P, Stonawska Z, Svoboda M (2005) Creep processes in pure aluminium processed by equal-channel angular pressing. Mater Sci Eng A 410-411:408–412. https://doi.org/10.1016/j.msea.2005.08.099

Sklenicka V, Dvorak J, Kral P, Svoboda M, Kvapilova M, Langdon TG (2012) Factors influencing creep flow and ductility in ultrafine-grained metals. Mater Sci Eng A 558:403–411. https://doi.org/10.1016/j.msea.2012.08.019

Sklenicka V, Kucharova K, Kvapilova M et al (2013) Creep in an electrodeposited nickel. J Mater Sci 48:4780–4788. https://doi.org/10.1007/s10853-013-7209-9

Tjong SC, Chen H (2004) Nanocrystalline materials and coatings. Mater Sci Eng R45:1–88. https://doi.org/10.1016/j.mser.2004.07.001

Van Petegem S, Brandstetter S, Schmitt B, Van Swygenhoven H (2009) Creep in nanocrystalline Ni during X-ray diffraction. Scripta Mater 60:297–300. https://doi.org/10.1016/j.scriptamat.2008.10.034

Voyiadjis GZ, Deliktas B (2010) Modeling of strengthening and softening in inelastic nanocrystalline materials with reference to the triple junction and grain boundaries using strain gradient plasticity. Acta Mech 213:3–26. https://doi.org/10.1007/s00707-010-0338-1

Wang CL, Lai YH, Huang JC, Nieh TG (2010) Creep of nanocrystalline nickel: a direct comparison between uniaxial and nanoindentation creep. Scripta Mater 62:175–178. https://doi.org/10.1016/j.scriptamat.2009.10.021

Wang YJ, Ishii A, Ogata S (2013) Entropic effect on creep in nanocrystalline metals. Acta Mater 61:3866–3871. https://doi.org/10.1016/j.actamat.2013.03.026

Yamakov V, Wolf D, Phillpot SR, Gleiyìter H (2002) Grain-boundary diffusion creep in nanocrystalline palladium by molecular-dynamics simulation. Acta Mater 50:61–73. https://doi.org/10.1016/S1359-6454(01)00329-9

Yang W, Ma XL, Wang HT, Hong W (2006) Deformation and diffusion in nano-grained metals. In: Sun QP, Tong P (eds) IUTAM symposium on size effects on material and structural behavior at micron- and nano-scales. Springer, Dordrecht. https://doi.org/10.1007/978-1-4020-4946-0_8

Yang X-S, Zhai H-R, Ruan H-H, Shi S-Q, Zhang T-Y (2018) Multi-temperature indentation creep tests on nanotwinned copper. Int J Plasticity 104:68–79. https://doi.org/10.1016/j.ijplas.2018.01.016

Yin W, Whang SH (2005) The creep and fracture in nanostructured metals and alloys. JOM 57:63–70. https://doi.org/10.1007/s11837-005-0066-5

Yin WM, Whang SH, Mirshams R, Xiao CH (2001) Creep behavior of nanocrystalline nickel at 290 and 373 K. Mater Sci Eng A 301:18–22. https://doi.org/10.1016/S0921-5093(00)01385-X

Zhang K, Weertman JR, Eastman JA (2004) The influence of time, temperature, and grain size on indentation creep in high-purity nanocrystalline and ultrafine grain copper. Appl Phys Lett 85:5197. https://doi.org/10.1063/1.1828213

Zhang S, Wang Y, Jiang H et al (2013) Constitutive model for plastic deformation of nanocrystalline materials with shear band. Meccanica 48:175–185. https://doi.org/10.1007/s11012-012-9592-8

Zhang W, Gao Y, Nieh TG (2019) Competing grain boundary and interior deformation mechanisms with varying sizes. In: Hsueh CH et al (eds) Handbook of mechanics of materials. Springer, Singapore. https://doi.org/10.1007/978-981-10-6884-3_75

Chapter 7
Superplasticity in Nanostructured Materials

7.1 Introduction

Superplasticity is the property of materials exhibiting very high ductility up to hundreds or thousands of elongation during tension (Padmanabhan et al. 2018a). This is exhibited by microcrystalline materials once grain size is below 10 μm at temperatures 0.5–0.6 T_m and strain rates in the range of 10^{-3} to 10^{-4} s^{-1} (Padmanabhan et al. 2018b). The constitutive equation describing the superplastic flow in microcrystalline materials is described by

$$\dot{\varepsilon} = A\frac{DGb}{kT}\left(\frac{b}{d}\right)^{p}\left(\frac{\sigma}{E}\right)^{n}$$

which is the strain rate, D is the coefficient of grain boundary diffusion, G is the shear modulus, b is the Burgers vector, k is the Boltzmann constant, T is the temperature of testing, d is the grain size, p is the grain size exponent (usually equal to 2), σ is the flow stress, and n is the stress exponent. A decrease in grain size leads to increase in superplasticity at a given temperature and strain rate (Chuvil'deev et al. 2010). Basing on these considerations, the development of SPD techniques allowed to produce bulk materials optimized for superplastic forming (Valiev 2000). The equation suggests that low-temperature superplasticity can also be achieved by enhancing grain boundary diffusion (Padmanabhan et al. 2018c). The diffusion coefficient increases exponentially with increasing deformation temperature and such a strong relationship between the diffusion and temperature is the main reason for the occurrence of superplasticity at high homologous temperatures. The enhancement of diffusion at the grain boundaries can be realized without increasing the temperature by engineering the grain boundary composition and segregation (Edalati et al. 2017).

© Springer Nature Switzerland AG 2021

P. Cavaliere, *Fatigue and Fracture of Nanostructured Materials*,
https://doi.org/10.1007/978-3-030-58088-9_7

7.1.1 Grain Boundary Sliding

During superplasticity, grain boundary sliding is the dominant deformation mechanism (Padmanabhan 2009). Given that grain boundary sliding during superplastic deformation is a thermally activated process, the macroscopic shear strain rate is given by

$$\dot{\gamma} = N_v A B \nu \exp\left[\frac{-\Delta G \tau_e}{kT}\right]$$

where N_v is the number of places per unit volume where thermally activated shear can occur, A is the area swept out per successful thermal fluctuation, m is the frequency of vibration, and ΔG is the Gibbs free activation energy, which is a decreasing function of the effective shear stress $\tau_e = \tau - \tau_0$, where τ is the applied stress and τ_0 a back stress or threshold stress. N_v is given by

$$N_v = \frac{\delta}{db^2}$$

where δ is the grain boundary width; given that $\Delta G = \Delta F - V\tau_e$ with ΔF Helmholtz free energy and $V = d^3$, the shear strain rate is given by

$$\dot{\gamma} = \frac{6\delta\nu_D}{d} \sinh\left(\frac{V\tau_e}{kT}\right) \exp\left(-\frac{\Delta F}{kT}\right)$$

7.1.2 Diffusion

For NC materials, owning to their fine grain size, large grain boundary area, and high self-diffusivity, superplasticity is expected at lower temperatures and/or higher strain rates (Dong et al. 2007). Before describing the dependence of the different mechanisms acting during deformation as a function of grain size, some assumptions about diffusion in nanostructured materials must be stated. Diffusion in nanostructured materials differs from that in microcrystalline regime because the crystallite interfaces provide paths of high diffusivity, whereas in more coarse-grained crystals, volume self-diffusion or substitutional atom diffusion is substantial generally only at temperatures greater than approximately half the melting temperature. Interface diffusion, in combination with a high fraction of interfaces, gives rise to modified physical properties of nanostructure solids, first of all enhanced ductility and grain boundary sliding at lower temperatures and higher strain rates with respect to the microcrystalline counterparts (Kawasaki et al. 2007). In nanocrystalline materials, diffusion is governed by the following equation:

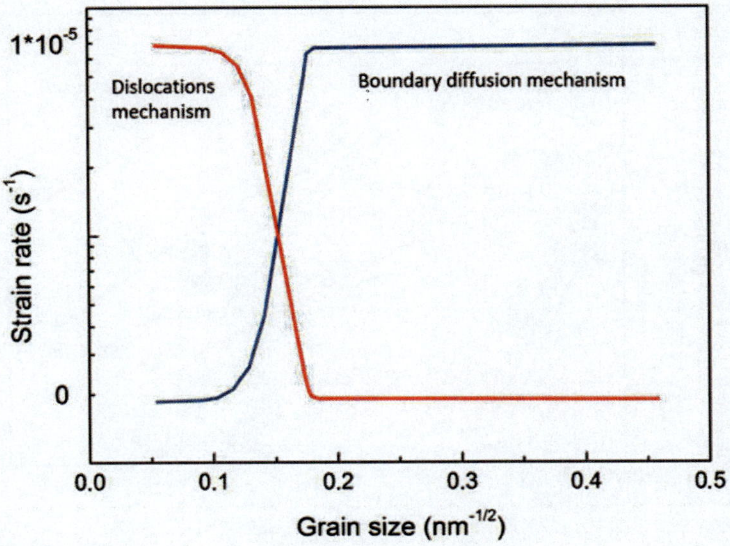

Fig. 7.1 Mechanisms affecting strain rate as a function of grain size

$$C \propto \frac{1}{\sqrt{D_{\mathrm{B}}t}} \exp\left(-\frac{x^2}{4D_{\mathrm{B}}t}\right)$$

where C is the concentration and D_{B} is the diffusion coefficient. By decreasing the grain size, dislocation-related mechanisms influence strain rate decreases, while boundary diffusion mechanisms influence increases (Fig. 7.1).

Some examples of how diffusivity varies with homologous temperature for some pure metals are shown in Fig. 7.2.

The interface diffusion in highly dense NC metals after cluster synthesis or severe plastic deformation is accompanied by strong crystallite growth. The effect of annealing temperature on crystallite growth for different metals is shown in Fig. 7.3.

A quantitative analysis shows that the rapid crystallite growth can be attributed to the high driving forces due to the small crystallite size, taking into account the dynamics of equilibrium-like grain boundaries.

Now, as a function of grain size different regimes can be underlined for grain boundary sliding (Sripathi and Padmanabhan 2014). Dislocation pileup is the main mechanism for $d > 10^{-6}$ m; the rate-controlling mechanism is

$$\tau = M\alpha Gb\left[\left(\rho_0(T,\gamma,\dot{\gamma})^{\frac{1}{2}} + \frac{\beta(T,\dot{\gamma})\gamma}{bd^{\frac{1}{2}}}\right]$$

where M and α are constants, and ρ_0 is the dislocation density. For $10^{-6} < d < 10^{-8}$ m, grain boundary shear promoted by the pileup of dislocations is considered to be the deformation mechanism. The rate-controlling mechanism is

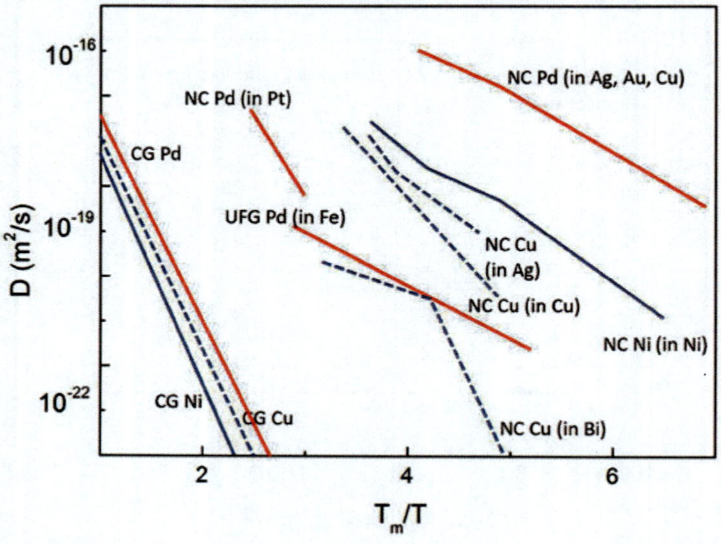

Fig. 7.2 Diffusivity for various metals from microcrystalline to nanocrystalline regime

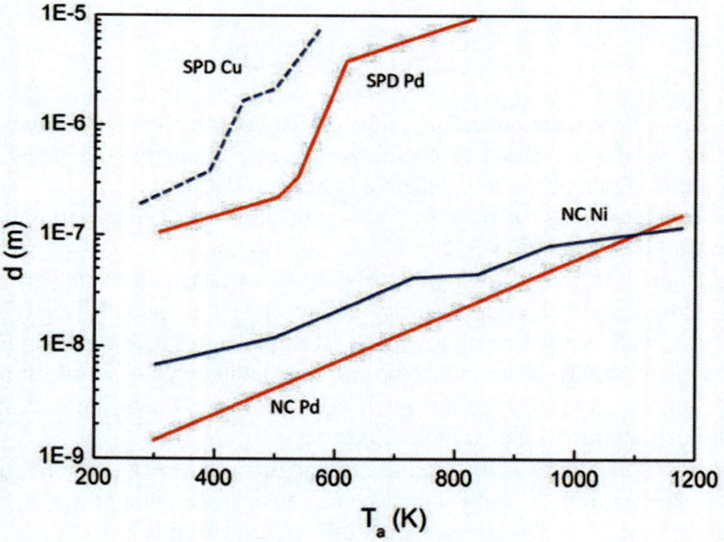

Fig. 7.3 Crystallite size variation with annealing temperature

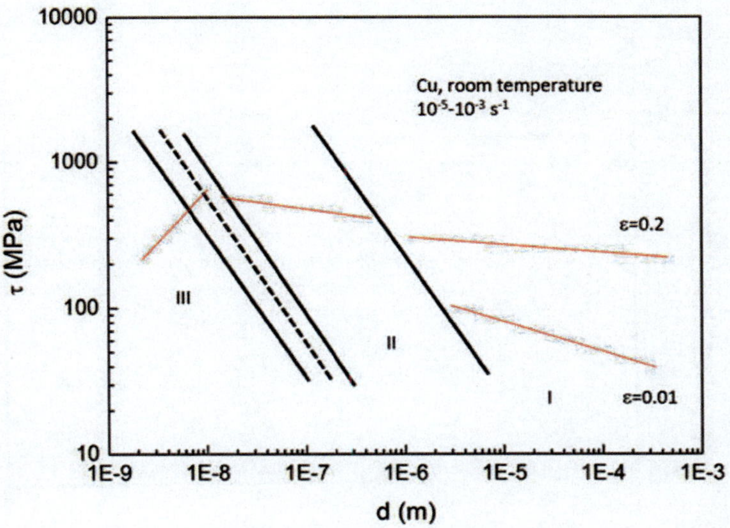

Fig. 7.4 Stress vs. grain size for pure copper in the different regimes (Meyers et al. 2006)

$$\tau = \alpha G b \rho^{\frac{1}{2}} + \left[\left(\Delta F^{*} - kT \left(\frac{\dot{\gamma}_0}{\dot{\gamma}} \right) \right)^{\frac{1}{2}} \left(\frac{Gb}{\pi V} \right)^{\frac{1}{2}} \right] d^{-\frac{1}{2}}$$

For $d < 10^{-8}$ m, grain boundary shear is the dominant mechanism. The rate-controlling mechanism is

$$\tau - \tau_0 = \left[\frac{kT}{V} \ln \left(\frac{\delta v_D}{\dot{\gamma}} \right) + \frac{\Delta F^{*}}{V} \right] + \frac{kT}{V} \ln d$$

where τ_0 is the threshold stress for sliding. The different regimes are summarized for pure Cu and represented in Fig. 7.4.

The negative slope in the regime III is representative of grain boundary sliding where the material behavior is completely different with respect to the regimes I and II. As d decreases, the contribution to plastic flow coming from grain boundary sliding increases. Anyway, this mechanism is not sufficient to explain the real behavior observed in nanostructured materials. For this reason it was introduced also the contribution of plastic accommodation; in this way grain boundary sliding can be represented by a viscous and a plastic accommodation term:

$$\tau_a = (\eta_i + \eta_d)\dot{\gamma} + \tau_p$$

where τ_a is the shear stress, η_i is an intrinsic grain boundary viscosity, η_d is a diffusional component of the accommodation process, and τ_p is a plastic deformation

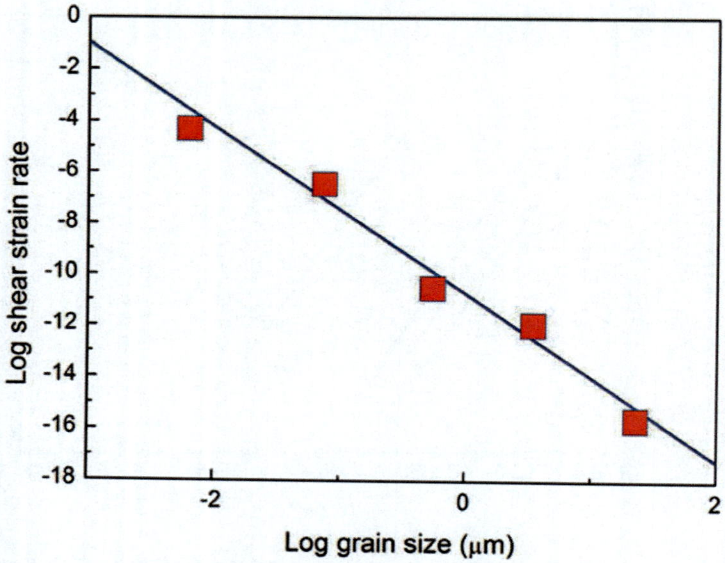

Fig. 7.5 Shear strain rate vs. grain size for pure Cu

term. Calculations on pure Cu led to the definition of the dependence of shear strain rate with grain size in accordance with all the mentioned contribution (Fig. 7.5).

7.1.3 Constitutive Relationships

It was found experimentally that the rate of flows in the superplasticity varies inversely with the grain size raised to a power of ~2. It was observed, also, that the superplastic regime is displaced to faster strain rates when the grain size is reduced. The constitutive equation to describe the superplasticity can be expressed by

$$\dot{\varepsilon} = A\frac{D_0Gb}{kT}\left(\frac{b}{D}\right)^p\left(\frac{\sigma - \sigma_0}{G}\right)^n \exp\left(-\frac{Q}{RT}\right)$$

where σ_0 is the threshold stress. During high-temperature deformation of SPD materials, it is fundamental to recognize that many nanostructured grains are not so stable as the temperature increases. For this reason, it is useful to have materials with stable precipitates in order to retain fine microstructure capable of exhibiting superplasticity at high temperature for a broad range of strain rates. As a matter of fact, various aluminum alloys exhibit very high elongations in a very broad range of strain rates as a function of the processing route and composition (Fig. 7.6).

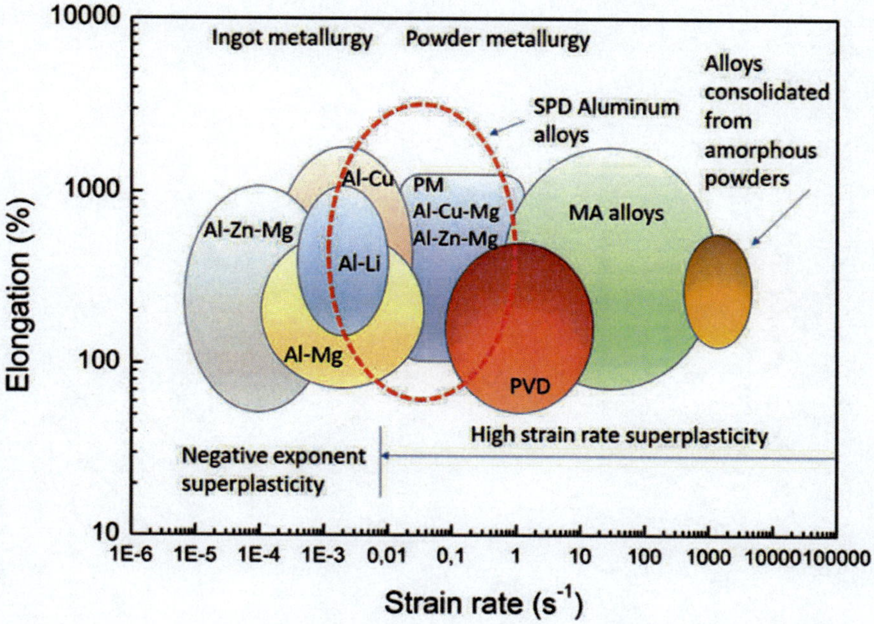

Fig. 7.6 Elongation as a function of strain rate for various aluminum alloys obtained via different processing routes

In the following the main superplastic properties of nanostructured materials will be described.

7.2 Superplasticity in SPD Materials

Superplasticity in micro-grained materials (*d* in the range of 1 μm–10 pm) is characterized by large elongations that are equal to or greater than 350%. These large elongations are usually attained at strain rates in the range of 10^{-3} to 10^{-2} s^{-1}. Such a strain rate range is too slow to be practical for commercial forming of structural materials, which is normally performed at strain rates >10 s^{-1}. As a result, extensive efforts were devoted to addressing this problem. These efforts have led to the emergence of high strain rate (HSR) superplasticity, which is very beneficial for commercial applications. The origin of the basic concept for HSR is the report that the ductility-strain rate curve of a micro-grained superplastic alloy shifts to higher strain rates with decreasing grain size. This finding has led to the following novel approach: HSR superplasticity can be induced via grain refinement.

New frontiers in superplasticity have been reached thanks to the development of bulk severely plastic deformed materials such as ECAPed and HPTed ones. The

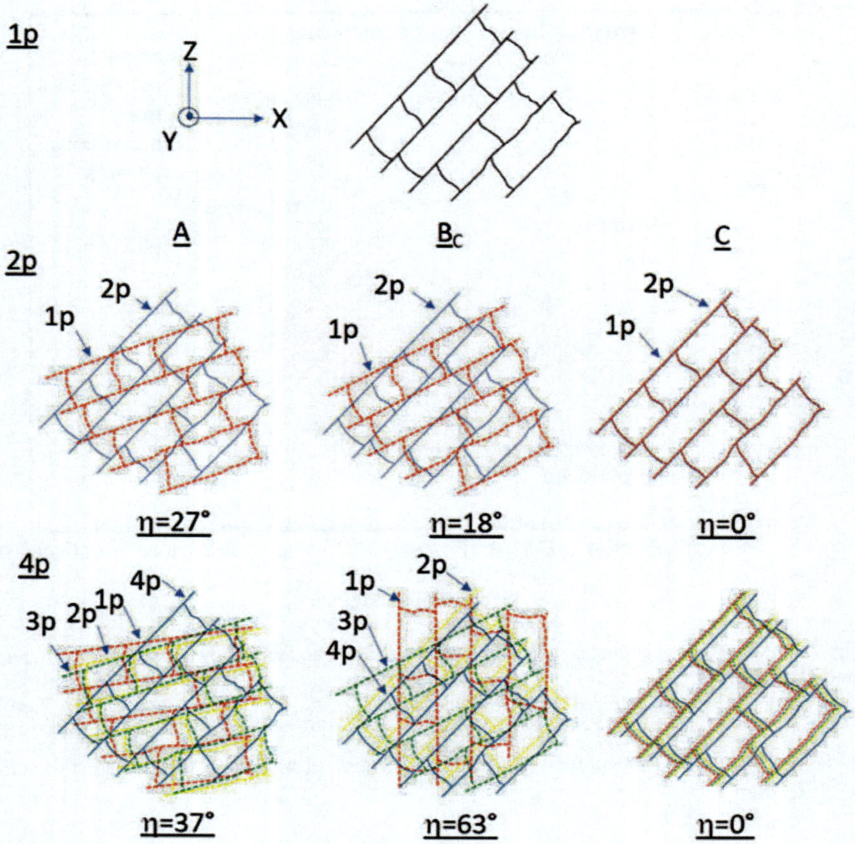

Fig. 7.7 Grain refinement during ECAP

schematic of grain refinement (for FCC metals) is illustrated in Fig. 7.7 for the main employed routes.

One of the major roadblocks in widespread implementation of nanocrystalline metals is connected to the thermal stability needed to consolidate them into bulk form. Grain boundaries in ultrafine-grained (UFG) materials processed by severe plastic deformation (SPD) are often called "nonequilibrium" grain boundaries. Such boundaries are characterized by excess grain boundary energy, presence of long-range elastic stresses, and enhanced free volumes (Sauvage et al. 2012). Ultrafine-grained (UFG) materials contain in their microstructure a very high density of grain boundaries (GB), which can play a significant role in the development and exhibition of novel properties. For this reason, UFG materials can be typically considered as interface-controlled materials. Unlikely to the nanocrystalline materials where grain boundary material can represent a significant, e.g., a percent or even larger, fraction of the whole volume, the volume fraction of GBs in an UFG material is less than 1%.

However, the structure, kinetic, and thermodynamic properties of GBs could be modified so significantly that they start to dominate some important material properties (Chokshi 2020).

7.2.1 Grain Development Behavior

The high grain boundary fraction gives a large driving force for coarsening and nanocrystalline materials exhibit limited thermal stability, with grain growth even being reported at room temperature for pure nanocrystalline metals. Grain sizes and grain boundaries (GBs) are the major microstructural features influencing the strength of nanocrystalline alloys (Smirnov et al. 2005). However, the thermal stability of nanocrystalline materials is generally low due to the large volume fraction (up to 49%) of high-energy GBs. Nanocrystalline materials tend to undergo rapid grain growth to eliminate GBs even at room temperature (RT). Introducing alloying elements into nanocrystalline materials is a promising approach to suppress grain growth and thereby improve their thermal stabilities. Furthermore, controlling the grain size at the nanoscale through alloying could significantly increase the mechanical strength of the materials because of the increased number of GBs that act as obstacles to sliding dislocations. Recently, increasing the GB stability of nanocrystalline metals, through GB relaxation and segregation, was found to be an effective approach to enhance strength (Demirtas and Purcek 2019).

The lack of thermal stability can be addressed in two ways: thermodynamic stabilization or kinetic stabilization of the microstructure. The former is achieved by reducing the excess grain boundary energy while the latter is achieved by reducing the grain boundary mobility. The effect of these two mechanisms can be summarized:

$$v = M \times P = M_0 \exp\left(-\frac{Q_m}{RT}\right) \times \frac{2\gamma}{r}$$

where v is the curvature-driven velocity of a grain boundary during grain growth, M is the grain boundary mobility, P is the driving force for grain growth, M_0 is a mobility constant, Q_m is an activation energy, γ is grain boundary free energy per unit area, and r is the radius of curvature. To obtain a stable microstructure, either M or P should be reduced, ideally to zero if stability is expected over long time periods. Kinetic approaches often rely on precipitation of second-phase particles or solute drag that reduces the grain boundary mobility by applying a pinning pressure on grain boundaries. The mobility term follows an Arrhenius function and extremely high temperatures can ultimately make kinetic approaches impractical. In thermodynamic approaches, γ should be reduced since the driving force for grain boundary migration is proportional to grain boundary energy. This is often done by doping grain boundaries with a solute that has a large atomic size mismatch with the solvent atoms. Grain boundary free energy is much less temperature dependent, and thermodynamic methods are philosophically more promising since they could stabilize a microstructure indefinitely.

Kinetic stabilization mechanisms stagnate grain growth with kinetic barriers. Without solute atoms, low-energy GBs tend to exhibit significantly less sliding than high-energy GBs (Du et al. 2011). Assuming solute segregates to the grain boundaries, solute drag can limit grain growth. Here, the grain boundary must carry the excess solute atoms as it moves, which counteracts the driving force for grain growth and consequently decreases the grain boundary velocity. The total drag force is described according to the following equation:

$$F_{\mathrm{b}}^{\mathrm{t}} = F_{\mathrm{b}}^{0} + F_{\mathrm{b}}^{\mathrm{d}}$$

where the total drag force ($F_{\mathrm{b}}^{\mathrm{t}}$) is equal to the sum of the inherent driving force for grain boundary migration without solute segregation (F_{b}^{0}), which is simply ν_b/M, and the solute drag force with solute segregation ($F_{\mathrm{b}}^{\mathrm{d}}$), which can be further defined as

$$F_{\mathrm{b}}^{\mathrm{d}} = \frac{\alpha C_{\infty} v_{\mathrm{b}}}{1 + \beta^2 v_{\mathrm{b}}^2}$$

where the solute drag force ($F_{\mathrm{b}}^{\mathrm{d}}$) is related to the drag force for unit concentration of solute per unit velocity of a moving boundary (α), bulk solute concentration (C_{∞}), grain boundary velocity (ν_b), and time required for solute atoms to diffuse one unit distance (β). Here α is approximately inversely proportional to the atomic diffusivity across the grain boundary, so a slower diffusing species will invoke a higher drag force:

$$F_{\mathrm{b}}^{\mathrm{t}} = \frac{v_{\mathrm{b}}}{M} + \frac{\alpha C_{\infty} v_{\mathrm{b}}}{1 + \beta^2 v_{\mathrm{b}}^2}$$

Another kinetic mechanism that has been shown to limit grain growth is Zener particle drag. This mechanism argued that particles sitting on grain boundaries similarly decrease the total grain boundary velocity. The simplifying assumptions for this analysis include the following: that the particles are randomly distributed, that they are inert with the matrix, that they do not move nor coarsen, and that the particles are spherical. The maximum drag force limiting grain boundary velocity is

$$F_{\mathrm{b}}^{\mathrm{p}} = \frac{3 f_{\mathrm{v}} \gamma_{\mathrm{b}}}{2r}$$

where the particle drag force ($F_{\mathrm{b}}^{\mathrm{p}}$) depends on the volume fraction of particles (f_{v}), the grain boundary energy, and the particle radius (r). If the particle drag force is equivalent to the intrinsic driving force, the limiting grain radius is

$$\overline{R} = \frac{4r}{3f_{\mathrm{v}}}$$

Another strategy to limit grain growth of nanocrystalline materials is to reduce the diffusivity of the solute and solvent atoms, either along or across the grain boundaries and through the lattice. The grain growth rate depends on the atomic diffusivity across the grain boundary. Thus, if heavy solute atoms can be selectively placed on the grain boundaries by solute segregation, the grain growth rate can be greatly diminished. Otherwise, heavy solute atoms located in the lattice can increase the activation energy, and consequently increase the onset temperature for grain growth. Chemical ordering has been cited as another kinetic stability mechanism to reduce diffusivity through the lattice or a grain boundary.

7.2.2 *Uniform Elongation*

Basically, these SPD techniques allow to obtain large tensile deformations at lower temperatures and higher strain rates. The first evidences were scientifically presented at the end of the 1980s for severely deformed aluminum alloys with high-angle grain boundaries (fundamental condition for high-temperature stability). Another fundamental aspect for achieving superplasticity in metals is the retardation of necking related to plastic instability during tension. A schematic of Zn-22Al prepared by HPT (Fig. 7.8) reveals how uniform elongation with very limited necking is experienced by the material up to 1800% deformation.

Studied performed on HPTed Ni_3Al intermetallics demonstrated that the achieved superplastic properties are mainly experienced through GBS. This was demonstrated by the aspect that showed limited dislocation activity in the material during deformation. Given the initial grain size of 50 nm (in the nanocrystalline range), this

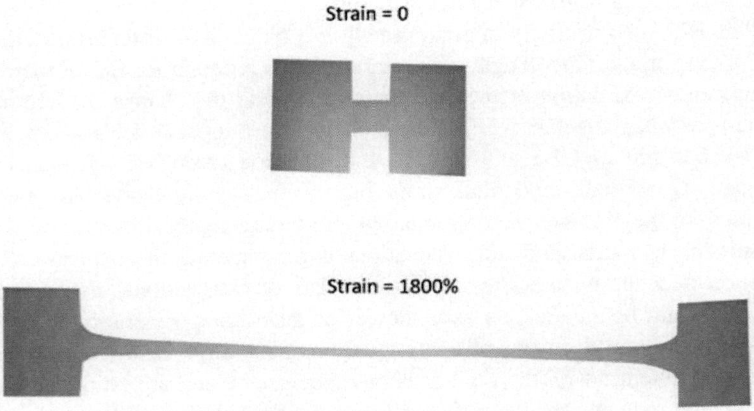

Fig. 7.8 Uniform elongation achieved in Zn-22Al after HPT at 200 °C

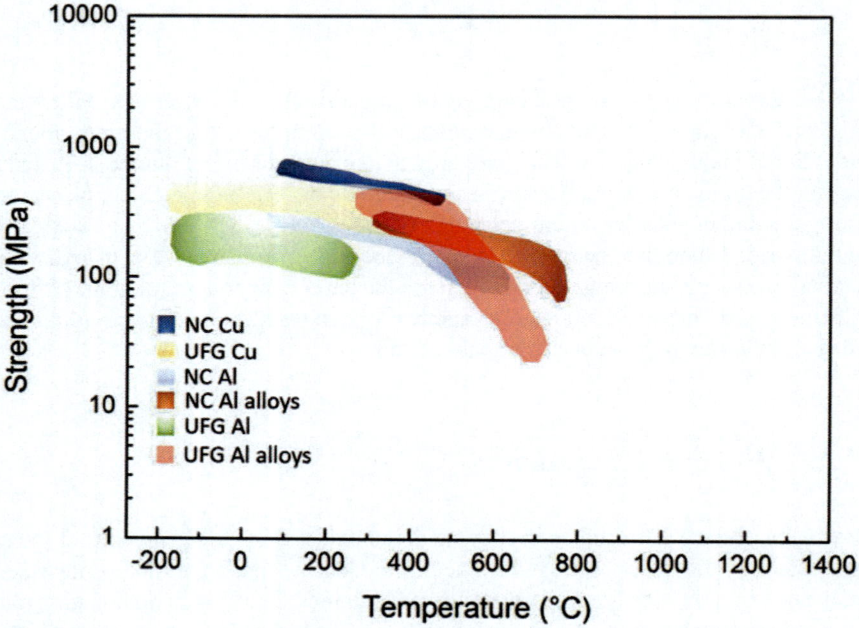

Fig. 7.9 Strength vs. temperature for ECAP materials

suggested that the superplastic flow in the nanocrystalline regime differs from that in the ultrafine regime. Data are presented on superplasticity of other nanocrystalline materials, and it is noted that their superplastic flow has a number of features as distinct from superplastic behavior of the alloys with micron-sized grains such as excessive strain hardening during tensile tests, high flow stresses, a correlation between microstructural instability (in terms of onset of grain growth) and superplastic behavior, and an absence of cavitation.

Obviously, the sensitivity to grain growth is a function of different materials. In Fig. 7.9 the variation of strength with temperature is reported for ECAP materials.

Apparently, exhibition of these features deals with the change of deformation mechanisms when, together with suppression of conventional lattice dislocation slip, such mechanisms as GBS and rotation and diffusion creep become increasingly important. Quasi-static and high strain rate testing are fundamentally different processes, so the plasticity accommodation will be inherently different. In general, the emission of partial and full dislocations is a competing mechanism and their activation depends on the stress levels. In case of conventional nanocrystalline materials, grain boundaries act as a source for dislocation generation which then runs freely and absorbs at the opposite grain boundary sink. Here, the dislocations are emitted, interact with the high-density nanoclusters, and are pinned at various sites, thereby reducing the mean free path for the propagation of dislocations. The motions of dislocations are restricted and there exists a stress required to unpin for

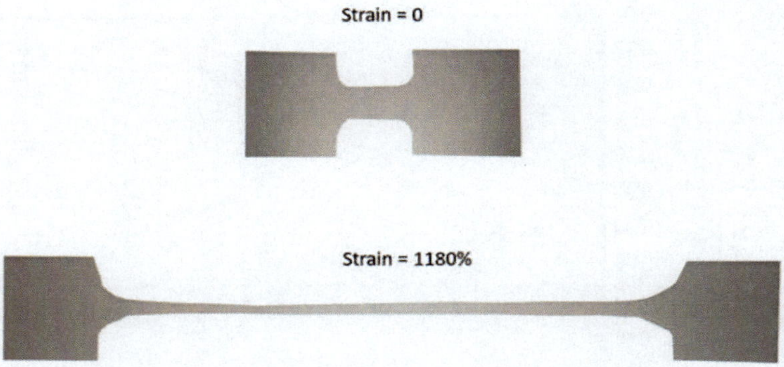

Fig. 7.10 Uniform elongation achieved in Al-Mg-Li-Zr after ECAP at 400 °C

the motion to resume. The other scenario includes the dislocation emission from the grain boundary, which travels to the opposite side and leaves behind two sides. The propensity of dislocation slip increases with the increase in testing temperature which helps in overcoming the energy barrier required for dislocation-based activity in these systems. However, the mean free path for the dislocation motion is dependent on the grain size and density of the nanoclusters that are embedded in the material system. In case of high-strain-rate conditions, nucleation of partial dislocations is favored at RT due to the high applied stress which surpasses the barrier for twinning-based deformation. Further, these nanoclusters pin the grain boundaries through Zener pinning mechanism, thereby arresting diffusion which is responsible for grain growth. Furthermore, these nanoclusters contribute towards strengthening.

In the 1990s, the first example of superplasticity at high strain rate (10^{-2} s^{-1}) in Al-Mg-Li-Zr revealed the broad potential of ECAP as an instrument for the production of superplastic materials (Fig. 7.10).

The superplastic deformed alloy showed a very uniform deformation up to 1180% strain. Further analyses showed that aluminum alloys containing Sc and Zr exhibited elongations >2000% thanks to the pinning activity of stable Sc-Zr precipitates. The superplastic behavior of these materials was revealed by values of the strain rate sensitivity in the range of 0.4–0.45. The nucleation of the Al_3Sc particles at the grain boundaries of the UFG metals before the recrystallization (grain boundary migration) onset ensures a high thermal stability of the structure and mechanical properties. High thermal stability of the fine-grained structure in the UFG Al-0.5 Mg-Sc alloys ensures high superplastic properties (Chuvil'deev et al. 2020).

The experimental evidences are in agreement with the theoretical calculations as shown in Fig. 7.11 (Turba et al. 2007; Pereira et al. 2017).

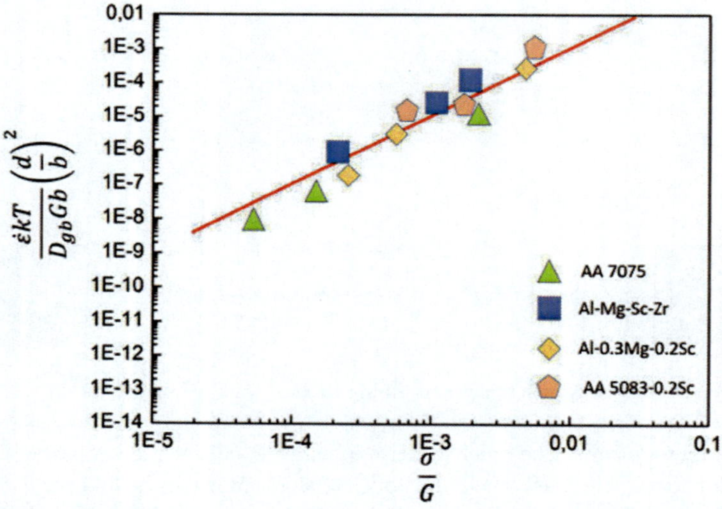

Fig. 7.11 The temperature- and the grain size-compensated strain rate versus normalized stress for various aluminum alloys

7.2.3 Mechanisms in UFGed Materials

The SPD conditions must be optimized, per each material, in order to induce perfect grain boundary sliding during deformation in order to achieve superplastic forming.

The dominant flow mechanism for superplasticity by GBS requires a grain size that is smaller than the equilibrium sub-grain size, λ (Kawasaki and Langdon 2018). The schematic is shown in Fig. 7.12.

In high-temperature creep with coarse-grained samples, the grains become divided into arrays of sub-grains having an average size of λ so that Fig. 7.12a corresponds to the condition where $d > \lambda$. Thus, in high-temperature creep GBS occurs through the movement of dislocations along the boundaries and this produces stress concentrations as at the triple point A, accommodating slip is initiated in the adjacent grain and these dislocations then move to the first sub-grain boundary at B where they climb into the boundary. Conversely, superplasticity requires a small grain size and it has been shown that it needs a grain size which is no larger than the average sub-grain size so that $d < \lambda$ as in Fig. 7.12b. The stress concentration at triple point C due to GBS is then accommodated by intragranular slip in the adjacent grain and now, in the absence of any sub-grain boundaries, the dislocations move across the grain and climb into the opposing grain boundary at D.

As a matter of fact, the elongations to failure are significantly enhanced by ECAP processing in comparison to the non-SPDed material. Second, the peak elongations are displaced to faster strain rates when the samples are subjected to larger numbers of passes in ECAP. In addition, superplastic behavior is demonstrated to be optimized by increasing the SPD temperature. This seems to be due to the optimal

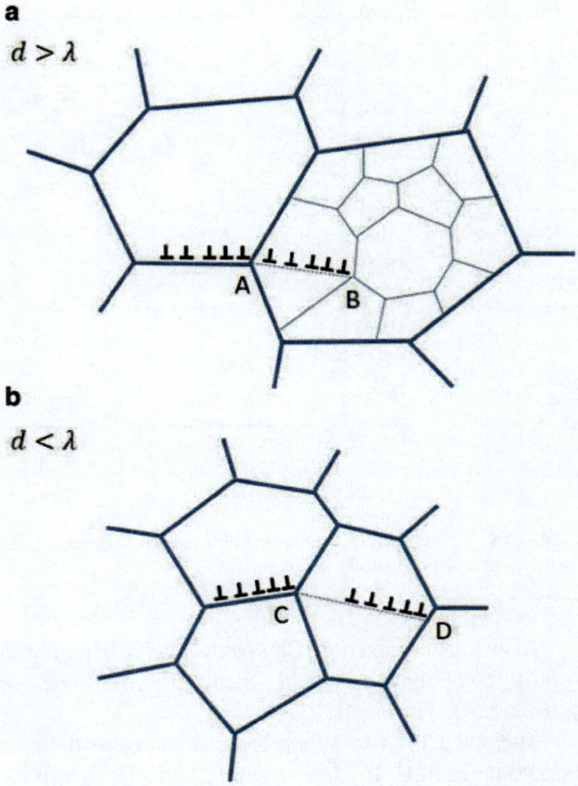

Fig. 7.12 Schematic illustration of a unified model for grain boundary sliding in (**a**) conventional creep when $d > \lambda$ and (**b**) superplasticity when $d < \lambda$

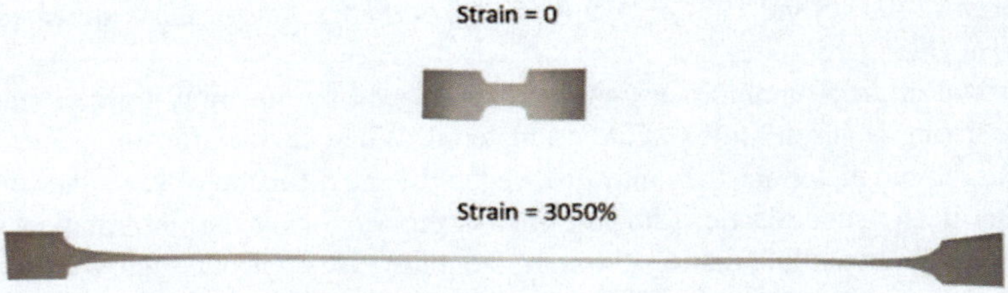

Fig. 7.13 Uniform elongation achieved in ZK60 Mg alloy after ECAP

second-phase precipitation allowing to obtain an exceptional grain stability. In addition, the higher temperature during SPD is believed to reduce textures that could be detrimental for the uniform elongation during superplastic deformation. By optimizing ECAP procedure in terms of temperature and processing angles, it was possible to record 3050% of strain in superplastic deformation of ZK60 Mg alloy. This is the highest recorded elongation for Mg alloys (Fig. 7.13).

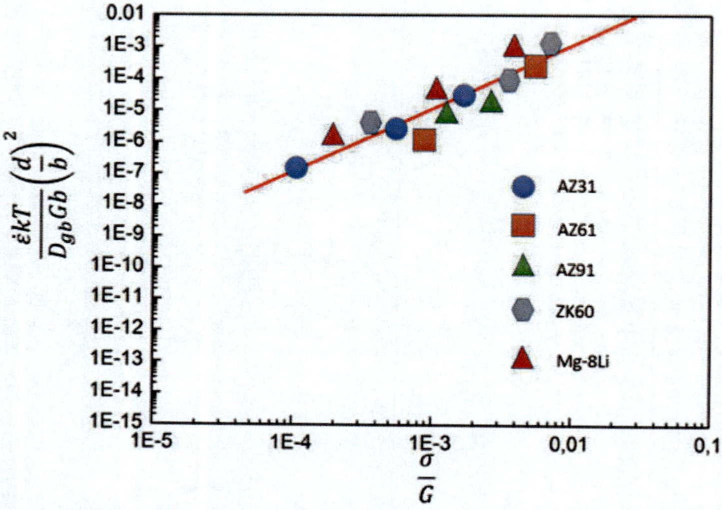

Fig. 7.14 The temperature- and the grain size-compensated strain rate versus normalized stress for various magnesium alloys

Also for optimized SPDed Mg alloys it is demonstrated that exceptional superplastic behavior is achieved when pure grain boundary sliding is the deformation mechanism (Fig. 7.14).

The main limit to superplastic forming of these materials is cavitation during straining. In order to observe such a behavior AA7034 was spray formed (grain size of 2 μm) and then ECAPed (300 nm grain size). After deformation in the same conditions (400 °C and 10^{-2} s^{-1}), the profiles of the fracture surfaces appeared very different (Fig. 7.15).

First of all, a pronounced necking is experienced by the spray-formed material. This necking is absent in the ECAPed material. When cavities form in superplastic nanostructured materials, they may grow either by the diffusion of vacancies into the cavities in the superplastic diffusion growth process or by the deformation in the surrounding crystalline lattice through the plasticity-controlled growth process. These two mechanisms lead to different cavity shapes because superplastic diffusion growth gives cavities that are essentially spherical, whereas in plasticity-controlled growth, the cavities pull out and become elongated along the tensile axis. The smaller cavities grow by diffusion growth and the larger cavities grow by plasticity-controlled growth. There are numerous clear demonstrations that the superplastic effect is achieved in these nanostructured materials at strain rates that are significantly faster than those in conventional micrometer-grained materials. Nevertheless, it is important to recognize that superplasticity can be achieved only in those materials where the ultrafine grain sizes introduced through processing remain small and reasonably stable at the temperatures needed to attain diffusion-controlled plastic flow. This means in practice that superplastic flow is not easily

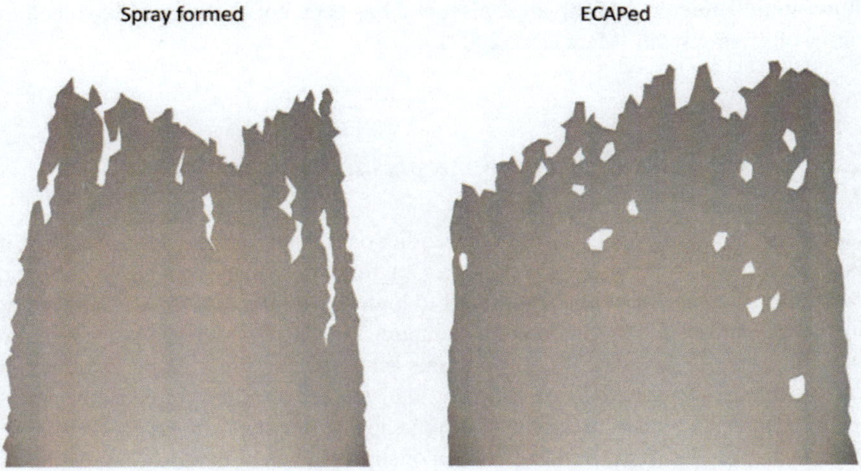

Fig. 7.15 Different fracture surfaces of AA7034 in as-sprayed and as-ECAPed conditions

achieved in pure metals or solid solution alloys where the grains grow rapidly when heated to high temperatures.

7.3 Superplasticity in Nanocrystalline Materials

As shown in the previous chapters, nanocrystalline materials often exhibit high strength, high hardness, and wear resistance. Anyway, in most cases, they are characterized by low tensile ductility limiting the range of applications. Recently, however, several examples of high ductility and superplasticity in NC materials was reported. A reduction in grain size can lead to a reduction in the superplastic temperature at constant strain rate, or an increase in the superplastic strain rate at constant temperature. Early speculation regarding enhanced superplasticity in nanocrystalline materials was based primarily on the grain size dependence of superplastic flow. The results with nanocrystalline materials show that a reduction in superplastic temperature has been achieved. However, even at lower temperatures, grain growth can be significant. The data show that the onset of nanocrystalline superplasticity coincides with the onset of microstructural instability (McFadden et al. 2001). From the first equation in the introduction, the grain size dependence also leads to an expectation of lower flow stresses in nanocrystalline materials compared to their microcrystalline counterparts. Nanocrystalline materials have also shown extensive strain hardening during superplastic deformation. In contrast, microcrystalline superplasticity is generally free of large-scale strain hardening. Because of the grain size dependence in the constitutive relationship for superplastic flow, strain hardening during superplasticity has been conventionally explained in terms of grain growth (Mara et al. 2007).

7.3.1 Grain Size Effect on Superplastic Behavior

According to the information mentioned earlier on SPDed material superplasticity, it is expected that as the grain size decreases from micrometer to nanometer, the superplastic region would be transposed to high strain rates (HSRs) or observed at low temperatures such as room temperature, i.e., HSR and/or low-temperature superplasticity is possible. The mechanisms acting during superplastic deformation of nanocrystalline materials are different with respect to other grain size regimes. This is due to the basic aspect that NC grains show higher flow stress for superplastic deformation. In NC regime, the activation of slip systems for the accommodation of deformation results to be much more difficult. As a matter of fact, in NC grains grain boundary sliding acts through triple junction migration. The basic theoretical view of the situation acting in NC materials related to slipping is shown in Fig. 7.16 (Mishra et al. 2001).

As shown in the previous chapters, partial dislocations are active in NC metals and alloys at low stress levels. In the case of GBS, dislocations accumulate at the triple junctions by increasing the energy that can be released through the emission of partial dislocations. In addition, the triple junction angles increase to accommodate the GBS. Atomic shuffling-accommodated GBS is also active in NC materials. This is known as stress-induced grain boundary diffusion. The global mechanism model

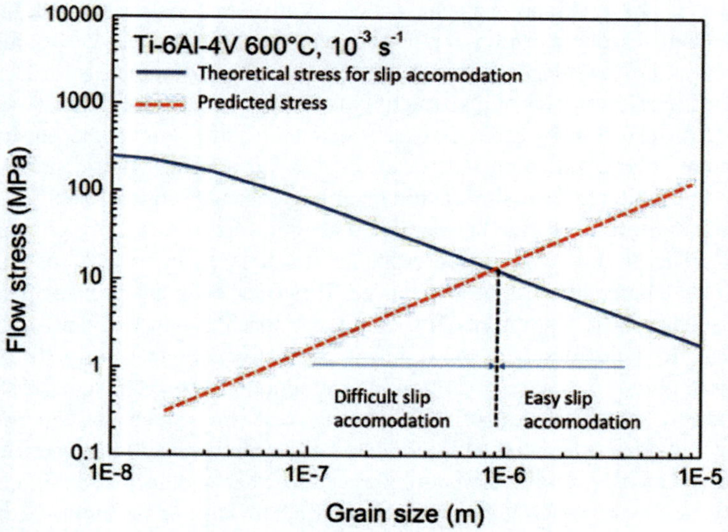

Fig. 7.16 Deformation by slip accommodation in Ti6Al4V

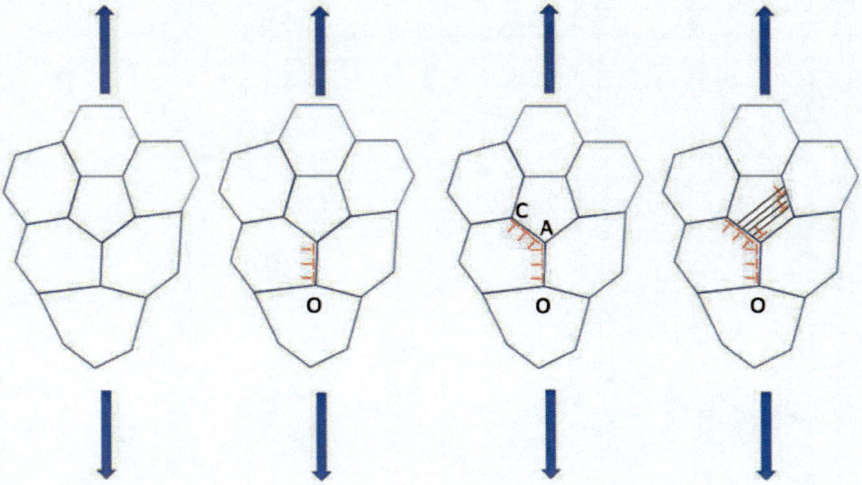

Fig. 7.17 Nanotwin formation during superplastic deformation

is shown in Fig. 7.17 where the sequential step of GBS, dislocation pileup, and twin formation are presented.

The piled-up dislocation splits into two types: immobile GB dislocation and mobile partial dislocation. This partial dislocation then moves cooperatively on every slip plane in a nanoscale region to generate nanotwin.

Ovidko (2005) analyzed the deformation mechanisms acting in UFG and NC grained materials for each grain size regime. Grain boundary-mediated mechanisms (grain rotation, GB diffusional creep, triple junction diffusional creep, GBS) act during superplastic deformation in the finest grains.

7.3.2 *Dislocation Behavior*

Superplasticity in nanocrystalline materials shows very high flow stresses and hardening in the first stage of deformation followed by softening. It is favored in the first stage thanks to the high strain rate sensitivity and by the suppression of nano-crack nucleation and propagation in the second stage. The mechanisms leading to the nano-crack formation in nanocrystalline materials during superplastic deformation were proposed by Ovidko and Sheinerman (2005); the schematic model is shown in Fig. 7.18.

Lattice dislocations are generated by the plastic deformation (Fig. 7.18a); they are adsorbed at the grain boundary splitting into GB dislocations (Fig. 7.18b); the model assumes that the splitting results in the formation of GB dislocations of two types: gliding and climbing GB dislocations with the Burgers vectors being parallel and perpendicular to the GB plane (Fig. 7.18c). The gliding dislocations move under the

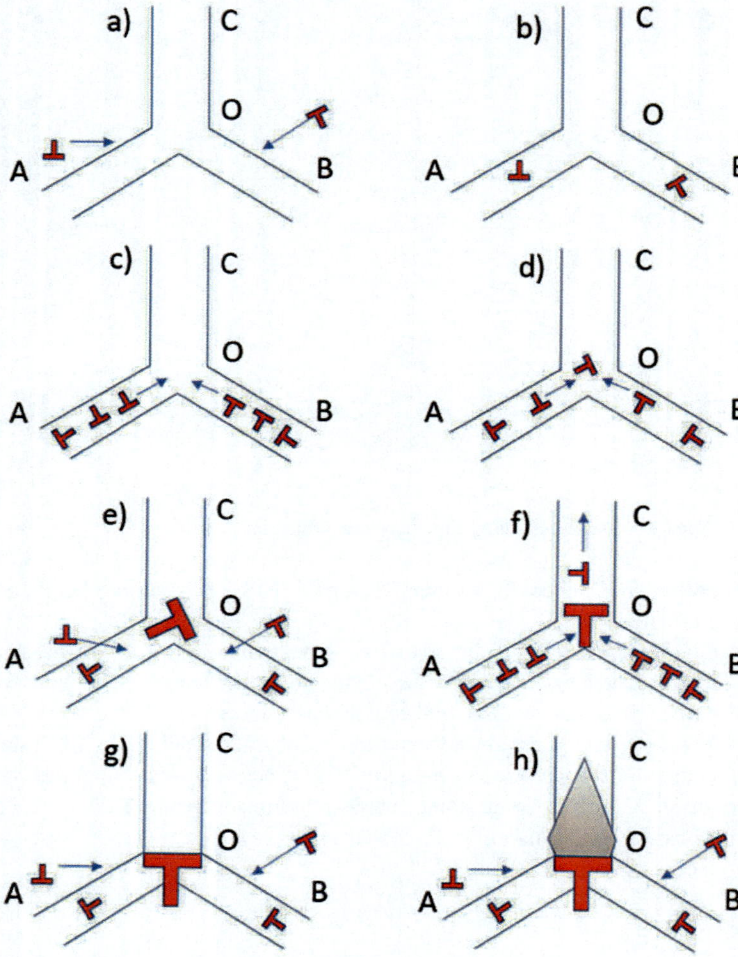

Fig. 7.18 Nano-crack formation mechanism at the twin boundary

action of the external stress along GB planes towards triple junctions. At a certain level of the stress, the gliding dislocations reach triple junctions where they converge (Fig. 7.18d); sessile dislocations form at the triple junction; by increasing the strain the process is repeated, thereby increasing the Burgers vector of the triple junction dislocation (Fig. 7.18e–g). The triple junction dislocation interacts with the GB dislocations leading to the hardening of the NC material. The Burgers vector of the triple junction dislocation increases leading to the formation of a "superdislocation." The triple junction superdislocations create tensile stresses that are capable of inducing the formation of nano-cracks in the vicinity of triple junctions (Fig. 7.18h) at some critical plastic strain.

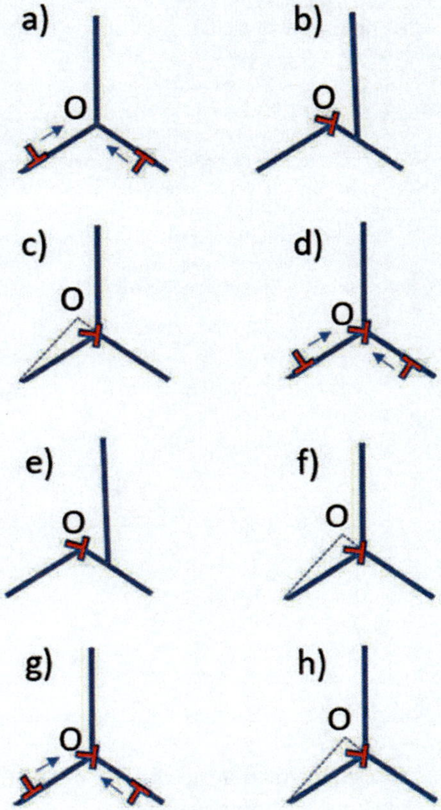

Fig. 7.19 GBS and defect transformations at triple junctions

Superplasticity acts in nanocrystalline materials via GBS as a dominant mechanism accompanied with intense diffusion and diffusion-controlled rotational deformation with triple junction migration. The schematic details are shown in Fig. 7.19.

Two grain boundary dislocations move towards the triple junction in O (Fig. 7.19a); the convergence in O gives rise to a sessile dislocations; the deformation of the upper grain leads to the migration of the triple junction (Fig. 7.19b); local grain boundary migration is exhibited in (Fig. 7.19c); the movement of two dislocations along grain boundary is repeated (Fig. 7.19d); the Burgers vector of the sessile dislocation is increased leading to shear in the upper grain (Fig. 7.19e); local grain boundary migration happens (Fig. 7.19f); there is continuous grain boundary sliding via the same mechanisms (Fig. 7.19g, h).

In this model, the storage of grain boundary dislocations causes the strengthening of NC materials in the first stage of deformation. This is very different from the case of microcrystalline materials where hardening is produced by the storage of lattice dislocations into the grain interior. Obviously, the excess of dislocation storage at triple junctions can lead to nano-crack formation. Anyway, the nano-crack nucleation can be limited in NC materials thanks to the diffusion-assisted rotation, and

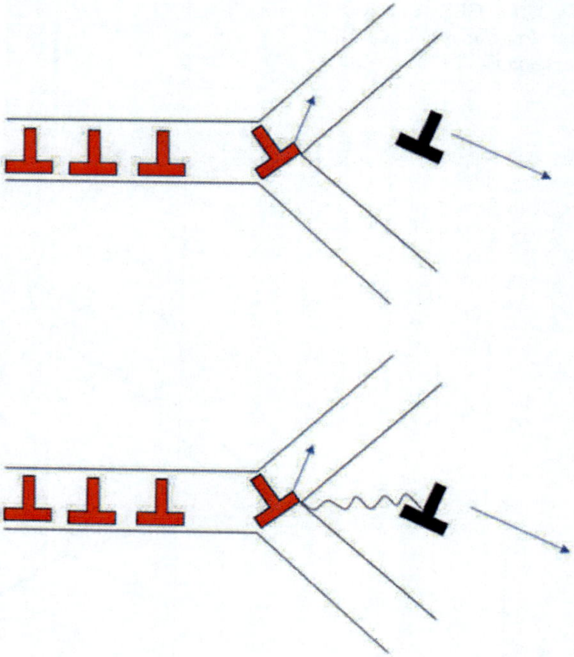

Fig. 7.20 Emission of lattice dislocation or partial dislocation at triple junction

local diffusion processes with the emission of lattice dislocations from triple junctions (Fig. 7.20).

The grain rotation occurs through diffusion-assisted climb of grain boundary dislocation. With this being a recovery mechanism, it allows for the softening and reduction of dislocation storage at triple junctions avoiding nano-crack nucleation. In addition, point defect diffusion at triple junctions provides additional softening with stress relaxation (Fig. 7.21).

Taking into account the model described in Fig. 7.18, the diffusion processes acting at the superplastic deformation temperature lead to the relaxation of the stresses induced by the formation of the "superdislocation." In certain conditions, the diffusion level allows for the suppression of nano-cracking at triple junctions; this aspect is fundamental for the high strain rate superplasticity of NC materials.

At this stage of superplastic deformation of nanocrystalline metals and alloys cooperative grain boundary sliding and grain rotation act as shown in Fig. 7.22 (Sergueeva and Mukherjee 2006).

This mechanism is associated not only with individual grain sliding and rotating relative to one another, but also with entire groups of grain sliding along commonly oriented grain boundaries, resulting in the large amount of mass transfer responsible for elongations in the hundreds of percent. In fact, while the sliding and rotation of individual grains are certainly an important aspect of superplasticity, it is difficult for such a mechanism to solely explain the elongations of hundreds of percent seen in superplastic materials. In CGBS, grains slide relative to one another along sets of

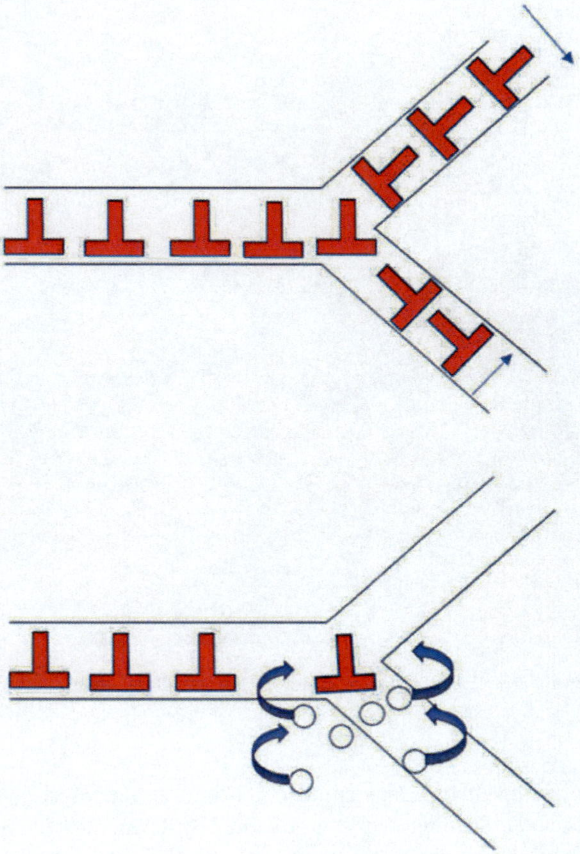

Fig. 7.21 Climbing grain boundary dislocations and vacancy diffusion

grain boundaries that have aligned themselves to form planes of long-range shear (Yu et al. 2014). Since the mass transport associated with a cooperative event is much greater than that carried by singular grains, CGBS can easily account for the large elongations seen in superplastic deformation. Of course, the alignment of the grain boundaries to form these "stringers," as the boundaries of long-range shear are commonly referred to, must occur by some process, which is where local grain rotation and/or dislocation accommodation must come into play. Sliding surfaces support the mechanism of cooperative grain boundary sliding; the schematic is shown in Fig. 7.23.

Generally, three models are presented in literature for theories of grain boundary sliding accommodated by dislocation sliding; they are summarized in Fig. 7.24.

During deformation in Ball-Hutchison model, grain boundaries are properly aligned and slide as groups. When the sliding is blocked by other grains, the local stresses would result in dislocation in the blocking grain, piling up at the opposite grain boundary until the back stress stops further generation of dislocation. The

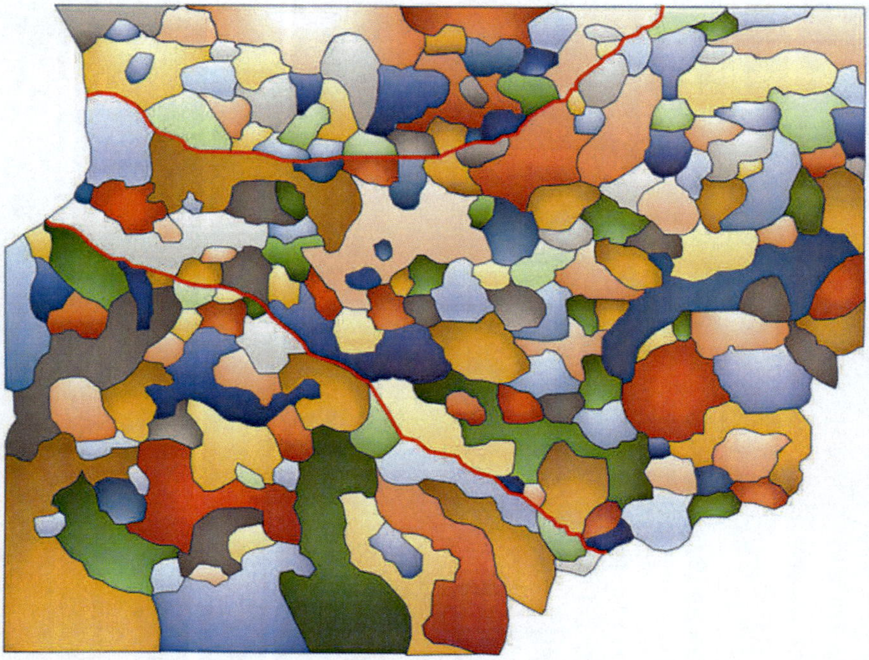

Fig. 7.22 Grain boundary sliding in Ni_3Al with starting grain size of 50 nm tensile deformed at 750 °C

piled-up dislocations could climb into and along the grain boundary. Thus, grain boundary sliding, which is governed by the kinetics of climb along grain boundary to annihilation areas, is feasible due to the constant replacement of the dislocation (Wang et al. 2018). The Mukherjee model elaborates dislocation-accommodated sliding mechanism of single grain. Dislocations are generated at scraggy surfaces of grain boundaries. The alignment mechanism of the generated dislocations is the same as that of the Ball-Hutchison model. Gifkins described grain boundary sliding and accommodation process as grain boundary dislocation motion and proposed the core-mantle model. Grain boundary sliding around grain triple junctions is accommodated by generation of new dislocation and their climbing along the boundaries. Dislocation motion is constrained to occur at grain boundary and mantle area rather than in the core of grain. According to the model, grains are able to slide by switching places in three-dimensional space.

Many experimental evidences show superplasticity in NC Ni-based materials. A summary of the recorded elongation to failure as a function of the starting grain sizes is shown in Fig. 7.25.

Here it is clear how also stable ceramic reinforcements can act as grain boundary pinning by reducing the grain growth and favoring the superplastic behavior of NC materials (Chan et al. 2004).

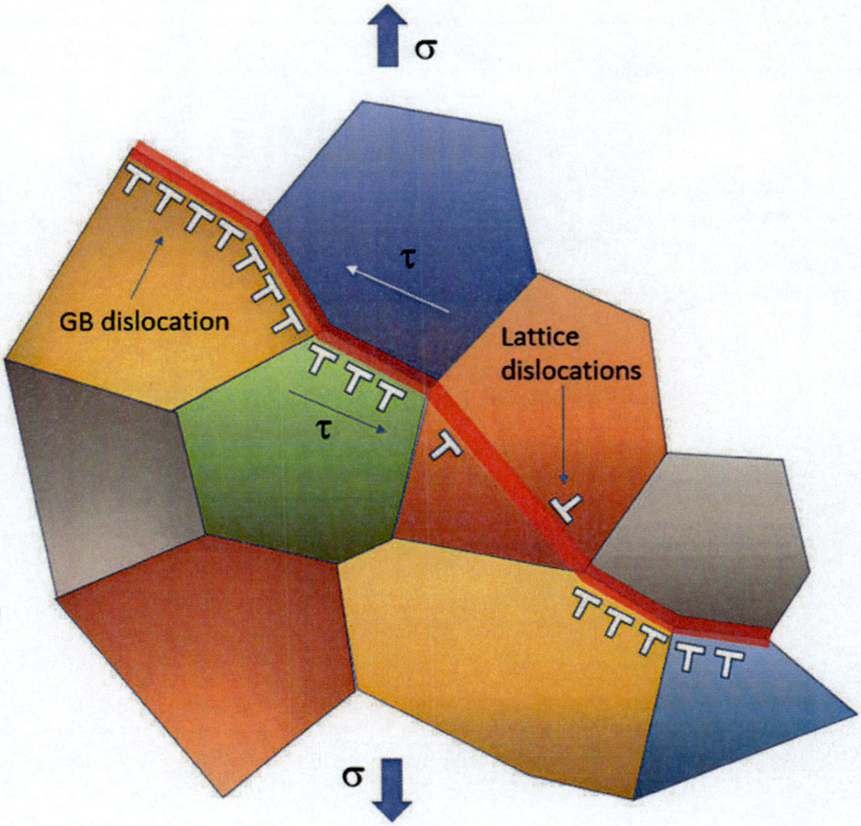

Fig. 7.23 Schematic of CGBS

Superplasticity is generally not observed under conventional quasi-static strain rates in pure metals because of rapid grain growth at elevated testing temperatures. Although there are several reports of superplasticity at low temperatures in nano-Ni, most of the data correspond to low strain rate superplasticity or strain rates in the vicinity of a transition to high strain rate superplasticity. Studies provide evidence for both low temperature and high strain rate superplasticity in nanocrystalline Ni-based materials (Prasad and Chokshi 2010). In conventional superplastic materials, there is usually a decrease in ductility at higher strain rates, corresponding to a decrease in strain rate sensitivity caused by a transition to intragranular dislocation creep. Failure in many superplastic materials is related to the nucleation, growth, and interlinkage of cavities; microstructural observations revealed concurrent cavitation in the fractured Ni samples. Based on an understanding of the kinetics of stress concentration and stress relaxation, it has been shown that cavity nucleation is favored at higher stresses.

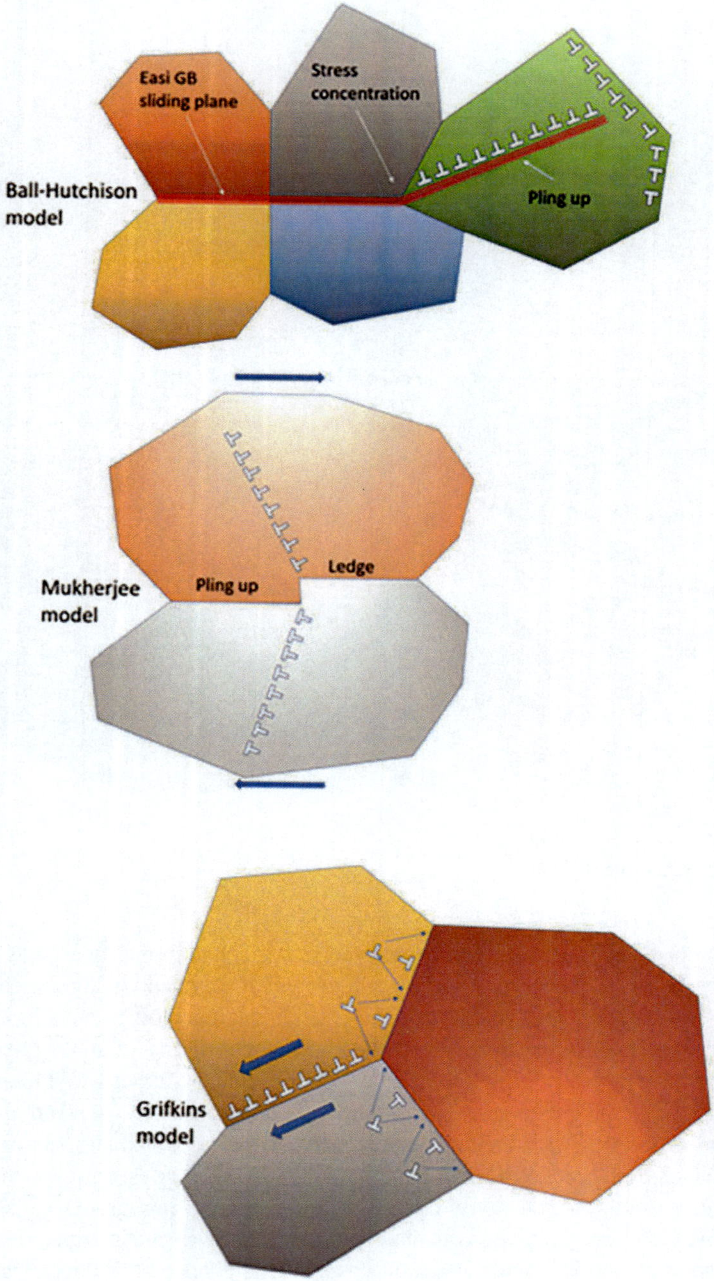

Fig. 7.24 Schematic of GBS models

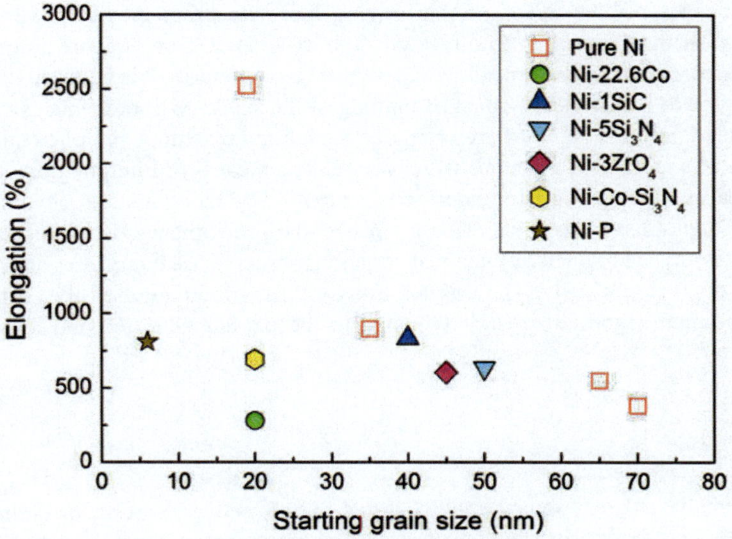

Fig. 7.25 Elongation to failure as a function of the initial grain size for various Ni-based materials

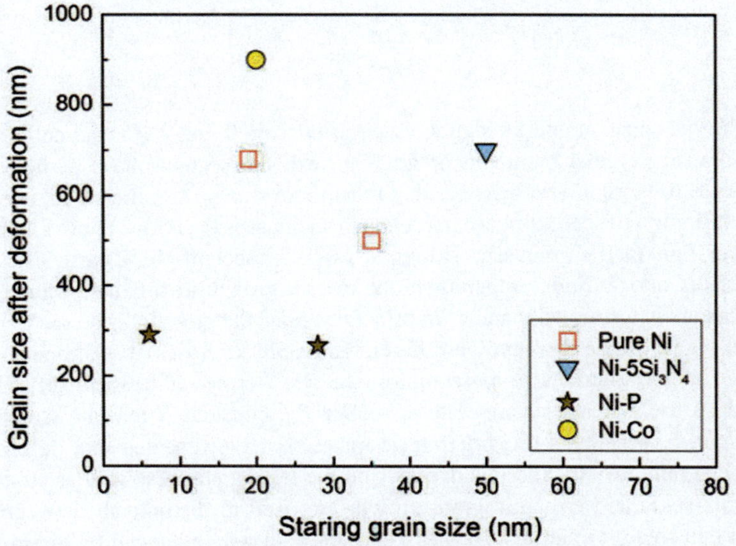

Fig. 7.26 Grain growth due to superplastic deformation temperature

Taking into account that the elongation values reported in Fig. 7.25 are the highest recorded per each material, it is very useful to monitor the grain size measured at the end of straining in order to analyze the grain growth during high-temperature superplastic forming (Fig. 7.26).

In nanocrystalline materials, the driving force for grain growth is very high, because of the large interfacial area per unit of volume. Consequently, significant grain growth can occur even at low superplastic temperatures. It is believed that Ni-P grain growth is impeded by the GB pinning of the stable Ni_3P precipitates.

Since fine grains tend to grow rapidly at high temperatures, the microstructural design of superplastic materials usually requires a means for limiting grain growth. This necessary microstructural design is accomplished using microduplex alloys or quasi-single-phase alloys with fine grain boundary precipitates, where microstructural stability is provided by chemical variations across phase boundaries and Smith–Zener drag, respectively. The original Smith–Zener formulation of drag suggested that the matrix grain size (d_m) is related to the particle size (d_p) and its volume fraction f:

$$\frac{d_m}{d_p} = \frac{2}{3f}$$

There have been several modifications to the above expression, taking into account the random distribution of second-phase particles, the curvature of grain boundaries encountering particles, and other factors. However, most studies lead to the general form:

$$\frac{d_m}{d_p} = \frac{B}{f^a}$$

where B is a constant and the exponent α varies from 0.3 to 1 (Prasad and Chokshi 2011). In the original treatment of grain growth the second-phase particles were considered to be rigid and incoherent, so that there was no significant change in the size distribution of second-phase particles; in such cases the above approach leads to a limiting final matrix grain size. However, when the second-phase particles can also grow or dissolve at high temperatures the overall growth of the matrix and second phase occurs in a coupled manner. In principle, since the growth of the second phase and the matrix phase can involve different activation energies, it is anticipated that the ratio of the matrix size to second-phase size (termed Z henceforth) will also depend on the annealing temperature, so that the constant B may be temperature dependent. Experiments on superplastic materials have revealed that deformation enhances grain growth, and this depends on the testing strain as well as strain rate; the terms static and dynamic grain growth are used to distinguish between grain growth due solely to thermal exposure and grain growth enhanced by deformation, respectively. Dynamic grain growth is usually interpreted in terms of the disturbance of triple grain junctions by grain boundary sliding, which enables faster grain growth. Geometric perturbation of Zener pinning can lead to deformation-enhanced grain growth, even in a system where the second phase is essentially inert. If grain boundary sliding can lead to distortions at triple junctions and accelerated grain growth it is possible that superplasticity can also lead to modifications in Z. In comparison with nano-Ni it is clear that the presence of Ni_3P particles restrains grain growth in the Ni-P alloy. Superimposition of the cumulative grain size distribution plots normalized by the mean grain sizes, for both static grain growth and dynamic grain growth, indicates that the normalized distribution is time (and strain) invariant, which is a characteristic of normal grain growth. The kinetics of normal grain growth are usually represented as

$$d^{N_g} - d_0^{N_g} = K_g t$$

where N_g is the grain growth exponent, and the temperature-dependent constant is

$$K_g = K_{0g} \exp\left(-\frac{Q_g}{RT}\right)$$

with K_{0g} being a pre-exponential term and Q_g being the activation energy for grain growth. Normal grain growth involves two topological processes of grain disappearance and grain switching. Superplastic flow also involves grain switching and grain boundary sliding, which can disturb triple junction geometries, both of which can enhance grain growth. Concurrent grain growth during superplastic deformation in the Ni-P alloy led to a greater number of grains containing intragranular Ni_3P precipitates, as some of the original intergranular precipitates detached from grain boundaries. The reduction in the number of particles restraining grain boundary mobility can lead to enhanced matrix grain growth. In addition, the reduction in the effective volume fraction of particles at grain boundaries should lead to a higher value of Z. The conventional approach in microduplex alloys considers grain boundary sliding involving multiple or two grains with the emission of intragranular dislocations which traverse across a grain and are recovered by a climb at the opposite grain boundary; the rate-controlling process is the climb of dislocations. In contrast to conventional microduplex alloys, the nanoduplex Ni-P alloy contained fine intragranular precipitates which impeded intragranular dislocation glide. In such cases, it is possible for superplasticity to be controlled by a climb process to overcome intragranular particles (Fig. 7.27).

The velocity of dislocation climb at the head of a pileup at an intragranular particle is given by

$$v_c = \frac{2D\Psi\sigma^2 b^3}{3\lambda GbkT}$$

where D is the appropriate diffusion coefficient, Ψ is the length of the dislocation pileup, and λ is the climb distance. The macroscopic strain rate necessary for dislocation climb is

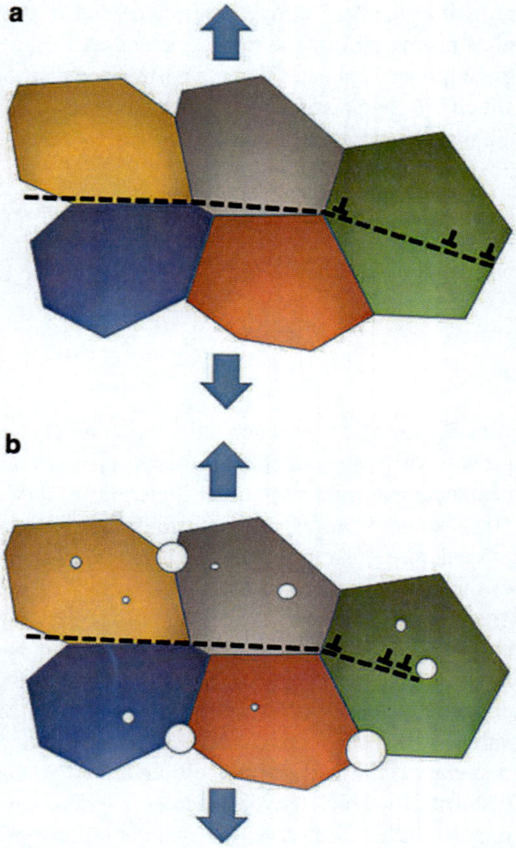

Fig. 7.27 Schematic illustration of grain boundary sliding and intragranular dislocation activity in (**a**) conventional superplastic alloys and (**b**) nano-Ni-P alloy with intragranular precipitates

$$\dot{\varepsilon} \propto \frac{D\Psi\sigma^2}{\lambda^2 d}$$

Conventional superplastic alloys frequently display steady-state flow with no significant changes in the flow stress. Under such conditions it may be reasonable to assume that flow localization is dictated by the value of strain rate sensitivity (m). It is well known that strain hardening can also contribute to the stability of tensile deformation in superplastic materials. Strain hardening is related to an accumulation of intragranular dislocations. The reduction in strain rate sensitivity under optimum conditions is accompanied by an increase in strain hardening, so that the large elongation to failure is attributed to the combined contributions of strain rate sensitivity and strain hardening. Cavity nucleation in superplastic materials may occur at grain boundary particles, grain boundary ledges, and triple junctions. The kinetics of cavity nucleation are given by

$$\dot{N} \propto \exp\left(-\frac{4\gamma^3 F_v}{\sigma^2 kT}\right)$$

where N, γ, and F_v are the nucleation rate, the grain boundary energy, and a shape factor, respectively. The above equation suggests that the cavity nucleation kinetics will be substantially more at the higher stress, which is consistent with the observation of greater cavitation at higher strain rates.

In NC Ni-Co a significant amount of twinning and dislocation is observed in the deformed specimen, and it is considered that the deformation twinning is an important accommodation mechanism of GBS at large deformation (Wang et al. 2006).

As a matter of fact, the data reported for NC materials showed the following characteristics: (1) the stress exponent (the inverse of strain rate sensitivity) was high and variable; (2) the grain size sensitivity was three not two; and (3) an activation energy was close to the activation energy for boundary diffusion but decreased with increasing applied stress. Consideration of these characteristics indicated that deformation of nanograined materials was not controlled by the same mechanism operating during superplastic flow in micro-grained and ultrafine-grained materials. Several mechanisms were proposed to account for the deformation behavior of nanograined materials. It was shown that these mechanisms were not satisfactory and that the predictions of a new mechanism based on deformation accommodated by boundary sliding (as previously mentioned) account for the deformation characteristics of nanograined materials.

The development of the model was based on the concept that plasticity in NC materials was the result of grain boundary sliding accommodated by the generation and motion of dislocations under local stresses, which were higher than applied stresses due to the development of stress concentrations. Specifically, it was assumed that as a result of sliding of a group of grains, the shear stress became concentrated at any grain, triple point, or protrusion that obstructed the motion of this group; that this high local stress can then generate dislocations in the blocking grain (or initiate voids); and that the generated dislocations move one by one to the opposite boundary where they can climb to their annihilation sites (no dislocation-ups).

By postulating that the creep rate was governed by the time for the climb of dislocation along the boundary until annihilation occurs, the following rate-controlling equation was derived:

$$\gamma = 9\left(\frac{b}{d}\right)^3\left(\frac{D_{gb0}}{b^2}\right)\exp\left(-\frac{Q_{gb}}{RT}\right)\left[\exp\left(\frac{\tau v}{kT}\right)-1\right]$$

where D_{gb0} is the frequency factor for grain boundary diffusion. The stress exponent for creep, n, exhibits high and variable values; for the applicable range of stresses, $n > 5$. Accordingly, it is expected that ductility in NC materials would be much lower than those characterizing micro-grained superplastic alloys for which $n = 2$ because ductility depends on $1/n$ ($n = 1/m$).

One of the most successful deformation mechanisms for superplasticity is the one based on the analysis in which dislocations glide in the interior of the blocking grain and then pile up at the opposite grain boundary, where they climb into and along the grain boundary. The smallest number of edge dislocations in a dislocation pileup is two dislocations. Under this condition, the equilibrium distance between two edge dislocations, L, is given by

$$\frac{L}{b} = 0.25\left(\frac{G}{\tau}\right)$$

The transition in creep behavior from superplastic behavior to NC behavior occurs when the dislocation pileup at the boundary reduces to no-dislocation pileup. Such a condition is met when the grain size, d, is less than L. On this basis, the transition from superplastic behavior to NC behavior occurs when

$$\frac{d}{b} = 0.25\left(\frac{G}{\tau}\right)$$

It resembles in form the equation which defines the transition from creep behavior controlled by dislocation climb ($n = 5$, $s = 0$, and $Q = Q_D$) to creep behavior characterizing superplastic flow ($n = 2$, and grain size sensitivity, $s = 2$, and $Q = Q_{gb}$), and which is represented by the following expression:

$$\frac{d}{b} = 10\left(\frac{G}{\tau}\right)$$

The behavior of nanograined materials is explained in terms of the emergence of a new deformation mechanism that is different from the mechanism operating during the deformation of micro-grained and ultrafine-grained materials. By considering the details of this new mechanism and the mechanism proposed for micro-grain superplasticity, the condition for a transition from the former to the latter is

$$0.25\left(\frac{G}{\tau}\right) < \frac{d}{b} < 10\left(\frac{G}{\tau}\right)$$

7.4 Conclusions

During superplasticity, grain boundary sliding is the dominant deformation mechanism. A decrease in grain size leads to increase in superplasticity at a given temperature and strain rate. Based on these considerations, the development of SPD techniques allowed to produce bulk materials optimized for superplastic formation. For NC materials, owing to their fine grain size, large grain boundary area, and high self-diffusivity, superplasticity is expected at lower temperatures and/or higher strain rates.

The superplastic flow in the nanocrystalline regime differs from that in the ultrafine regime. Data are presented on superplasticity of nanocrystalline materials, and it is noted that their superplastic flow has a number of features as distinct from superplastic behavior of the alloys with micron-sized grains such as excessive strain hardening during tensile tests, high flow stresses, a correlation between microstructural instability (in terms of onset of grain growth) and superplastic behavior, and an absence of cavitation. Obviously, the sensitivity to grain growth is a function of different materials. The SPD conditions must be optimized, per each material, in order to induce perfect grain boundary sliding during deformation in order to achieve superplastic formation. Superplastic behavior is demonstrated to be optimized by increasing the SPD temperature. This seems to be due to the optimal second-phase precipitation allowing to obtain an exceptional grain stability. In addition, the higher temperature during SPD is believed to reduce textures that could be detrimental for the uniform elongation during superplastic deformation. When cavities form in superplastic nanostructured materials, they may grow either by the diffusion of vacancies into the cavities in the superplastic diffusion growth process or by the deformation in the surrounding crystalline lattice through the plasticity-controlled growth process. These two mechanisms lead to different cavity shapes because superplastic diffusion growth gives cavities that are essentially spherical, whereas in plasticity-controlled growth, the cavities pull out and become elongated along the tensile axis. The smaller cavities grow by diffusion growth and the larger cavities grow by plasticity-controlled growth. It is important to recognize that superplasticity can be achieved only in those materials where the ultrafine grain sizes introduced through processing remain small and reasonably stable at the temperatures needed to attain diffusion-controlled plastic flow. This means in practice that superplastic flow is not easily achieved in pure metals or solid solution alloys where the grains grow rapidly when heated to high temperatures.

As the grain size decreases from micrometer to nanometer, the superplastic region would be transposed to high strain rates (HSRs) or observed at low temperatures such as room temperature, i.e., HSR and/or low-temperature superplasticity is possible. The mechanisms acting during superplastic deformation of nanocrystalline materials are different with respect to other grain size regimes. This is due to the basic aspect that NC grains show higher flow stress for superplastic deformation. In NC regime, the activation of slip systems for the accommodation of deformation results to be much more difficult. The diffusion processes acting at the superplastic deformation temperature lead to the relaxation of the stresses induced by the formation of the "superdislocation." In certain conditions, the diffusion level allows for the suppression of nano-cracking at triple junctions; this aspect is fundamental for the high strain rate superplasticity of NC materials.

At this stage of superplastic deformation of nanocrystalline metals and alloys cooperative grain boundary sliding and grain rotation act. In nanocrystalline materials, the driving force for grain growth is very high, because of the large interfacial area per unit of volume. Consequently, significant grain growth can occur even at low superplastic temperatures. Since fine grains tend to grow rapidly at high temperatures, the microstructural design of superplastic materials usually requires a means for limiting grain growth. This necessary microstructural design is accomplished using microduplex alloys or quasi-single-phase alloys with fine grain boundary precipitates, where microstructural stability is provided by chemical variations across phase boundaries and Smith–Zener drag, respectively.

References

Chan KC, Wang CL, Zhang KF (2004) Low temperature and high strain rate superplasticity of Ni-1 mass%SiC nanocomposite. Mater Trans 45(8):2558–2563. https://doi.org/10.2320/matertrans.45.2558

Chokshi AH (2020) Grain boundary processes in strengthening, weakening, and superplasticity. Adv Eng Mater 22:1900748. https://doi.org/10.1002/adem.201900748

Chuvil'deev VN, Shchavleva AV, Nokhrin AV et al (2010) Influence of the grain size and structural state of grain boundaries on the parameter of low-temperature and high-rate superplasticity of nanocrystalline and microcrystalline alloys. Phys Solid State 52:1098–1106. https://doi.org/10.1134/S1063783410050422

Chuvil'deev VN, Shadrina IS, Nokhrin AV, Kopylov VI, Bobrov AA, Gryaznov MY, Shotin SV, Tabachkova NY, Chegurov MK, Melekhin NV (2020) An investigation of thermal stability of structure and mechanical properties of Al-0.5Mg–Sc ultrafine-grained aluminum alloys. J Alloys Compd 831:154805. https://doi.org/10.1016/j.jallcom.2020.154805

Demirtas M, Purcek G (2019) Room temperature superplasticity in fine/ultrafine grained materials subjected to severe plastic deformation. Mater Trans 7:1159–1167. https://doi.org/10.2320/matertrans.MF201922

Dong Y, Li Z, Sun J (2007) A model to explain extensive superplasticity in polycrystalline materials. J Mater Sci 42:7977–7980. https://doi.org/10.1007/s10853-007-1886-1

Du N, Qi Y, Krajewski PE et al (2011) The effect of solute atoms on aluminum grain boundary sliding at elevated temperature. Metall Mater Trans A 42:651–659. https://doi.org/10.1007/s11661-010-0326-z

Edalati K, Masuda T, Arita M, Furui M, Sauvage X, Horita Z, Valiev RZ (2017) Room-temperature superplasticity in an ultrafine-grained magnesium alloy. Sci Rep 7:2662. https://doi.org/10.1038/s41598-017-02846-2

Kawasaki M, Langdon TG (2018) Superplasticity in ultrafine-grained materials. Rev Adv Mater Sci 54:46–55

Kawasaki M, Figueiredo RB, Xu C et al (2007) Developing superplastic ductilities in ultrafine-grained metals. Metall Mater Trans A 38:1891–1898. https://doi.org/10.1007/s11661-006-9000-x

Mara NA, Sergueeva AV, Mara TD, McFadden SX, Mukherjee AK (2007) Superplasticity and cooperative grain boundary sliding in nanocrystalline Ni_3Al. Mater Sci Eng A463:238–244. https://doi.org/10.1016/j.msea.2006.08.123

McFadden SX, Valiev RZ, Mukherjee AK (2001) Superplasticity in nanocrystalline Ni_3Al. Mater Sci Eng A319–321:849–853. https://doi.org/10.1016/S0921-5093(01)01098-X

Meyers MA, Mishra A, Benson DJ (2006) Mechanical properties of nanocrystalline materials. Prog Mater Sci 51:427–556. https://doi.org/10.1016/j.pmatsci.2005.08.003

Mishra RS, Stolyaroov VV, Echer C, Valiev RZ, Mukherjee AK (2001) Mechanical behavior and superplasticity of a severe plastic deformation processed nanocrystalline Ti–6Al–4V alloy. Mater Sci Eng A298:44–50. https://doi.org/10.1016/S0921-5093(00)01338-1

Ovidko IA (2005) Superplasticity and ductility of superstrong nanomaterials. Rev Adv Mater Sci 10:89–104

Ovidko IA, Sheinerman AG (2005) Suppression of nanocrack generation in nanocrystalline materials under superplastic deformation. Acta Mater 53:1347–1359. https://doi.org/10.1016/j.actamat.2004.11.026

Padmanabhan KA (2009) Grain boundary sliding controlled flow and its relevance to superplasticity in metals, alloys, ceramics and intermetallics and strain-rate dependent flow in nanostructured materials. J Mater Sci 44:2226–2238. https://doi.org/10.1007/s10853-008-3076-1

Padmanabhan KA, Prabu SB, Mulyukov RR, Nazarov A, Imayev RM, Chowdhury SG (2018a) Mechanics of superplastic deformation and assessment of superplastic behavior. In: Superplasticity. Engineering materials. Springer, Berlin. https://doi.org/10.1007/978-3-642-31957-0_2

Padmanabhan KA, Prabu SB, Mulyukov RR, Nazarov A, Imayev RM, Chowdhury SG (2018b) Theories of superplasticity. In: Superplasticity. Engineering materials. Springer, Berlin. https://doi.org/10.1007/978-3-642-31957-0_8

Padmanabhan KA, Prabu SB, Mulyukov RR, Nazarov A, Imayev RM, Chowdhury SG (2018c) Structural superplasticity in relatively higher melting temperature materials—experimental. In: Superplasticity. Engineering materials. Springer, Berlin. https://doi.org/10.1007/978-3-642-31957-0_4

Pereira PHR, Huang Y, Langdon TG (2017) Examining the thermal stability of an Al-Mg-Sc alloy processed by high-pressure torsion. Mater Res 20(S1):39–45. https://doi.org/10.1590/1980-5373-MR-2017-0207

Prasad MJNV, Chokshi AH (2010) Superplasticity in electrodeposited nanocrystalline nickel. Acta Mater 58:5724–5736. https://doi.org/10.1016/j.actamat.2010.06.047

Prasad MJNV, Chokshi AH (2011) Microstructural stability and superplasticity in an electrodeposited nanocrystalline Ni–P alloy. Acta Mater 59:4055–4067. https://doi.org/10.1016/j.actamat.2011.03.029

Sauvage X, Wilde G, Divinsky S, Horita Z, Valiev RZ (2012) Grain boundaries in ultrafine grained materials processed by severe plastic deformation and related phenomena. Mater Sci Eng A540:1–12. https://doi.org/10.1016/j.msea.2012.01.080

Sergueeva AV, Mukherjee AH (2006) Crystalline plasticity of nanocrystalline materials at elevated temperatures. Rev Adv Mater Sci 13:1–5

Smirnov BI, Shpeizman VV, Nikolaev VI (2005) High strength and superplasticity of nanocrystalline materials. Phys Solid State 47:840–844. https://doi.org/10.1134/1.1924842

Sripathi S, Padmanabhan KA (2014) On the experimental validation of a mesoscopic grain boundary sliding-controlled flow model for structural superplasticity. J Mater Sci 49:199–210. https://doi.org/10.1007/s10853-013-7693-y

Turba K, Malek P, Cieslar M (2007) Superplasticity in a Zr and Sc modified AA7075 aluminium alloy produced by ECAP. Kokove Mater 45:165–170

Valiev RZ (2000) Superplastic behaviour of micro- and nanograined materials. In: Lépinoux J, Mazière D, Pontikis V, Saada G (eds) Multiscale phenomena in plasticity: from experiments to phenomenology, modelling and materials engineering, NATO science series (Series E: Applied sciences), vol 367. Springer, Dordrecht. https://doi.org/10.1007/978-94-011-4048-5_38

Wang GF, Chan KC, Zhang KF (2006) Low temperature superplasticity of nanocrystalline electrodeposited Ni–Co alloy. Scripta Mater 54:765–770. https://doi.org/10.1016/j.scriptamat.2005.11.024

Wang X-G, Li Q-S, Wu R-R, Zhang X-Y, Ma L (2018) A review on superplastic formation behavior of Al alloys. Adv Mater Sci Eng. https://doi.org/10.1155/2018/7606140

Yu M, Fang Q, Feng H et al (2014) Effect of cooperative grain boundary sliding and migration on dislocation emitting from a semi-elliptical blunt crack tip in nanocrystalline solids. Acta Mech 225:2005–2019. https://doi.org/10.1007/s00707-013-1039-3

Chapter 8
Mechanical Properties of Thin Films and Coatings

8.1 Introduction

Thin films of one or more materials deposited onto substrates, on other thin films, or on their own can have properties which their thicker, micro-grained counterparts could never achieve. Their structure and functionality can be catered by changing the processing methodology and technique by which they are made. This allows thin films to attain a variety of useful properties which can in turn be used in a variety of applications.

Thin films are applied in many industries such as nano/microelectromechanical systems (NEMS/MEMS) (Zorman 2017), microelectronics, and optics (Spengen et al. 2007). One of the main issues related to the development of these devices is the material degradation and failure, especially related to the mechanisms developing in a system where the grain size is comparable to the film thickness. Metal thin films are usually fabricated by employing chemical or physical methods such as electrodeposition, chemical vapor deposition (CVD), plasma-enhanced CVD, atomic layer deposition (ALD), electron beam evaporation, or magnetron sputtering (Dobrzański et al. 2015).

Sputtering is a physical vapor deposition technique in which highly accelerated ions and electrons are made to strike a solid source. This bombardment causes atoms from the solid source to be ejected into the gas phase. Once in the gas phase, atoms are free to interact and settle on surfaces they come in contact with. This method can be precisely controlled by tailoring the amount of material ejected from the source (the sputter yield) by varying the source material, bombarding ion energies and masses and source temperature. Because of the kinetics of particle-particle interaction, virtually any material can be used as a sputtering source. Microstructure can be controlled through the use of masks, movement of the source or target, and energies related to the deposition impact. This technique is operable at a range of temperatures, creates uniform particle distribution over relatively large areas, and is comparatively cost effective.

© Springer Nature Switzerland AG 2021

P. Cavaliere, *Fatigue and Fracture of Nanostructured Materials*,
https://doi.org/10.1007/978-3-030-58088-9_8

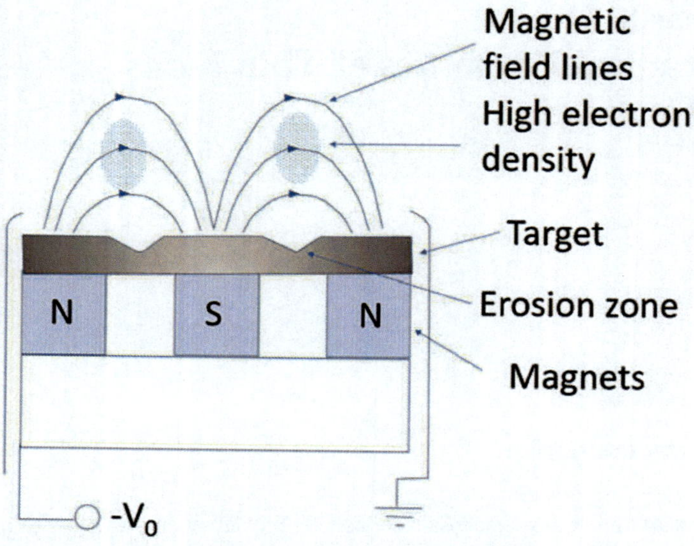

Fig. 8.1 Magnetron sputtering schematic

Magnetron sputtering is probably the most widely used variant of DC sputtering (Musil 2006). Some of the advantages of magnetron sputtering is one or two orders of magnitude higher ion current (i.e., higher deposition rates) and reduced operating pressure (i.e., higher energy of deposited atoms) compared to simple DC sputtering. In DC magnetron sputtering, permanent magnets are arranged in an appropriate configuration behind the target plate (Fig. 8.1).

The magnetic field lines penetrate the target and form a closed path on its front surface. The parallel component of the magnetic field strength with respect to the target surface is typically a few hundred gauss measured on the target front surface. Electrons launched slightly off the target normal will initially execute a helical motion along the magnetic field emanating normal to the target. Encountering the region of the parallel component of the magnetic field, also denoted the magnetron component, the electrons are forced to drift in an orbit back to the target. By solving the equations of motion for the electrons one finds that the electrons follow cycloidal trajectories near the target along the space confined by the magnetic field lines. Thus, in the presence of the magnetic field, the secondary electrons make more ionizing collisions close to the target and thereby increase the flux of bombarding ions (i.e., higher ion current), resulting in a higher deposition rate. The significantly increased ionization efficiency makes it possible to reduce the operating pressure and still maintain a stable discharge and a reasonable high ion current. Typical magnetron sputtering pressure is 10^{-2} Pa, which results in a mean free path of the sputtered atoms in the range of 1–100 cm. Thus, compared to simple DC sputtering, the sputtered atoms collide less with the gas atoms, whereby the loss of sputtered atoms to the chamber walls is lowered (increasing the deposition rate). Furthermore, the sputtered atoms preserve most of their initial kinetic energy before hitting the substrate, which is of great importance to the resulting film microstructure. Besides the increased plasma density and a lower operating pressure, the magnetic field also prevents electrons emitted at the target from bombarding the substrate, thereby limiting high substrate heating effects. Generally, the deposition rate in a magnetron deposition system is proportional to the DC power dissipated in the magnetron. The significantly increased deposition rate attainable by magnetron sputtering (compared to simple DC sputtering) is often very desirable. For example, the impurity level in the deposited samples is lower (at a given background pressure) due to a higher flux of target atoms with respect to impurity atoms onto the substrate (Battaile and Hoyt 2005). Further, a high deposition rate makes magnetron sputtering attractive for industrial applications. However, a drawback of magnetron sputtering is the creation of an erosion crater on the target, denoted as the "race track," where the density of secondary electrons is highest due to the confinement by the magnetic field. This irregular erosion results in a typical target material utilization of only 20–30%. Recently, target utilizations of about 50% have been achieved by optimizing the shape of the magnetic field by use of profiled magnets.

8.1.1 Microstructural Evolution in Thin Films

The properties of the deposited thin films are in direct relation to their structure—this structure spans surface characteristics, to interior structure, to defects and bonding. These structural properties must be investigated in order to determine both the properties of the material system and the effectiveness of the deposition technique in achieving the desired results with repeatable precision (Pantano et al. 2012).

The microstructural evolution of the films is characterized by interphase formation. The interphases possess a certain disorder which governs the free energy of the system. The general relationship for the Gibbs free energy is

$$G = G_0 + \sum_i A_i \gamma_i$$

where G_0 is the bulk Gibbs energy, and A and γ are the interface area and the energy, respectively.

Taking into account also the contribution of defects leading to an increase in the energy

$$G = G_0 + \sum_i A_i \gamma_i + \sum_j G_{Sj}$$

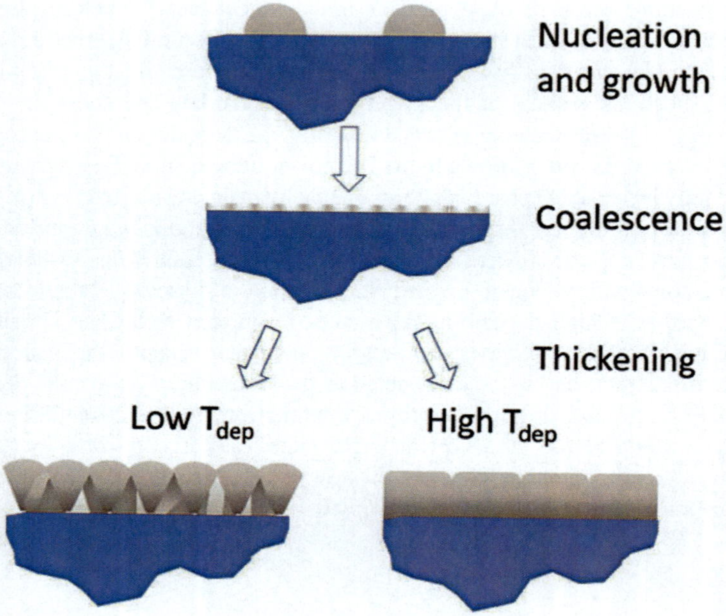

Fig. 8.2 Growth mechanisms during magnetron sputtering

The main mechanisms taking place during the deposition are nucleation and growth of isolated islands, island coalescence, and formation of the continuous film (thickening); the schematic is shown in Fig. 8.2.

The way the film structure evolves during the various processes is of course very dependent on the exact processing conditions: choice of materials, deposition rate, deposition and annealing temperatures, and ion bombardment (Frolov et al. 2000). They influence the diffusion and consequently the thin-film structure. As a matter of fact, nucleation rate increases as the deposition rate increases and temperature decreases. In order to obtain a nanocrystalline microstructure, high nucleation rate is required. Once energy minimization conditions are satisfied, growth occurs. After reaching a certain dimension, islands start to impinge on each other. When two islands come in touch, they start to form the grain boundary. If the diffusivity is low (low temperature), the formed grain size is stable. When a continuous film is formed, its thickness increases as the deposition continues. The new deposited atoms can form new grains (if the thermodynamic of the system allows this) or contribute to epitaxial growing (Ievlev 2004). Grain growth occurs in polycrystalline materials to decrease the interfacial energy and hence the total energy of the system.

8.1.2 Grain Evolution

Since nanocrystalline materials have a highly disordered large interfacial component, the driving force for grain growth in nanocrystalline materials is expected to be especially high (Ovid'ko 2000). As the unique properties of nanocrystalline materials are derived from their fine grain sizes, grain growth, which may deteriorate the material, is a crucial aspect of the thermal stability of nanocrystalline materials (Hultman and Mitterer 2006). Grain growth occurs through grain boundary migration (Gutkin and Dynkin 2012). Its driving force is due to the pressure difference formed by a curved grain boundary, which is proportional to γ_{gb}/r, where γ_{gb} is the grain boundary energy and r is the radius of curvature of the grain (proportional to the grain size). The rate of grain boundary migration is

$$\frac{dD}{dt} = k\frac{\gamma_{gb}}{r}$$

The grain size dependence on the temperature is given by

$$D^2 - D_0^2 = K(T)t$$

where D_0 is the initial mean grain size, and the rate constant $K(T)$ is

$$K = K_0 \exp\left(-\frac{Q}{k_b T}\right)$$

The previous equations refer to the so-called normal grain growth. In many cases the dependence is given by

$$D^{\frac{1}{n}} - D_0^{\frac{1}{n}} = K_1(T)t$$

For nanocrystalline materials $n < 0.5$. Taking into account the pinning effect of defects

$$\frac{D_0 - D}{D_m} + \ln\frac{D_m - D_0}{D_m - D} = K_2(T)$$

where D_m is the maximum grain size resulting due to the pinning forces. Resistance to grain growth in nanocrystalline materials results from insufficient driving force due to structural factors such as narrow grain size distribution, equiaxed grain morphology, low-energy grain boundary structures, and relatively flat grain boundary configurations (Suryanarayana and Koch 2000).

8.2 Mechanical Properties of Thin Films

The modification of the surface properties of a material can be achieved by the application of a thin film on the surface of an existing material. This application of thin films opens up a world of possibility for the creation of materials having specific and controlled surface properties while having different properties for the interior of the material. Mechanical properties form the last major group of properties which have important applications in thin films. The hardness of a material can be modified by the deposition of a thin film having a different surface hardness (i.e., harder or softer) than the material on which it is deposited. This process is commonplace in tools which are coated with hard materials in order to increase wear and abrasion resistance associated with their use. Another way to alter the mechanical property of a material is by the coating of material having different adhesion or friction properties.

8.2.1 Thin-Film Strength

Failure properties of NC metals, including thin films, are size dependent and cannot be extrapolated from bulk counterparts and are still an open challenge (Pelleg 2013). Therefore, it is of great importance to understand failure behavior of NC thin films including the sources of plasticity in nano-sized grains. The properties of thin films differ significantly from those of their bulk counterparts due to the dimensional constraints imposed by the film thickness and/or grain size. With decreasing film thickness, the dislocation motion is more pronouncedly confined which leads to an increase of the film strength. Furthermore, when the film thickness and/or grain size decreases down to the nanoscale, grain boundary (GB) instability under cyclic loading is frequently observed.

Very thin films have a great technological importance in the modern industry in addition to conventional bulk nanostructured materials (Antolovich and Armstrong 2014). In the analyses of thin films, the size scale is fundamental because of the microstructural features involved with their deformation (Fig. 8.3) (Aliofkhazraei 2011).

The experimental evidence for this behavior can be shown for NC Cu thin films (Fig. 8.4).

The unique properties of NC FCC metals clearly suggest a departure in plastic deformation mechanisms from the conventional transgranular dislocation processes in CG metals. As the grain size becomes smaller the dislocation-mediated processes become increasingly more difficult and grain boundary-assisted mechanisms become increasingly more important. Limited available sources for dislocation processes lead to activation of alternative deformation mechanisms including grain boundary (GB) sliding, grain rotation, grain growth, and twinning (Han et al. 2014).

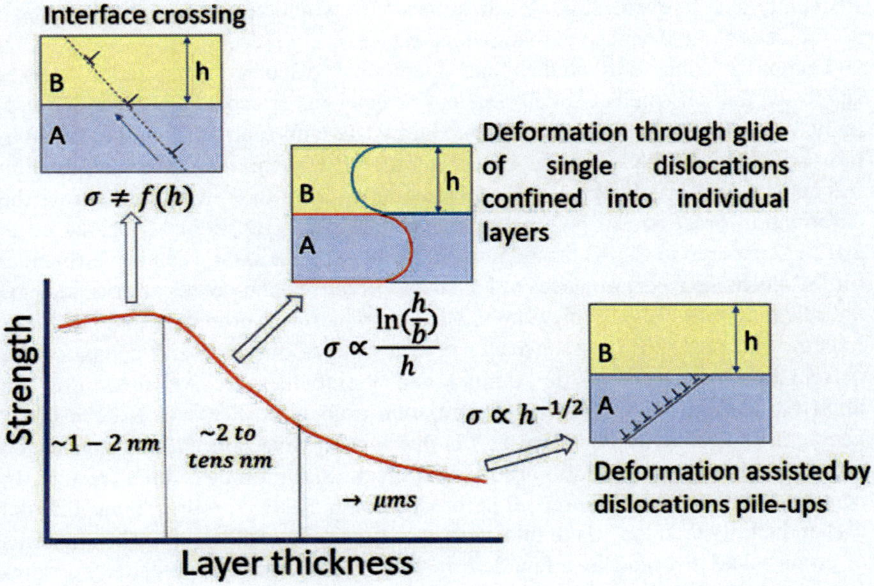

Fig. 8.3 Effect of size scale on the deformation of thin films

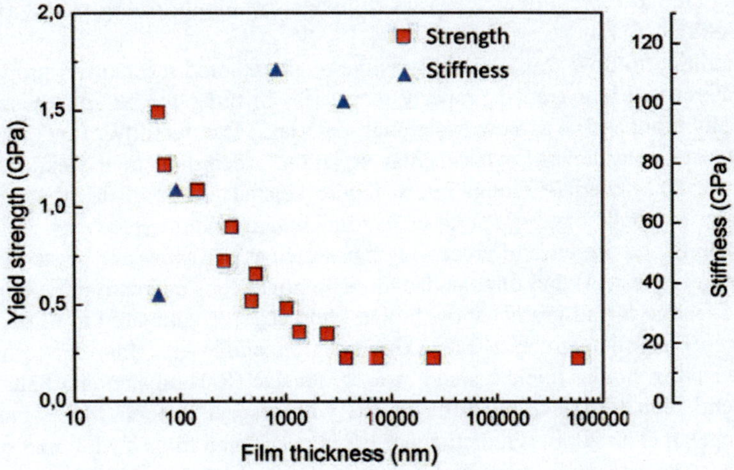

Fig. 8.4 Strength and stiffness vs. thin-film thickness for NC Cu

Cyclic deformation behaviors have been found to be greatly affected by the microstructure instability. Evident effects of cyclic softening during cyclic deformation are related to dynamic grain growth. The abnormally coarsened grains frequently lead to cyclic strain localization which would accelerate fatigue damage accumulation and decrease the fatigue resistance.

Computer simulation studies and theoretical calculation prediction studies suggested that GB-mediated deformation mechanisms are able to perform individually, or alternatively cooperate and accommodate with each other in the deformation processes of NC metals. CGBS means that more than one deformation mechanisms are involving in and accommodating with the GB sliding during the deformation processes in NC metals rather than pure GB sliding (Argibay et al. 2017). Compared to CGBS mechanism, frequent pure GB sliding can be detrimental for the mechanical performances of NC metals because nano-cracks always nucleate at triple junctions (TJs) during pure GB sliding which will promote rapid NC metal fracture (He et al. 2017). As a matter of fact, studies observed and compared the deformation behaviors of the thinner and the thicker NC-Au films (30 nm thick vs. 100 nm thick). For the deformation processes of 30 nm NC-Au films, nano-cracks frequently nucleated at TJs due to excessive pure GB sliding, which rapidly evolved into some sub-propagating cracks and promoted rapid fracture. In contrast, multi-slips were observed in the 100 nm thick NC-Au films, indicating the higher ductility than the 30 nm thickness ones (Espinosa and Prorok 2003). In terms of proposed CGBS models, a few deformation mechanisms can potentially accommodate with GB sliding. With these accommodated mechanisms, fewer nano-cracks would nucleate during the GB sliding processes. The possible cooperation between GB sliding and other mechanisms is schematically shown in Fig. 8.5.

Grain rotation, grain coalescence, stress-driven GB migration, and intergranular shearing are most likely to create grain alignment condition for GB sliding (Gutkin et al. 2008).

In addition to flow stress, ductility is the most explored mechanical property of NC materials. It is a crucial property to possess in order for NC materials to be practically competitive as new functional materials. The ductility of NC metals is subjected to considerable experimental variation. According to the experimental results for NC materials, elongation to failure depends on geometrical constraints (thin film vs. bulk material), microstructural heterogeneities (such as grain size distribution), and presence of processing flaws such as porosities and contamination. As a matter of fact, Al thin films with a more homogeneous microstructure (identical Schmid factors for all grains) show higher yield stresses compared to films with a heterogeneous microstructure but exhibit significantly less Bauschinger effect, despite having similar thickness and mean grain size (Rajagopalan and Saif 2011). In general, thin films exhibit different BE with respect to their MC counterpart (Rajagopalan et al. 2008). Predictions of BE in passivated films have found support in various experimental studies that have revealed early yielding in thin metal films on substrates during thermomechanical cycling. Early yielding in these passivated films has normally been attributed to the presence of stored dislocation energy, which assists in reverse plastic deformation during unloading. Energy gets stored during the forward deformation because the dislocations are prevented from exiting the film by the passivation layer, resulting in dislocation pileups or misfit dislocation segments being deposited at the film/passivation layer interface. In the absence of a

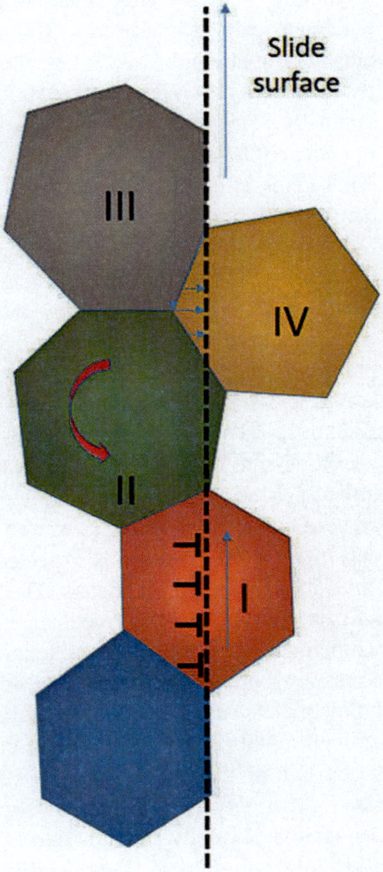

Fig. 8.5 Schematic illustrations of the proposed CGBS model, GB sliding could be accommodated by several deformation mechanisms. Grain I: intergranular shearing. Grain II: grain rotation and grain migration. Grains III and IV: pure grain migration

passivation layer, dislocations are free to exit the film and hence it is accepted that unpassivated films should not show early yielding. Even recent experiments that provided direct evidence of BE in passivated thin metal films did not reveal any BE in similar unpassivated films. Unpassivated free-standing metal films, but with smaller thicknesses and grain sizes compared to films examined in the above studies, exhibit a distinct BE during unloading. The mechanism responsible for BE in unpassivated films seems vastly different from that in passivated films. In passivated films, blockage of dislocations by the passivation layer leads to BE. In unpassivated films the selective plastic relaxation of larger/favorably oriented grains, coupled with elastic accommodation in smaller/unfavorably oriented grains, appears to be the cause.

Plastic instability is also another major limiting factor, leading to localized shear band formation due to limited strain hardening capacity. However, introduction of a high density of nanoscale growth twins can significantly enhance the strain hardening capacity (Zhang et al. 2008).

In order to increase the linear range of operation of thin film-based devices, films with nanoscale grain size ($d < 100$ nm) have been employed which have resulted in outstanding strength. Moreover, the need for planar and near-stress-free thin films resulted in layer-by-layer deposition processes that reduce the film grain size dramatically. The reduced grain size leads to increased yield strength due to grain boundary (GB) strengthening; that is, GBs act as obstacles to dislocation motion (Hall-Petch effect) but the increased volume fraction of GBs and triple points in the nanocrystalline face-centered cubic (FCC) metals can lead to enhanced diffusion-based creep and relaxation processes that finally reduce the material yield strength for grain sizes smaller than ~20 nm. The beneficial increased yield strength due to nanocrystallinity is eclipsed by increased strain rate sensitivity and creep rates compared to films with larger grain sizes. Nanocrystalline metals have large volume fractions occupied by GBs that control dislocation nucleation, pinning and annihilation, and GB-mediated creep while affecting the relative contribution of thermal and stress-driven inelastic processes. These often competing mechanisms result in increased rate sensitivity which is not uniform across time scales: diffusion-controlled processes are important at slow-applied strain rates and may dominate in the case of very small grain sizes. The competition between GB and intragranular deformation mechanisms is expected to control rate sensitivity and activation volumes in nanocrystalline metals. The dominant (rate controlling) deformation mechanisms in nanocrystalline metals are strongly dependent on the state of material, the temperature, and the applied loading rate. It is also noteworthy that the rate-controlling physical processes behind dislocation-mediated plasticity in nanocrystalline metals subjected to high strain rates could be different from those for coarse-grained metals. It was suggested that the dislocation cell size (average dislocation spacing) lmin calculated from the Taylor relation becomes comparable to the grain size of nanocrystalline metals. Thus, it is unlikely that forest hardening-based dislocation plasticity seen in coarse-grained FCC metals could occur in their nanocrystalline counterparts. On the other hand, a number of studies have also suggested that nucleation and annihilation of dislocations can take place at GB ledges, triple junction points, and other GB imperfections which may also act as stress raisers. Thus, nucleation and annihilation of GB dislocations could become the rate-controlling processes in nanocrystalline metals. It has also been suggested that GB dislocation emission is a thermally activated mechanism. Dislocations emitted at a GB could then be pinned at other GB locations and the ensuing thermally activated depinning of dislocations could become a rate-controlling process.

Many scientific evidences show that thin films exhibit uniform necking during deformation (Fig. 8.6).

The large stresses result in grain boundary decohesion and nano-void formation which coalesce with the main crack (Bufford et al. 2016), leading to intergranular fracture (in contrast to the typical transgranular cracking in coarse-grained metals).

The tensile fracture is governed by void coalescence and propagation (Fig. 8.7).

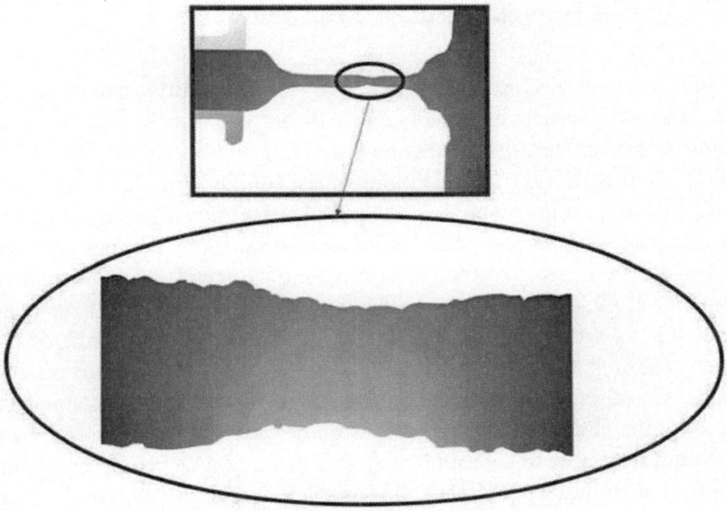

Fig. 8.6 Uniform necking in Al thin film with 200 nm thickness

Fig. 8.7 Fracture of Al thin film 200 nm thickness

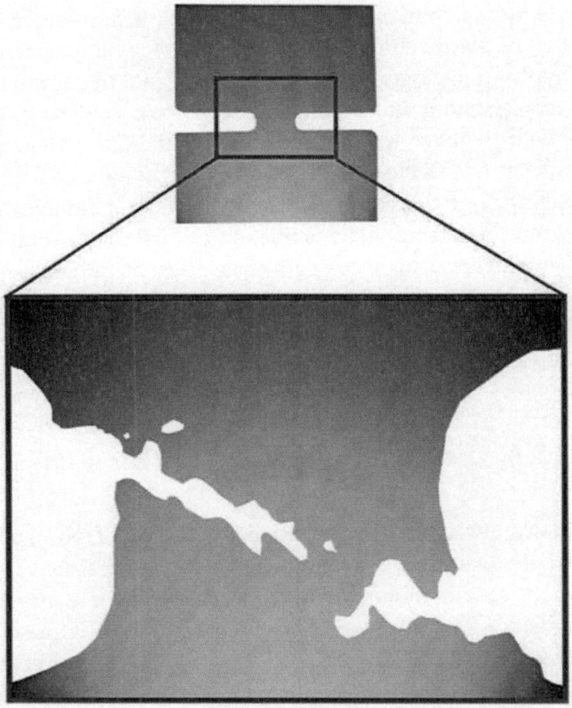

8.3 Fatigue of Nanostructured Thin Films

It was specified how mechanical properties of thin films differ from those of bulk materials. These differences are mainly due to the large surface-to-volume ratio since the microstructure of the surface will have fundamental influence on the thin-film behavior (Larsen et al. 2003). The continuing trend of miniaturizing materials has led to a strong demand for understanding the complex fatigue properties of thin films at small length scales (including grain size *d* and film thickness *h*) both for scientific aspects and in the interest of industrial reliability. Recently, a number of typical studies of length-scale effects on mechanical fatigue behaviors have been conducted in metallic thin films with *h* or *d* range spanning from microns to submicrons. All the experimental results show that, since the geometric and microstructural characteristic dimensions of the materials are in the range of microns to nanometers, the constraints of the characteristic dimensions on dislocation activities and the effects of surface and interfaces in the thin films result in the fatigue behaviors of thin films different from their bulk counterparts. At large length scales (*d* and/or $h > 1$ μm), the fatigue deformation is accommodated by extended dislocation structures (accompanied with extrusions/intrusions) and shows weak length scale-dependent crack nucleation and propagation. At submicron length scales, which are too small for the formation of extended dislocation structures and inhibit the localized accumulation of plastic strain within grains, the fatigue processes are presumably controlled by the constraints on motion of individual dislocations and the interface-mediated damage and grain boundary (GB) process become more prevalent. The changes in fatigue damage with length scales (*d* and/or *h*) suggest that the fatigue mechanism transits from dislocation-mediated extrusion formation to crack formation-controlled behavior, and it can be attributed to the inhibition of dislocation mobility and the limited availability and activation of dislocation sources on the small length scale. Thus there is a distinct length-scale effect on fatigue damage morphology and fatigue lifetime, which reflects the change of length scale-related fatigue mechanisms (Zhang et al. 2011).

8.3.1 Fatigue Mechanisms in Thin Films

Many attractive abovementioned features of NC metals, such as high strength, would likely be unavailing if the fatigue resistance of these materials does not meet certain minimum acceptable levels for particular demands. Fatigue crack initiation in conventional coarse grain pure FCC metals is triggered by extrusion-mediated surface roughness. This roughness occurs at the intersection of free surface and persistent slip bands (PSBs) and is a direct consequence of cyclic microplasticity. In the saturation stage of the cyclic loading, strain is highly localized near PSBs. Ultimately cyclic irreversibilities along PSBs trigger the formation of protrusions. The corners of protrusions, with irreversible slip-unslip, are the preferential sites for fatigue crack initiation. This sequence of events is regarded as conventional for fatigue crack nucleation in pure monocrystalline and coarse grain FCC metals. One of the most comprehensive theories of surface relief formation is based on the hypothesis that the origin of irreversibility for the motion of PSBs is dislocation pair annihilation and generation of a vast number of vacancies within PSBs. The PSB-based models are unlikely to apply for NC metallic thin films, as the typical micron-size PSBs are too large to form within nanograins. This is only true when the structure of the grains is stable; otherwise, cyclic stress-assisted grain growth could lead to conventional fatigue mechanisms by PSB formation in coarsened grains. Due to high strain rate sensitivity, NC metals can also creep at room temperature and therefore it is difficult to distinguish time-dependent degradation from cycle-dependent deformation.

Owing to small dimensions of the materials and the constraint of the substrate, new experimental methods have been developed to evaluate fatigue properties accurately by considering three factors as (1) ultralow load cell, high-resolution displacement sensor, and a data acquisition system; (2) reliable clamping without damaging the samples; and (3) accurate method to determine the fatigue failure point because films generally fracture before the substrate (van Spengen et al. 2017). To overcome the difficulty in the determination of the fatigue failure point, several methods were developed based on change in the physical or mechanical properties of the fatigued samples, including electrical resistance method, mechanical energy loss method, continuous stiffness measurement, and direct monitoring of fatigue damage formation (Luo et al. 2019). In general, mechanical fatigue testing methods can be roughly classified into uniaxial cyclic tensile loading, dynamic bending mode, and resonance loading. Films are deposited onto those flexible substrates with dog-bone or rectangle shapes. Fatigue life is determined by the electrical resistance method, the mechanical energy loss method, or the method to directly monitor fatigue damage formation. The uniaxial tensile cyclic tests are generally applicable for thin films on flexible substrates.

Traditionally, the physical origin of the initiation of fatigue damage in bulk FCC materials is intimately related to cyclic strain localization in typical persistent slip bands (PSBs), leading to the formation of surface extrusions and intrusions, and crack initiation along these extrusion/intrusions. The PSBs with a so-called ladder or wall structure consist of dislocation walls with dense multipolar bundles and screw dislocations in the channel. PSB wall spacing is around 1.3 μm and the wall thickness ranges from 30 to 250 nm. When decreasing the characteristic length of the material below the critical size, materials would show obvious length scale-dependent fatigue properties, indicating different fatigue damage mechanisms with scaling down the length scales. Above the film thickness and the grain size of around 3 μm, typical bulk-like fatigue damage behavior was observed, in that self-organized dislocation structures (such as dislocation cell and PSB walls) formed within the grains and extrusions formed at the surfaces (Fig. 8.8).

As the film thickness or the grain size decreases to 1–3 μm, extrusions become smaller and rarer. Diffuse, cell-like dislocation structures and small groups of tangled dislocations are observed. At even smaller film thicknesses and grain sizes

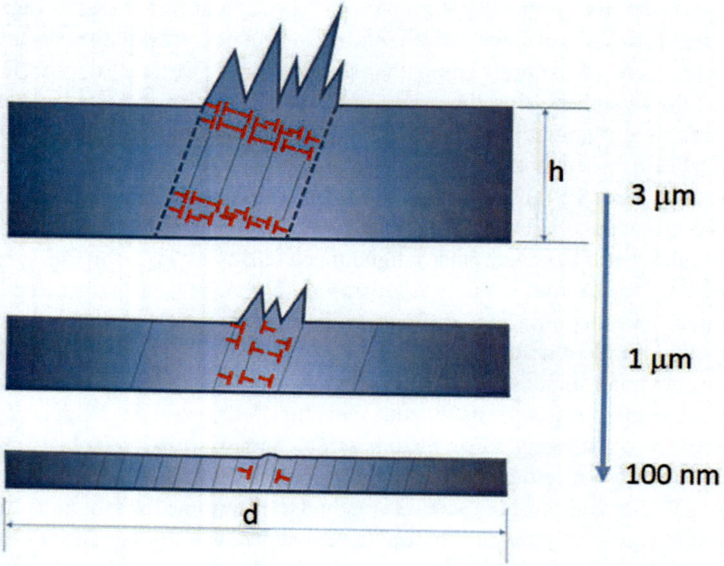

Fig. 8.8 Fatigue extrusions at the surfaces of the fatigued thin films with different thickness

between 1 μm and 100 nm, only individual dislocations are observed, and extrusions are almost completely suppressed and replaced by cracks along GBs and TBs at the submicron scale. The most frequently observed and discussed types of TBs in face-centered cubic (FCC) metals include (1) the low-energy coherent TB (CTB) across which a mirror image of atoms can be seen and (2) the incoherent TB (ITB) with high boundary energy where there is a partial dislocation on each slip plane. Due to the low GB energy, the CTB is the most favorable TB during grain growth (Luo et al. 2017). The production of growth twins during the grain growth is to reduce boundary energy; so it is reasonable to have dominant generation of growth twins with the low-energy CTB rather than the high-energy ITB. Thus, the newly formed twins in the coarsened grains are apparently not growth twins; they are deformation twins instead. The high stress at the crack tip caused not only grain growth but also deformation twins.

8.3.2 *Size Effect*

The variation in extrusion dimensions with the film thickness and the grain size can be evaluated in terms of the extrusion widths:

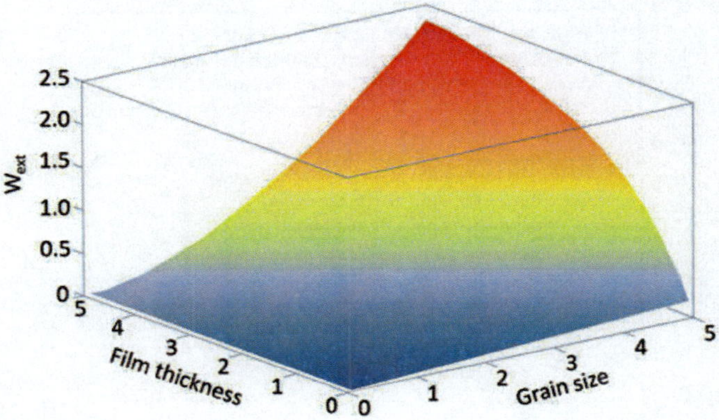

Fig. 8.9 Variation of fatigue extrusion width (Wext) as a function of film thickness and grain size

$$W_{\text{ext}} = \frac{\frac{{}_{\varepsilon}d}{Nb\sin\lambda} - 1}{2\ln\left(\frac{d}{b}\right)\left(\frac{2}{d} + \frac{\sin\phi}{h}\right)}$$

where ε_{pl} is the applied plastic strain, d is the grain size, h is the film thickness, N is the number of dislocations that have traversed the slip plane, b is the Burgers vector, and λ and ϕ are the angles of the slip direction and the slip plane normal with the out-of-plane direction of the film, respectively. There is a clear trend that the fatigue extrusion dimension gradually shrinks with decreasing film thickness or grain size (Fig. 8.9).

With further decreasing of the film thickness and grain size below 100 nm, the stress for full dislocation motion gradually exceeds that for partial dislocations; thus partial dislocation and twinning behaviors instead of the full dislocation take over in the nanoscale films. In addition, GBs usually become unstable and grain growth happens under fatigue loading. Luo and Zhang (2017) quantitatively investigated the effects of the film thickness on grain growth mechanisms and corresponding fatigue damage behaviors in the nanocrystalline Au films with a film thickness ranging from 20 to 930 nm. In thicker films (90 nm), abnormal grain growth happened and exhibited a bulk-like damage behavior. Fatigue cracks preferentially initiated in the abnormally grown grains with well-developed strain localization. When decreasing the film thickness to several nanometers, grain growth tended to be locally uniform. Fatigue damage in the thinner film was associated with GB-related behaviors, such as intergranular cracking, grain coarsening, as well as deformation twinning. They found that the formation of nanotwins was an effective way to assist limited grain coarsening following a fundamental process that the mutual formation of nanotwins in two neighboring grains changes the local grain orientation and dissociates the GBs into new segments, which become more mobile. Such limited grain growth did contribute to certain plastic dissipation and the newly formed nanotwins played a role in the crack deflection, which effectively enhanced the

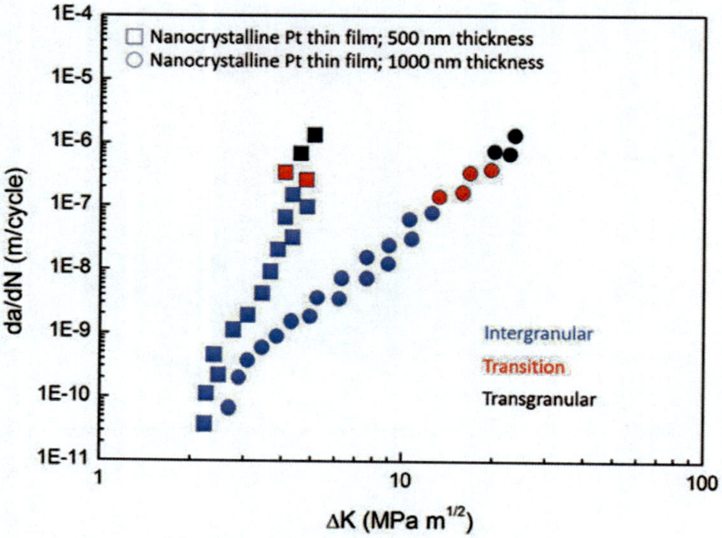

Fig. 8.10 Crack initiation and growth behavior of nanostructured Pt thin films

resistance to the fatigue crack growth. At the nanoscale, interface-dominant cracking behavior is also the predominant fatigue damage mechanism.

Recently, length scale-dependent fatigue mechanisms have been closely investigated. A noncommon type of fatigue deals primarily with scenarios exclusive to nanocrystalline or thin-film metals in which grains are no longer large enough to contain dislocations or surface layer-mediated crack initiation and growth is complicated by the increased surface-to-interior ratio in thin films. Obviously, only experimental testing of free-standing nanocrystalline metallic films can help shed light on the deformation mechanism.

Basically, the nanostructured thin-film thickness has a very noticeable effect on the fatigue behavior of materials (Fig. 8.10). Recent studies have shown that the thinner metal films exhibit longer fatigue lifetime than the thicker metal films (Wan et al. 2016). It is believed that the increase in fatigue lifetime with decreasing film thickness is associated with the increase in yield stress and the gradual transition of fatigue damage mode from typical fatigue extrusions/intrusions at the micrometer scale to the damage along the grain boundaries (GBs) at the submicron scale. As the length scale (grain size or film thickness) further decreases to the nanometer scale, GBs become so unstable that grain growth/coarsening always occurs under cyclic loading. In addition, partial dislocations or twinning behavior instead of full dislocations would be the dominant deformation mechanism, because the critical stress to nucleate a full dislocation will be larger than that for a partial dislocation or a twin when the length scale is below a critical scale.

In most polycrystalline metals there is a progressive loss of crack tip constraint as the monotonic and cyclic plastic zones increase in size, relative to the specimen thickness, and the fracture toughness is enhanced. In very thin sheets, however, this loss of constraint causes a reduction in fracture toughness relative to thicker material forms and the fracture surface in mode I fatigue loading changes from being flat to inclined through the thickness. Metallic thin films have low fracture toughness and inferior resistance to crack propagation when compared to conventional coarse-grained bulk form of the same metal (Buehler 2008). Atomistic simulations on NC Ni (grain sizes ranging from 5 to 10–12 nm) revealed that the amplified stresses ahead of the crack tip trigger intragranular dislocations that would otherwise not occur in such small grains.

As the film thickness or grain size is reduced, the fatigue life under a constant applied strain is increased. It was demonstrated that in copper films of less than 1 μm thickness, extrusions are absent and cracks initiate at boundaries and defects, indicating that the local accumulation of cyclic plastic strain becomes increasingly difficult with a decrease in film thickness. This change in the behavior of fatigue damage is attributed to the inhibition of dislocation mobility and the limited availability and activation of dislocation sources on thin films. Veins, slips, or dislocation cells do not form in films of less than 1 μm thickness during the course of fatigue. A decrease in thickness and grain size inhibits the localized accumulation of plastic strain within grains (such as at extrusions/intrusions and extended dislocation structures), and promotes the formation of cracks at twin and grain boundaries during fatigue. This effect is a likely cause of the increase in fatigue life (Hu et al. 2013).

In NC Cu thin films there is a significant effect of length scale on Nf: larger $\Delta\varepsilon$ is required to cause failure in thinner type I films except for the thinnest one. The applied $\Delta\varepsilon$ required for causing failure increases with decreasing h (h in the range of 100–700 nm). Typically, global or local plastic strains are the driving force for fatigue failure. Therefore, it is not surprising that an increase in $\Delta\varepsilon$ is required for causing failure since it corresponds to an increase in σ_y under monotonic loading. It indicates that the fatigue resistance in present Cu films is controlled by the constraint effect on the dislocation mobility. By scaling the characteristic dimensions, the fatigue damage transits from extrusion dominated to crack dominated, owing to the deformation mechanism transition from dislocation glide to grain boundary-mediated deformation.

8.3.3 Fracture Behavior

The extraordinarily thin films and their textured nanoscale grain morphologies mean that the usual assumptions, such as isotropy, are not applicable and trends in fracture toughness and fatigue susceptibility are often inverted from what is found in larger scale forms and morphologies. Pt thin films show a very different crack growth behavior as a function of the film thickness. In addition, the transition from intergranular to transgranular fracture mode is observed for both the films (Meirom et al. 2012). A central challenge to understanding how fatigue damage accumulates in nanograined films has been the limited stability of the structure of the films. In

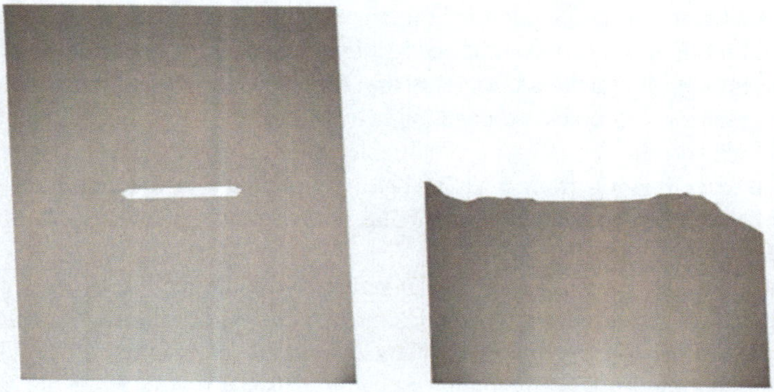

Fig. 8.11 Fracture profile of Pt thin film before and after fatigue tests

many metallic films (e.g., Ni, Cu, and Ag) the grains spontaneously coarsen ("self-annealing"), a phenomenon that can also be driven by stress (Meirom et al. 2011).

Micrographs of a notched tensile specimen of Pt during testing and immediately after failure (Fig. 8.11) show ductile tearing at the notch (flat portions on either side of the notch), fracture occurred via a shear mechanism that proceeded on surfaces inclined to the loading direction (45° slanted portions on either end of the flat portions).

It is perfectly in line with the performed MD simulations (Fig. 8.12).

Although a grain boundary (GB), as an effective barrier to dislocation motion, plays a key role in enhancing yield strength of polycrystalline metals through reducing grain size down to nanometer scales, GBs in nanograined metals usually become so unstable that grain growth characterized by GB migration/grain coarsening always occurs. Grain growth is most commonly associated with thermally activated GB migration, but grain growth in NC metals could be induced mechanically. Several mechanisms are currently correlated to mechanically induced grain growth such as grain rotation and agglomeration, GB migration, cooperative mechanism, and so on. It is suggested that the grain rotation is a significant factor only for very small grains and at very high temperatures, and the growth mechanism is associated with gradual GB dissociation caused by dislocation motion. For GB migration, the motion of a low-angle GB (LAGB) is generally related to the collective motion of the individual dislocations in these boundaries, while the migration of a high-angle GB (HAGB) is mainly described by the shuffling model, the DSC (displacement shift complete) model, and the shear coupling models. In the DSC model, the HAGB migration is attributed to the motion of the secondary GB dislocations, which result in a combined GB migration and sliding. Furthermore, a more generalized formulation was proposed to describe the shear-migration coupling of ordinary GBs with noncoincidence relationships and irrational habit planes. Upon plastic deformation of the NC FCC-structured metal, Shockley partial dislocations (PDs) are usually emitted from GBs, leading to frequent twin nucleation.

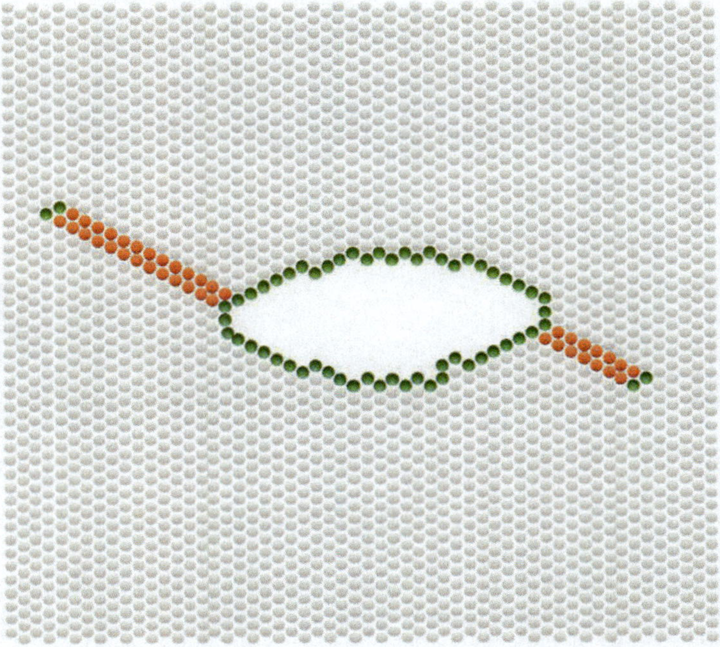

Fig. 8.12 MD simulation of crack propagating in NC thin film

Besides, molecular dynamics simulation and experimental observations suggest that twins can form by GB motion or dissociation.

Under mechanical loading, nanocrystalline metals show unique behaviour, among the most common of which are high strength, mechanically induced grain growth, and twin formation (Fig. 8.13).

The formation of nanotwins is an effective way to assist grain coarsening, following a fundamental process that the mutual formation of nanotwins in two neighboring grains changes the local grain orientation and dissociates the grain boundary into new segments, which become more mobile (Luo et al. 2013). The two nanograins before cyclic loading (Fig. 8.13a) start to show nanotwin formation leading to local dissociation of GB (Fig. 8.13b, c). Successive twin formation within two grains further splits the GB (Fig. 8.13d). Undissociated segments of GB move into the neighbor grain (Fig. 8.13e). Coupled with the GB motion, two grains finally coalesce into one grain with multiple twins (Fig. 8.13f).

Advanced material properties can be obtained by optimization of the amount and character of these interfaces in materials. To this end, materials exhibiting nanoscale twins, so-called nanotwinned materials, are shown schematically in Fig. 8.14.

Nanotwin acts as a barrier for dislocation motion and a new source for dislocation nucleation and interactions (Andrievski 2016). There is also a significant size dependence of mechanical properties on the twin lamellar spacing much like the grain size dependence of the strength in NC metals.

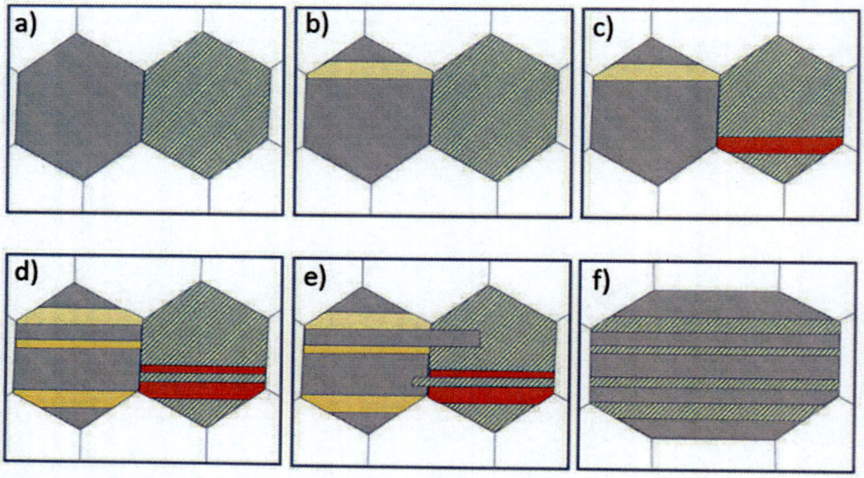

Fig. 8.13 Nanotwin-assisted grain growth in thin films

This special microstructure can be synthesized by phase transformation, by deformation, or by growth. Growth twins are typically produced by electro- or sputter deposition. Nanotwinned microstructures can exhibit ultrahigh strength. At the same time, nanotwinned metals can retain a certain degree of ductility and can show significant work hardening in contrast to nanocrystalline metals. Further, the fracture toughness and fatigue properties of nanotwinned materials could be improved by increasing the twin density and maintaining a constant grain size. For industrial application, not only the mechanical but also the thermal stability of these structures is crucial. Deformation-induced detwinning could dramatically reconfigure microstructure. Some nanocrystalline deformation phenomena, like stress-induced grain boundary migration, would lead to subtle changes in the boundary network that could nonetheless accumulate during cyclic loading. Unfavorably, distinct detwinning was observed during heat treatments in pure FCC metals (Shute et al. 2011).

In Ni-W and Ni-Mo thin films, the twin density increases with increasing solute contents (Fig. 8.15).

The decrease of stacking fault energy is accompanied with an increase of twin probability. For very low solute contents, the stacking fault energy for the Ni-W solid solutions decreases more rapidly with solute content than for the Ni-Mo solid solutions, but for larger solute contents it depends less than for the Ni-Mo solid solution on solute content. At the same time, the thermal stability of the twins increases with increasing solute content (Fig. 8.16).

The diffusion coefficient is smaller for the Ni-W solid solutions for all solute contents. Moreover, the diffusion coefficient for the Ni-W solid solutions decreases more pronouncedly with increasing solute content. The occurrence of W segregation during annealing is demonstrated. This segregation mostly occurs at grain

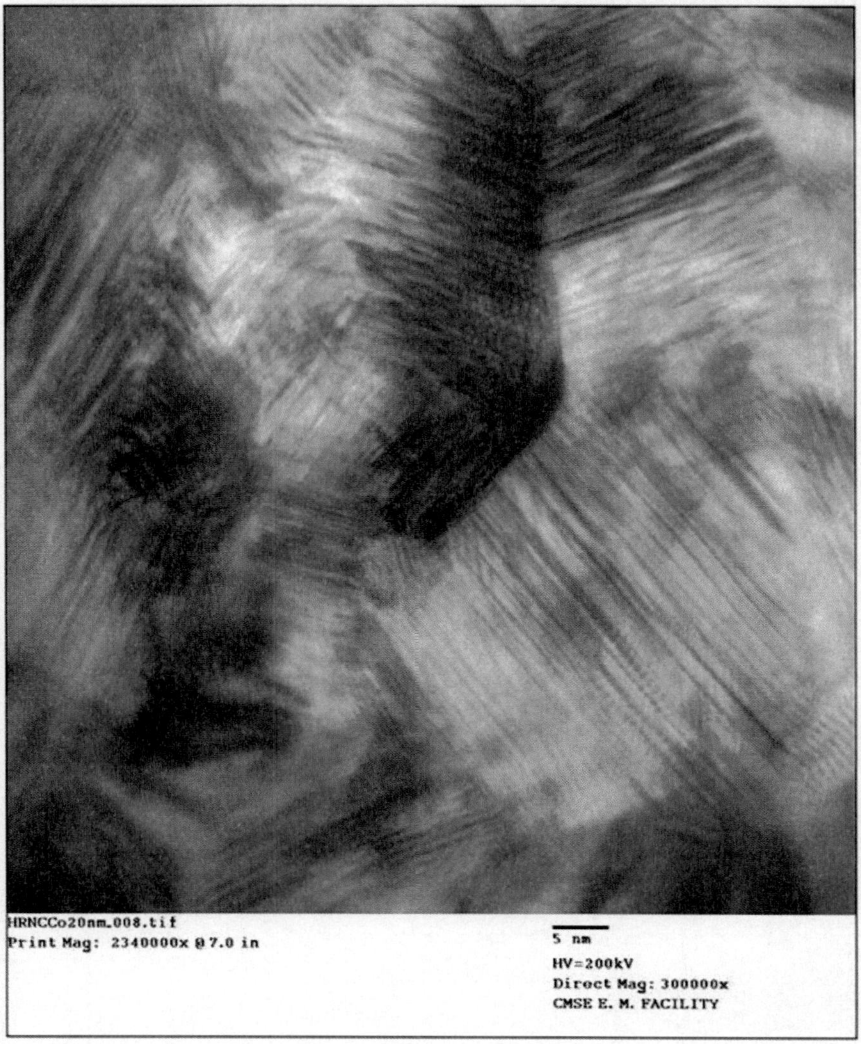

Fig. 8.14 Nanotwin formed in electrodeposited NC material

boundaries and twin faults and stabilizes the twin-faulted, nanocrystalline structure. Films with W contents <15% show distinct detwinning upon annealing.

Like the GBs, the coherent twin boundaries (TBs) are also effective in blocking the dislocation motion. Thus, creating nanoscale twins in micro- and nanocrystals can also improve the yield strength of alloys. For that, the development of nanotwinned nanocrystalline HEAs is attracting considerable attention (Huo et al. 2019). Compared to coarse-grained samples, the nanotwinned films showed remarkable fatigue resistance, demonstrated by history-independent and near-stable fatigue

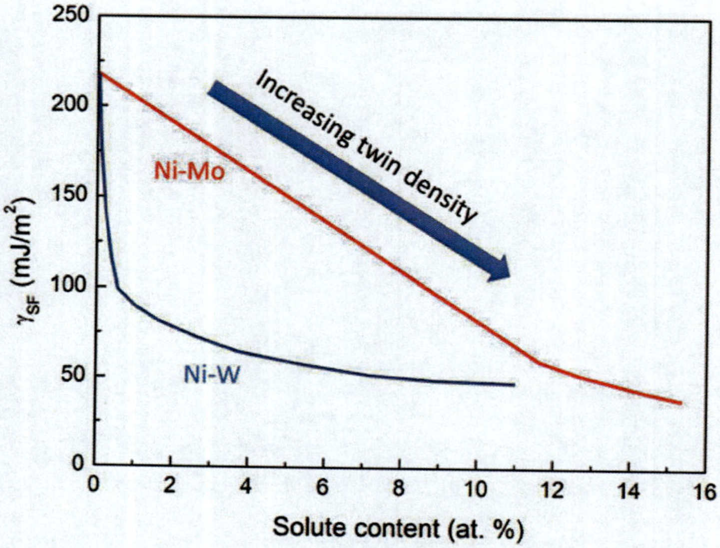

Fig. 8.15 Stacking fault energy and twin density variation with solute content in Ni-W and Ni-Mo thin films

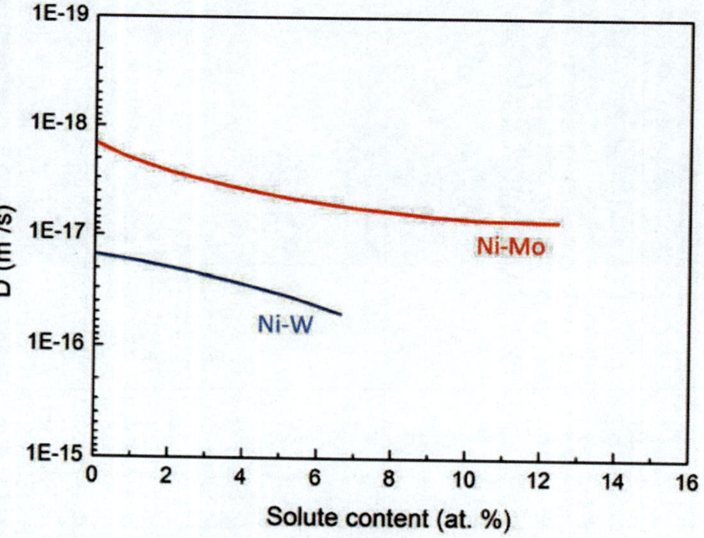

Fig. 8.16 Diffusion coefficient variation with solute content in Ni-W and Ni-Mo thin films

responses. Under the influence of the textures of the films, the nanotwin planes are oriented in parallel to the loading direction during each fatigue cycle. The extremely stable correlated necklace dislocations moved back and forth along the twin

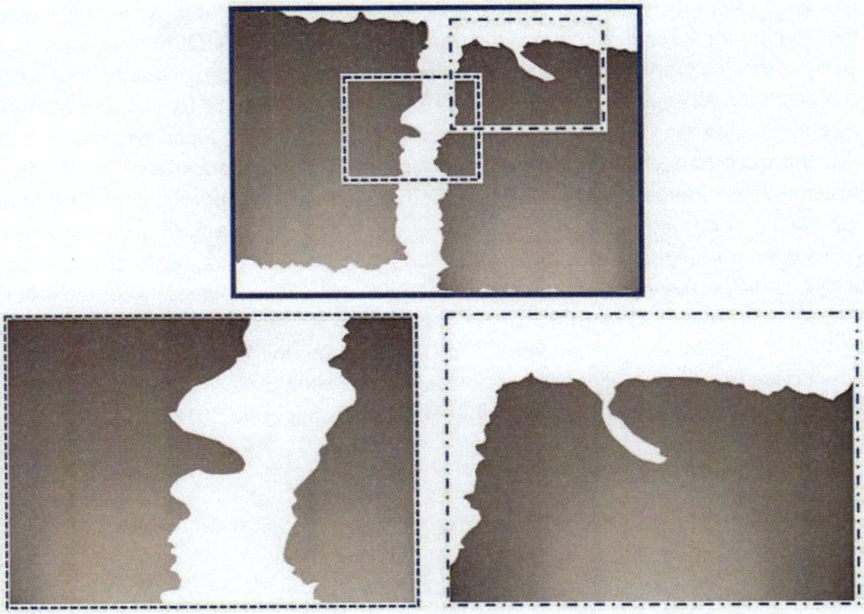

Fig. 8.17 Fracture surface of the ultrathin Au NC film

boundaries during cyclic tests, which preserved the slip system as well as the coherency and stability of twin boundaries.

In situ transmission electron microscopy provides dynamic observations of the physical behavior of materials in response to external stimuli such as temperature, gas environment, stress, and electric or magnetic fields (Ross and Minor 2019). In situ TEM fatigue tests of ultrathin Au nanocrystalline films showed enhanced ductility along the fracture surface (Fig. 8.17).

The 100 nm thick, 1.5 mm wide nanocrystalline Au specimens have a tensile strength of 1.2–1.3 GPa, and a total elongation to failure of 2.5%. The specimens exhibit ratcheting behavior under tension–tension cyclic loading in near-stress-controlled conditions. Nanoscale fatigue cracks were observed at nearly 7000 cycles for σ_{max} 0.95 GPa (Hossenian and Pierron 2013).

Adjusting GBs is a universal strategy to inhibit grain growth in NC metals. Simulation results have suggested a promising and simple approach by alloying NC metals with immiscible elements. Immiscible elements tend to segregate at GBs, reducing their free energy and thus the capillary driving force for grain growth. The segregated elements can also reduce GB mobility through the solute drag effect. In addition, immiscible elements often form precipitates that can additionally pin GBs. These effects are effective in preventing grain growth in NC metals, but alloying has only been found in a few empirically discovered systems (Zhang et al. 2014). GB segregation or GB doping should affect the mechanical properties and deformation behaviors of NC metals. Molecular dynamics simulations have revealed a significant

GB strengthening by solute segregation, in the case of solutes with larger size mismatch with the matrix atoms. One proposed mechanism for GB doping strengthening is that the activation of grain boundaries, such as GB sliding and GB shuffling, can be effectively suppressed when the GB energies are reduced by GB segregation. Accompanying the GB doping strengthening, the ductility of doped NC metals will also be affected significantly. As mentioned above, NC metals have insufficient dislocation accumulation capability. GB activities make an important contribution to the plastic deformation of NC metals. It is likely that the plastic deformation of a nanocrystalline metal will directly alter the grain boundary character distribution, which is in sharp contrast to conventional grain boundary engineering. In the latter, the applied strain serves primarily to accumulate strain energy that is useful for driving subsequent recrystallization that reconfigures the boundary network. In a nanocrystalline metal, it has been hypothesized that even modest plastic deformation can alter boundary character rather dramatically (Bober et al. 2018).

8.3.4 Grain Boundary Structure

When GB activities are suppressed by solute segregation, it seems that the ductility of NC metals will be further reduced.

The addition of low-solubility elements that segregate to grain boundaries (GB) can stabilize a nanocrystalline material against grain growth, suppress grain boundary-mediated plasticity, and improve the fatigue characteristics of the material. Zr has been found to strongly segregate to GBs in nanocrystalline Cu, and is effective at stabilizing it at elevated temperatures. This segregation has also been shown to lead to the formation of amorphous intergranular films under certain annealing conditions (Borovikov et al. 2019). Studies performed on ordered nanocrystalline- and AIF-containing Cu-Zr thin films showed that ordered grain boundary sample experienced grain growth and dislocation activity at the crack initiation site (Fig. 8.18).

The Cu-Zr alloy experiences very little grain growth, only coarsening to an average grain size of 45 nm due to a reduction in grain boundary energy with Zr segregation. The high-temperature annealing treatment used to induce Zr segregation is also useful for promoting AIF formation.

An exciting concept for the design and control of interfacial properties is complexions, defined as interfacial structures that are in thermodynamic equilibrium and have a stable, finite thickness (Rajgarhia et al. 2010). Similar to bulk phases, complexions can be described with thermodynamic parameters and can even undergo phase-like transitions in response to alterations of external variables such as temperature, pressure, and chemistry. The interfacial potential between the adjoining grains can be calculated as a function of the grain boundary thickness, composition, and interatomic distance assuming exponentially decaying interfacial forces. This results in an oscillating free energy versus complexion thickness relationship where energy minima occur when the complexion thickness is an

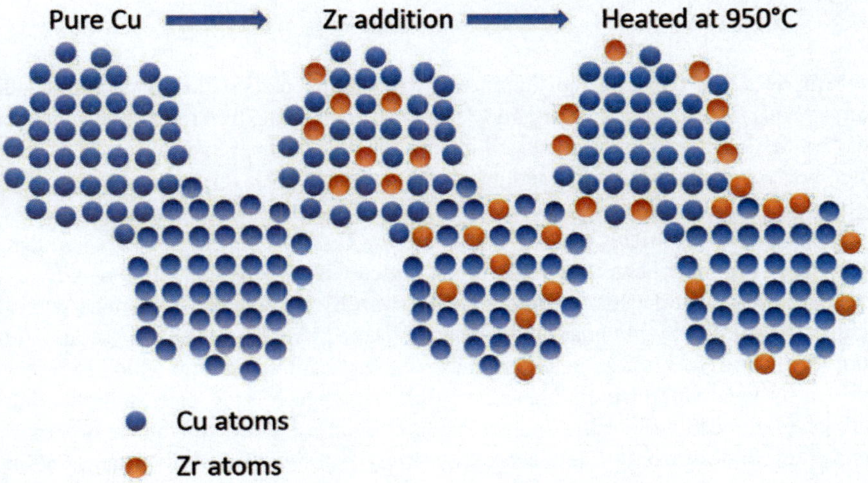

Fig. 8.18 Cu-Zr NC alloy evolution

integer number of bond lengths. This means that energetically favorable grain boundary structures with characteristic thicknesses can occur. Different complexion types have been shown to dramatically influence material behavior and have been deemed the root cause of several previously unexplained phenomena. By preferentially incorporating certain complexion types within the grain boundary network of nanocrystalline metals, dramatic changes to grain boundary structure and composition can be achieved, resulting in dramatically changed material behaviors. The different types are generally known as (1) clean grain boundaries (CGBs), (2) ordered grain boundary complexions (OGBCs), and (3) amorphous intergranular films (AIFs). CGBs have no alloying elements at the grain boundaries. In contrast, OGBCs and AIFs are formed because of solute segregation to grain boundaries (Turlo and Rupert 2018). Dislocation emission is easier in the OGBC and AIF samples than in the CGB specimen. The critical stress for dislocation emission demonstrates an inverse correlation with the excess free volume at the interface, meaning the increase in the free volume due to alloying element segregation to the grain boundary complexions is responsible for the softening effect. In contrast, the flow stress for dislocation propagation is larger for the OGBC and AIF samples than for the CGB. This strengthening effect is caused by a combination of local ledges in boundary structure and local stress variation caused by dopant segregation. Dopant segregation and grain boundary complexion formation can make the dislocation pinning effect stronger, creating areas of reduced stress near the grain boundary interfaces.

AIFs form when the energy needed to create an amorphous region of a given thickness with two new crystalline-amorphous interfaces is less than the energy of the original grain boundary:

$$\Delta G_{\mathrm{amorph}} \cdot h + 2\gamma_{\mathrm{cl}} < \gamma_{\mathrm{gb}}$$

where $\Delta G_{\mathrm{amorph}}$ refers to the volumetric free energy penalty for an undercooled amorphous film at a given binary alloy composition, h is the film thickness, γ_{cl} is the excess free energy of the new crystalline-amorphous interfaces, and γ_{gb} is the excess free energy of the original crystalline grain boundary. AIF formation is a form of grain boundary premelting, which is prohibitively difficult to observe in pure metals but accessible for highly doped boundaries. AIFs have been shown to impart extremely desirable behaviors when incorporated within nanocrystalline metals.

Here AIF is produced by complexions driven by alloy thermodynamics with a combination of grain boundary dopants and elevated temperatures required, and did not form through another process such as solid-state amorphization which is purely driven by local composition. Nanocrystalline metals have very high strengths, but are generally also very brittle and exhibit little ductility, with elongations or strain-to-failure values of only a few percent typically reported. This brittle nature often restricts the usage of nanocrystalline materials in technological applications. The high strength comes from the fact that traditional, intergranular dislocation plasticity mechanisms are shut off at these extremely fine crystallite sizes, requiring dislocations to be nucleated at an interface, propagated across the nanograin, and then absorbed at the opposite grain boundary. The repeated dislocation absorption required for such a physical mechanism may be the cause of the brittle nature of nanocrystalline metals, as atomistic simulations show that a single absorption event leads to high local stresses at the grain boundary while multiple absorption events result in crack nucleation. It was also recently demonstrated that solute segregation can affect a GB's ability to emit dislocations. Emission of dislocations from GBs is a key deformation mechanism for many nano/polycrystalline materials, so segregating solutes that stabilize the microstructure against coarsening can also increase the yield strength. The effect of segregation on dislocation nucleation depends on the nature of the disorder at GBs. If disorder is complete and the GB does not retain any dislocation embryos, the yield stress is increased because of the lack of dislocation nucleation sites. Partial disorder can lead to dramatic decreases in the yield stress if some dislocation embryos remain within the GB, because the atomic shuffling in amorphous regions near the dislocation sources aids in dislocation emission.

Nano-cracks formed within the ordered grain boundary sample interior and grew until a bridging grain detached, connecting the internal nano-crack to the notch to allow the start of crack propagation. The AIF-containing sample had no grain growth at crack initiation and instead had distributed plastic activity surrounding the notch region (Schuler et al. 2019). The ordered grain boundary sample demonstrated unsteady, accelerating crack growth characterized by the formation of an extensive nano-crack network interspersed with grain bridges. In contrast, the AIF-containing sample experienced steady, constant-rate crack growth with distributed plastic activity preceding the crack tip (Fig. 8.19).

The evenly distributed plastic activity in the AIF-containing sample indicates that the grain boundaries were better able to mitigate dislocation damage, whereas the

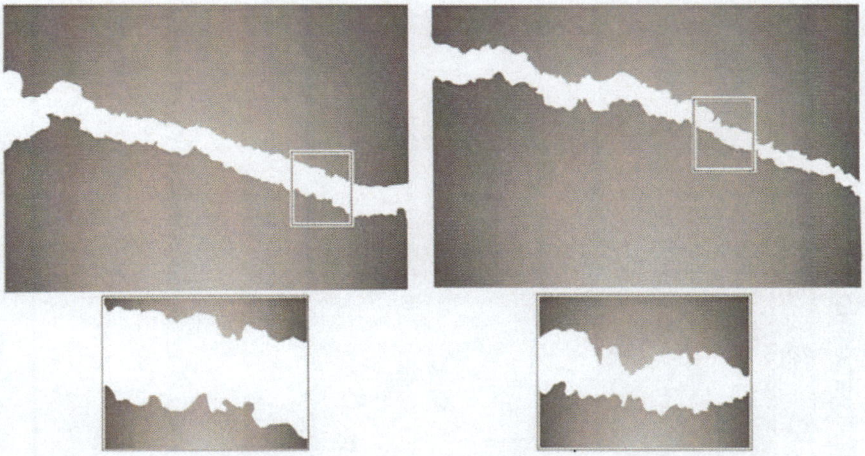

Fig. 8.19 Fracture surface of the ordered grain boundary (left) and amorphous intergranular film (AIF) containing samples after fatigue failure (right)

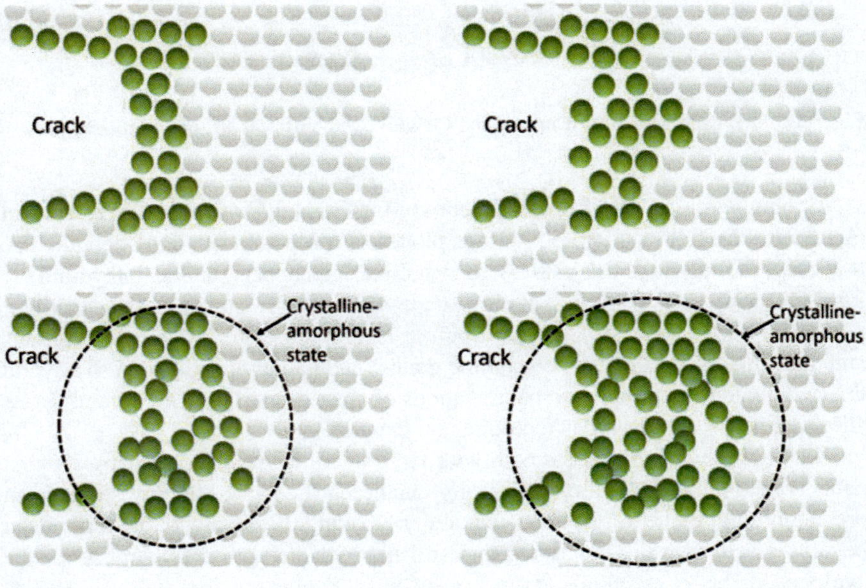

Fig. 8.20 Crack propagating in crystalline and amorphous microstructure

ordered grain boundary sample had extensive nano-cracking and highly localized plasticity. The MD simulation of a crack moving towards amorphous state is shown in Fig. 8.20.

Movement in a crystalline-amorphous structure shields the atom at the crack tip and impedes the increase of its stresses at this small concentrated region.

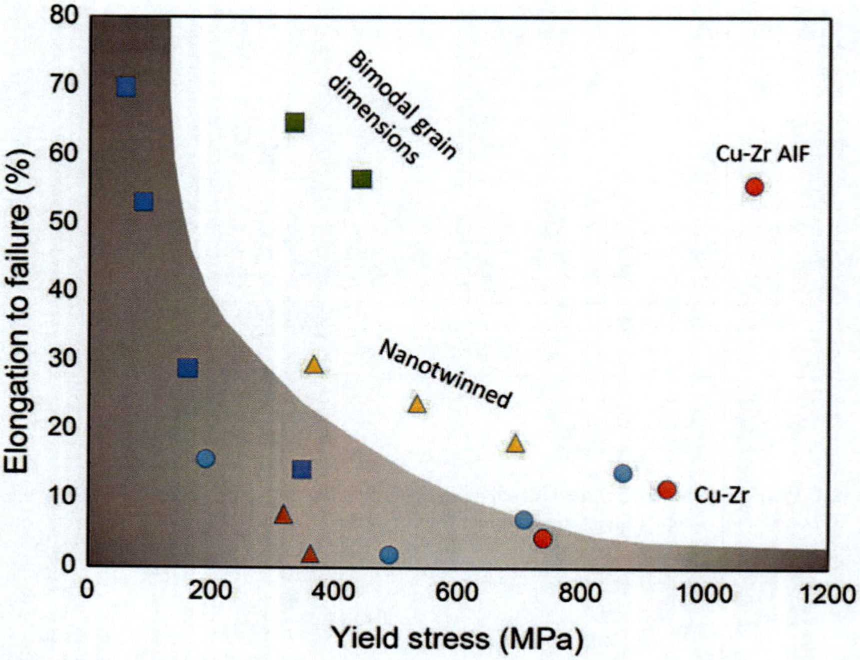

Fig. 8.21 Strength-ductility for many Cu and Cu alloys with different NC microstructure

A compilation of data from the literature for Cu and Cu alloys is presented in Fig. 8.21, where elongation to failure is plotted as a function of yield strength. The properties of traditional Cu, both microcrystalline and nanocrystalline, fall within the gray envelop, demonstrating a clear trade-off between strength and ductility. Advanced alloys, such as Cu with a bimodal grain structure (micron-scale grains embedded in a matrix of nanocrystalline grains) and nanotwinned Cu, push beyond this and demonstrate improved combinations of these properties. The Cu and Cu-Zr alloys with AIF are shown as red points on the plot.

The observation of high strength with large ductility fits a recent narrative that nanocrystalline materials are intrinsically ductile on the nanoscale, but suffer from premature failure due to strain localization or cracking. If this premature failure can be suppressed, then the traditional trade-off between strength and ductility can be avoided (Mishra et al. 2009).

Simulations on NC Al (grain size: 5 nm) ahead of a "blunt" crack (or nano-notch, with a root radius of 5 nm) highlighted the role of grain rotation (similar to the Ashby–Verrall mechanism) ahead of the notch, which resulted in a homogeneous (gradient free) stress field and led to the conclusion that NC metals may be insensitive to stress concentration. In situ TEM observations revealed intergranular crack growth via nucleation and growth of nano-voids at triple junctions and/or grain boundary nano-cracks, and their coalescence to the main crack. Crack growth in thin

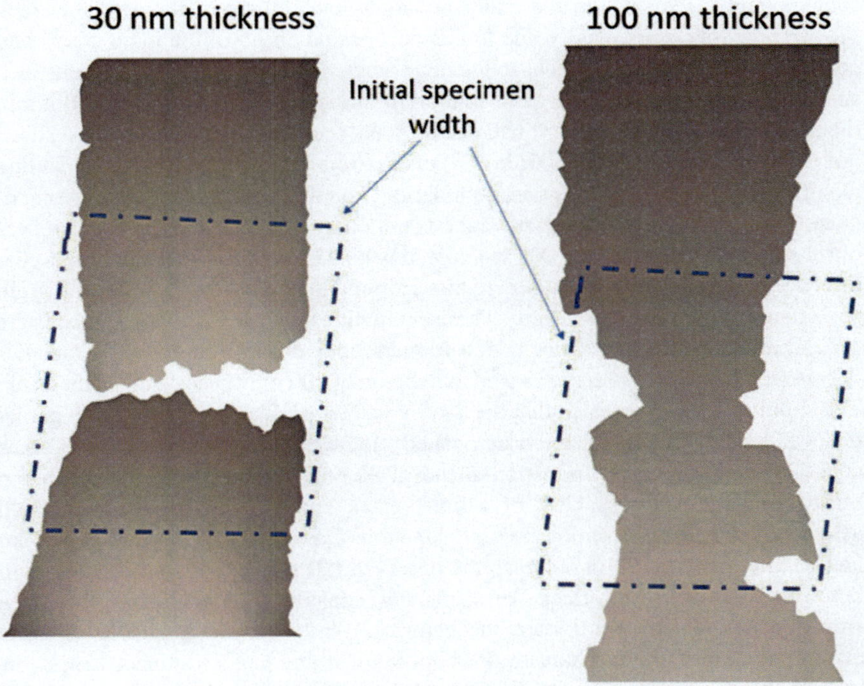

Fig. 8.22 Fracture of Au thin films with different thickness

films no longer follows the classic mechanisms, observed in bulk metals when film thickness decreases to 100 nm and below (Chang 2018). As a matter of fact, experiments performed on 20 and 100 nm thickness Au thin films showed the fractures reported in Fig. 8.22.

The 100 nm thickness film shows a more pronounced necking. In the 30 nm thickness film an intergranular crack propagated from the right edge through the entire specimen's width, along a plane roughly perpendicular to the loading axis. The intergranular crack propagates mainly via the growth of nano-cracks that form due to GB decohesion ahead of the main crack and that eventually coalesce with the main crack beginning of the holding steps during which large amounts of dislocation activities can be observed in a localized region ahead of the main crack. In contrast, the crack growth process of the 100-nm-thick Au specimens is distinctly different from that of the 30-nm-thick specimens. Although intergranular crack growth also prevails, GB decohesion does not occur ahead of the main crack tip. Instead, the main crack extends in a continuous manner along the GBs as a result of significant GB dislocation activities leading to grain boundary sliding. So, significant differences in the intergranular crack growth processes between the two types of specimens that can only result from the effects of thickness (30 nm vs. 100 nm) and grain size distribution must be highlighted. For the 30-nm-thick specimens having an average grain size of 40 nm and grains no larger than 100 nm in diameter, the crack growth process occurs mainly due to GB decohesion ahead of the main crack, and coalescence of these nano-cracks to the main crack. In contrast, for the 100-nm-thick specimens having an average grain size of 70 nm and grains as large as 350 nm in diameter, intergranular crack growth occurs mainly as a result of sustained (i.e., even after a displacement hold of 30 min) GB dislocation activities leading to GB sliding along the large (>200 nm) grain boundaries. For these thicker specimens, nano-crack formation ahead of the main crack is not observed. The crack growth process observed in the 30-nm-thick Au specimens is consistent with nucleation and coalescence of nano-cracks, and with atomistic simulations performed on NC Ni (grain sizes ranging from 5 to 10–12 nm). These simulations revealed that large stresses are present ahead of the crack tip, which triggers both dislocation activities and GB decohesion, which we observe as well in a region of 200 nm ahead of the main crack. The simulations also revealed that the nano-void formation can be assisted by partial dislocation activities, and that it is associated with vacancy cluster formation (Farkas et al. 2005). Nano-crack formation ahead of blunted cracks due to the combined effects of large stresses ahead of a blunt crack and the superposed stress field associated with GB dislocations at triple junctions was largely explained by Ovid'ko and Sheinerman (2012). In contrast, the observed GB sliding-assisted crack growth behavior for the 100-nm-thick Au specimens constitutes a significant difference from the previously documented mechanisms. While the crack growth mode is also intergranular, the mechanism does not involve any nano-void/nano-crack formation ahead of the main crack, which may be explained by either a lower driving force for GB decohesion or a higher resistance against it (Meyers et al. 2009). Larger grain sizes decrease the likelihood of nano-crack formation because of the decreased stress levels farther away from the crack tip. It is therefore possible that the larger grains in the 100-nm-thick Au specimens do not provide the necessary conditions to form nano-cracks. GBs may be more resistant to decohesion in the thicker films. Both thickness and grain size distribution could therefore affect the formation of nano-cracks ahead of the main crack tip, and alternative crack growth mechanisms become dominant if nano-crack formation does not occur. The motion of GB dislocations clearly dictated the intergranular stable crack growth via GB sliding (Hosseinian et al. 2018).

8.4 Conclusions

The properties of the deposited thin films are in direct relation to their structure—this structure spans surface characteristics, to interior structure, to defects and bonding. The way the film structure evolves during the various processes is of course very dependent on the exact processing conditions: choice of materials, deposition rate, deposition and annealing temperatures, and ion bombardment. They influence the diffusion and consequently the thin-film structure.

The properties of thin films differ significantly from those of their bulk counterparts due to the dimensional constraints imposed by the film thickness and/or grain size. With decreasing film thickness, the dislocation motion is more pronouncedly confined which leads to an increase of the film strength. Furthermore, when the film thickness and/or grain size decreases down to the nanoscale, grain boundary (GB) instability under cyclic loading is frequently observed.

Many scientific evidences show that thin films exhibit uniform necking during deformation. The large stresses result in grain boundary decohesion and nano-void formation which coalesce with the main crack, leading to intergranular fracture (in contrast to the typical transgranular cracking in coarse-grained metals). The tensile fracture is governed by void coalescence and propagation.

At large length scales (d and/or $h > 1$ μm), the fatigue deformation is accommodated by extended dislocation structures (accompanied with extrusions/intrusions) and shows weak length scale-dependent crack nucleation and propagation. At submicron length scales, which are too small for the formation of extended dislocation structures and inhibit the localized accumulation of plastic strain within grains, the fatigue processes are presumably controlled by the constraints on motion of individual dislocations and the interface-mediated damage and grain boundary (GB) process become more prevalent. The changes in fatigue damage with length scales (d and/or h) suggest that the fatigue mechanism transits from dislocation-mediated extrusion formation to crack formation-controlled behavior, and it can be attributed to the inhibition of dislocation mobility and the limited availability and activation of dislocation sources on the small length scale. Thus there is a distinct length-scale effect on fatigue damage morphology and fatigue lifetime, which reflects the change of length scale-related fatigue mechanisms.

As the film thickness or grain size is reduced, the fatigue life under a constant applied strain is increased. It was demonstrated that in films of less than 1 μm thickness, extrusions are absent and cracks initiate at boundaries and defects, indicating that the local accumulation of cyclic plastic strain becomes increasingly difficult with a decrease in film thickness. This change in the behavior of fatigue damage is attributed to the inhibition of dislocation mobility and the limited availability and activation of dislocation sources on thin films. Veins, slips, or dislocation cells do not form in films of less than 1 μm thickness during the course of fatigue. A decrease in thickness and grain size inhibits the localized accumulation of plastic strain within grains (such as at extrusions/intrusions and extended dislocation structures), and promotes the formation of cracks at twin and grain boundaries during fatigue. This effect is a likely cause of the increase in fatigue life.

References

Aliofkhazraei M (2011) Size effect in physical and other properties of nanostructured coatings. In: Nanocoatings. Engineering materials. Springer, Berlin. https://doi.org/10.1007/978-3-642-17966-2_6

Andrievski RA (2016) Twinned boundaries in nanomaterials. Rev Adv Mater Sci 46:104–120
Antolovich SD, Armstrong RW (2014) Plastic strain localization in metals: origins and consequences. Prog Mater Sci 59:1–160. https://doi.org/10.1016/j.pmatsci.2013.06.001
Argibay N, Chandross M, Cheng S et al (2017) Linking microstructural evolution and macro-scale friction behavior in metals. J Mater Sci 52:2780–2799. https://doi.org/10.1007/s10853-016-0569-1
Battaile CC, Hoyt JJ (2005) Modeling the effects of impurities on microstructure formation in nanocrystalline nickel thin films. JOM 57:71–73. https://doi.org/10.1007/s11837-005-0120-3
Bober DB, LaGrange T, Kumar M, Rupert TJ (2018) Pronounced grain boundary network evolution in nanocrystalline Cu subjected to large cyclic strains. J Mater Res 34(1):35–47. https://doi.org/10.1557/jmr.2018.334
Borovikov V, Mendelev MI, King AH (2019) Effects of grain boundary disorder on dislocation emission. Mater Lett 237:303–305. https://doi.org/10.1016/j.matlet.2018.11.114
Buehler MJ (2008) Deformation and fracture mechanics of geometrically confined materials. In: Atomistic modeling of materials failure. Springer, Boston, MA. https://doi.org/10.1007/978-0-387-76426-9_8
Bufford DC, Stauffer D, Mook WM, Amanulla SAS, Boyce BL, Hattar K (2016) High cycle fatigue in the transmission electron microscope. Nano Lett 16(8):4946–4953. https://doi.org/10.1021/acs.nanolett.6b01560
Chang SY (2018) In-situ nanomechanical testing in electron microscopes. In: Hsueh CH et al (eds) Handbook of mechanics of materials. Springer, Singapore. https://doi.org/10.1007/978-981-10-6855-3_53-1
Dobrzański LA, Gołombek K, Lukaszkowicz K (2015) Physical vapor deposition in manufacturing. In: Nee A (ed) Handbook of manufacturing engineering and technology. Springer, London. https://doi.org/10.1007/978-1-4471-4670-4_29
Espinosa HD, Prorok BC (2003) Size effects on the mechanical behavior of gold thin films. J Mater Sci 38:4125–4128. https://doi.org/10.1023/A:1026321404286
Farkas D, Van Petegem S, Derlet P, Van Swygenhoven H (2005) Dislocation activity and nano-void formation near crack tips in nanocrystalline Ni. Acta Mater 53:3115–3123. https://doi.org/10.1016/j.actamat.2005.02.012
Frolov GI, Zhigalov VS, Mal'tsev VK (2000) Temperature effect on structural transformations in nanocrystalline cobalt films. Phys Solid State 42:334–336. https://doi.org/10.1134/1.1131208
Gutkin MY, Dynkin NK (2012) Dislocation-disclination models of grain boundary migration in ultrathin nanocrystalline films. Phys Solid State 54:798–807. https://doi.org/10.1134/S1063783412040105
Gutkin MY, Mikaelyan KN, Ovid'ko IA (2008) Grain growth and collective migration of grain boundaries during plastic deformation of nanocrystalline materials. Phys Solid State 50:1266–1279. https://doi.org/10.1134/S1063783408070135
Han X, Zhou J, Liu H et al (2014) The dominant deformation mechanism of nanocrystalline materials with the finest grains: grain boundary sliding or grain boundary migration? J Nanopart Res 16:2200. https://doi.org/10.1007/s11051-013-2200-6
He T, Xiao W, Zhang Y et al (2017) Effect of cooperative grain boundary sliding and migration on dislocation emission from a branched crack tip in deformed nanocrystalline solids. Int J Fract 206:1–10. https://doi.org/10.1007/s10704-017-0193-3
Hosseinian E, Gupta S, Pierron ON, Legros M (2018) Size effects on intergranular crack growth mechanisms in ultrathin nanocrystalline gold free-standing films. Acta Mater 143:77–87. https://doi.org/10.1016/j.actamat.2017.10.004
Hossenian E, Pierron ON (2013) Quantitative in situ TEM tensile fatigue testing on nanocrystalline metallic ultrathin films. Nanoscale 5:12532. https://doi.org/10.1039/c3nr04035f
Hu TC, Wang YT, Hsu FC, Sun PK, Lin MT (2013) Cyclic creep and fatigue testing of nanocrystalline copper thin films. Surf Coat Technol 215:393–399. https://doi.org/10.1016/j.surfcoat.2012.08.089

Hultman L, Mitterer C (2006) Thermal stability of advanced nanostructured wear-resistant coatings. In: Cavaleiro A, De Hosson JTM (eds) Nanostructured coatings. Nanostructure science and technology. Springer, New York. https://doi.org/10.1007/978-0-387-48756-4_11

Huo W, Fang F, Liu X, Tan S, Xie Z, Jiang J (2019) Fatigue resistance of nanotwinned high-entropy alloy films. Mater Sci Eng A739:26–30. https://doi.org/10.1016/j.msea.2018.09.112

Ievlev VM (2004) Thin-film nanostructures: substructural aspects. Inorg Mater 40:S71–S90. https://doi.org/10.1007/s10789-005-0048-6

Larsen KP, Rasmussen AA, Ravnkilde JT, Ginnerup M, Hansen O (2003) MEMS device for bending test: measurements of fatigue and creep of electroplated nickel. Sensors Actuators A103:156–164. https://doi.org/10.1016/S0924-4247(02)00306-0

Luo X-M, Zhang GP (2017) Grain boundary instability dependent fatigue damage behavior in nanoscale gold films on flexible substrates. Mater Sci Eng A 702:81–86. https://doi.org/10.1016/j.msea.2017.07.006

Luo X-M, Zhu X-F, Zhang G-P (2013) Nanotwin-assisted grain growth in nanocrystalline gold films under cyclic loading. Nat Commun 5:3021. https://doi.org/10.1038/ncomms4021

Luo X-M, Li X, Zhang G-P (2017) Forming incoherent twin boundaries: a new way for nanograin growth under cyclic loading. Mater Res Lett 5(2):95–101. https://doi.org/10.1080/21663831.2016.1203365

Luo X-M, Zhang B, Zhang G-P (2019) Fatigue of metals at nanoscale: metal thin films and conductive interconnects for flexible device application. Nano Mater Sci 1:198–207. https://doi.org/10.1016/j.nanoms.2019.02.003

Meirom RA, Alsem DH, Romasco AR, Clark T, Polcawich RG, Pulskamp JS, Dubey M, Ritchie RO, Muhlstein CL (2011) Fatigue-induced grain coarsening in nanocrystalline platinum films. Acta Mater 59:1141–1149. https://doi.org/10.1016/j.actamat.2010.10.047

Meirom RA, Clerk TE, Muhlstein CL (2012) The role of specimen thickness in the fracture toughness and fatigue crack growth resistance of nanocrystalline platinum films. Acta Mater 60:1408–1417. https://doi.org/10.1016/j.actamat.2011.11.015

Meyers MA, Traiviratana S, Lubarda VA et al (2009) The role of dislocations in the growth of nanosized voids in ductile failure of metals. JOM 61:35. https://doi.org/10.1007/s11837-009-0025-7

Mishra AC, Thakur AK, Srinivas V (2009) Effect of deposition parameters on microstructure of electrodeposited nickel thin films. J Mater Sci 44:3520–3527. https://doi.org/10.1007/s10853-009-3475-y

Musil J (2006) Physical and mechanical properties of hard nanocomposite films prepared by reactive magnetron sputtering. In: Cavaleiro A, De Hosson JTM (eds) Nanostructured coatings. Nanostructure science and technology. Springer, New York. https://doi.org/10.1007/978-0-387-48756-4_10

Ovid'ko IA (2000) Interfaces in nanostructured films and coatings. In: Chow GM, Ovid'ko IA, Tsakalakos T (eds) Nanostructured films and coatings, NATO science series (Series 3: High technology), vol 78. Springer, Dordrecht. https://doi.org/10.1007/978-94-011-4052-2_20

Ovid'ko IA, Sheinerman AG (2012) Generation and growth of nanocracks near blunt cracks in nanocrystalline solids. Eur J Mech A Solid 33:39–47. https://doi.org/10.1016/j.euromechsol.2011.11.001

Pantano MF, Espinosa HD, Pagnotta L (2012) Mechanical characterization of materials at small length scales. J Mech Sci Technol 26:545–561. https://doi.org/10.1007/s12206-011-1214-1

Pelleg J (2013) Mechanical behavior in the micron and submicron/nano range. In: Mechanical properties of materials. Solid mechanics and its applications, vol 190. Springer, Dordrecht. https://doi.org/10.1007/978-94-007-4342-7_8

Rajagopalan J, Saif MTA (2011) Effect of microstructural heterogeneity on the mechanical behavior of nanocrystalline metal films. J Mater Res 26(22):2826–2832. https://doi.org/10.1557/jmr.2011.316

Rajagopalan J, Han JH, Saif MTA (2008) Bauschinger effect in unpassivated freestanding nanoscale metal films. Scripta Mater 59:734–737. https://doi.org/10.1016/j.scriptamat.2008.06.010

Rajgarhia RK, Spearot DE, Saxena A (2010) Behavior of dopant-modified interfaces in metallic nanocrystalline materials. JOM 62:70–74. https://doi.org/10.1007/s11837-010-0184-6
Ross FM, Minor AM (2019) In situ transmission electron microscopy. In: Hawkes PW, Spence JCH (eds) Springer handbook of microscopy, Springer handbooks. Springer, Cham. https://doi.org/10.1007/978-3-030-00069-1_3
Schuler JD, Barr CM, Heckman NM et al (2019) In situ high-cycle fatigue reveals importance of grain boundary structure in nanocrystalline Cu-Zr. JOM 71:1221–1232. https://doi.org/10.1007/s11837-019-03361-7
Shute CJ, Myers BD, Xie S, Li SY, Barbee TW Jr, Hodge AM, Weertman JR (2011) Detwinning, damage and crack initiation during cyclic loading of Cu samples containing aligned nanotwins. Acta Mater 59:4569–4577. https://doi.org/10.1016/j.actamat.2011.04.002
Spengen W, Modlinski R, Puers R, Jourdain A (2007) Failure mechanisms in MEMS/NEMS devices. In: Bhushan B (ed) Springer handbook of nanotechnology, Springer handbooks. Springer, Berlin. https://doi.org/10.1007/978-3-540-29857-1_52
Suryanarayana C, Koch CC (2000) Nanocrystalline materials - current research and future directions. Hyperfine Interact 130:5. https://doi.org/10.1023/A:1011026900989
Turlo V, Rupert TJ (2018) Grain boundary complexions and the strength of nanocrystalline metals: dislocation emission and propagation. Acta Mater 151:100–111. https://doi.org/10.1016/j.actamat.2018.03.055
van Spengen WM, Modliński R, Puers R, Jourdain A (2017) Failure mechanisms in MEMS/NEMS devices. In: Bhushan B (ed) Springer handbook of nanotechnology, Springer handbooks. Springer, Berlin. https://doi.org/10.1007/978-3-662-54357-3_40
Wan HY, Luo XM, Li X, Liu W, Zhang GP (2016) Nano twin-enhanced fatigue resistance of ultrathin Ag films for flexible electronics applications. Mater Sci Eng A676:421–426. https://doi.org/10.1016/j.msea.2016.09.010
Zhang X, Anderoglu O, Hoagland RG et al (2008) Nanoscale growth twins in sputtered metal films. JOM 60:75–78. https://doi.org/10.1007/s11837-008-0123-y
Zhang JY, Zhang X, Liu G, Wang RH, Zhang GJ, Sun J (2011) Length scale dependent yield strength and fatigue behavior of nanocrystalline Cu thin films. Mater Sci Eng A528:7774–7780. https://doi.org/10.1016/j.msea.2011.06.083
Zhang P, Zhang JY, Li J, Liu G, Wu K, Wang YQ, Sun J (2014) Microstructural evolution, mechanical properties and deformation mechanisms of nanocrystalline Cu thin films alloyed with Zr. Acta Mater 76:221–237. https://doi.org/10.1016/j.actamat.2014.04.041
Zorman CA (2017) Materials aspects of micro- and nanoelectromechanical systems. In: Bhushan B (ed) Springer handbook of nanotechnology, Springer handbooks. Springer, Berlin. https://doi.org/10.1007/978-3-662-54357-3_7

Chapter 9
Contact Fatigue and Crack Behavior of Nanostructured Metal Alloys and Composites

9.1 Introduction

In the past two decades, NC metals and alloys have attracted dramatically increasing attentions because of their promising structural properties, such as the ultrahigh strength and superior wear resistance (Murty et al. 2013). The unlubricated sliding of metals is important in many mechanical applications covering a wide range of sliding velocities. The understanding of the sliding behavior of metals at high sliding velocities is important in mechanical parts such as bearings, brake liners, gears, and friction clutches, and also in some microelectromechanical systems.

Ultralow wear rates are attributed to those materials' volumetric wear rate less than 10^{-8} mm^3 N^{-1} m^{-1} (Curry et al. 2018). In order to obtain these properties in metals, the deformation due to the cyclic contact stresses must be mitigated. Now, reduced grain size is the very effective method to increase hardness in metals and alloys with direct evidences in the increment of wear resistance (Hoornaert et al. 2009; Leyland and Matthews 2000). Once the surface stresses do not overcome the flow stress typical of sliding contact, the wear behavior is a fatigue-dominated mechanism (Argibay et al. 2010). Fatigue wear is related to the formation of surface cracks due to cyclic contact stresses. In this case, wear is characterized by particle formation via cohesive failure. Now, when microcrystalline-grained materials show dislocation-mediated plasticity, nanocrystalline metals and alloys can present grain growth leading to cracking (Boyce and Padilla 2011). Improving the thermomechanical stability of nanocrystalline alloys can mitigate stress-driven microstructure evolution at elevated contact stresses, and ultimately suppress delamination wear.

The friction coefficient vs. wear rate map for tribological materials is shown in Fig. 9.1.

Tribological studies show significant reductions in the coefficient of friction and wear rate for NC Ni (grain size ~10 nm) produced by electroplating compared to its microcrystalline (MC) counterpart (Guidry et al. 2009). Similar evidence has been

© Springer Nature Switzerland AG 2021
P. Cavaliere, *Fatigue and Fracture of Nanostructured Materials*,
https://doi.org/10.1007/978-3-030-58088-9_9

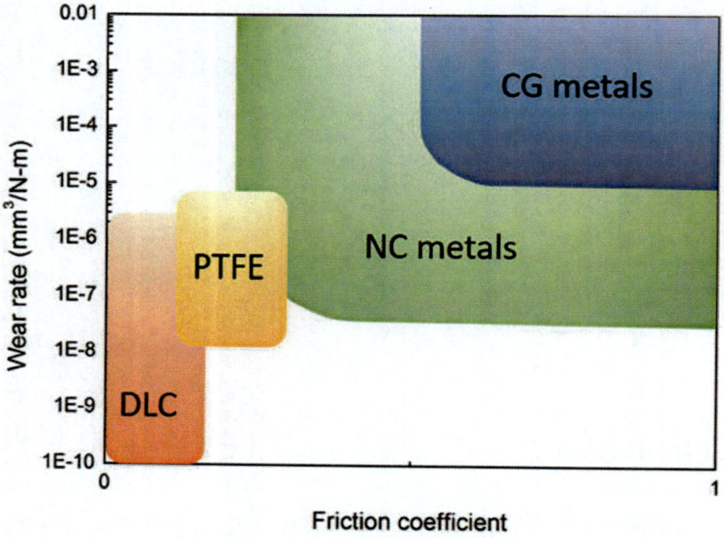

Fig. 9.1 Property map of tribological materials

reported in the literature and has been attributed to the higher hardness of the NC structure. It is evident that the significant increase in strength (hardness) arising from the grain size reduction in the nano domain is expected to impact the mechanical processes at asperity contacts that are critical in determining wear behavior.

Tribocorrosion is the degradation of materials due to the simultaneous effects of electrochemical corrosion and mechanical wear in a tribological contact (Meng et al. 2020). The total material removal rate differs from sum of the wear rate measured in the absence of corrosion and the corrosion rate observed in the absence of wear. For example, passive coatings would be subject to active corrosion due to the annihilation of protective passive film in wet sliding contact, when the frictional movement in corrosive medium is continuous (Fathollahzade and Raeissi 2014).

Because nanocrystalline metals are known to deform through unique grain boundary-mediated mechanisms instead of dislocation-dominated processes, their subsurface microstructures could in fact evolve in manners completely different to their coarse-grained counterparts.

9.2 Wear Mechanisms in Nanostructured Materials

The tribological behavior of sliding surfaces is a very complex system; in such an analysis parameters such as elasticity as elastic modulus, plasticity as hardness or yield strength, and ductility as fracture toughness result to be very important but lead to very complex models (Dupont and Sansoz 2010). The tribocontact has been studied on macro-, micro-, and nano-level with a very deep range of mechanical properties used to describe the frictional and wear behavior. For the deep understanding of frictional and wear sliding mechanisms the interaction between solid and solid asperities needs to be studied. Such phenomena are actually very less discovered at nanoscale levels in which the material strength increase with the grain size reduction and the influence of asperity geometry result to be strongly pronounced (Ma et al. 2013). The advance in the evolution of characterization techniques such as atomic force microscopy and nanoindentation allows to obtain very useful information on sliding and wear behavior from nano- to microscale levels (Wen et al. 2017). Nanotribological studies result to be fundamental in the development of sliding models at very fine levels, for the understanding of nanostructure behavior used in magnetic storage systems, microelectromechanical systems (MEMS), and other industrial applications. At such a dimensional scale, the frictional problems are much more dependent on the surface interactions and the lubricant effect must be analyzed from a different point of view (Cavaliere and Prete 2010). The normal indentation test has limited application in predicting tribological response. On the other hand, the scratch test, in which a hard indenter is slid across the surface of the material, provides a tool to test materials under conditions of controlled abrasive wear (Nazemian and Chamani 2019). It is routinely employed in practice to compare hardness and abrasive resistance of surfaces, to extract information relating to mechanisms of deformation and material removal, and to study delamination of coatings. Developments in instrumentation now provide the means to monitor load–depth response in normal as well as in sliding contact across several length scales. In addition, it is possible to observe friction evolution through continuous measurement of tangential loads along the scratch, and obtain residual scratch profiles using high-precision profilometers and/or an atomic force microscope (AlMotasem et al. 2017). In the scratch test, the indenter is moved first normally into the surface to make contact at a specified normal load or depth, following which it is moved along a chosen scratch direction. The "scratch hardness" HS is defined as the average pressure of contact assuming total loss of contact at the back of the indenter. Since the scratch test closely simulates controlled abrasive events, its use has been limited as a practical tool in industrial applications such as studying delamination of coatings, ranking materials in order of their abrasive resistance, and characterizing the mechanisms of deformations for various tribological purposes (Fu et al. 2012).

9.2.1 Wear Characterization

In general, the indentation and scratch hardness values depend on the indenter geometry/dimensions and on the indenter–material contact area. By employing different indenters, it was possible to measure very precisely the plastic properties of electrodeposited metals with different grain sizes. The use of spherical indenters allows to simulate an axisymmetric condition especially at very low loads. All the experiments were accompanied with indentations performed with a Berkovich

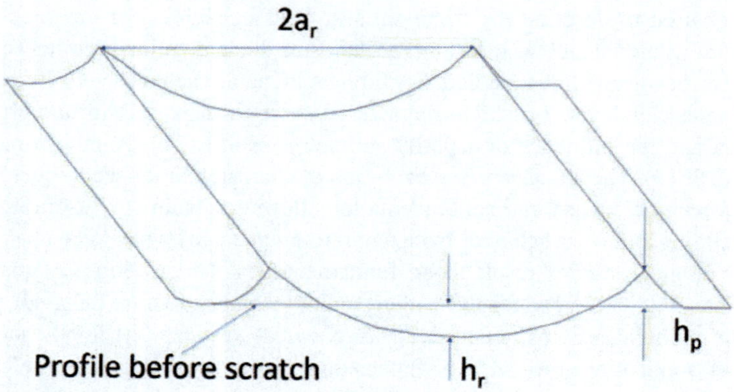

Fig. 9.2 Residual scratch profile of the spherical indenters

indenter for comparison. The schematic representation of the residual scratch profile is shown in Fig. 9.2.

Here, $2a_r$ is the residual width, h_r the residual penetration depth, and h_p the residual pileup height. With P being the applied load, the scratch hardness can be measured as

$$H_s = \frac{2P}{\pi a_r^2}$$

while the overall frictional coefficient is given by

$$\mu_{tot} = \frac{F_t}{P} = f(\mu_a, \mu_w)$$

with μ_a being the contribution due to the friction for the normal contact and μ_w the contribution due to the work of plastic deformation.

For a given value of P, h_r, and F_t and the material properties for the indenter, the frictional sliding process can be fully determined. In such an analysis, the pileup height h_p/h_r, significantly varying with material properties, assumes a fundamental importance. The analysis of the material response can be modeled following the Hooke's law and the von Mises yield criterion with isotropic power law hardening. The dependence of the true stress σ on the true strain ε is commonly expressed as

$$\sigma = \begin{cases} E\varepsilon \text{ for } \sigma \leq \sigma_y \\ R\varepsilon^n \text{ for } \sigma \geq \sigma_y \end{cases}$$

where E is the Young's modulus, R a strength coefficient, n the strain hardening exponent, and σ_y the initial yield stress at zero offset strain. By considering the load–displacement indentation curves, it is convenient to consider the total strain as decomposed in its elastic and plastic component:

$$\varepsilon = \varepsilon_y + \varepsilon_p$$

For $\sigma = \sigma_y$

$$E\varepsilon_y = R\varepsilon_y^n$$

For $\sigma > \sigma_y$ it is possible to replace the R value with the elastic properties obtaining

$$\sigma = \sigma_y \left(1 + \frac{E}{\sigma_y}\varepsilon_p\right)^n$$

From profilometric analysis it is possible to measure the plastic strain due to the frictional sliding obtaining the complete constitutive relationships for the material under investigation. In addition, for a fixed geometry, the strain hardening exponent can be measured from the indentation data by analyzing the stress-strain response of the material. The influence of the strain hardening behavior on the pileup was evaluated from frictional sliding results obtained for nanocrystalline nickel and alloys of different grain sizes demonstrating that the pileup height decreases with increasing strain hardening exponent. In the case of electrodeposited pure Ni in the NC and UFG regimes, a significant increase was shown in friction coefficient with decreasing mean grain size. Such a behavior becomes less evident by increasing the applied load and the penetration depth due to the different strain hardening of the material which increases with decreasing mean grain size. It was also demonstrated that the indentation response is strongly influenced by the surface finishing. The hardness decreases with increasing indentation load in those materials exhibiting low roughness levels while it results to be constant in those materials with higher surface roughness.

As mentioned before, delamination is one of the most important characteristics to be monitored in the case of coatings (Musil 2012). Especially in the case of thin coatings, nanoindentation and nanoscratch tests are fundamental instruments for the characterization of those coatings. Experiments performed on Ni-P coatings in both nanocrystalline and amorphous states allowed for the analyses of many mechanisms acting in the materials during loading and scratch. First of all, there is strong experimental evidence that the elastic modulus has a main effect on the wear resistance (Leyland and Matthews 2000). In addition, as the material hardness is increased, higher wear properties are expected. At the beginning of the indentations, the experienced strain is purely elastic under very low compressive stress; after this, the curve deviates from the Hertzian behavior because of the accumulation of the shear stress at the indentation tip, leading to the initiation of plastic deformation. For amorphous microstructures, this behavior is due to the formation and extension of shear bands; on the contrary, in crystalline materials it is due to the nucleation and sliding of dislocations (Hadipuor et al. 2019). In amorphous Ni and Ni alloys, plastic deformation acts as a flow of thin layer located in the shear plane. The material immediately close to these bands deforms only elastically. Once the alloy is heat

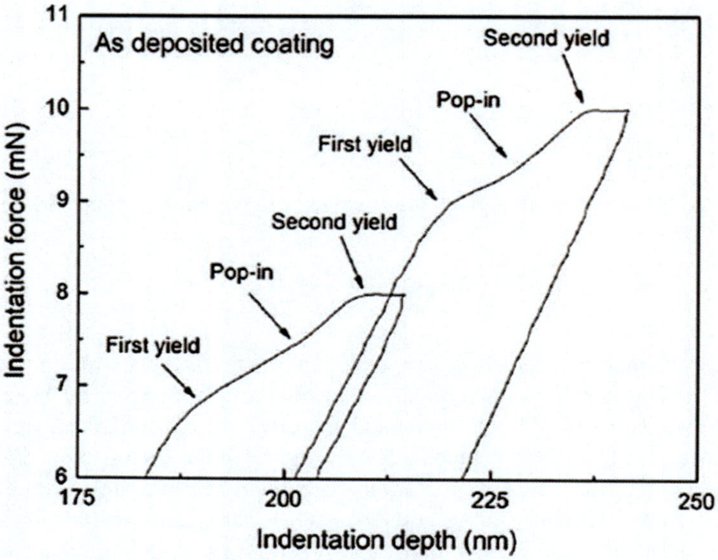

Fig. 9.3 Multistep indentation of Ni-P amorphous coating

treated, the microstructure becomes crystalline, but the high hardening is experienced due to the effect of phosphide precipitation. This also leads to the increase of the fracture toughness, indicating that ductile rupture governs failure in the heat-treated coatings. The main mechanisms, leading to the strength increase, have been recognized in the formation of well-defined high-angle boundaries, also leading to the decrease of interior defect density. Now, as previously indicated for amorphous microstructures, deformation behavior is due to the formation and extension of shear bands; on the contrary, in crystalline materials it is due to the nucleation and sliding of dislocations (Chang et al. 2006). A magnification of the indentation curves performed in multistep tests for the amorphous Ni-P showed different stages during the deformation (Fig. 9.3).

These stages (pop-in) are associated with the emission of individual shear bands underneath the indenter tip. These bands then appear as a series of steps around the periphery of the indentation. Each pop-in event corresponds to the formation of shear bands that quickly accommodate the applied strain. Second, when multiple unloading-reloading tests were conducted during an indentation experiment at the same loading, a peculiar hardening behavior was observed. Each unloading segment is noted to be completely reversible upon subsequent reloading, indicating that the elastic deformation indeed occurred. More interestingly, the onset of yielding upon each reloading is always noted to take place at a higher load than the load immediately before the previous unloading, apparently suggesting a hardening effect (Fig. 9.4).

This stage can be defined as the first yield. No pop-in or multiple yields are revealed in the crystallized material; in fact, it is expected that the deformation leads

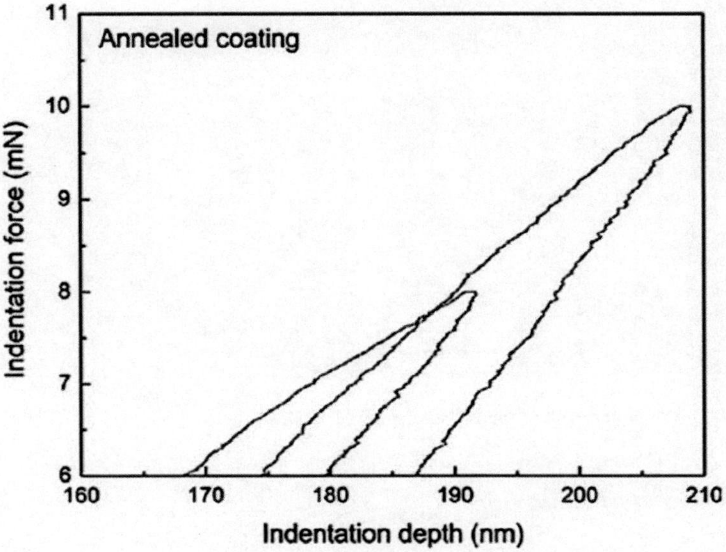

Fig. 9.4 Multistep indentation of Ni-P NC coating

to crack formation at the indentation tip without shear band formation. In amorphous Ni and Ni alloys, plastic deformation acts as a flow of thin layer located in the shear plane (Shen et al. 2012). The material immediately close to these bands deforms only elastically. Once the alloy is heat treated, the microstructure becomes crystalline, but the high hardening is experienced due to the effect of phosphide precipitation. This also leads to the fracture toughness increase, indicating that ductile rupture governs failure in the heat-treated coatings. As a matter of fact, for the heat-treated crystalline material, cracks form at the indent tips for the as-deposited amorphous coating; the characteristic patterns inside the indent are not cracks but overlapping layers of displaced material because of the indenter (Fig. 9.5).

Owing to the volume-conserving nature of the plasticity, deformation in elastic-perfectly plastic solids, like amorphous alloys, occurs by the pileup of the material against the faces of the indenter. Because of the inhomogeneous nature of plastic deformation in amorphous alloys or constraint of the surrounding elastic material, the unstable extension of dominant shear bands was suppressed, so the pileup of the material is seen as discrete steps. The pileup of the material, in the form of circular shear bands, was also found for nanoindentation into the surface of the amorphous Ni-P coatings.

Fig. 9.5 Indentation aspect of the NC and amorphous Ni-P

9.2.2 *Scratch Behavior*

Heat treatment improves the adhesion properties because of the positive action of nickel phosphide precipitation. Once they are uniformly distributed, they act as a barrier, preventing the coagulation of nickel grains promoting the adhesion. This is because the mean precipitate diameter leads to a bowing mechanism interaction between the particles and the dislocations. By analyzing the SEM micrographs of the scratch track, it was revealed that the first critical load with the appearance of the first hair cracks (inclined to 30°) is at about 19 N (green line in Fig. 9.6), which corresponding microstructure is shown in Fig. 9.7.

The hairy cracks visible in the center of the scratch groove are due to the tensile stresses in front of the sliding indenter. The coating, in the NC state, is susceptible to adhesive wear typical of a mild wear regime. The microcracks obviously increase in density and length by increasing the vertical load (Fig. 9.8).

The cracks inside the track, due to the compressive stress of the indenter, become remarkable at a vertical load of 30 N for an indentation depth of 19 μm. After this load also the tensile component of the force starts to produce cracks parallel to the scratch direction up to a load of 45 N when only cracks perpendicular to the scratch direction seem to be predominant and macrocracks are transferred to the substrate (Fig. 9.9).

Based on the observations of the model described, no bending cracks are observed for these loads revealing that the substrate provides still sufficient mechanical support. Only after this load, depressions start to appear, revealing that coating delamination occurs because excessive strain of the coating is experienced at that stage of the scratch. At these loads, the coating is subjected to a strain of 5% which is significantly above the ductility of Ni-P coatings. At low loads, arc cracks are visible in the scratch track; this reveals adhesive failure up to 30 N, and then the linear cracks reveal that the failure becomes cohesive. These wing-shaped delaminations at

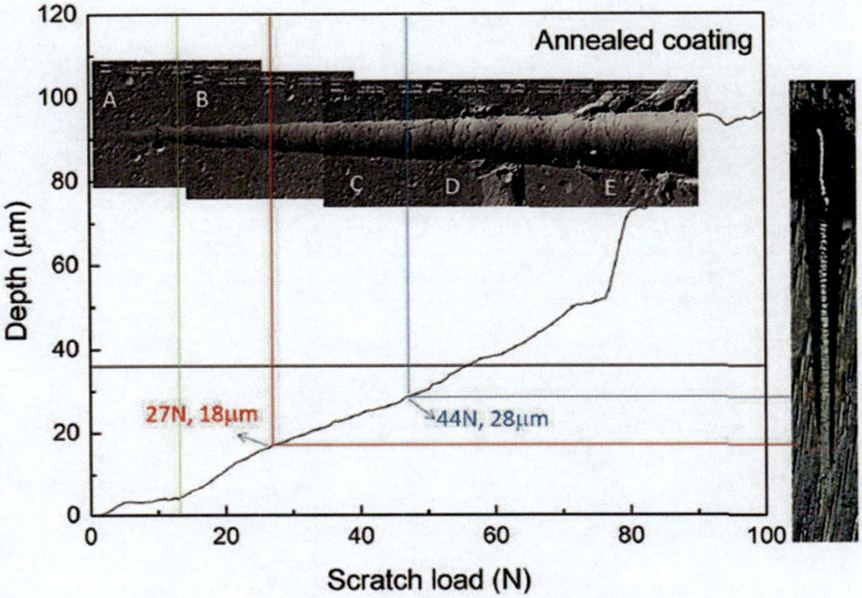

Fig. 9.6 Scratch curve with the indication of the substrate and the vertical load

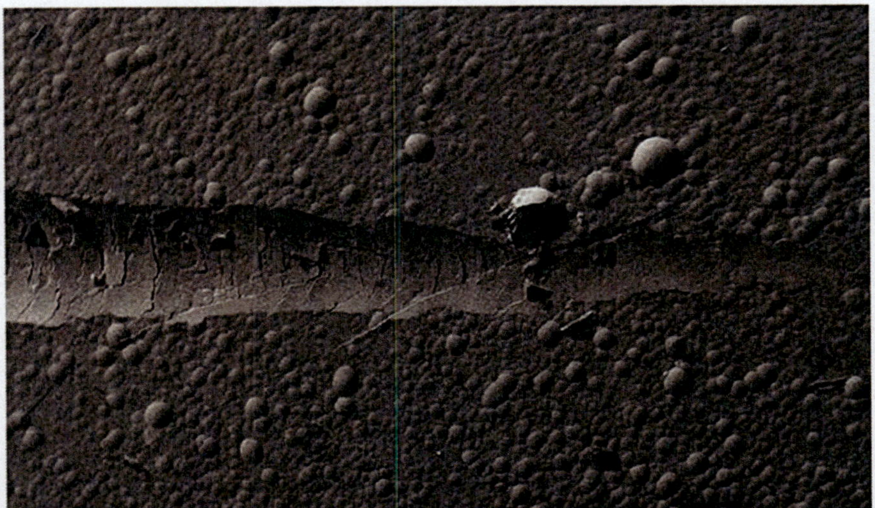

Fig. 9.7 First appearance of microcracks

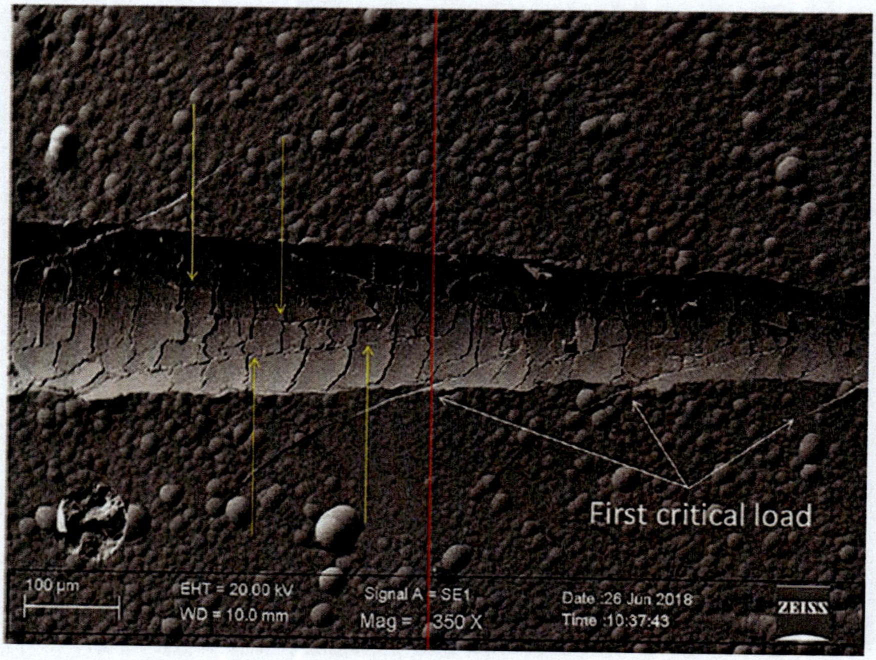

Fig. 9.8 Microcracks and compressive cracks

the edge of the groove are due to very complex multiaxial stress fields resulting from the indenter movement and pressure and from the bulged coating.

9.3 Fretting in Nanostructured Materials

Many scientific evidences reveal that tribological properties of nanostructured materials have been paid more attention and many experimental results have shown a significant enhancement in friction and wear properties of nanostructured materials (Gao et al. 2012).

Fretting is described as the oscillatory relative tangential movement which contacting surfaces may experience as a result of vibration or cyclic stressing of one of the surfaces (Pinchuk et al. 2015). It is often the origin of catastrophic failures or loss of functionality in many industrial applications, including bolted mechanical joints, stacks of objects in transport, and electrical connectors in vibrating machinery. Fretting is a common problem resulting in increased contact resistance and possible failure. Initially, before the fretting problems occur, the contacts will form so-called a-spots, metal-to-metal contacts with low contact resistance, by plastically deforming surface asperities. Yet, after some time fretting failure takes place, and may be divided into fretting wear, fretting corrosion, and fretting fatigue. At the wear

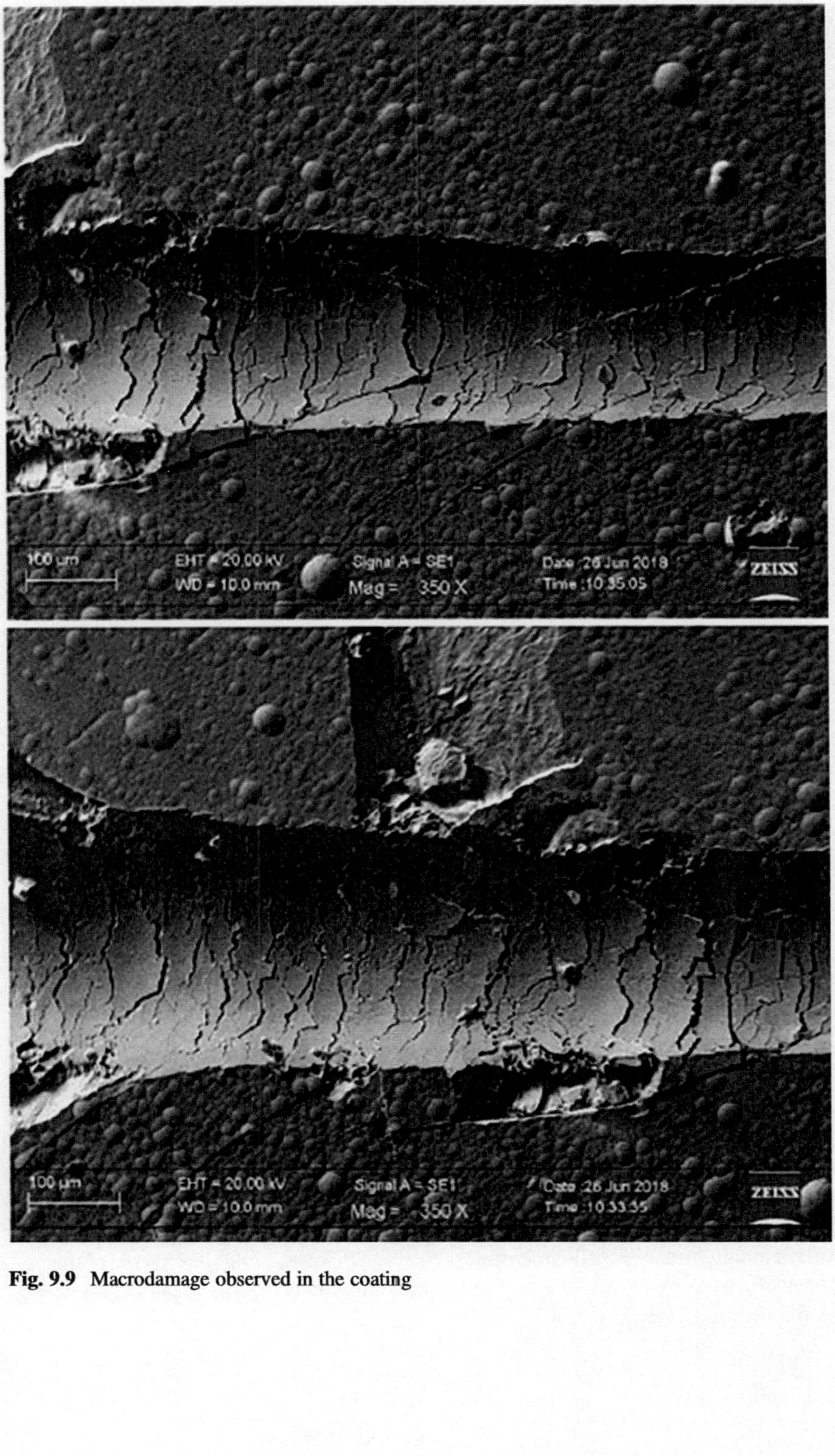

Fig. 9.9 Macrodamage observed in the coating

stage, contact surfaces are slowly worn down due to the friction caused by vibrations, although not significantly influencing the conductivity and only filling some of the cavities.

The damage due to fretting wear can accelerate fatigue failure of components by creating crack initiation sites on the surface. In order to prevent fretting wear, modification of the surface to improve the tribological properties is necessary. To this end, it has been a great challenge to develop an effective surface modification technique for fretting applications (Amanov et al. 2012). Surface modification techniques such as ion implantation (II), laser beam quenching (LBQ), shot peening (SP), laser shot peening (LSP), plasma nitriding (PN), plasma immersion ion implantation (PIII), surface mechanical attrition treatment (SMAT), and ultrasonic nanocrystal surface modification (UNSM) have already been identified as effective methods to enhance the ability of materials to resist fretting wear.

In recent decades, how to improve the fretting wear resistance of a material has always been a huge challenge for material scientists due to its importance in industrial application. The movement may be either the result of external vibration (fretting wear) or the consequence of one of the members of the contact being subjected to a cyclic stress (fretting fatigue). Whatever the consequences, both fretting types may give rise to service failure due to the production of debris or the initiation and propagation of fatigue cracks. Reciprocating movements as short as 0.1 μm in amplitude, which is coupled with high number of reciprocating cycles, can cause failure of components even at very low levels of stress.

9.3.1 Fretting Mechanisms

It has been shown that if the applied coating exhibits low-friction characteristic, there could be a significant reduction in the fretting fatigue according to the following equation:

$$S_{\mathrm{fr}} = S_0 + 2\mu P_0\left[1 - \mathrm{e}^{\left(-\frac{I}{k}\right)}\right]$$

where S_{fr} is the fretting fatigue strength (MPa), S_0 is the fatigue strength in the absence of fretting (MPa), μ is the coefficient of friction, P_0 is the contact pressure (MPa), I is the reciprocating amplitude (μm), and k is a constant, typically $k = 3.8$ (μm), and the reciprocating amplitude greater than 25 (μm) makes the exponential term negligible. From this equation, it can be observed, in general, that the fretting wear resistance can be enhanced by the application of low-friction contact material.

Except in the case of brittle materials, it was shown that a specific superficial layer is formed during the very first cycles of fretting loading. This layer is called the tribologically transformed structure or TTS. Understanding the mechanisms of formation of the TTS is a key step in the modeling of WIF. The mechanisms acting in the materials are shown in Fig. 9.10.

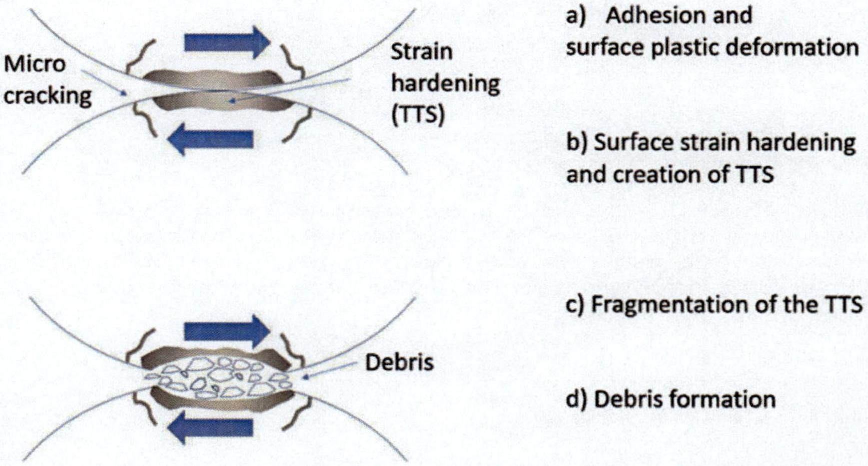

Fig. 9.10 Mechanism formation of the TTS

Fig. 9.11 Debris evolution during sliding cycles

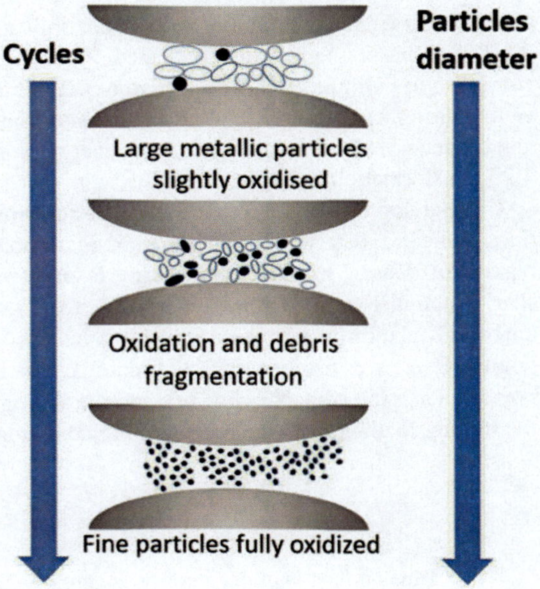

Obviously, the debris evolves during continuous straining (Fig. 9.11).

Sliding contact entails significant compression and shear stresses at asperity contacts that are expected to be accommodated by entirely different mechanical deformation mechanisms in MC and NC metals and alloys. Under the repeated loading, large plastic strains are expected in the former resulting in work hardening but also wear by contact fatigue. The schematic of the different zones is shown in Fig. 9.12.

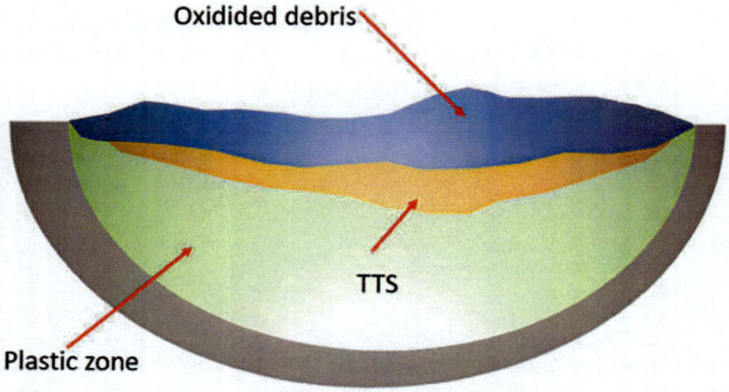

Fig. 9.12 Cycled sliding surface zones

Cracks could be observed within the TTS; shape and direction differed for the several groups of alloys. For instance, TTS seemed to burst in the case of steels while longer cracks parallel to the surface were observed for titanium alloys. These differences in crack morphology were possibly related to the crystallography of the TTS. Measurement always indicated a great increase in hardness in the TTS as compared to bulk material, irrespective of the initial microstructure revealing hardening during cyclic sliding. When increasing the number of cycles, the sliding amplitude, and the normal force, more energy is dissipated and then it favors the TTS transformation.

Comparison between experiments on various microstructure and TEM observations indicates that plastic deformation is responsible for TTS formation. The maximum density in dissipated energy is observed at the center of the fretting scar, which also corresponds to the point of the wear and TTS analysis. A simple analytical relation is expressed which relates the local dissipated energy on such points (E_{d0}) as a function of the friction coefficient (μ), the maximum contact pressure (p_0), the contact radius (a), and the sliding amplitude (D). To differentiate the fretting loading from the reciprocating condition a sliding ratio is introduced:

$$e = \frac{D}{a}$$

If $e < 1$ the contact is under fretting, whereas when $e > 1$ reciprocating sliding is in operation. Such a sliding variable influences the dissipated energy. Under gross slip conditions ($e < 1$)

$$E_{d0} = 2a\pi p_0 \left[e\left(1 - e^2\right)^{\frac{1}{2}} + \arcsin\,(e) \right]$$

For reciprocating conditions ($e > 1$), the maximum density of dissipated energy is constant:

$$E_{d0} = 2a\pi p_0$$

The local energy approach permits quantifying the different steps of the wear processes which consist, primarily, of initiating the TTS layer from plastic strain accumulation. After TTS stabilization, wear extends in depth through several fronts of progression defined first by a subsurface plastic domain, a constant TTS layer, and surface wear. The TTS generation kinetics next to the plastic domain is assumed to be similar to TTS surface degradation in order to maintain a constant TTS thickness. Reaching such a steady-state wear process, a linear evolution is usually observed between the wear volume and the accumulated dissipated energy (Sauger et al. 2000).

Experiments performed on NC Cu (10 nm on the surface up to 100 nm grain size at a thickness of 25 μm) showed different mechanisms as a function of the fretting conditions (Zhang et al. 2008). Under a low load, discrete oxide debris layers on the plate and the ball are formed during fretting test, because of a small quantity of remained oxide debris at the contact area. When the load is increased, more oxide debris would be generated during the fretting test since the wear volume increases with an increase of the load. Moreover, the contact temperature on the solid surface subjected to relative slip could increase with an increasing load, which would significantly promote tribo-oxidation reaction. So under loads in excess of a critical value, a continuous oxidation layer should be formed between tribo-pair. In this case, the friction between Cu and the counterpart ball is transformed completely to the friction between Cu oxide layer on the plate and the ball, which thus results in an increased friction coefficient. Since the adhesion between the transfer layer and the ball is poor, material transfer onto the counterpart and detachment from it would take place alternatively. Consequently, when the loads exceed a critical value, there occurs a notable increase in the friction coefficient and the wear volume. It is reasonable to find that for frequencies exceeding 50 Hz, the contact between the Cu plate and the ball is also replaced completely by a Cu oxide and Cu oxide tribological pair, leading to an abrupt increase of the friction coefficient and wear volume, completely similar to the situation under high loads. This behavior can be related to the increasing wear loss and contact temperature, which can be encouraged by the frequency. When the frequency increases to 175 Hz, wear debris should be eliminated from the worn track so quickly because of the high rate of debris ejection that it even has no time to be oxidized or transferred to the ball surface. It is suggested that high contact temperature as a result of high frequency will lead to the softening of the materials to some extent, which should also partly account for the sharp increase of the friction coefficient and wear volume at a frequency of 175 Hz.

9.3.2 Size Effect

The deformation mechanisms are expected to be controlled by scale effects and be fundamentally different. Thus, NC materials offer an opportunity to impact the the wear behavior through exploration of their mechanical deformation under sliding contact. Sliding contact induces significant compression and shear at asperities, leading to large plastic deformations even at low-applied loads (Qi et al. 2009). With increasing strain, nano-crystallization has been envisioned to progress through a sequence of dislocation emission, formation of elongated microbands, dislocation cells, dislocation cell blocks, and equiaxed submicro- and nanocrystal grains. The final grain size is reached when the grain size merges with the sub-grain size, the smallest available length scale within which dislocations are absorbed. The reduction in the grain size is also limited by the rate of recovery during processing. At certain strain level, dislocations annihilate and self-assemble to form small-angle grain boundaries, reducing the dislocation density. Thus, a high dislocation density is required for substantial grain size refinement. The final grain size is determined by the competition between deformation and recovery. Thus, high-melting-point elements are expected to produce smaller grain sizes. The smallest grain size for Ni produced by mechanical attrition has been found to be around 12 nm. Besides nanocrystallization occurring during sliding wear, the subsurface region is simultaneously subjected to alternate tensile and compression stresses leading to wear by fatigue. The latter process continuously interrupts the nano-crystallization process resulting in delamination cracking and spalling of material portions. Indeed, development of a high dislocation density in conjunction with triple grain boundary interaction can lead to crack nucleation below the surface disrupting the nanocrystallization processes and resulting in a high wear rate (Kolomeichenko and Kuznetsov 2014). A distinction should be made, however, between nanocrystallization of MC metals during sliding wear and nano-crystallization induced by surface SPD. It should be noted that the latter metals exhibit improved tribological performance.

Deformation-induced coarsening is also observed and attributed to grain boundary migration. Grain coarsening has also been observed in NC Ni during scratch experiments using a nanoindenter. Deformation-induced coarsening in NC Ni involves a very thin surface layer that is continuously supported by the underlying high-hardness nanostructured substrate. Such high hardness has been found to diminish damage accumulation in nanostructured metals and increase wear resistance. Thus, due to the particular substructures two different mechanisms operate: wear/fatigue followed by removal of coarse pieces of material in the MC Ni and ultrafine abrasion wear in the NC Ni.

Some other evidences show that the microstructure is unstable during cyclic sliding in NC Ni and NC Ni-W showing grain coarsening and hardening in grains starting from 3 to 10 nm mean grain size and measuring 20 nm after sliding (Rupert and Schuh 2010).

Also nanotwinning has an influence on the sliding resistance of nanostructured materials. In Cu, the density increase of nanotwins leads to high resistance to surface damage and microstructure modification (Singh et al. 2011). Recent studies have also revealed that nanoscale twins introduced in the microstructure lead to improved mechanical properties in terms of not only higher strength, but also retention of reasonable ductility, higher rate sensitivity of deformation, and greater resistance to both fatigue crack initiation and growth. These effects can be further enhanced through refining nanotwin (NT) spacing. However, high-quality pure nanotwinned specimens have thus far been produced successfully only in thin-film form. This can impose severe constraints on potential applications of nanotwinned materials for bulk structural components (Singh et al. 2012). Design against wear damage is a topic of considerable technological interest as it is one of the most common ways of material loss in most engineering applications. Excessive material removal due to repeated frictional sliding and rubbing between contacting surfaces can lead to device dysfunction in microelectromechanical systems, failure of engineering components such as ball bearings, and adverse immune response in the human body to metal-based biological implants. There have been only limited studies done on the wear response of nanograined (NG) and NT materials in general. In NG Ni, wear damage has been shown to decrease with decreasing grain size, but beyond the Hall-Petch breakdown point increased material loss was observed with grain refinement. However, degraded wear response with strengthening has been observed in NG Ni-B alloy film produced by electrodeposition and NG iron produced by rolling. Grain coarsening under repeated sliding has been documented for NG Ni as well as for NG Ni-W (Mahidashti et al. 2018). Moreover, a recent experimental study on repeated frictional sliding of NT Cu within ultrafine grains averaging 450 nm in size confirmed that higher density NT Cu exhibited greater resistance to both microstructure changes and surface damage after a frictional sliding pass; however, after many repeated sliding passes, copper samples with different NT densities were found to exhibit similar surface hardness and microstructure. This behavior bears a striking resemblance to the uniaxial strain-controlled fatigue behavior of medium-to-high stacking fault energy (SFE) metals, in which a uniform steady-state structure evolves after repeated mechanical loading as a result of rearrangements in defect structure facilitated by cyclic deformation. The DPD process improves hardness while providing improved tribological properties compared to CG Cu. The introduction of nanograins and nanotwins in DPD nano-Cu appears to aid in reducing the material removal under repeated frictional sliding (Yan et al. 2019). The friction coefficients of DPD nano-Cu and CG Cu during repeated sliding were found to be similar in spite of significant differences in their initial hardness values. This may be attributed to the significantly higher strain hardening capability of CG Cu than that of DPD nano-Cu. Under repeated frictional sliding, DPD nano-Cu softens and CG Cu hardens, and eventually both specimens exhibit similar values of hardness within the deformation-affected zone.

9.4 Conclusions

It is evident that the significant increase in strength (hardness) arising from the grain size reduction in the nano domain is expected to impact the mechanical processes at asperity contacts that are critical in determining wear behavior. Tribological studies show significant reductions in the coefficient of friction and wear rate for NC metals produced by electroplating compared to its microcrystalline counterpart.

In the case of electrodeposited pure metals in the NC and UFG regimes, a significant increase in friction coefficient was shown with decreasing mean grain size. Such a behavior becomes less evident by increasing the applied load and the penetration depth due to the different strain hardening of the material which increases with decreasing mean grain size. It was also demonstrated that the indentation response is strongly influenced by the surface finishing. The hardness decreases with increasing indentation load in those materials exhibiting low roughness levels while it results to be constant in those materials with higher surface roughness.

It has been shown that if the applied coating exhibits low-friction characteristic, there could be a significant reduction in the fretting fatigue. Sliding contact entails significant compression and shear stresses at asperity contacts that are expected to be accommodated by entirely different mechanical deformation mechanisms in MC and NC metals and alloys. Under the repeated loading, large plastic strains are expected in the former resulting in work hardening but also wear by contact fatigue.

Also nanotwinning has an influence on the sliding resistance of nanostructured materials. The density increase of nanotwins leads to high resistance to surface damage and microstructure modification. Recent studies have also revealed that nanoscale twins introduced in the microstructure lead to improved mechanical properties in terms of not only higher strength, but also retention of reasonable ductility, higher rate sensitivity of deformation, and greater resistance to both fatigue crack initiation and growth.

References

AlMotasem AT, Bergström J, Gåård A et al (2017) Atomistic insights on the wear/friction behavior of nanocrystalline ferrite during nanoscratching as revealed by molecular dynamics. Tribol Lett 65:101. https://doi.org/10.1007/s11249-017-0876-y

Amanov A, Cho I-S, Kim D-E, Pyun Y-S (2012) Fretting wear and friction reduction of CP titanium and Ti–6Al–4V alloy by ultrasonic nanocrystalline surface modification. Surf Coat Technol 207:135–142. https://doi.org/10.1016/j.surfcoat.2012.06.046

Argibay N, Bares JA, Keith JH, Bourne GR, Sawyer WGG (2010) Copper-beryllium metal fiber brushes in high current density sliding electrical contacts. Wear 268(11–12):1230–1236. https://doi.org/10.1016/j.wear.2010.01.014

Boyce BL, Padilla HA (2011) Anomalous fatigue behavior and fatigue-induced grain growth in nanocrystalline nickel alloys. Metall Mater Trans A 42:1793–1804. https://doi.org/10.1007/s11661-011-0708-x

Cavaliere P, Prete P (2010) Tribomechanisms of pure electrodeposited Ni at ultra-fine and nano-scale level. Wear 268:1490–1503. https://doi.org/10.1016/j.wear.2010.02.027
Chang S-Y, Lee Y-S, Hsiao H-L et al (2006) Mechanical properties and deformation behavior of amorphous nickel-phosphorous films measured by nanoindentation test. Metall Mater Trans A 37:2939–2945. https://doi.org/10.1007/s11661-006-0175-y
Curry JF, Babuska TF, Furnish TA, Lu P, Adams DP, Kustas AB, Nation BL, Dugger MT, Chandross M, Clark BG, Boyce BL, Schuh CA, Argibay N (2018) Achieving ultralow wear with stable nanocrystalline metals. Adv Mater 30:1802026. https://doi.org/10.1002/adma.201802026
Dupont V, Sansoz F (2010) Multiscale modeling of contact-induced plasticity in nanocrystalline metals. In: Dumitrica T (ed) Trends in computational nanomechanics, Challenges and advances in computational chemistry and physics, vol 9. Springer, Dordrecht. https://doi.org/10.1007/978-1-4020-9785-0_6
Fathollahzade N, Raeissi K (2014) Electrochemical evaluation of corrosion and tribocorrosion behavior of amorphous and nanocrystalline cobalt-tungsten electrodeposited coatings. Mater Chem Phys 148:67–76. https://doi.org/10.1016/j.matchemphys.2014.07.013
Fu L, Yang J, Bi Q et al (2012) Sliding wear behaviour of nanocrystalline Fe88Si12 alloy under low load and speed. Tribol Lett 48:329–335. https://doi.org/10.1007/s11249-012-0028-3
Gao N, Wang CT, Wood RJK et al (2012) Tribological properties of ultrafine-grained materials processed by severe plastic deformation. J Mater Sci 47:4779–4797. https://doi.org/10.1007/s10853-011-6231-z
Guidry DJ, Lian K, Jiang JC, Meletis EI (2009) Tribological behavior of nanocrystalline nickel. J Nanosci Nanotechnol 9:4156–4163. https://doi.org/10.1166/jnn.2009.M25
Hadipuor A, Rahsepar M, Hayatdavoudi H (2019) Fabrication and characterisation of functionally graded Ni–P coatings with improved wear and corrosion resistance. Surf Eng 35(10):883–890. https://doi.org/10.1080/02670844.2018.1539295
Hoornaert T, Hua ZK, Zhang JH (2009) Hard wear-resistant coatings: a review. In: Luo J, Meng Y, Shao T, Zhao Q (eds) Advanced tribology. Springer, Berlin. https://doi.org/10.1007/978-3-642-03653-8_257
Kolomeichenko AV, Kuznetsov IS (2014) Tribotechnical properties of electrospark coatings of amorphous and nanocrystalline iron alloys. J Frict Wear 35:501–504. https://doi.org/10.3103/S1068366614060087
Leyland A, Matthews A (2000) On the significance of the H/E ratio in wear control: a nanocomposite coating approach to optimized tribological behavior. Wear 246(1–2):1–11. https://doi.org/10.1016/S0043-1648(00)00488-9
Ma G, Yang J, Liu Y et al (2013) Friction and wear behavior of nanocrystalline nickel in air and vacuum. Tribol Lett 49:481–490. https://doi.org/10.1007/s11249-012-0089-3
Mahidashti Z, Aliofkhazraei M, Lotfi N (2018) Review of nickel-based electrodeposited tribo-coatings. Trans Indian Inst Metals 71:257–295. https://doi.org/10.1007/s12666-017-1175-x
Meng Y, Xu J, Jin Z et al (2020) A review of recent advances in tribology. Friction 8:221–300. https://doi.org/10.1007/s40544-020-0367-2
Murty BS, Shankar P, Raj B, Rath BB, Murday J (2013) Unique properties of nanomaterials. In: Textbook of nanoscience and nanotechnology. Springer, Berlin. https://doi.org/10.1007/978-3-642-28030-6_2
Musil J (2012) Hard nanocomposite coatings: thermal stability, oxidation resistance and toughness. Surf Coat Technol 207:50–65. https://doi.org/10.1016/j.surfcoat.2012.05.073
Nazemian M, Chamani M (2019) Experimental investigation and finite element simulation of the effect of surface roughness on nanoscratch testing. J Mech Sci Technol 33:2331–2338. https://doi.org/10.1007/s12206-019-0432-9
Pinchuk VG, Buyanovskiy IA, Korotkevich SV (2015) Kinetics of microstructure and selective mechanism of fracture of metal surface layer under friction. Inorg Mater Appl Res 6:355–362. https://doi.org/10.1134/S2075113315040206

Qi Z, Jiang J, Meletis EI (2009) Wear mechanism of nanocrystalline metals. J Nanosci Nanotechnol 9:4227–4232. https://doi.org/10.1166/jnn.2009.M37

Rupert TJ, Schuh CA (2010) Sliding wear of nanocrystalline Ni–W: structural evolution and the apparent breakdown of Archard scaling. Acta Mater 58:4137–4148. https://doi.org/10.1016/j.actamat.2010.04.005

Sauger E, Fouvry S, Ponsonnet L, Kapsa P, Martin JM, Vincent L (2000) Tribologically transformed structure in fretting. Wear 245:39–52. https://doi.org/10.1016/S0043-1648(00)00464-6

Shen YF, Liu WN, Sun X et al (2012) Plastic deformation in amorphous Ni-P coatings. Metall Mater Trans A 43(5):1610–1620. https://doi.org/10.1007/s11661-011-0989-0

Singh A, Dao M, Lu L, Suresh S (2011) Deformation, structural changes and damage evolution in nanotwinned copper under repeated frictional contact sliding. Acta Mater 59:7311–7324. https://doi.org/10.1016/j.actamat.2011.08.014

Singh A, Tao NR, Dao M, Suresh S (2012) Repeated frictional sliding properties of copper containing nanoscale twins. Scripta Mater 66:849–853. https://doi.org/10.1016/j.scriptamat.2012.02.017

Wen W, Becker AA, Sun W (2017) Determination of material properties of thin films and coatings using indentation tests: a review. J Mater Sci 52:12553–12573. https://doi.org/10.1007/s10853-017-1348-3

Yan J, Lindo A, Schwaiger R et al (2019) Sliding wear behavior of fully nanotwinned Cu alloys. Friction 7:260–267. https://doi.org/10.1007/s40544-018-0220-z

Zhang YS, Han Z, Lu K (2008) Fretting wear behavior of nanocrystalline surface layer of copper under dry condition. Wear 265(3–4):396–401. https://doi.org/10.1016/j.wear.2007.11.004

Chapter 10
Effect of Environment on Microstructure and Mechanical Properties of Nanostructured Metal Alloys and Composites

10.1 Introduction

As largely shown in the previous chapters, nanostructured materials into the nanometer range have been shown to result in unique and improved properties as compared to their conventional polycrystalline counterparts. An enhancement in hardness, ductility, fatigue behavior, coercivity, wear, and corrosion resistance was often observed from the reduction in the grain size towards the UFG and NC regimes. For many applications such as protective coatings and electrical contact components, corrosion damage of these nanostructures can result in the development of cracks or pores and ultimately malfunction of the components. The enhanced properties of the nanostructured material are controlled by their microstructure which is generally seen in the increase of their grain boundaries and triple junctions. Grain size reduction in NC materials considerably improves the corrosion performance for a wide range of electrochemical conditions. This is mainly due to the elimination of the localized attack at grain boundaries, which is one of the most detrimental mechanisms of degradation in polycrystalline materials. Several explanations have been put forward for this effect: the solute dilution effect by grain size refinement, crystallographic texture changes with decreasing grain size, and grain size-dependent passive layer formation.

One of the prime advantages of electrodeposition is that there are many parameters such as current density, solution concentration, pH, and addition of additives to the plating bath that could be adjusted to obtain suitable nanostructured deposits. Factors accelerating the nucleation rate and inhibiting grain growth are favorable to the formation of ultrafine-grained deposits. A large cathodic overpotential is important in achieving a high nucleation rate, which results in the formation of crystalline nuclei that are small in size and also in large quantities. The reduction of grain size can be attributed to the higher overpotential which is related to the crystalline nucleus formation probability W expressed by the following equation:

© Springer Nature Switzerland AG 2021

P. Cavaliere, *Fatigue and Fracture of Nanostructured Materials*,
https://doi.org/10.1007/978-3-030-58088-9_10

$$W = A \exp\left(\frac{-b}{\eta_k^2}\right)$$

where A and b are constants. The overpotential increases with the increasing current density as described in the following equation:

$$\eta_k = a + b \log i$$

where a and b are constants and i is the current density. Therefore, with increasing current density reduction in grain size could be achieved. The current density should be controlled and held within a suitable range with respect to bath composition and temperature. Insufficient or low current density will result in poor coverage of recesses and a low plating rate. On the other hand, the presence of an excessive current will cause other difficulties such as stress and high traces of impurities, and also may produce a burned plate. The effect of current density on the properties of nanocrystalline Ni-Co-W alloy was reported and the results obtained showed that, with increased current density, both the Co and W contents decreased. This resulted in better corrosion resistance of these nanocrystalline alloys (Farzaneh et al. 2010). The presence of different ion concentrations could also lead to different properties of electrodeposited coatings. With a fixed concentration of cobalt ions, the iron content in nanostructured CoFe alloy deposits increased gradually with increasing Fe^{2+} concentration in the electrolyte (Nik Masdek and Alfantazi 2012). The increase in iron content in the electrodeposits resulted in a refinement of grain size and thus increased the microhardness of these electrodeposited nanocrystalline alloy coatings. Maintaining pH of the plating solution is crucial in providing stability of the bath during deposition process. The optimal pH value has been reported to be in the range of 2–6. Very high pH values could cause the formation of hydroxide precipitates of the metal ions and thus results in unstable bath solutions. Meanwhile, low pH values result in excessive hydrogen evolution. Control of the deposition bath temperature is vital for the consistent performance of any deposition process. Small deviation of the bath temperature is sufficient to harm plating quality and deposition rates. Hence, bath temperature is one of the primary parameters that have a huge impact on the properties of deposited nanomaterials. The deposition temperature was observed to significantly change the microstructure, morphology, and magnetic properties of electrodeposited films. The composition and grain size change considerably with increasing temperature. In order to ensure a nanostructured grain size, the grain growth of existing grains could also be suppressed by certain inhibiting molecules. Therefore, adding organic inhibitors to the electrolyte has been a common practice during the electrodeposition process. These organic additive molecules are believed to adsorb in a reversible way on active sites of the electrode surface and thus block the active sites and reduce crystallite growth. In addition, the absorbed organic molecules also hinder the surface diffusion of adatoms. This results in fewer metal adatoms reaching growth sites and promotes formation of new nuclei. In general, these additive molecules decomposed at the cathode surface promote nucleation and impede grain growth, which refines the grain size further by at least two orders of magnitude.

The electrochemical corrosion behavior of nanocrystalline materials is expected to be different from that of polycrystalline materials due to the presence of higher volume of grain boundaries and triple junctions (more active sites available), and higher diffusivity of the alloying element. Other parameters such as solute atoms, impurities, grain size, surface defects, and precipitate distribution could lead to the difference of corrosion resistance between nanocrystalline and conventional materials.

Environmentally assisted cracking (EAC) is a very critical materials science problem that concerns many technological areas. This type of corrosion is initiated at the microscopic level and is complicated due to the combination of chemistry (reaction caused by corrosive agents) and mechanics (varying load). As EAC is generally related to the segregation of impurity elements to defects (mainly grain boundaries), the symptoms of risk may not be apparent from the exterior of the metal components: hence EAC remains latent and gives no sign of warning until the failure occurs. Under certain conditions, the cracks or failure of materials happen at stress far below their tensile strength and this corrosion incident is known as environment-assisted cracking (EAC). Broadly, environment-assisted cracking includes a group of physical phenomena such as stress corrosion cracking (SCC), corrosion fatigue (CF), and hydrogen embrittlement (HE). EAC is mainly caused by corrosive environments and mechanical load. Although the dissolution and diffusion of impurity particles responsible for corrosion may proceed without the assistance of stress, still the stress plays a major role in promoting crack initiation and growth. More specifically, stress-corrosion cracking may be facilitated by static or residual stresses above some threshold value and corrosion fatigue can be initiated by fluctuating or cyclic stresses (Chabok et al. 2015). Similar to intergranular corrosion, EAC is related to the dissolution of solutes, especially the ones with corrosive effects into host materials. In particular, the presence of defects such as vacancies and grain boundaries induces higher solubility of impurity atoms, and makes materials more prone to cracking as a consequence. Stress corrosion may induce brittle cracks even in ductile materials.

10.2 Corrosion of Nanostructured Materials

The environment as well as the nature of the material also affects the electrochemical response of these fine grain materials (Andrievskii 2013). The intercrystalline surface area fraction may be closely related to the corrosion properties since the intercrystalline defects are considered as preferential dissolution sites. If the intercrystalline surface area fraction is considered to be related to the cathode-to-anode area ratio, nanocrystalline materials are believed to show improved performance against corrosion than materials with microcrystalline structures. If the corroding metal is assumed to be equivalent to a short-circuited cell where energy is dissipated during consumption of a cathodic reagent there will be no external source of electrons. When the area of the anodic and cathodic site is large, the current densities need not be equal. However, when there is same surface area between anodic and cathodic sites as for nanocrystalline materials, the anodic current density can equal the cathodic current density in absolute value. Thus, reduction in grain size will reduce the penetration current density and result in uniform and better localized corrosion resistance due to the highly distributed corrosion current. It is expected that nanocrystalline materials exhibit a higher corrosion rate because of the high volume of grain boundaries, dislocations, and triple points that serve as anodic sites. Although nanostructure may accelerate the corrosion rate by producing more electrochemical reaction sites between the large amount of grain boundaries and the matrix, the high density of grain boundaries and lattice defects could also provide more nucleation sites for formation of protective passive films which, in return, results in a high fraction of passive layers and high corrosion resistance.

Potentiodynamic and potentiostatic polarization tests were used to analyze the corrosion performance of nanocrystalline nickel in deaerated 2 N H_2SO_4. The typical active-passive-transpassive behavior was obtained for nanocrystalline nickel electrodeposits with grain sizes ranging from 500 to 32 nm (Shiraman et al. 2007), which was similar to the polycrystalline nickel, although the passive current density was seen to be one magnitude higher. The increased electrochemical response for the nanocrystalline Ni was due to the higher dissolution rate corresponding to the increased intercrystalline region that provided more active sites for corrosion activity. However, this passivation current density between the samples was negligible at high potentials.

The corrosion of both microcrystalline and nanocrystalline Ni with grain sizes ranging from 2 μm to 16 nm was studied in three different solutions (10% NaOH, 3% NaCl, and 1% H_2SO_4 solutions). The lowest passive current was observed for nanocrystalline Ni with grain size of 16 nm in NaOH while the corrosion current of nanocrystalline Ni in NaCl was reported to be ten times lower as the grain size decreased. The results indicate that the corrosion response of nanocrystalline materials depends not only on their grain sizes, but also on the environment where the material is being exposed (Andrievski and Khatchoyan 2016a, b).

Corrosion behavior of nanostructured Ni deposits was compared with bulk nickel in an H_2SO_4 solution. Results confirmed that a higher number of defects on the passive film resulted in easier and more diffusion of nickel cations through the defected film and thus created a higher passive current density. A significant shift of zero current potential towards the noble direction was observed for nanostructured deposits as compared to bulk nickel due to the difference in grain size and catalysis of hydrogen processes (Fig. 10.1).

The effect of grain size and solute segregation for both polycrystalline and nanocrystalline Ni revealed uniform corrosion morphology with high density of evenly distributed corrosion pits of less than 2 μm deep on the nanocrystalline Ni coatings. Passive film formed on the nanostructured specimen is more defective than that formed on the polycrystalline specimen, while the thickness of the passive layer on both specimens was similar. This resulted in a more uniform breakdown of the

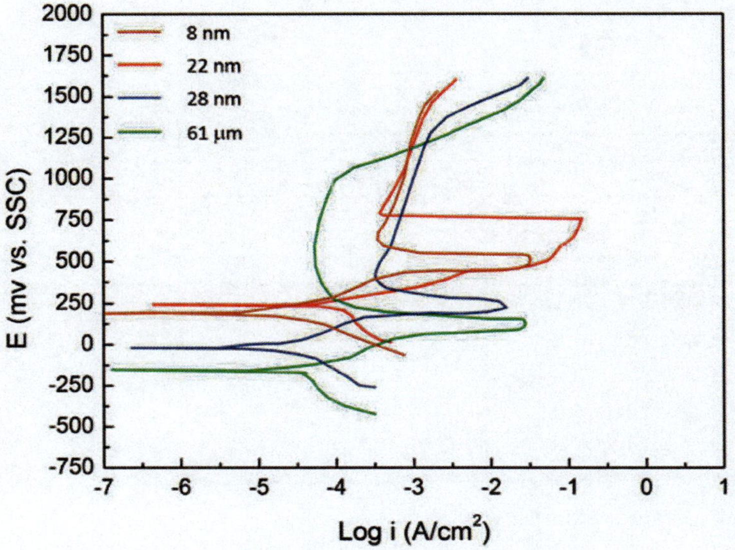

Fig. 10.1 Potentiodynamic polarization curves of nickel of different grain sizes

passive film. Meanwhile, a contrast on the corrosion morphology was observed for the coarse-grained Ni which exhibited severe localized corrosion mainly along grain boundaries and triple junctions. The breakdown of the passive film occurs first at the grain boundaries and triple junctions rather than the crystal surface, leading to preferential attack at these defects. The general corrosion was somewhat enhanced compared to conventional coarse-grained nickel; however, the nanostructured nickel was much more immune to localized attack which often can lead to catastrophic failure.

As a matter of fact, in nanocrystalline-electrodeposited alloys, the GB structure is fundamental in governing the SCC. Depending on its circumstance, EAC failure can be either transgranular (ignoring grains) or intergranular (following grain boundaries). The latter case may resemble intergranular corrosion, as tensile stress can induce crack initiation along grain boundaries. In Fig. 10.2, the fracture behavior of pure NC Ni during SCC is shown; the analyses belong to MD simulations.

The SCC behavior of solute-enriched NC Ni is shown in Fig. 10.3.

Electrochemical theories are based on the postulate that cracks progress along continuous localized paths such as grain boundaries, under segregation of chemically reactive elements. Due to the difference in chemical potential, those paths are more susceptible to corrosion and become sources of stress concentration. Sufficient stress might lead to mechanical fracture and as a result the metal tears apart. Since the GB is a planar discontinuity where two single crystals meet, it serves as an ideal channel for the migration of atoms. Additionally, these boundaries have low coordination and are prone to mechanical damage as they are more brittle, compared to the bulk crystal. The presence of impurity elements may modify the GB energy and

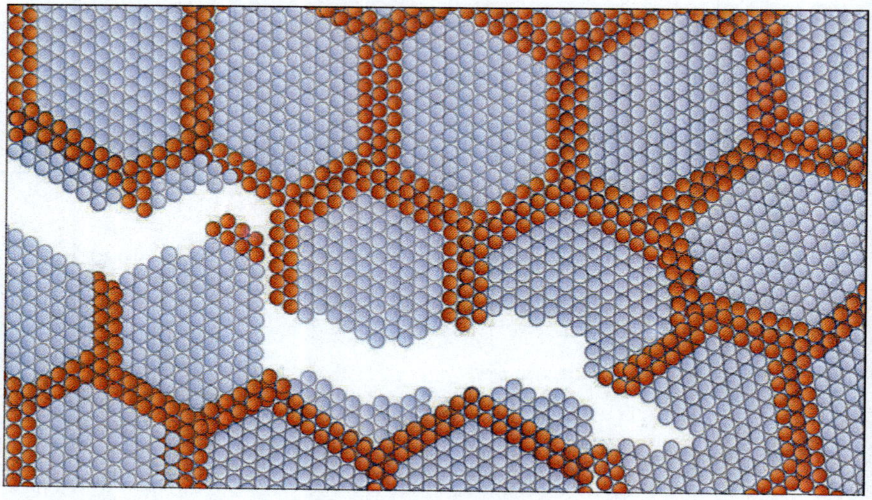

Fig. 10.2 Fracture of SCCed pure NC Ni

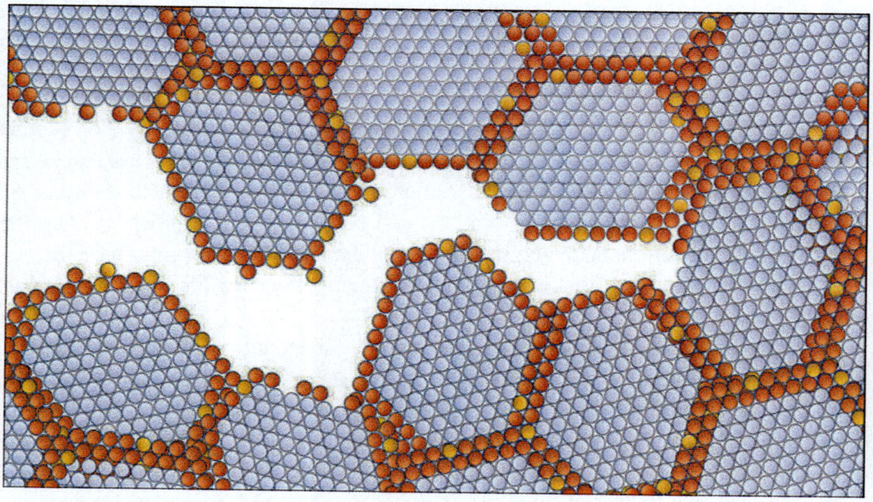

Fig. 10.3 Fracture of SCCed solute-enriched NC Ni

local crystal structure, including substantial changes in interatomic distances. First of all, pure NC Ni shows intergranular fracture indicating that grain boundaries are not degraded during SCC; on the other hand, solute addition (mainly diffused at the high-energetic grain boundary) leads to GB embrittlement.

Nanocrystalline materials have a large volume fraction of grain boundaries, many of which are special low-sigma CSLs formed during electrodeposition. According to the Σ3 regeneration model, the propensity for Σ3 boundary formation becomes greater as more preexisting grain boundaries have Σ3 character. This predominance of HAGBs would also favor dislocation-accommodated grain boundary sliding during deformation. This accommodation process could prevent or delay the formation of voids at triple junctions during deformation. Thus, ED NC metals with a large percentage of these special boundaries have great potential for improved salient mechanical properties in addition to excellent corrosion resistance. Corrosion resistance has been demonstrated to be significantly superior in several ED NC metals and alloys as compared to their coarse-grained counterparts. The susceptibility to localized corrosion was also found to be lower for ED NC Ni. These attributes of superior corrosion resistance of ED NC Ni can be ascribed to the large volume fraction of naturally occurring coherent low-sigma CSL boundaries. This characteristic could have far-reaching implications for engineering NC materials since increasing the relative frequency of these special boundaries not only augments corrosion resistance but should also enhance salient mechanical properties as a result of dislocation-accommodated grain boundary sliding (Roy et al. 2008).

The corrosion properties of nanocrystalline iron (Fe) were also investigated to determine the effect of surface nano-crystallization produced from pulse electrodeposition. The results verified the beneficial effect of grain size reduction on the formation of a superior protective passive film on the coating surface and thus significantly enhanced the corrosion resistance of the nanostructured Fe deposits.

The transport of ions in passive films takes place by migration and/or diffusion of defects (grain boundaries, linear dislocations, vacancies, or interstitials), so more protective passive layer is formed on nanocrystal-electrodeposited coating. This may be attributed to the widening of the energy band by decreasing the grain size.

Wang et al. (2006) reported on both the positive and negative effects of the fine grain size on the corrosion performance of nanocrystalline Co in different test solutions. The positive effect of high volume fraction of grain boundaries due to the small grain size of nanocrystalline Co produced by double-pulse current electrodeposition could be observed in alkaline solutions (NaOH and NaCl). Formation of a protective passive film was enhanced by the high density of grain boundaries leading to a much higher corrosion resistance as compared to coarse grain Co coatings. On the other hand, negative effects were present for the corrosion performance of nanocrystalline Co in HCl solutions. The grain boundaries were preferential attack sites when exposed to high-acidity environments causing detrimental corrosion behavior. The results here were similar as observed for nanocrystalline Ni in both acidic and alkaline solutions.

Adding alloying elements is believed to improve the corrosion behavior of nanostructured alloy coatings. This is likely because the diffusivity of alloying and impurity elements in nanocrystalline materials is much higher as compared to polycrystalline materials which leads to better corrosion resistance. It is reported that the boundary diffusivity of nanocrystalline materials is about three orders of magnitude larger than that in conventional polycrystals.

This positive effect of alloying can be seen on Co-P nanocrystalline alloys, where the addition of P leads to a remarkable increase in corrosion resistance in an acidic aqueous medium. A positive shift in corrosion potential (59 mV) was seen for nanocrystalline Co-1.1 wt%P which significantly reduced the anodic dissolution rates as compared to pure nanostructured Co deposits. In addition, at open circuit potential, total interfacial impedance of nanocrystalline Co-1.1 wt%P was significantly larger than that of nanocrystalline Co as observed from EIS measurements.

The corrosion rate of the nanocrystalline Ni-P alloy was also considered to be significantly enhanced due to the enrichment of P on the surface coating which inhibited the dissolution rate. Dissolution was limited by the formation and absorption of the hypophosphite anion on the alloy surface that forms a barrier layer between the alloy and the electrolyte and prevents H_2O from reaching the alloy surface. Various experimental results have shown that the nanostructured sleeve is intrinsically resistant to intergranular attack and intergranular stress corrosion cracking due to the P alloying. The material was also found to be resistant to pitting attack and only slightly susceptible to crevice corrosion.

Meanwhile, alloying nanostructured Ni with W has been found to be beneficial in several applications, for example, as a barrier coating in electronics and for wear protection in engineering components, owing to their superior strength and thermal stability. However, it was reported that the competing effects of W content and grain size evidently govern the corrosion behavior of the nanocrystalline Ni-W alloy. While an increase in W content has been noted to promote the formation of a mixed oxide film that increases corrosion resistance, it also leads to finer grains providing more active sites for corrosion reaction.

The addition of Cr nanoparticles affected the electrochemical behavior, particularly pitting corrosion of the electrodeposited nanocrystalline Ni. Numerous pits occurred for the 4.5 wt% Cr-nanocrystalline Ni although, with an increase in Cr addition, the formation of pits was effectively prevented. High Cr content (10.9 wt.%) increased pitting corrosion resistance with fewer pits present on the coating surface. This is due to the rapid formation of a dense protective Cr-oxide-rich passive film during the polarization test on the entire surface of the nanostructured Ni deposits. The results indicate that there exists a critical co-deposition content of Cr nanoparticles where, in this case, the critical value is presumed to be close to 11 wt.% Cr.

Ni alloying was seen to significantly enhance the surface morphology of the deposited Cu alloy. The fine grain size of Cu-Ni deposits proved to be the main reason in increasing the corrosion performance of these alloys in NaCl solutions.

In another study, the corrosion of an electrodeposited Ni-Co alloy with different amounts of Co in a 3.5% NaCl solution was investigated. The authors reported that both the crystal structure and surface morphology of the deposits significantly changed while varying the amounts of Co. A single FCC crystal structure was present for pure Ni deposits while a mixed structure (FCC + HCP) was observed for deposits with cobalt content of 50 wt.%. Further increase in cobalt was attributed to the change of phase structure to full HCP phase. The surface morphology of Ni-Co alloys changed from regular polyhedral crystallites to spherical cluster for a cobalt content of 50 wt.% and further increase resulted in increase in cluster size. Finally, for cobalt content of 80 wt.% and beyond there was a change to acicular crystallite morphology. The highest corrosion resistance was obtained from the

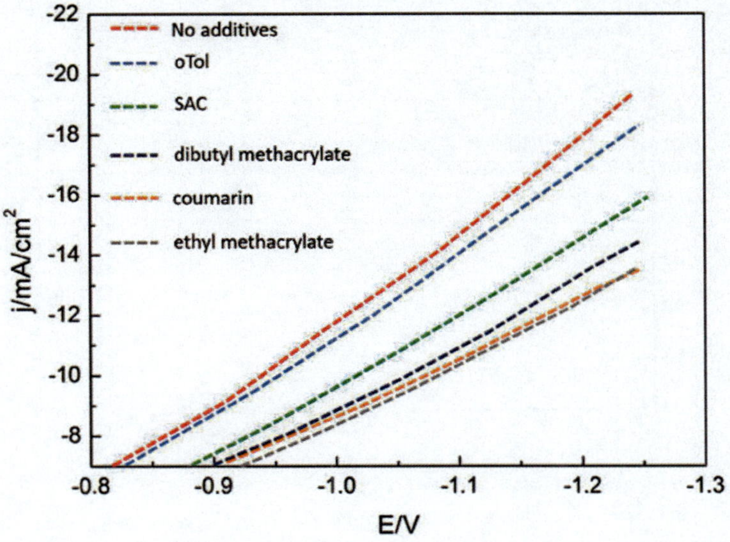

Fig. 10.4 Cathodic polarization curves of Co-W electrodeposition from electrolytes with different additives

Ni-20 wt.% Co alloy as compared to other Ni-Co alloys, pure Ni, and plain cobalt as observed from the polarization studies. The better corrosion performance of this alloy was due to the presence of both FCC and HCP crystal structure.

Corrosion of nanocrystalline alloys was also seen to be affected by organic additives used during the electroplating of these nanostructured deposits. Different types of organic additives (saccharin (SAC), *o*-toluene sulfonamide (oTOL), and methacrylate reagent) on the cathodic process of electrodeposited Co-W thin films were studied and it was found that the shape of the potentiodynamic polarization curves was similar for all curves, although a shift to more negative potential and decrease in current density were observed, depending on the presence and nature of the organic additives (Fig. 10.4).

From Fig. 10.4 we can see that the negative shift on the potential increased in the following order: oTOL, SAC, dibutyl methacrylate, coumarin, and ethyl-methacrylate.

Recent studies on CoFe nanocrystalline alloys show a marked grain refinement of the alloy with Fe content (Fig. 10.5).

The gradual increase obtained initially at lower iron content and the drop at higher iron content may be attributed to the change in the phase structure from a HCP- to BCC-phase structure as iron is further increased in the deposits. As can be seen, the hardness increases and reaches a maximum value of about 620 HV at around 18 wt. % Fe content which coincides with the BCC phase from deposits with an average grain size of 20 nm. Another important factor affecting the behavior of the microhardness of these nanocrystalline CoFe in this study may be attributed to the grain size effect. Initially, the increase at low Fe content in the deposits could be the

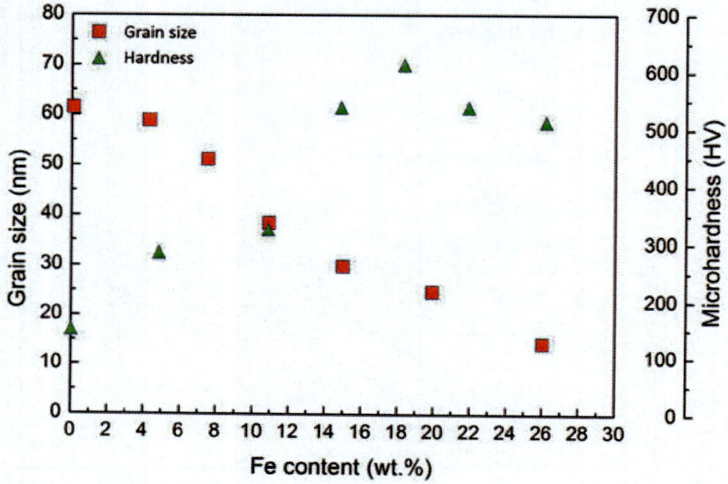

Fig. 10.5 Grain size and hardness variation with Fe content in CoFe-electrodeposited alloys

result from the refinement in grain size as the Fe increases in the alloy coatings. The increase in strength is due to the solid solution hardening with an increase in Fe of the alloy. The formation of an ordered Ni_3Fe in nanocrystalline NiFe due to faster grain boundary diffusion was proposed as the main factor for the increase in strength. On the other hand, the decrease in hardness at higher Fe content may be due to several different reasons. One of the possible reasons causing this deviation from the Hall-Petch equation is that the solid solution hardening is no longer effective for deposits with higher than 25 wt.% Fe. Furthermore, this softening effect may have resulted from the refinement of the grain size due to the significant increase of the intercrystalline volume fraction associated with the fraction of triple junction.

Now, grain boundary diffusion is a particularly important and interesting topic from the viewpoints of technology and basic understanding of diffusion processes in solids. It plays a crucial role in different processes such as sintering, creep, grain growth, or solid-state reactions, all of them affecting the properties of nanostructured materials. GB diffusion is crucial for corrosion in metals and transport properties of oxides. A simple model for diffusion is shown in Fig. 10.6.

The GB has a constant thickness δ, with the diffusion coefficient (D_{gb}) in the GB D_g within the grain.

The general equation for diffusion process is

$$\frac{\partial C_g}{\partial t} = D_{gb}\frac{\partial^2 C_g}{\partial y^2} + \frac{2D_g}{\delta}\frac{\partial C_g}{\partial x}\bigg|_{x=\frac{\delta}{2}}$$

wherein the solution for the present model is given by

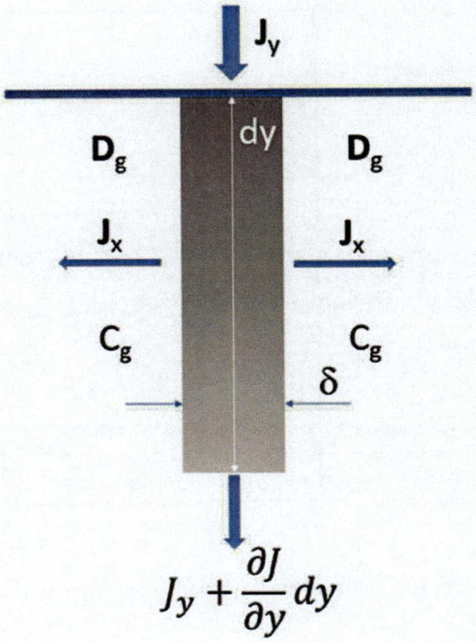

Fig. 10.6 GB diffusion model

$$C_g(\eta, \xi, \beta) = C_0 \operatorname{erfc}\left(\frac{\eta}{2}\right) + \frac{C_0 \eta}{2\pi^{\frac{1}{2}}} \int_1^{\Delta} \frac{d\sigma}{\sigma^{\frac{3}{2}}} \exp\left(-\frac{\eta^2}{4\sigma}\right) \operatorname{erfc}\left[\frac{1}{2}\left(\frac{\Delta - 1}{\Delta - \sigma}\right)\left(\frac{\sigma - 1}{\beta} + |\xi|\right)\right]$$

The solution is of integral form with the integral being responsible for GB diffusion. The first term of the solution, obviously, represents diffusion within the grain. The dimensionless quantities η and ξ are, in fact, dimensionless coordinates and given by

$$\eta = \frac{y}{\sqrt{D_g t}}$$

and

$$\xi = \frac{x \pm \delta/2}{\sqrt{D_g t}}, \text{ if } |x| \geq \frac{\delta}{2}$$

The meaning of the dimensionless parameter β is more complex. It characterizes the diffusion process in the sense that a large β-value means that the diffusion along the GBs is greatly pronounced in comparison with bulk diffusion; that is, the penetration depth along the GB is much larger. It varies over the orders of magnitude

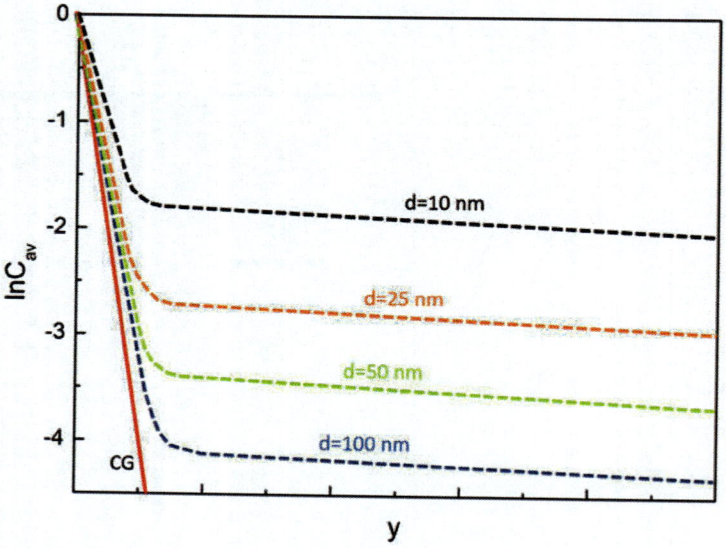

Fig. 10.7 Diffusivity as a function of grain size

from several thousands to tens as a function of the grain boundary dimensions. It is also related to the inclination angle between an isoconcentration line and the GB. Finally, it is high not only when Δ is large, but also when t is short. The analytical form of β is

$$\beta = \frac{(\Delta - 1)\delta}{2\sqrt{D_g t}}$$

In the very beginning of the diffusion process the atoms (traces) move through the GB without a significant contribution to the bulk. Nanocrystalline materials have β-values of the order of several thousands. So, considering C_{av} average concentration obtained by integrating C_g along the direction perpendicular to the grain boundary, the diffusion as a function of grain size is shown in Fig. 10.7.

Considering the previous equations, the segregation effect is quantified firstly considering the grain boundary volume fraction g:

$$D_{eff} = gD_{gb} + (1 - g)D_g$$

And taking into account the segregation factor s

$$D_{eff} = sgD_{gb} + (1 - sg)D_g$$

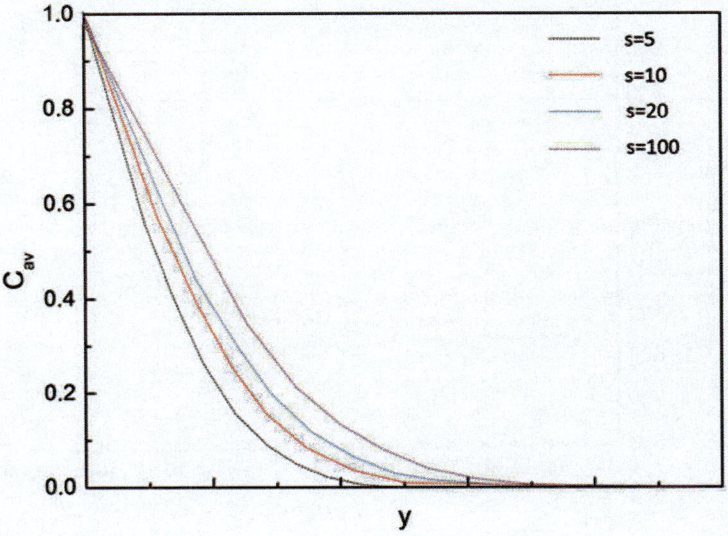

Fig. 10.8 Diffusivity in a ten-grain size as a function of segregation

That is again related to the grain size leading to a modification of the diffusion kinetic. Figure 10.8 shows the variation of C_{av} as a function of the segregation factor for a constant grain size of 10 nm.

Figure 10.9 illustrates the potentiodynamic polarization behavior of nanocrystalline Co and CoFe alloys with different iron contents in deaerated 0.1 M H_2SO_4 with a scan rate of 1 mV s^{-1}.

Typical anodic and cathodic polarization curves were observed. All samples exhibit active dissolution without any distinctive transition to passivation within the applied potential range. It is also apparent that Co coating exhibited a mass transfer limit which reached at about 0 mVSCE while CoFe alloy coatings exhibited a much lower limit. The effect of alloying iron led to an increase in the kinetics of the anodic metal dissolution which moved the corrosion potential (E_{corr}) to a more negative value and increased corrosion current.

Experiments performed on NC Ni-Fe (with starting grain size of 20 nm) subjected to annealing in order to induce grain growth showed that the material undergoes localized corrosion when immersed in 3.5 wt.% NaCl solutions during potentiodynamic-polarization experiments. The possible formation of Fe_2O_3 deposits surrounds the pits for the as-received NC material. The cracking phenomenon is most likely associated with the coupling effects of possible residual stresses and detrimental species, such as chlorides or hydrogen, on the pit core (Fig. 10.10).

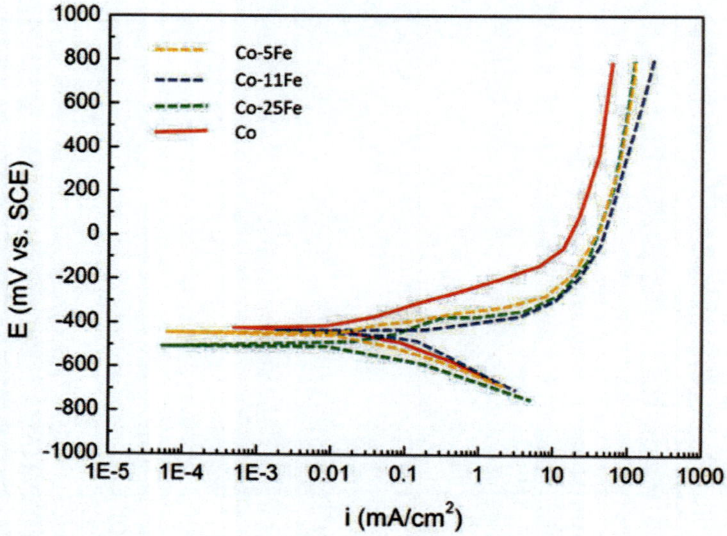

Fig. 10.9 Potentiodynamic polarization curves of CoFe with various iron contents

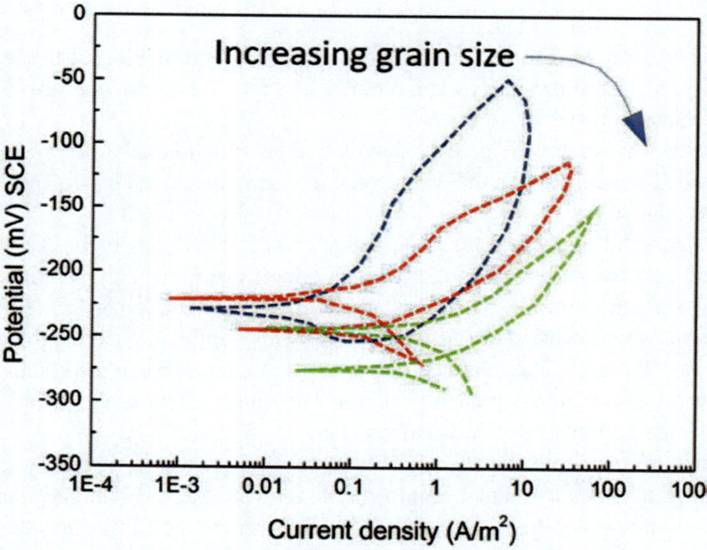

Fig. 10.10 Potentiodynamic polarization curves of NC Ni-Fe as varying the grain size

Overall, the corrosion behavior of the as-deposited nanocrystalline material is the best compared to those reported for the aged NC materials. The reason is believed to be the lack of surface area available for the completion of cathodic reactions by the consumption of electrons released from the oxidation of metal to metal ions in the aged NC materials.

10.3 SCC in Nanostructured Materials

The performance and lifetime of materials are often severely limited by corrosion under stress loads. Most critical here is premature and catastrophic failure of materials resulting from chemically influenced corrosion. The basic requirements for the operation of structural systems exposed to corroding conditions under stress loads are safety and reliability. Safe and reliable operation is endangered by the uncertainties in stress-corrosion cracking (SCC). Understanding the behavior of fine-grained materials that will be used in structures, and that are subjected to loading (static and cyclic) while exposed to aggressive environments, is of significant importance.

Overall, most stress-assisted cracking lies within two categories: stress-corrosion cracking and corrosion fatigue. Stress-corrosion cracking is attributed to a sustained-loading condition where the material, before the introduction of stresses, is susceptible to localized corrosion. The material can produce cracks at grain boundaries or pits. Cracking along grain boundaries is referred to as intergranular cracking. Cracks that emanate from pits, once these pits have reached a critical size, are most likely associated with the formation of atomic hydrogen as a by-product during metal hydrolysis. The increase in the local hydrogen relative to the bulk electrolyte creates a potential difference gradient. Hence, the mass transport of detrimental species, such as chlorides to the pit, is permitted. Additionally, the increase in the atomic hydrogen within the pit decreases the pH to cause acidification. At a critical pit depth, a crack (or cracks) initiates from the pits to cause SCC. Whether the culprit site is attributed to chemistry inhomogeneities, surface flaws, or pits, the addition of a stress to corrosion introduces a more complex region of activities. Thus, cracking occurs, resulting from the lattice dilation of diffused hydrogen into interstitial sites within the metallic lattice. The dilation causes a local condition of plane strain, which requires an applied stress equal to at least three times the yield strength to promote the crack growth and subsequent failure. The fracture behavior is a brittle one due to the large increase in the yield strength from the more macroscopic yield strength. The passivity region is marked by the formation of a protective film within a specific potential range. The transition from the active to the passive regions is one of the most susceptible areas to cracking since active processes and oxide formation processes are in competition. Likewise, the transition from the passivity to pitting is also vulnerable to cracking, since the depassivation and oxide formation occur simultaneously.

The resulting combined action of loading and aggressive environment often results in stress corrosion cracking and corrosion fatigue, both ensuing in damage that significantly reduces a material's fracture resistance. Since reduced fracture resistance can result in serious or catastrophic failure, a study of the environmental degradation mechanisms and fatigue properties of fine-grained materials is clearly needed (Sharma et al. 2015). Both forms of stress corrosion account for the influences of stresses on a material in a given electrolyte, and plausible crack initiation phenomena can be explained with the linear elastic fracture mechanics. However, there are some subtleties. SCC occurs under conditions of constant stresses. The fracture mechanics parameter, K, the stress intensity factor, can be used to quantify the stress in which corrosion will not enhance or expedite the propagation rate of a given crack size. The stress intensity factor associated with the enhanced behavior is known as the critical stress intensity factor, K_c, which is usually lower than that observed in air. However, corrosion fatigue is more complex in that the alternating stress is involved. When studying the behavior of corrosive media during crack propagation, ΔK_{ICF}, the critical stress intensity factor range due to corrosion can serve as a marker for this type of distinction. Stress intensity factor ranges below ΔK_{ICF} are mostly dominated by mechanical stresses, while values above ΔK_{ICF} are generally dominated by electrochemical influences. Further studies have established a relationship to determine the susceptibility of a material to corrosion fatigue by knowing the fatigue endurance limit in both corrosive and ambient environments. If the ratio of the corrosion fatigue endurance limit to that observed in air is less than 1, the material is likely to perform poorly in the corrosive media. Materials exhibiting ratios greater than 1 are deemed immune to corrosion under fatigue. Corrosion fatigue mechanisms for metals in the active and passive regimes differ. For metals in the passive regime, a perturbation in the passive film must occur, such as film rupture, before corrosive species can cause dissolution of the freshly exposed surface. Metals in the active regime experience dissolution more readily due to the inability to maintain a protective layer. Hence, the subjection to corrosion fatigue exacerbates the corrosion behavior observed in the presence of aggressive solutions. Given a sufficient time, corrosion species attack the bare metal exposed by the formation of extrusions and intrusions on the material surface. The mechanism commonly adopted is that by which dissolution of the freshly exposed material occurs, and the ease of plastic deformation or the localized surface plasticity is more favorable. The corrosion fatigue of materials can be categorized into two types: type 1—pitting, and type 2—embrittlement. The major distinction between the two cases is whether the material state is homogeneous or inhomogeneous and whether the material can maintain a protective film. The response to the addition of stress is affected by the material state. Among the embrittlement postulates, two approaches are acceptable for the deterioration of a material. These two categories are known as the slip-step anodic dissolution and hydrogen embrittlement. In both cases, the crystal structure of the material is weakened and experiences dissolution. However, the mechanism by which slip-step anodic dissolution occurs is different than the mechanism for hydrogen embrittlement. In hydrogen embrittlement, local acidification occurs due to the saturation of atomic hydrogen. For slip-step anodic dissolution, the atomic hydrogen buildup is due to the local entrapping of the corrosive electrolyte, which produces hydrogen at a faster rate as opposed to having hydrogen readily available from the bulk electrolyte.

Modeling of corrosion fatigue phenomena presents certain challenges since every material environment system is a synergistic effect, coupling the material nature, surface reactions, and stress conditions. The most employed model considers the superposition approach, which assumes that the operating mechanisms under sustained loading (stress-corrosion cracking) and corrosion fatigue hold true. A model to determine the crack growth rate was derived according to the following equation:

$$\left(\frac{da}{dt}\right)_{\mathrm{SCF}} = \left(\frac{da}{dt}\right)_{\mathrm{F}} + \left(\frac{da}{dt}\right)_{\mathrm{SC}}$$

Assuming that the rate of the growth of the pit radius is proportional to Faraday's law, the following equation can be written:

$$\frac{da}{dt} = \frac{MI}{\eta \rho F}$$

The crack growth rate, *da/dt*, is proportional to the quotient of the molecular weight, *M*; current, *I*; valence electrons, *n*; density, *ρ*; and Faraday's constant. Then the radius as a function of time is determined by the separation of variables and integrating the equation where a_{co} is a constant representing the critical stable nucleus size to precede the crack propagation:

$$a = a_0 + \frac{MI}{\eta \rho F} t$$

Once a critical pit size is reached, the fatigue-crack growth rate follows the relation

$$\frac{da}{dN} = C_{\mathrm{F}}(\Delta K - \Delta K_{\mathrm{th}})^{n_{\mathrm{C}}}; \Delta K = \beta \Delta \sigma \sqrt{a}$$

The parameters in the equation describe the relationship between the fatigue-crack growth rate, *da/dN*, and the difference between the applied stress intensity factor range and the near-threshold stress intensity factor range, where crack growth is not permissible. The stress intensity factor range can also be expressed according to the linear elastic fracture mechanics relationship, where the stress intensity factor range is proportional to the stress range, $\Delta\sigma$, and the square of the crack length, *a*. The CF and β parameters are the proportionality constants, and n_{C} is the fatigue exponent.

A transition criterion for both the stress intensity factor range for the critical pit size, ΔK_{Pit}, and the fatigue-crack growth rate, $(da/dt)_{\mathrm{fcg}}$, is governed according to the following:

$$\Delta K_{\mathrm{pit}} \geq \Delta K_{\mathrm{th}}$$

And

$$\left(\frac{da}{dt}\right)_{\text{fcg}} \geq \left(\frac{da}{dt}\right)_{\text{pit}}$$

where ΔK_{th} represents the near-threshold stress intensity factor range and $(da/dt)_{\text{pit}}$ represents the rate of the pit growth before the onset of cracks emanating from pits, respectively.

Although the grain size effect on mechanical properties has been largely demonstrated, the grain refinement effect on SCC should be deeply investigated. It seems that the effect of grain size reduction on corrosion resistance is mostly positive in stainless steels and aluminum alloys, whereas the effect is marginal in copper and titanium alloys (Asabe et al. 2017). The limited available literature on the effect of grain size on stress corrosion cracking (SCC) indicates an increasing resistance to SCC with decreasing grain size. There is some evidence that nanoscale alloys can exhibit superior corrosion properties, especially in conditions where corrosion cracking, wear corrosion, and localized corrosion are important (Tellkamp and Lavernia 1999). Nanoscale alloys can be processed to yield a homogeneous microstructure, which makes the material less susceptible to failure by grain pullout or grain dropping. In addition to the high strength, high temperature, and increased fatigue properties reported for these materials, corrosion properties are of particular interest. Nanomaterials may improve performance in applications where bulk materials are used in structures, components, and machinery and where service conditions subject the material to loading (static and cyclic). However, this combined action of loading and aggressive environment often results in stress corrosion cracking (SCC) and ensuing damage that significantly reduces a material's fracture resistance (Sharma and Ziemian 2008).

SCC of nanocrystalline materials demonstrated that nanocrystalline nickel electrodeposited on an austenitic stainless steel pipe has supreme resistance to SCC in high-pressure and high-temperature water. Intergranular propagation distance of SCC by stochastic approach with respect to varying grain size and grain boundary character distribution (GBCD) by supposing that an intergranular SCC is blunted at low-energy, coincidence site lattice (CSL) boundaries was analyzed. Grain size reduction and control of GBCD can improve the resistance to intergranular SCC.

Different nano-crystallization processes could introduce different structures including the degree of grain refinement, volume fraction of grain boundaries, and density of dislocations, all of which influence the characteristics, including thickness, composition, and electronic structure of the passive film on surface, and subsequently the corrosion resistance (Zheng et al. 2012; Gupta and Birbilis 2015).

Grain boundaries formed during SPD are often characterized by a nonequilibrium state, where grain boundaries have extrinsic dislocations by absorbing lattice dislocations. Thus, grain boundaries have higher grain boundary energy than that in equilibrium state. Thus, the post-ECAP annealing is expected to enhance the resistance to intergranular degradation such as intergranular corrosion and SCC, which are closely associated with grain boundary energy and structures. Severely deformed

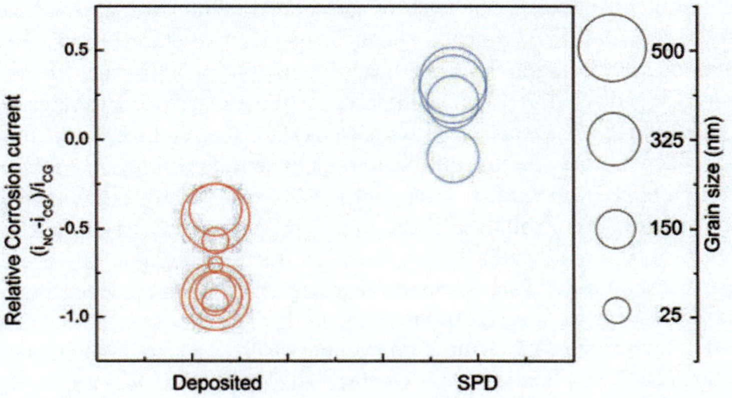

Fig. 10.11 Relative corrosion in nanostructured Cu produced via different routes

nanocrystalline materials are likely to have a preponderance of RHAGBs, while the deposited materials should be characterized by a high special fraction. This in turn should manifest as a difference in corrosion behavior, namely that deposited nanocrystalline material should be more corrosion resistant than that prepared by SPD (Fig. 10.11).

Based on UFG studies, RHAGBs are more susceptible to grain boundary sliding so promoting ductility. The high RHAGB fraction is also likely to leave the microstructure more vulnerable to coarsening because RHAGBs are more prone to thermal migration than are low-energy boundaries.

Liao et al. (2012) found that fine-grained Mg-Al alloys exposed to 0.1 M NaCl solution had superior corrosion resistance when compared to micro-grained hot-extruded alloys of the same composition. They attributed the enhanced corrosion performance to the enhanced passivity of the oxide film generated on the surface.

The corrosion fatigue strength of a UFG Al–7.5 Mg alloy was investigated and compared to its conventional counterpart 5083 H111. The UFG Al–7.5 Mg alloy was observed to have superior fatigue limit than the conventional alloy in air. It was shown that under cyclic loading, the fatigue limit of the UFG alloy was drastically reduced when exposed to a marine environment. At high cyclic stresses the UFG alloy had a superior fatigue resistance compared to the conventional 5803 alloy.

The presence of stress along with a sensitized microstructure leads to severe susceptibility towards intergranular stress corrosion cracking (IGSCC). The nature of precipitation along the grain boundaries and segregation in the matrix dictates the mechanism leading to SCC. Changes in grain cohesion and electrochemical properties due to solute segregation can alter fracture toughness and corrosion of metallic alloys by orders of magnitude (e.g., hydrogen embrittlement and stress-corrosion cracking). Interfacial chemistry also affects GB energy and mobility, and through this grain coarsening as well as stability of nanostructures. Local solute decoration can lead to the nucleation of second phases, formation of complexions, or selective melting of GBs (Herbig et al. 2014).

The continuous precipitation leads to accelerated failure through anodic dissolution in the presence of detrimental chemical species like chloride ions. In case of discontinuous precipitation, multiple mechanisms for crack propagation have been suggested (Argade et al. 2013). The importance of the nature of deformation slips in the materials on the SCC characteristics is well established and the crack propagation has large dependence on the dislocation–particle interaction. A microstructure exhibiting a planar slip tends to have higher SCC susceptibilities in comparison to a microstructure that exhibits a wavy slip. The higher repassivation kinetics of submicron-sized grains increased the resistance for pit initiation; however, for the long-duration exposure in saline environments, pit propagation has been reported to show concerns. López et al. (2006) reported higher SCC resistance for fine-grained austenitic stainless steel microstructure in comparison to coarse-grained microstructure in 30% NaCl solution at 90 °C. The increased SCC resistance was attributed to faster passivation–repassivation kinetics associated with fine-grained microstructure.

A higher overall resistance against SCC and corrosion fatigue can be obtained after grain reduction via equal-channel angular pressing although the general liquid corrosion assessed electrochemically by the average dissolution rate at a given potential is somewhat higher for the severely pre-strained material. Stress-corrosion cracking (SCC) and corrosion fatigue (CF) result from the combined effects of static or cyclic stresses and a deleterious environment. Both phenomena are related to the environment-assisted cracking. Thus, two stages of damage should be distinguished: crack initiation and propagation. SCC in ECAP Cu can be explained in principle within the framework of the film rupture and slip-dissolution model if one adopts that most plastic deformation is confined to the grain boundary and grain boundary-affected region. It has long been recognized that many ECAP materials are prone to shear banding during tensile or cyclic deformation. Besides, it was convincingly shown that namely the nonequilibrium grain boundaries produced during ECAP play a key role in shear banding which occurs along the grain boundaries in the plane of maximum shear stresses. The shear banding breaks the passive film on the surface, exposing therefore the grain boundary region to aggressive species. This results in the intergranular crack nucleation and propagation. Regarding corrosion fatigue, the following scenario has been proposed (Fig. 10.12).

The plastic strains and internal stresses accumulated during the first steady stage of cycling trigger intensive cyclic softening. This softening is caused by dislocation recovery and abnormal grain growth. As a result, a nonuniform structure consisting of fine grain matrix and coarse grains embedded here and there is created. The PSB-like dislocation structure forms in the enlarged grains and the surface relief typical for PSBs in CG materials emerges. The PSBs break the passive surface film and interact with grain boundaries, which are substantially weakened by the selective corrosive attack that promotes preferential intergranular cracking. The microscopic observations, on different stages of fatigue, confirm the proposed scenario. The beginning of cyclic softening is associated with rapid grain coarsening and abnormal grain growth, which occurs at the expense of the pre-deformed matrix. Then, when the grain size becomes large enough (of the order of several micrometers

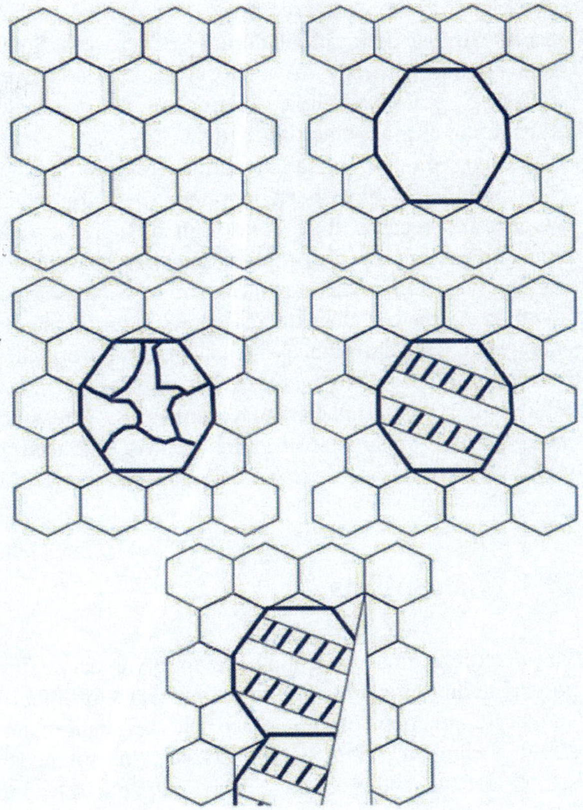

Fig. 10.12 Fatigue structure evolution and intergranular damage due to corrosion-deformation interaction in the grain boundary (GB) region

or tens of micrometers) the ordinary dislocation sources come into force in theses enlarged grains, dislocation multiplication occurs, typical cell structure forms, and then a ladderlike dislocation configuration is observed. The dislocation wall pattern becomes similar to what is commonly observed in PSBs of Cu. The appearance of PSB-like structure in numerous enlarged grains and strain localization associated with PSBs can account for the secondary saturation (Vinogradov et al. 2002).

Sensitivity to SCC of nanocrystalline Cu-10%Zn alloys fabricated by ECAP was higher than that of coarse grain counterpart in classical ammonia vapor. It seems that cracks propagated intergranularly as has long been acknowledged in the environment. The tarnish-film rupture has been considered as the main mechanism, where brittle oxide film on the surface is cracked in the crack tip followed by further oxidation and cracking at the crack tip. SCC propagated intergranularly, since growth of oxide film is faster at grain boundaries than that in grain interior. On the other hand, when tarnish film does not form, SCC tends to propagate transgranularly. Grain boundary energy varies depending on the misorientation between two crystals at equilibrium state. Thus, sensitivity to intergranular SCC is associated with misorientation of the grain boundary. Nonequilibrium grain boundaries, however, have higher grain boundary energy, irrespective of the misorientation because they absorb extrinsic dislocations during SPD process. Small grain size and overall high grain boundary energy may help a crack to advance in straight manner without "selecting" grain boundaries in front of the crack path. The crack initiation and propagation at grain boundaries under lower stress levels than yield stress might be associated with grain boundary sliding (Miyamoto et al. 2008). The nonequilibrium grain boundaries have been reported to exhibit grain boundary sliding at room temperature because they have high diffusivity. Grain boundary sliding at the crack tip may cause film rupture at lower stress level. Further studies on SPDed Cu-Zn revealed that the time to fracture by SCC increased with decreasing grain size down to 1 μm but then decreased with further decreases in grain size into the submicron scale. In other words, there was a critical grain size above which the susceptibility began to increase, and this grain size matched the grain size that divides the two Hall-Petch relationships. Stress-corrosion cracks propagated intergranularly regardless of grain size. SPD-induced grain boundaries have a high sensitivity to chemical reaction and intergranular SCC (Asabe et al. 2017).

10.4 Conclusions

The electrochemical corrosion behavior of nanocrystalline materials is different from polycrystalline materials due to the presence of higher volume of grain boundaries and triple junctions (more active sites available), and higher diffusivity of the alloying element. Other parameters such as solute atoms, impurities, grain size, surface defects, and precipitate distribution could lead to the difference of corrosion resistance between nanocrystalline and conventional materials. Grain size reduction in NC materials considerably improves the corrosion performance for a wide range of electrochemical conditions. This is mainly due to the elimination of the localized attack at grain boundaries, which is one of the most detrimental mechanisms of degradation in polycrystalline materials. Several explanations have been put forward for this effect: the solute dilution effect by grain size refinement, crystallographic texture changes with decreasing grain size, and grain size-dependent passive layer formation.

In nanocrystalline electrodeposited alloys, the GB structure is fundamental in governing the SCC. Depending on its circumstance, EAC failure can be either transgranular (ignoring grains) or intergranular (following grain boundaries). The latter case may resemble intergranular corrosion, as tensile stress can induce crack initiation along grain boundaries.

Adding alloying elements is believed to improve the corrosion behavior of nanostructured alloy coatings. This is likely because the diffusivity of alloying and impurity elements in nanocrystalline materials is much higher as compared to that in polycrystalline materials which leads to better corrosion resistance. The transport of ions in passive films takes place by migration and/or diffusion of defects (grain boundaries, linear dislocations, vacancies, or interstitials), so more protective passive layer is formed on nanocrystal-electrodeposited coating. This may be attributed to the widening of the energy band by decreasing the grain size.

References

Andrievski RA, Khatchoyan AV (2016a) Mechanical actions effect upon nanomaterials. In: Nanomaterials in extreme environments, Springer series in materials science, vol 230. Springer, Cham. https://doi.org/10.1007/978-3-319-25331-2_4

Andrievski RA, Khatchoyan AV (2016b) Nanomaterials behavior in corrosion environments. In: Nanomaterials in extreme environments, Springer series in materials science, vol 230. Springer, Cham. https://doi.org/10.1007/978-3-319-25331-2_5

Andrievskii RA (2013) The role of nanoscale effects in the interaction between nanostructured materials and environments. Prot Met Phys Chem Surf 49:528–540. https://doi.org/10.1134/S207020511305002X

Argade GR, Kumar N, Mishra RS (2013) Stress corrosion cracking susceptibility of ultrafine grained Al–Mg–Sc alloy. Mater Sci Eng A 565:80–89. https://doi.org/10.1016/j.msea.2012.11.066

Asabe T, Rifai M, Yuasa M, Miyamoto H (2017) Effect of grain size on the stress corrosion cracking of ultrafine grained Cu-10 wt% Zn alloy in ammonia. Int J Corros. https://doi.org/10.1155/2017/2893276

Chabok A, Dehghani K, Ahmadi Jazani M (2015) Comparing the fatigue and corrosion behavior of nanograin and coarse-grain IF steels. Acta Metall Sin (Engl Lett) 28:295–301. https://doi.org/10.1007/s40195-014-0196-2

Farzaneh MA, Raeissi K, Golozar MA (2010) Effect of current density on deposition process and properties of nanocrystalline Ni-Co-W alloy coating. J Alloys Compd 489(2):488–492. https://doi.org/10.1016/j.jallcom.2009.09.092

Gupta RK, Birbilis N (2015) The influence of nanocrystalline structure and processing route on corrosion of stainless steel: a review. Corros Sci 92:1–15. https://doi.org/10.1016/j.corsci.2014.11.041

Herbig M, Raabe D, Li YJ, Choi P, Zaefferer S, Goto S (2014) Atomic-scale quantification of grain boundary segregation in nanocrystalline material. Phys Rev Lett 112:126103. https://doi.org/10.1103/PhysRevLett.112.126103

Liao J, Hotta M, Yamamoto N (2012) Corrosion behavior of fine-grained AZ31B magnesium alloy. Corros Sci 61:208–214. https://doi.org/10.1016/j.corsci.2012.04.039

López HF, Cisneros MM, Mancha H, García O, Pérez MJ (2006) Grain size effects on the SCC susceptibility of a nitrogen steel in hot NaCl solutions. Corros Sci 48:913–924. https://doi.org/10.1016/j.corsci.2005.02.017

Miyamoto H, Vinogradov A, Hashimoto S (2008) Susceptibility to stress corrosion cracking in ammonia of nanostructured Cu-10wt% Zn alloy produced by severe plastic deformation. Mater Sci Forum 584–586:877–892. https://doi.org/10.4028/www.scientific.net/MSF.584-586.887

Nik Masdek NR, Alfantazi AM (2012) Nanocrystalline cobalt-iron alloy: synthesis and characterization. Mater Sci Eng A 550:388–394. https://doi.org/10.1016/j.msea.2012.04.092

Roy I, Yang HW, Dinh L, Lund I, Earthman JC, Mohamed FA (2008) Possible origin of superior corrosion resistance for electrodeposited nanocrystalline Ni. Scripta Mater 59:305–308. https://doi.org/10.1016/j.scriptamat.2008.03.033

Sharma MM, Ziemian CW (2008) Pitting and stress corrosion cracking susceptibility of nanostructured Al-Mg alloys in natural and artificial environments. J Mater Eng Perform 17:870–878. https://doi.org/10.1007/s11665-008-9215-7

Sharma MM, Tomedi JD, Parks JM (2015) A microscopic study on the corrosion fatigue of ultra-fine grained and conventional Al–Mg alloy. Corros Sci 93:180–190. https://doi.org/10.1016/j.corsci.2015.01.020

Shiraman KR, Ganesh Sundara Raman S, Seshadri SK (2007) Corrosion behaviour of electrodeposited nanocrystalline Ni–W and Ni–Fe–W alloys. Mater Sci Eng A 460–461:39–45. https://doi.org/10.1016/j.msea.2007.02.055

Tellkamp V, Lavernia E (1999) Processing and mechanical properties of nanocrystalline 5083 Al alloy. Nanostruct Mater 12:249–254. https://doi.org/10.1016/S0965-9773(99)00110-5

Vinogradov A, Miyamoto H, Mimaki T, Hashimoto S (2002) Corrosion, stress corrosion cracking and fatigue of ultra-fine grain copper fabricated by severe plastic deformation. Ann Chim Sci Mat 27(3):65–75

Wang L, Zhang J, Gao Y, Xue Q, Hu L, Xu T (2006) Grain size effect in corrosion behavior of electrodeposited nanocrystalline Ni coatings in alkaline solution. Scripta Mater 55(7):657–660. https://doi.org/10.1016/j.scriptamat.2006.04.009

Zheng ZJ, Gao Y, Gui Y, Zhu M (2012) Corrosion behaviour of nanocrystalline 304 stainless steel prepared by equal channel angular pressing. Corros Sci 54:60–67. https://doi.org/10.1016/j.corsci.2011.08.049

Index

© Springer Nature Switzerland AG 2021
P. Cavaliere, *Fatigue and Fracture of Nanostructured Materials*,
https://doi.org/10.1007/978-3-030-58088-9

D

H

I

O